STRATEGIC MANAGEMENT IN THE AVIATION INDUSTRY

Strategic Management in the Aviation Industry

Edited by
WERNER DELFMANN
University of Cologne, Germany

HERBERT BAUM
University of Cologne, Germany

STEFAN AUERBACH
Lufthansa Consulting GmbH, Germany

SASCHA ALBERS
University of Cologne, Germany

Co-published by Kölner Wissenschaftsverlag

ASHGATE

© Werner Delfmann, Herbert Baum, Stefan Auerbach and Sascha Albers 2005

All rights reserved. No part of this publication may be reproduced, stored in a retrieval system, or transmitted in any form or by any means, electronic, mechanical, photocopying, recording or otherwise without the prior permission of the publisher.

Werner Delfmann, Herbert Baum, Stefan Auerbach and Sascha Albers have asserted their right under the Copyright, Designs and Patents Act, 1988, to be identified as the editors of this work.

Published by
Ashgate Publishing Limited
Gower House
Croft Road
Aldershot
Hampshire GU11 3HR
England

Ashgate Publishing Company
Suite 420
101 Cherry Street
Burlington, VT 05401-4405
USA

Ashgate website: http://www.ashgate.com

British Library Cataloguing in Publication Data
Strategic management in the aviation industry
   1.Aeronautics, Commercial - Management 2.Strategic planning
   I.Delfman, Werner
   387.7'068

Library of Congress Cataloging-in-Publication Data
Strategic management in the aviation industry / edited by Werner Delfmann ... [et al.].
       p. cm.
  Includes index.
  ISBN 0-7546-4567-3
 1. Aircraft industry--Management. I. Delfmann, Werner.

  HD9711.A2S77 2005
  387.7'068'4--dc22
                                                                 2005013252

Reprinted 2006, 2008

ISBN-13: 978-0-7546-4567-2

Printed and bound in Great Britain by MPG Books Ltd, Bodmin, Cornwall

Contents

Preface	X
List of Figures	XI
List of Tables	XV
Introduction – Moving Targets: Strategic Trends in the Aviation Sector	1
Werner Delfmann, Herbert Baum, Stefan Auerbach and Sascha Albers	

Part I — Passenger Airlines

1 Competition Policy in European Aviation Markets — 19
   Carsten Bermig

2 Business Models in the Airline Sector – Evolution and Perspectives — 41
   Thomas Bieger and Sandro Agosti

3 Consolidating the Network Carrier Business Model in the European Airline Industry — 65
   Stefan Auerbach and Werner Delfmann

4 From Production-Orientation to Customer-Orientation – Modules of a Successful Airline Differentiation Strategy — 97
   Kim Flenskov

5 STRATEGIC AIRLINE POSITIONING IN THE GERMAN
LOW COST CARRIER (LCC) MARKET   119
THORSTEN KLAAS AND JOACHIM KLEIN

6 LOW COST CARRIERS IN SOUTHEAST ASIA:
A PRELIMINARY ANALYSIS   143
MARK GOH

7 THE FUTURE OF CONTINENTAL TRAFFIC PROGRAM:
HOW LUFTHANSA IS COUNTERING COMPETITION
FROM NO-FRILLS AIRLINES   165
CHRISTOPH KLINGENBERG

8 COMPETITIVE DYNAMICS THEORY – APPLICATION TO AND
IMPLICATIONS FOR THE EUROPEAN AVIATION MARKET   185
CAROLINE HEUERMANN

PART II — ALLIANCES

9 COOPERATION AND INTEGRATION AS STRATEGIC OPTIONS
IN THE AIRLINE INDUSTRY – A THEORETICAL ASSESSMENT   223
THOMAS FRITZ

10 INTERFACE MANAGEMENT IN STRATEGIC ALLIANCES   255
MATTHIAS GRAUMANN AND MARCUS NIEDERMEYER

11 SYNERGY ALLOCATION IN STRATEGIC AIRLINE ALLIANCES   275
BJÖRN GÖTSCH AND SASCHA ALBERS

12 THE EFFECT OF HIGH-LEVEL ALLIANCE FORMATION ON THE
   PROFITABILITY OF PARTICIPATING AIRLINE CARRIERS   305
   JENS RÜHLE

13 MARRIAGES AND DIVORCES: STRATEGIC ALLIANCES
   IN THE NETWORKED ECONOMY – THE CASE OF
   AIR NEW ZEALAND   325
   KATHRYN PAVLOVICH

PART III — AIRPORTS

14 FORCES DRIVING INDUSTRY CHANGE – IMPACTS FOR
   AIRPORTS' STRATEGIC SCOPE   345
   ROBERT J. AARONSON

15 THE AIRPORT OF THE FUTURE AGAINST THE BACKDROP OF
   DRAMATIC CHANGES IN THE AVIATION SECTOR   361
   MICHAEL GARVENS

16 INTERNATIONALIZATION STRATEGIES FOR AIRPORT
   COMPANIES   377
   BENJAMIN KOCH AND SVEN BUDDE

17 EVALUATING AIRPORT COOPERATION AND
   ACQUISITION STRATEGIES   409
   ANDREA PAL AND WOLFGANG WEIL

18 THE ROLE OF INTERMODAL TRANSPORTATION IN AIRPORT MANAGEMENT: THE PERSPECTIVE OF FRANKFURT AIRPORT   427

HANS G. FAKINER

PART IV — AIR CARGO

19 REFLECTING THE PROSPECTS OF AN AIR CARGO CARRIER   451

ANDREAS OTTO

20 DRIVERS OF ALLIANCE FORMATION IN THE AIR CARGO BUSINESS   473

PETER GRÖNLUND AND ROBERT SKOOG

21 AIRFREIGHT DEVELOPMENT SUPPORTING THE STRATEGY OF GLOBAL LOGISTICS COMPANIES   489

RENATO CHIAVI

22 INTEGRATOR NETWORK STRATEGIES AND PARAMETERS OF AIRPORT CHOICE IN THE EUROPEAN AIR CARGO MARKET   517

BENJAMIN KOCH AND ANDREAS KRAUS

23 ESCAPING THE AIR CARGO BAZAAR: HOW TO ENFORCE PRICE STRUCTURES   539

KORNELIA REIFENBERG AND JAN REMMERT

24 ASSESSING AIR HUBS: THE HAYES-WHEELWRIGHT FRAMEWORK APPROACH   557

SIN-HOON HUM AND SWEE-KOON TAN

PART V — AIRPORT AND AIRLINE STRATEGY IMPACTS

25 THE IMPACT OF AIRPORTS ON ECONOMIC WELFARE 585
    HERBERT BAUM

26 LEARNING FROM THE AIRLINES? A COMPARATIVE ASSESSMENT OF COMPETITIVE STRATEGIES FOR RAILWAY COMPANIES 605
    CHRISTIAN KAUFHOLD AND SASCHA ALBERS

ABOUT THE EDITORS   *637*
LIST OF AUTHORS   *641*
INDEX   *655*

PREFACE

This book evolved from a long-standing collaboration of the editors at the University of Cologne. When Werner Delfmann and Herbert Baum decided to set up an airline and airport-related course for both of their majors (transport economics and logistics/strategic management), bringing together economics and management topics and viewpoints, the close involvement of airline representatives was a major cornerstone in the development of the course. Thus, the involvement of Lufthansa Consulting in such an early phase of the course development process as well as in teaching was simply a logical consequence, since both departments had a rich history of successful joint projects with this Lufthansa spin-off. The intensive discussions with students as well as among ourselves, together with the perception that a forum for strategic management issues in this industry, despite their undisputed and frequently hailed relevance, is still widely underrepresented with regard to current questions and specific research, finally led us to take on this book project.

Many of the authors involved have held talks and presentations in the frame of our courses, or even welcomed us to their sites with our students. We thank them for their commitment and participation. Thanks also go to a variety of people who have helped us in collecting, reviewing and improving the chapters in this volume, as well as producing this interesting book, we hope, for both airline and airport scholars and practitioners. First and foremost we have to thank Björn Götsch who has skillfully supported the coordinative and administrative processes which are related to such a voluminous book publication in a highly professional manner. Special thanks are also owed to the referees who carefully read and commented on every paper we received and thereby significantly supported and facilitated our editing tasks. The reviewers were Rowena Arzt, Torsten Geißler, Björn Götsch, Sören Grawenhoff, Caroline Heuermann, Thorsten Klaas, Juliane Kloubert, Kai Krause, Christiane Müller-Rostin, Heiko Peters, Markus Raueiser, Markus Reihlen and Jens Rühle.

Furthermore, we thank Michael Becker, Katrin Linzbach, Linda Horn and Heike Kirch for their extensive formatting and spell-checking efforts.

Werner Delfmann, Herbert Baum, Stefan Auerbach, Sascha Albers

List of Figures

Figure 3-1: Operating Results of AEA Airlines 1990-2003. 67
Figure 3-2: Development of Market Shares in Europe 70
Figure 3-3: The Integrated Transport System of Network Airlines 74
Figure 3-4: Traffic Segments and Unit Margins of Network Carriers 76
Figure 3-5: S-Curve-Effect at Alitalia. 78
Figure 3-6: Traffic Segments 85
Figure 3-7: Players in the Future Airline Industry 86
Figure 3-8: Hub-connectivity at International Airports in 2004 87
Figure 3-9: Comparison of Hub-structures at Frankfurt and Atlanta 89
Figure 4-1: Aircraft Utilization, Seat Range 132-180 103
Figure 4-2: Five-Year Net Margin Average 1997-2002 104
Figure 4-3: Market Capitalization Million US$ (22nd August 2003) 105
Figure 4-4: Number of Fractional Shares and Business Jet Deliveries 1986-2002 106
Figure 5-1: Segments of the Airline Market 122
Figure 5-2: Strategic Triangle of the LCC Business 126
Figure 5-3: Strategic Positioning in the LCC Market 129
Figure 5-4: Logotype of Germanwings 132
Figure 5-5: Germanwings House of Corporate Strategy 133
Figure 5-6: Germanwings Strategic Triangle 134
Figure 6-1: AirAsia's Route Network 150
Figure 6-2: ValuAir's Route Network 152
Figure 6-3: NokAir's Route Network 154

List of Figures

Figure 7-1: Comparing Ryanair's Average Costs per Offered Seat Kilometer in cts with Three European Hub & Spoke Carriers, 2001...........168

Figure 7-2: Development of Traffic between Germany and GB from 2002 to 2003 170

Figure 7-3: Holding Delay Before and After Depeaking in Frankfurt...........180

Figure 8-1: A Competitive Dynamics Model 188

Figure 8-2: Multimarket Contact and Competitive Intensity.........195

Figure 8-3: Contingencies in Multimarket Competition................198

Figure 9-1: Airline Value Chain 232

Figure 10-1: Interface Management in the Star Alliance................260

Figure 10-2: Interface Management in the WOW Alliance............262

Figure 10-3: Efficiency Hypotheses concerning Self-coordination Measures.269

Figure 10-4: Efficiency Hypotheses concerning External Coordination Measures 270

Figure 11-1: Positive and Negative Potential Synergies (examples)...............284

Figure 11-2: Four-phase Synergy Allocation Model 287

Figure 11-3: Deduction of Fair Shares – Overview.........292

Figure 11-4: Matrix of Pairwise Comparisons (level 1) 294

Figure 11-5: Matrix of Pairwise Comparisons (level 2) 295

Figure 11-6: Aggregated Calculation of RWs (level 3)...............296

Figure 11-7: Required Money Transfers 297

Figure 12-1: Unadjusted Profitability Index: For All Carriers 314

Figure 12-2: Profitability Index – Base Year 1989 for All Carriers 316

Figure 12-3: Correlation between US-GDP and Overall Operating Profit......317

Figure 12-4: US, Hong Kong, and UK GDP Growth 317

Figure 12-5: GDP-Adjusted Index: Base Year 1989 for All Carriers..............318

Figure 13-1: Lord of the Rings Themed 747 326

Figure 13-2: Timeline of AirNZ Alliances 332

List of Figures

Figure 13-3: AirNZ Global Partnerships 338
Figure 14-1: Factors Impacting the Airport Industry.......................................346
Figure 15-1: Catchment Area of Cologne Bonn Airport 370
Figure 16-1: The Air Transport Chain (Passenger Transport).........................379
Figure 16-2: The Value Chain of the Airport 380
Figure 17-1: Airport Privatization Projects 1995-2004 412
Figure 17-2: Alliance Categories...413
Figure 17-3: Strategic Alliance Life Cycle..423
Figure 18-1: Market Potential of the Major European Airports 430
Figure 18-2: AIRail Market Shares 436
Figure 18-3: Shifts among Modes of Transportation (main modes of access and egress) 437
Figure 18-4: Growth of the Catchment Area 438
Figure 19-1: Estimated Development of the Air Cargo Market 454
Figure 19-2: World-wide Demand for Air Cargo ex Origin............................455
Figure 19-3: Transport Chain for Airfreight..457
Figure 19-4: Development of the Air Cargo Yield, 1985-2001.......................459
Figure 19-5: Segment Growth 460
Figure 19-6: Operating Results of Lufthansa Cargo 1997-2002......................464
Figure 19-7: Product Advertisement "Live/td"...467
Figure 21-1: Strategic Framework of Global Logistics Companies 490
Figure 21-2: Consumer Prices (annual change in percent) 494
Figure 21-3: Indexed World Airfreight Cargo Yields 500
Figure 21-4: World Growth, Trade and Airfreight Development....................504
Figure 21-5: Top Ten Airfreight Lanes in Market Shares and their Forecasted Annual Growth Rates (CAGR 2000-2020)...............510
Figure 22-1: Transported Air Cargo Volumes in Europe (in million Tons)....521
Figure 22-2: Flight Times from the Central European Hubs (minutes)..........532

List of Figures

Figure 22-3: TNT's European Single-Hub Air Network 532

Figure 22-4: DHL's European Multi-Hub Air Network 532

Figure 22-5: Operational Pattern at DHL's Main Hub 533

Figure 22-6: Operational Pattern at DHL's Sub-Hubs 533

Figure 22-7: Flight Times from a Potential Sub-Hub in Leipzig/Halle, Germany (minutes) 536

Figure 23-1: Reasons for Price Wars and Historical Examples 540

Figure 23-2: Customization vs. Standardization 546

Figure 23-3: The Rate Card Concept 558

Figure 23-4: Example Weekday Pricing 552

Figure 23-5: Price Differences between Rate Cards 553

Figure 26-1: Strategic Decision Domains for the Comparative Analysis 608

Figure 26-2: Western Airline's Route Network *prior to* US Deregulation 610

Figure 26-3: Western Airline's Route Network *after* US Deregulation 611

List of Tables

Table I-1:	Timeline of Deregulation Processes in the US, Europe and Asia	3
Table I-2:	The Airbus A380 Orderbook as of December 2004	10
Table 2-1:	Stages of Air Transport Development	42
Table 2-2:	The Evolution of Strategic Planning Units against the Background of the Evolution of Strategy Theory	46
Table 2-3:	Business Models Used by Airlines – A First Example	49
Table 2-4:	Strategic Success Factors of Business Models in Airline Business	55
Table 2-5:	Transformation of Airline Business Models	58
Table 2-6:	Performance of Business Models	61
Table 3-1:	Competition on the Cologne-Berlin Route	75
Table 4-1:	Low-fare Carrier Online Sales	101
Table 4-2:	Online Majors Travel Booking Revenues	101
Table 5-1:	Action List of Key Success Drivers in the LCC Business	128
Table 5-2:	Airline Service-Quality Components	131
Table 6-1:	Dollar Value Percentage Comparison between US Airlines and SIA	147
Table 6-2:	Summary of Company Profile	156
Table 9-1:	Influencing Competitive Position through Cooperation/Integration	236
Table 9-2:	Comparison of Results from the Theoretical Analyses	247
Table 10-1:	Measures of Interface Management	264
Table 11-1:	Typology of Alliance Compensation Rules	279
Table 11-2:	Relative Weights on the Different Levels (level 1 – 3)	293

Table 11-3:	Ordinal Rating Scale	293
Table 12-1:	Alliance Groupings Taken for the Analysis	309
Table 12-2:	Performance of Selected Alliances	319
Table 16-1:	Comparison of Empirical Case Study Results	397
Table 16-2:	Success Factors for Airport Internationalization Projects	402
Table 17-1:	Options for Private Sector Participation in Airports	422
Table: 18-1:	Criteria for the Standardisation of Intermodal Products/Services	434
Table 21-1:	World Real GDP Impact on Airfreight (FTK)	505
Table 21-2:	Forecast for Regional Real GDP Growth	507
Table 21-3:	Forecasts	509
Table 22-1:	Integrators' Infrastructure Requirements to Airports (Examples)	530
Table 22-2:	Timing Scheme of DHL's Hub Operations	534
Table 24-1:	Major Air Hubs in Asia	563
Table 25-1:	Direct, Indirect and Induced Employment and Income Effects for the Eextension-scenario and the Non-extension-scenario	592
Table 25-2:	Data Collection Program for the Empirical Determination of the Impacts of Airports as a Location Factor	593
Table 25-3:	Impacts of Frankfurt Airport as a Location Factor (2015)	594
Table 25-4:	Quality Attributes of Labour	596
Table 25-5:	Employed Persons Resulting from Construction, Operation and Location Factor Effects (in 2015) Divided into Quality Criteria (Status Quo)	598
Table 26-1:	Main Features of Incumbents' and Newcomers' Airline Strategies	620

INTRODUCTION

MOVING TARGETS: STRATEGIC TRENDS IN THE AVIATION SECTOR

WERNER DELFMANN, HERBERT BAUM, STEFAN AUERBACH AND
SASCHA ALBERS

1   Increasing Strategic Scope 2
2   Trends in the Aviation Industry 4
3   The Chapters in this Book 13
4   References 14

Summary:
The aviation industry changed over the last two decades. Deregulation and liberalization of air transport markets and renegotiation of existing bilateral agreements caused the traditional flag carriers to face stronger competition and led to turbulent times in deregulated markets. After a phase of consolidation in the 1990s the industry is again confronted with major changes and new developments. These are not only caused by the spread of deregulation over the globe, but are accompanied by technological breakthroughs which are likely to change the international traffic patterns, powerful new competitors both among the network carriers and the low cost airlines, as well as new freedoms and requirements on the side of airports and cargo transportation. This chapter outlines these changes, reflects on their impact and subsequently puts the other chapters collected in this volume into context.

1 Increasing Strategic Scope

The aviation industry has undergone a phase of transition in which the actors involved have consecutively been granted a greater degree of autonomy as regulatory bonds have been released successively (see table I-1).

In general, the development and impact of deregulation on the market conditions and the strategies of airlines and airports follow a typical iterating pattern of turbulence and consolidation. Deregulation in all cases implies that the relevant actors' freedom is increased, which on the one hand reduces certainty and predictability of their environment, but on the other hand allows for Schumpeterian processes of entrepreneurial actions and competition to unfold.

Whereas airlines were the first to encounter the importance of strategies and strategic decision-making in a competitive environment, this is only slowly beginning to transcend to airports and their organizations. Most airports were and still are seen as public utilities and are in public ownership. In times of fiscal problems of their public owners, airport privatization is a popular and widely followed means to ensure infrastructure expansion and maintenance.[1] However, even though privatization of airports comes with an increased need to thoroughly evaluate infrastructure investments with regard to their rates of return in financial terms rather than with regard to their overall economic or regional benefits,[2] business-like and strategic mindsets still have to be established among most airport operating companies. In certain areas of their activities, such as ground-handling, competition has already slowly set in. Thus, whereas airlines in the meantime basically have all major parameters of competition at their disposal (e.g. price differentiation, product features and production plans and equipment used, advertising and marketing campaigns, customer segmentation), with regulations limiting their use only occasionally, airports and airport operating companies are still (and for good reasons) seen as local monopolies and limited in their actions. However, as has been portrayed before, their strategic scope has also expanded, requiring detailed analyses and recommendations for their needs.

---

[1] See Graham (2001).
[2] See Baum's chapter 25 in this volume.

North America	Europe	Asia-Pacific
1978: Deregulation (US Airline Deregulation Act)	Traditional, strong regulatory regimes	
1978-1990: Increasing competition in the US (market entries)[3]		
1984: New Canadian Air Policy, initiating deregulation in Canada	1987-1993: Intra-EU deregulation: three packages, increased competition among the flag carriers, start of low cost competition in selected markets, consolidation in national markets, no trans-border consolidation	
1990: Concentration process in Canada set in soon after deregulation (formation of Canadian Airlines)		1989/90: Australia domestic deregulation; Air New Zealand privatized
		Renegotiation of bilateral agreements, e.g. Singapore – Australia; China –EU
1990-2000: Consolidation of the US industry; concentration of trunk carriers		1996: Australia-New Zealand Single Aviation Market Agreement
Since 2000: Increased competition by spread of low cost carriers	Since 2000: Increased competition by spread of low cost carriers, only slow consolidation[4]	2000: New Zealand and Australia adopting open-skies policies
		Spread of low cost carriers in liberal economies (e.g. Australia, New Zealand, Singapore, Thailand)

Table I-1: Timeline of Deregulation Processes in the US, Europe and Asia

---

[3] Between 1978 and 1987, the number of US airlines grew from 36 to 123. However, the real number of entries was significantly higher. In the same period, about half of the existing carriers and two thirds of the newly founded disappeared in approximately 160 bankruptcies and mergers from the market. See Knieps (1987), p. 34 and Borenstein (1992), p. 46.

[4] E.g. the acquisition of KLM by Air France, and Swiss by Lufthansa.

2 Trends in the Aviation Industry

Doganis identifies nine major trends in the aviation industry for the 21st century.[5] The further spread of deregulation of international air service agreements, which will ultimately result in an international open skies regime, has meanwhile taken further shape. The increased relaxation of airline ownership restrictions has not yet seen any significant advancement. Despite rather liberal regimes in the US and the EU, minimum domestic shareholder quotas are still required by law and seem to persist at least in the near future. However, despite only slow advancements with regard to the ownership clauses, the trend towards increased concentration among airlines, also on an international scale, is taking shape. The takeover (merger) of KLM by Air France is the first major equity-based consolidation in the European airline industry and is widely seen as a necessary first chance to eliminate the persistent overcapacities in the market. Lufthansa's acquisition of Swiss which has just received the two airlines' boards' approval is the second major move towards European consolidation.[6] However, it was only after a long period of demise (first of Swissair, then of Swiss) that the proud Swiss agreed to abandon their flag carrier. Belgium's Sabena has received a second life as SN Brussels Airline before being merged with Virgin Express, but ailing Alitalia is still backed by stately financial injections. The increasing dominance of only a handful of airline alliances continues with Skyteam exhibiting a breathtaking speed in catching up with Oneworld and even Star.

Privatization will also continue, both with regard to airline as well as airport operators. During the last ten years, formerly state-owned European flag carriers, such as British Airways, Lufthansa, Air France, KLM, or Iberia have been privatized. In the United States, airlines traditionally have been owned by the private sector. However, the vast majority of the world's airlines still remain publicly owned and in many cases only subsidies prevent them from bankruptcy. But since public budgets are globally tightening, the privatization of these state assets can help to fill the gaps. Currently the Mexican State Agency Cintra plans to privatize Aeromexico and Mexicana, and the Hungarian privatization Agency ÁPV Rt. is looking for investors for its struggling airline Malév.

---

[5] See Doganis (2001).
[6] See e.g. Flottau (2005).

Tremendous investment programs in airport infrastructure are required simply to keep pace with the projected growth in air travel demand with an increasing role for the private sector as governments tend to reduce their level of support and funding for airport projects. These factors are literally changing the face of the world's airports in terms of the risks and opportunities present requiring an appropriate response. Today, only an estimated four percent of the world's commercial airports are genuinely managed or owned by the private sector. However, an expanded private sector role is inevitable. Its evolution is seen in varying stages around the world, in opening up ground handling to additional competition, outsourcing activities such as facility management, developing long-term management contracts, initiating public/private joint ventures, Build Operate Transfer (BOT) project financing, and in the case of a few airports so far, outright equity offerings. Whilst unpredicted events like the terrorist attacks of September 11th, Djerba and Bali, the SARS crisis and the aftermath of the Iraq war may have stifled demand in the short term, the long term is expected to require action to improve, modernize, and coordinate the air transport infrastructure to cope with volume growth.

Privatization of airports can be observed around the globe. In Asia for instance, the Indian government plans to proceed with privatization airports of New Delhi, Mumbai, Chennai, and Calcutta.[7] A Greenfield airport is going to be developed at Hyderabad on a build-own-operate (BOO) basis by a joint venture of Malaysian Airport Holdings and Indian construction firm GMR. The airport of Bangalore is transferred to a consortium led by Siemens and unique Zurich airport by a BOT-concession. In Australia, the Southern Cross Consortium, led by Hochtief Airports and Macquarie Bank, won the bid for a 99-year lease to Sydney's Kingsford Smith Airport. In China, Beijing International Airport was privatized by an Initial Public Offering (IPO) and the Chinese government announced that it will permit foreign investors to hold a majority stake in civil airport joint ventures, rather than being limited to the current 49 percent. Japan announced plans to privatize, via public share offerings, three major airports: Narita (New Tokyo International), Kansai, and Chubu/Nagoya.

Airport privatization in Europe started with the IPO of the BAA plc in 1987. The airport companies of Copenhagen (Copenhagen Airports A/S), Vienna (Flughafen Wien AG), Zurich (Flughafen Zürich AG) and Frankfurt (Fraport AG) are other examples for successful stock exchange listings. Another group of European airports have been (partially) transferred to the private sector by trade sales such as Dusseldorf, Hamburg, Bristol, East Midlands, Newcastle or Turin.

---

[7] The expansion and modernization of these airports should be transferred to the private sector via build-operate-transfer (BOT) concessions.

The privatization of Budapest, Prague, and the French regional airports are announced for the near future.

In South Africa, the private airport company ACSA manages 11 airports. In Tunisia, the planned new Enfidha Airport near Tunis should be developed under a BOT-concession. The Worldbank's private sector arm IFC has recently announced the privatization of the airports of Madagascar.

In Canada and the United States, privatization of airports is not a widespread phenomenon. In Canada, a majority interest in the country's only privately owned commercial airport was acquired by YVR, the outsourcing division of Vancouver Airport. In particular in Latin America, concessions are a common means to develop airport infrastructure, for instance in Bolivia, Mexico, Argentina and most recently the Ecuadorian airports of Quito and Guayaquil.

Increased competition with low fare airlines results in the continuity of the downward trend in fare levels, requiring in turn that airlines need to redouble efforts of cost reduction in their operations. Potentially helpful, at least at first sight, is the trend of disintermediation in the sales channel with travel agents and also global distribution systems potentially threatened by alternative, direct internet sales by airlines. The only partially restricted growth in air traffic will further emphasize the inadequacy of aviation infrastructure and, despite the depicted privatization efforts, will lead to more and more airports suffering congestion. However, air traffic control authorities will have to tackle the issues of increasing air transport as well. This directly leads to a final trend which will undoubtedly be reflected also in the capacity and traffic control issue, which is the increasing importance and concern of environmental issues of air transport.

These trends are still valid and have already in the last four years been illustrated by a variety of examples and policy decisions. From a strategic management perspective, however, we would like to add a couple of issues which have emerged since, or the consequences of which have been sharpened since 2001. These are (1) the further spread of deregulation to Asia and the consequences this evokes on the international airline alliance structures, (2) the recent technological advancements in aircraft technology with new quantum leaps with regard to size and reach of new aircraft, and (3) the emergence of new, financially strong competitors in intercontinental traffic in the near east.

2.1 Deregulation in Asia-Pacific

For a considerable period of time the deregulation of the domestic air transport markets in Australia and New Zealand in the late 1980s and early 1990s were the only examples for a significant move towards liberalization in the Asia-Pacific region (see table I-1 above).

In the past, moves to liberalization in Asia were largely focused on long-haul routes rather than within a region. Asian flag carriers considered their neighbouring airlines as their major competitors. Hence, their public owners were reluctant to relax local access provisions. Compared to trade relations with Europe and the US, intra-Asian trade was of a smaller scale and did not provide any impetus to change the regulatory system. Furthermore, Asia has been suffering infrastructural problems: half of the region's airports are bursting at the seams. Infrastructure deficiencies have given the region an ideal out against accusations of tardiness in the global liberalization stakes. In Asia it is called the Japan excuse: we would love to open our skies but we are just too full. Congestion in the air is also a barrier. Air corridors into and within Asia-Pacific are congested, a problem which will only be resolved by a new generation air navigation system. However, in the recent past the pace is accelerating.

Singapore has for a long time been happy to offer virtual free skies because it is the region's major hub, feeding off, through and stopover traffic. The more airlines fly in, the more profitable is the aviation business for Singapore and Singapore Airlines. New Zealand is pro-liberalization but, geographically isolated as it is, can only gain from freer access to Asia-Pacific markets, offering little in return. Thailand has forged a near open skies agreement with the US, after canceling the old bilateral and waiting for Washington to return to the table.

The Philippines too has a liberalized agreement with the US, although it was signed in the face of bitter protests from Philippine Airlines, which complained the government was removing its protection and handing over the farm to the Americans - a point of view supported by most analysts. India, like several other countries, has opened up its domestic market to competition and has a new, relatively open, bilateral agreement with the US New Delhi's policy on foreign participation domestically is inconsistent, but internationally it is clear cut: the government fears real open skies will damage Air India irrevocably. Japan's ongoing wrangle with the US has become one of the sagas of aviation history and its stance against open skies is backed by most neighbours. Asia-Pacific governments still regard their carriers, whether state-owned or privatized, as strategic assets critical to economic development, which must be protected.

Interestingly, China in particular has evolved as a frontrunner in deregulation. The Chinese government has introduced sweeping reforms to the country's air services regime to improve the local airline industry's competitiveness and drive economic growth. China's protectionist policies of the past were aimed at helping its state-owned carriers, but the authorities have moved to introduce liberal reforms because aviation is seen as a key driver of the economy.[8] The Civil Aviation Administration of China (CAAC) is considering an open skies policy for other cities apart from the three major gateways of Beijing, Guangzhou and Shanghai.

In the past 15 months, the CAAC signed more liberal air services agreements with several countries, including Australia, Pakistan, Thailand and the US. In most cases, the Chinese have granted fifth-freedom rights for cargo and in some instances – Pakistan, Thailand and the US – also fifth-freedom rights for passenger services. Pakistan, for example, gained rights to operate passenger flights beyond China to Tokyo and Los Angeles. Thai-designated carriers, meanwhile, are permitted to operate beyond China to destinations in South-East Asia.[9] The reason for these steps is that the fast growing demand for air services cannot be met by Chinese airlines alone.

In 2000-01, the CAAC initiated the building of three major airline groupings. The three leading carriers Air China, China Eastern Airlines and China Southern increased their scale by taking over smaller, state-owned carriers and implement massive re-fleeting programmes. Competition in China's domestic market is also set to intensify over the next 18 months because the Chinese authorities have decided to allow private carriers to operate domestically. At least four carriers have applied to the CAAC to operate domestic services, including Okay Airways in Tianjin, Jetwin Air Cargo in Shenzhen, Spring Autumn Airlines in Shanghai and United Eagle Airlines in Chengdu.

While there is the threat of increased competition from the private sector, the Chinese authorities are easing restrictions so legacy carriers can gain more access to private capital. China Eastern Airlines and China Southern Airlines have already completed IPOs and Air China is preparing to float late this year or early next year. Currently, private ownership – including foreign ownership – is

---

[8] One of the boldest moves so far has been the announcement in January 2005 of an open skies regime for the southern Chinese provincial island of Hainan. Under the new regime, foreign airlines can operate as many passenger or cargo services as they wish to Haikou Meilan airport in the north of the province and Sanya Phoenix in the south of the island, as well as to third countries. Foreign carriers are also permitted to stop in Hainan and travel on to any city in China, with the exception of Beijing, Guangzhou and Shanghai.

[9] These reforms are significant because in the past the CAAC was reluctant to grant fifth-freedom rights because state-owned carriers would have to compete more with foreign carriers.

capped at 49%, but this may change. CAAC may consider easing private and foreign ownership limits but there is no formal policy yet.

2.2 Technological Quantum Leap

In January 2005, Airbus Industries celebrated the roll-out of its double-deck A380 in Toulouse. After more than ten years of development, the aviation industry is awaiting the A380's maiden flight. After a 14-month testing and certification period, Singapore Airlines will be the first carrier to put the new type into commercial service in April 2006 (see table I-2 for the current A380 orderbook).

Compared to the largest commercial passenger aircraft in service by now, the Boeing 747-400 with a seating capacity of 400-480 (typical three-class) and a maximum take-off weight (MTOW) of 385 tons, the A380 exceeds these numbers by far. The seating capacity ranges from 550 (three-class-configuration) to 800 seats (one-class-charter) depending on the configuration, MTOW amounts to 540 tons (A380-100).[10] The enormous capacity of the aircraft provoked an intense discussion about whether there are hub-to-hub markets for such a giant aircraft or if a fragmentation towards more point-to-point services will prevail in the future.

Airbus' rival Boeing based its strategy on the hypothesis that the future of long-haul growth lies in higher frequency point-to-point services rather than the traditional hub system. Hence, Boeing is promoting its concept of the new mid-size aircraft 787 programme. The features of this aircraft concept are not shere capacity, but lower unit cost, mostly because of lower fuel consumption. Airbus strongly believes that future growth will be achieved through a mix of both fragmentation and hub consolidation. Accordingly, the Boeing Global Market Outlook suggests that aircraft with over 400 seats such as the 747 and A380 will capture only 4% of the market over the next 20 years.[11] Airbus argues that the present concentration of 80% of 747 operations at just 37 airports, growing congestion and the establishment of global airline alliances, the logic of which is

---

[10] Most of the carriers which placed an order for the A380 do not publish exact details on configuration. Only Emirates has indicated that a regular two-class configuration would have in the region of 650 seats, while on lower-density, ultra-long-range flights - such as Dubai-Sydney and Dubai-New York - the A380 would be configured for 480-490 passengers.

[11] See Boeing (2004), p. 12.

based on hub-to-hub operations, indicate the need for greater aircraft and are thus put forward as reasons behind developing the A380. Other key drivers are given as environmental pressures and operational efficiency demands.

Customer	Model	Firm	Options	Delivery
Emirates	A380-800	41	10	2006
	A380-800F		2	2008
Qantas Airways	A380-800	12	12	2006-11
Qatar Airways	A380-800	2*	2	2009
Virgin Atlantic	A380-800	6	6	2007-9
Air France	A380-800	10	4	2007-9
ILFC	A380-800	5		2007-12
	A380-800F	5		2007-12
Singapore Airlines	A380-800	10	15	2006-7
Lufthansa	A380-800	15		2007-15
FedEx	A380-800F	10	10	2008-11
Malaysia Airlines	A380-800	6		2007-8
Korean Air	A380-800	5	3	2007-9
Etihad Airways	A380-800	4*		2007-8
Thai Airways	A380-800	6*		2008-9
Total		**139**	**62**	

Table I-2: The Airbus A380 Orderbook as of December 2004[12]

The fact that Boeing has not received any order for a B747-400 passenger aircraft since November 2002 may indicate that there is a need for a new and innovative type. Currently, there is no reliable, quantitative information on unit cost of B787 aircraft available. With regard to the unit cost savings of the A380 operation compared to a B747-400, announced as a quantum leap in efficiency from a clean sheet design, translates into improvements of 15% in direct operating cost per seat on a 11,000km stage length and 20% per ton, while providing space for 35% more passengers. But a key maturity target is a massive

---

[12] Source: ATI. *Commitments announced, firm contracts not yet signed.

24% reduction in direct operating costs as well as 99% operational reliability within two years of entry into service through the application of new systems, fuel-efficient engines and lightweight materials, leading to lower fuel burn and reduced maintenance costs according to Airbus.[13]

In the air cargo business, the A380 is a step up from the B747 freighter version (B747-400F). Direct operating costs per tonne would be 20% lower, measured over a 7,400km sector at 80% gross volume. Compared with a typical 116t payload carried by the 747-400F, the difference between the two aircraft becomes still more evident. According to Airbus, the A380 will be able to carry that particular payload 13,500km, opposed to the 7,800km range of the Boeing aircraft.

The introduction of a new aircraft type of the dimension of an A380 causes infrastructure adaptations, such as aircraft stands, loading bridges and terminal space requirements to accommodate expected peak volumes. According to Airbus industries, over 60 airports will have to handle the A380 by 2010. Fifteen, mostly in the Far East, are ready today.

2.3 Emergence of New Network Players

The rise and success of low cost carriers entering the major airline markets caught the attention of researchers and airline managers alike for the last five years and remain an important issue on airline and airport managers' agendas (as witnessed by the respective chapters in this book). However, in addition to the rise of the LCC, and besides the worldwide regulatory changes and the technological advancements just depicted, there is still a new challenge on the horizon, which has yet been widely neglected in the literature. The Persian Gulf emirates' economies are reinforcing their development efforts in third sector industries, especially in tourism. To these ends, expenditures and investments into the tourism sector in the United Arab Emirates, Oman, Qatar, and Bahrain significantly increased over the last few years. Efforts have been undertaken to raise attractiveness by spectacular new hotel and leisure facilities, supported by a gearing up of marketing efforts and contact to international media. Consequently,

---

[13] In the media a discussion on weight problem with the A380 arose, which may cause penalties for Airbus. Airbus Industies admits that the A380 is approximately 2% over the weight at which some airlines have ordered the aircraft, Airbus officials stated that the overweight will be eliminated by entry into service. Although this additional weight appears to be of only minor importance, it equates to a payload reduction of around 50 passengers.

the attention which is paid to airlines as an economic growth factor and essential part in tourism development has increased in these countries as well. In addition to the traditional carriers Gulf Air and Qatar Airways, Dubai's Emirates Airlines was founded 20 years ago as a response to established Gulf Air's refusal to increase connections to Dubai,[14] and it is only two years ago that Etihad Airways was founded as Abu Dhabi's official airline.[15] This development of the traditionally oil fueled economies on the western gulf coast is of course legitimate and follows the tracks of successful Asian economies' development paths, using their airlines as an important vehicle to these ends (e.g. Singapore).[16]

The Gulf airlines are on aggressive expansion paths and are trying to pull traffic not only to their countries, but way beyond, with their home airports designed and destined to become global hub airports. The sheer dimension of the entrants' investment into aircraft (most visibly among them is Emirates with their firm oders of 41 A380 aircraft, see table I-2 above) and ground infrastructure emphasizes their ambitions, which is even aggravated by their government and oil money backing. However, it also illustrates the size of traffic volume that is aimed to be diverted from inter-continental traffic which hitherto is routed especially via Europe and Asia. This approach, of course, is at odds with traditional European and Asian network carriers, which see their traffic threatened and contested by these new aggressive entrants. The Wall Street Journal remarked lately with regard to Emirates as the most visible of these airlines that it "has emerged as one of the fastest growing and most feared competitors in the global airline industry [...] The big fear is that Emirates and its Persion Gulf neighbors will shake up international aviation in the same way no-frills airlines have done on shorter routes on the US and Europe."[17] If this fear is legitimate, or if the Gulf's airlines have taken on a too ambitious approach aiming at the world and finally fighting mainly themselves in their proximate home bases remains to be seen. The overall performance Emirates exhibited over the last decade, its successful recent expansion moves, e.g. in entering and shaking up the trans-Tasman market, however, suggest that potent competitors emerge in the top league of the global airline industry.

The aviation industry is a dynamic industry, perhaps one of the most dynamic industries of all. Major driving forces which stir up the competition in an industry are in constant change here: Technological progress reshuffles the odds

---

[14] See Gostelow (1987) and Sull et al. (2005).
[15] Furthermore, the first low cost carrier in the region was founded in 2003 by another emirate of the UAE: Air Arabia is owned by Sharjah (Kerr, 2003; McSheehy, 2004).
[16] See Bowen (2000).
[17] See Michaels (2005).

INTRODUCTION 13

with regard to efficient production structures and scale economies, the regulatory environment is all but stable and predictable, new competitors emerge whereas old vanish constantly, rivalry among incumbents varies in intensity but is usually significant. Customers are well informed about the product and have easy electronic access in comparing conditions and prices, providing a rather transparent market environment. Researchers and managers need to keep track at least, but better still remain ahead of these developments – even if this means that they are aiming at moving targets in their respective professions.

3 The Chapters in this Book

With this book we provide a comprehensive review of the major branches of the aviation sector in these turbulent times, including cargo airlines and airports. It is divided into five parts. The first part concentrates on recent developments and major strategic issues encountered by passenger airlines. Due to the importance of the regulatory framework (even though deregulated), we commence this section with an assessment of the European Commission's competition policy (Bermig). Subsequently, distinct business models and strategic options in the passenger sector are identified and discussed (Auerbach and Delfmann; Bieger and Agosti) before a variety of chapters focus on the low cost carrier market, currently probably the most important strategic challenge traditional airlines as well as airports are confronted with. Accordingly, we bring together three perspectives on the LCC business, the first by a representative of a successful German low cost airline (Klaas and Klein), the second takes a more remote perspective from a consultant's point of view (Flenskov), whereas in the third an example of the intended reaction of an incumbent airline is portrayed (Klingenberg). Hereafter, a special outlook and assessment of the LCC situation in Southeast Asia is provided (Goh), before a theoretical chapter offers insights into a possible understanding of the situation among LCC by taking a competitive dynamics perspective on the European airline market and derives strategy recommendations (Heuermann).

The second part is dedicated to the special importance alliances play in the world airline industry. A variety of different topics are addressed – from alliance organization (Graumann and Niedermeyer) and benefit allocation among the partners in an alliance (Götsch and Albers) to performance effects of airline alliances (Rühle). Furthermore, a theoretical assessment of the advantage of alliances with regard to alternative options (i.e. mergers) is provided (Fritz), and the illustrative case of Air New Zealand is theoretically assessed and discussed (Pavlovich).

Airports and their strategic challenges are discussed in the third part of the volume. An overview of the current trends and forces in the airport industry is provided first (Aaronson), before a successful strategic turnaround of an airport company is portrayed (Garvens). In a more general sense, and focusing on the airport operating company, two major strategic decision domains are discussed in the following chapters – internationalization and acquisition strategies (Pal and Weil, Koch and Budde). Finally, a major competitive parameter of airports – the connection of the airport to other transport modes (Fakiner) – is introduced and intensively discussed.

In the fourth part, air cargo carriers represent the center of the discussion. A general discussion of the strategic situation of a leading cargo carrier introduces this section (Otto) before special attention is given to the alliance option among cargo carriers (Grönlund and Skoog). A historical review on the importance air cargo plays in the perception of a major global logistics company (Chiavi) as well as a comparative assessment of challenges to the traditional air cargo model (Koch and Kraus) follow. Pricing as a major competitive parameter for air cargo airlines is assessed (Reifenberg and Remmert) before special attention is given to the role of airports in air cargo operations (Hum and Tan).

An interesting final part closes the book which relates the aviation industry to its environment. Airports and their impact on the region they are located in with regard to economical benefits and costs are assessed (Baum). Despite their long history of regulation, the airline industry was among the first transport modes in Europe which was deregulated, with trucking and railroads lagging behind. The final chapter (Kaufhold and Albers) therefore discusses the strategic options airlines face and their transferability on the (meanwhile also deregulated) European railroad sector.

4 References

BOEING (2004): Current Market Outlook 2004, Seattle.

BORENSTEIN, S. (1992): The Evolution of US Airline Competition, in: *Journal of Economic Perspectives* 6(2), Spring, pp. 45-73.

BOWEN, J. (2000): Airline Hubs in Southeast Asia: National Economic Development and Nodal Accessibility, in: *Journal of Transport Geography* 8(1), pp. 25-41.

DOGANIS, R. (2001): *The Airline Business in the 21st Century.* London, New York: Routledge.

FLOTTAU, J. (2005): Lufthansa Takes Over Swiss For Up To EUR300 Million, in: *Aviation Daily* (March 23) 359(55), p. 1.

GOSTELOW M. (1987): Can Emirates Fill the Gulf?, *Director* 40(12), pp. 69-72.

GRAHAM, A. (2001): *Managing Airports - An International Perspective,* Oxford et al.: Butterworth Heinemann.

KNIEPS G. (1987): *Deregulierung im Luftverkehr*, Tübingen: Mohr Siebeck.

MICHAELS D. (2005): Flying Sheik: From Tiny Dubai, an Airline With Global Ambition Takes Off, in: *Wall Street Journal*, 11 January 2005, p. A.1.

SULL, D.N., GHOSHAL, S., MONTEIRO, F. (2005): The Hub of the World, in: *Business Strategy Review* 16(1), pp. 35-40.

Part I

Passenger Airlines

1
COMPETITION POLICY IN EUROPEAN AVIATION MARKETS

CARSTEN BERMIG

1	Introduction	20
2	European Competition Rules Applicable to Air Transport	21
3	Competition Policy Objectives in Aviation Markets	23
4	Enforcement of Competition Rules in European Aviation Markets	25
5	Conclusion and Policy Outlook	37
6	References	38

Summary:
This chapter outlines legal and economic aspects of current European competition and state aid policy in European aviation markets. It describes the liberalization process, the applicable competition rules and competition policy objectives. With regard to competition enforcement, the chapter illustrates assessment criteria concerning airline alliances, predatory pricing, travel agents' commissions and state aid. Finally, a policy outlook on the Commission's regulatory review projects is given.

1 Introduction

The most spectacular growth over the last 20 years among all transport sectors has occurred in aviation, where increasing competition between network carriers has been given added impetus by the emergence of new, low cost carriers. The wider choice of frequencies and fares has spawned a mass market and this, coupled with a quantum leap in the number of destinations served, has cemented greater cohesion and sense of European identity among the traveling population. In terms of passenger kilometers, air transport has increased by over 7% a year and air traffic movements within the old EU15 have increased five fold.[1] Taking into account the impact of enlargement, it is confidently predicted that air transport will have doubled its market share by 2010. All this is a direct result of EU liberalization.

The liberalization process[2] has already triggered a process of consolidation and restructuring in the European aviation sector. Further liberalization of international air transport will bring about even more challenges to an industry that is already in transition.

In its ruling of 5 November 2003, the European Court of Justice concluded that the current "Open Skies" bilateral agreements signed by eight different Member States infringed EU law on two main aspects ("open-skies judgement").[3] The Court confirmed the principle that Community competence for international relations is established wherever internal EC rules have been agreed (here the Internal Market in air transport). It also considered that the clause on the ownership of airlines incorporated in these bilateral agreements ("nationality clause") is contrary to the rules on the right of establishment (Article 43 of the EC Treaty). In essence, a very large number of bilateral agreements has to be re-negotiated in order to introduce the so-called "Community carrier principle".[4]

---

[1] See European Commission (2003).
[2] On the history of the liberalization process see Doganis (2001), pp. 19-54.
[3] Judgements of the Court of Justice of the European Communities in cases C-466/98, C-467/98, C-468/98, C-469/98, C-471/98, C-472/98, C-475/98 and C-476/98 against the United Kingdom, Denmark, Sweden, Finland, Belgium, Luxembourg, Austria and Germany.
[4] See Regulation (EC) No 847/2004 of the European Parliament and of the Council of 29 April 2004 on the negotiation and implementation of air service agreements between Member States and third countries, OJ L 157, 30.4.2004, p. 7.

The introduction of the Community carrier principle in air services agreements concluded between EU Member States and third countries will certainly affect industrial organization in European aviation. Thus, it can be expected that the internal EU aviation market will see further restructuring and consolidation efforts by way of mergers and alliances, rendered possible by liberalization and caused by growing globalization in the international airline industry, i.e. the extension in scope of global alliances. Moreover, the increasing importance of the low cost sector is putting pressure on traditional network carriers to cut costs, in particular in the field of distribution (for example the growing importance of direct internet booking) and Computer Reservation Systems (CRSs). Finally, congestion problems at European airports have an impact on the allocation of slots, airport charges and environmental regulation.

2   European Competition Rules Applicable to Air Transport

Competition in a market economy is the driving force for economic welfare to be obtained by favoring an efficient resource allocation and technical progress. It moreover provides for containing economic power.[5] From the consumers' point of view competition is a simple and efficient means of guaranteeing high level products and services in terms of the quality and price. In order to be effective, competition assumes that the market is made up of suppliers who are independent of each other, each subject to the competitive pressure exerted by the others.

Agreements which restrict competition are prohibited (Article 81 of the EC Treaty). This is for example the case of price-fixing agreements and cartels between competitors. On the other hand, some horizontal agreements between competitors might bring concrete benefits to consumers, e.g. airlines alliances. To this latter type of agreements Article 81(3) is applicable, if the companies involved can demonstrate that the agreement cumulatively fulfills four conditions.[6]

---

[5]   See Neumann (2001), p. 45.

[6]   The four conditions are: "improving the production or distribution of goods or promoting technical and economic progress", "while allowing consumers a fair share of the resulting benefits", "[without] imposing on the undertakings concerned restrictions which are not indispensable" and "[without] affording such undertakings

Firms in a dominant position may not abuse of that position (Article 82 of the EC Treaty). This is for example the case for predatory pricing aiming at driving another carrier out of a certain route.[7]

Since 1 May 2004 a new merger Regulation is in force. The economic rationale underlying the merger control regulation[8] is that competition is a means to achieving efficient market outcomes and increased economic welfare. Consequently, distortions to the structure of a market, such as large-scale mergers that may "significantly impede effective competition in the common market or in a substantial part of it, in particular by the creation or strengthening of a dominant position",[9] ought to be screened and if necessary prevented by the Commission. However, mergers may not always be harmful for competition. When motivated by the companies' desire to become more efficient and competitive, mergers may contribute to the very process of optimal resource reallocation and improve the competitive performance of affected markets. Merger-specific efficiency gains can offset price increases or other anti-competitive effects caused by the creation or strengthening of a dominant position resulting from a merger.[10]

State aid that distorts intra-Community competition is prohibited by the EC Treaty (Article 87 and 88 and for transport also Article 73). State aid can frustrate free competition, seriously disturb the most efficient allocation of resources and posing a threat to the smooth running of the Internal Market. In many cases, the granting of state aid reduces economic welfare by preventing the

---

the possibility of eliminating competition in respect of a substantial part of the products in question".

[7] The European Commission is empowered by the EC Treaty to apply these prohibition rules of Article 81 and 82. It may impose substantive fines for their violations. Since the entry into force of Council Regulation (EC) No 1/2003 on 1 May 2004 (OJ L1, 4.1.2003), all National competition authorities are also empowered to apply fully the provisions of the Treaty in order to ensure that competition is not distorted or restricted. National courts may also apply directly these prohibitions so as to protect the individual rights conferred to citizens by the Treaty. Moreover, Regulation 1/2003 now also applies to air transport between the EU and third countries, which previously was not the case. Although there was no doubt that competition rules also applied also to long-haul routes, the Commission lacked the effective enforcement powers in this field. Therefore the assessment of a number of international alliances obliged the Commission to separate procedurally the intra-Community routes from third country routes.

[8] See Council Regulation (EC) No 139/2004 of 20 January 2004 on the control of concentrations between undertakings, OJ L 24, 29.1.2004, pp. 1-22.

[9] See Article 2 of Council Regulation (EC) No 139/2004.

[10] See De La Mano (2002).

most efficient resource allocation and weakens the incentives for firms to improve efficiency. Aid may also enable the less efficient to survive at the expense of the more efficient, delaying structural change and hindering productivity growth and competitiveness. Another basic reason for having a system of state aid control is "the risk of a subsidy race where Member States might outbid each other"[11] which would result in wasting tax payers' money. Accordingly, by giving certain carriers an advantage or air services favored treatment to the detriment of other carriers or services, it seriously disrupts competitive forces in European aviation markets.

But in certain cases, the Commission will apply the exceptions allowed by the Treaty and have regard to potential welfare enhancing effects of state aid and authorize aid where it is justified for example in case of market failure[12] or for regional development. In the aviation sector state aid is mainly granted for rescue and restructuring of ailing airlines, public service obligations or more recently for regional development purposes.

3 Competition Policy Objectives in Aviation Markets

Competition policy in this sector is underpinned, first, by the Lisbon process, one requirement of which is that liberalization of sectors such as transport should be "speeded up"; and, second, by the Cardiff process, which set in motion the integration of environmental objectives – chiefly sustainable development – into EU policy generally. Concretely, this means pursuing the liberalization agenda as a means of encouraging people to abandon their cars in favor of public transport and of persuading freight customers to shift from road haulage to more environmentally friendly modes; and using antitrust enforcement to ensure that progressive and pro-competitive market integration is maintained and not undermined by restrictive or exclusionary business practices.

Competition policy should not stand in the way of pro-competitive restructuring and efficiency enhancing co-operation between carriers. However, the Commission cannot allow industry agreements or unilateral conduct by carriers

---

[11] See Sinnaeve (1999), p. 13.
[12] For example positive and negative externalities or public service obligations. For an extensive discussion of market failures which are relevant to the analysis of state aid see European Commission (1999), pp. 25-31.

that are intended or have the effect of purely defending established positions on the market without additional efficiencies or consumer benefits being created.

Anti-trust enforcement also aims at ensuring a level playing field for competing business models (for instance low cost model versus network carriers) and for new entrants, i.e. access to infrastructure or slots. Furthermore, it monitors restrictive effects of certain business practices, e.g. loyalty programs or predatory pricing.

Due to the complexities related to changes in the regulatory framework brought about by the open-skies judgment, the Commission will increase its attention for global alliances. In this context it will also play an active role in the re-negotiations of bilateral air service agreements between Member States and third countries with a view to bringing them in line with Community law, particularly as regards Internal Market and competition rules.[13] In order to prevent the risk of conflicting decisions of different authorities, the Commission will further enhance co-operation with main third country authorities (such as the US, Canada, Australia, etc.).

Pro-competitive effects of measures taken in the market for air transport do not reach the end-consumer if there is a lack of competition in related downstream and upstream markets. Competition policy in European aviation therefore also monitors the structure and conduct of actors in related markets, e.g. the level of airport charges or potential anti-competitive effects of computer reservation systems.

The control of state aid focuses on the effects of competition of aid granted by Member States to undertakings. The EU system of state aid control – where the Commission scrutinizes and approves state aid measures by Member States – aims at minimizing all economic inefficiencies attached to the granting of aid. Indeed over the years, state aid control has been instrumental in attenuating many of the worst effects and impact of state aid. At the European Council Summits in Lisbon and later in Stockholm, Member States recognized that in order for Europe to become more competitive, increase productivity and deliver sustainable economic growth, with more and better jobs and greater social cohesion, it is necessary to improve its competitive position not only by ensuring that European society itself is both dynamic and knowledge-based, but also by reducing levels of state aid and redirecting it towards horizontal policy objectives of Community interest, such as research and development, economic and social cohesion or environmental protection. Nevertheless, some Member States are

---

[13] See Recital 16 and Article 1 of Regulation 847/2004, OJ L 157, 30.4.2004.

continuing to award aid that is particularly distortive of competition, such as rescue and restructuring aid to airlines.

4 Enforcement of Competition Rules in European Aviation Markets

This section outlines recent developments of competition enforcement in the aviation sector and describes the relevant competition assessment criteria, notably as regards airline alliances, predatory pricing, travel agents' commissions and state aid.

4.1 Alliances

The European airline industry has seen an increasing number of international airline alliances in recent years that has a significant impact on competition. The Commission takes a broadly positive approach to international airline alliances and mergers. Alliances can bring benefits to the economy from cost savings, as well as to consumers in the form of service improvements such as seamless services, improved schedules or reduced fares.[14]

However, "while rarely stated publicly as an objective when airline alliances are formed, there can be little doubt that airline executives see alliances, especially when they involve code sharing and capacity rationalization as a way of reducing or limiting competition. The reduction of effective competition is likely to be most marked in route-specific or regional alliances and least clear-cut in global alliances where routings via competitors' hubs may be a feasible alternative for long-haul passengers."[15]

As this statement suggests, there is a risk that the benefits of an alliance will be achieved at the expense of restricting or eliminating competition in certain markets. The Commission therefore usually tries to ensure continued competition on all routes affected by alliances by imposing a set of remedies that have the effect of facilitating market entry.

---

[14] See Gremminger (2003), p. 75.
[15] See Doganis (2001), p. 79.

Even if each alliance case[16] must be assessed on its own merits and every situation tends to be different, the competitive analysis carried out by the Commission follows a general approach. The approach consists of four steps: the market definition, the identification of competitively affected markets, the analysis of the conditions of Article 81 (3), notably benefits, entry barriers and elimination of competition, and finally the imposition of effective remedies.[17]

4.1.1 Market Definition

To establish the definition of the relevant market in air transport cases, the Commission applies the so-called 'point of origin/point of destination' (O&D) pair approach. According to this approach, every combination of a point of origin and a point of destination should be considered to be a separate market from the customer's viewpoint. To establish whether there is competition on an O&D market, the Commission looks at the different transport possibilities in that market, that is, not only at the direct flights between the two airports concerned, but also other alternatives[18] to the extent that they are substitutable to these direct flights. The Commission also investigates whether passengers traveling on unrestricted tickets are part of a different market from passengers with restricted tickets. These two groups of passengers are labeled as "time-sensitive" and "non-time sensitive" respectively.

Network carriers, operating a so-called hub and spoke system, usually argue that the market definition used in air transport should take into account that the airline industry is characterized by network competition among airlines alliances. Imagine however a consumer wishes to fly from a concrete point of origin to a certain point of destination. If no choice between airlines on this particular O&D

---

[16] The following part mainly outlines the competitive analysis of airline alliances under Article 81. It should be noted, however, that the competitive analysis of a merger is largely similar to that of an alliance.

[17] For an overview on the current practice regarding market definition, competition assessment and remedies in aviation merger and alliance cases of European competition authorities see ECA (2004). See also an OECD publication on the same topic: OECD (2000).

[18] These alternatives may be direct flights between the airports whose respective catchment areas significantly overlap with the catchment areas of the airports concerned at each end (airport substitution), indirect flights between the airports concerned, or other means of transport such as road, train or sea (inter-modal substitution). Whether one of those alternatives is substitutable to the direct route depends on a multiplicity of factors, such as the overall travel time, frequency of services and the price of the different alternatives and can only be decided on a route-by-route basis.

pair exists, the fact that airlines compete world-wide in the development of their respective networks is of no importance to the consumer. Thus, while the emphasis on network competition reflects the view of the supply side, from a demand side perspective it is necessary to analyze the effects of the co-operation primarily under the O&D pairs approach.

While network competition issues have little influence on the established market definition of the Commission, certain competition effects resulting from networks (e.g. frequent flyer programs, corporate discounts or limited access of third carriers to feeder traffic) may have to be considered as 'market entry barriers' in the context of assessing the alliance partners' market power on the respective routes.

4.1.2 Identification of Competitively Affected Markets

The competitive assessment of airline alliances is complex not only due to the high number of potentially affected O&D markets but also by the network nature of the industry. In practice, competition could take place between routes as well as on a specific route. When trying to identify the markets affected by an alliance, the Commission typically categorizes two broad types of affected markets: overlap and non-overlap markets. The first type deals with routes where the alliance partners operated independent services prior to the alliance (actual competition) and the second with routes on which only one alliance partner offered services before the co-operation, but where the non-operating partner could be considered a potential entrant (potential competition).

Competition assessment on overlap markets: as regards intra-European alliance cases the Commission in principle only assesses direct-overlap markets. For long-haul traffic the situation is insofar different as both direct and certain indirect routes may belong to the same relevant market.

Competition assessment on non-overlap routes: the Commission applies an economic approach towards potential competition, relying on the notion of a real commercial possibility of entry.[19] An airline will in principle only be considered as a potential competitor on a specific route if that route is either directly linked to one of its hubs or sufficiently frequented by local traffic to allow market entry on point-to-point basis, while taking into account the operational requirements of the respective business models.

---

[19] Joint cases T-374/94 etc. European Night Services and Others v. Commission [1998] ECR II-3141.

4.1.3 Market Power, Benefits, Entry Barriers and Elimination of Competition

Airline alliances usually involve a high degree of business integration (e.g. route coordination, revenue sharing, joint-pricing and marketing) which ends any actual and potential competition between alliance partners. In addition, alliances normally have, at least on some routes (in particular on the hub-to-hub routes), rather high market shares. Thus alliance agreements are usually caught by Article 81 with respect to all routes where the co-operation leads to a restriction of competition.

The Commission in principle accepts that alliances may contribute to improving the production and distribution of transport services and promote technical and economic progress. The Commission further accepts that connecting passengers can enjoy the various types of alliance benefits, such as wider choice of destinations, seamless service, and lower fares. It is however vital that alliance benefits are passed on to passengers. Benefits are more likely to be passed on the more complementary the respective networks are and consequently the less overlap routes exist. However, more difficult is to establish the benefits to be expected for point-to-point passengers on the routes where the parties were actual or potential competitors before. The Commission therefore puts a higher emphasis on receiving clear evidence on the expected benefits to point-to-point passengers on these routes.

Whether competition is likely to be eliminated depends mainly on the market position of the alliance, on the relevant markets and on the existence of significant entry barriers, hence whether the parties will enjoy market power, which might allow them to act independently of their competitors and customers. Airline alliances often lead, in particular on the overlap routes, to high market shares. The Commission does not consider high market share as such as a sign of significant market power and that competition is likely to be eliminated. On the other hand the Commission is well aware of "the imperfection of contestability of airline markets",[20] i.e. as regards potential competition the Commission relies on the notion of a real commercial possibility of entry rather than theoretical considerations.

In the airline industry, however, high market shares often come together with high entry barriers. In airline alliance cases the Commission usually faces various types of entry barriers. The most important are: regulatory barriers, such as government pricing restrictions for indirect flights or the unavailability of

---

[20] See Baumol (1989), p. 25; see also Hanlon (1999), pp. 41-42.

necessary traffic rights, slot shortages at congested airports, increased frequencies resulting from the cooperation, network effects resulting from joint frequent flyer, travel agency or corporate customer discounts or reduced third carrier access to transfer passengers, 'behavioral' barriers arising from possible predatory pricing or predatory frequency increases[21] aiming at squeezing new entrants.

Whether market power resulting from high market share and/or entry barriers is likely to eliminate competition needs to be primarily established on a route-by-route analysis. Of particular competition concern are usually barriers resulting from slot shortages at congested airports, because these are directly marginalizing the threat of entry of potential competitors on any route out of the airport concerned. Moreover it might also be necessary to assess certain network effects across all affected routes in order to identify market power arising from the overall strength of an alliance at their respective hubs or vis-à-vis certain customer groups. Absent significant entry barriers, the Commission analyzes to what extent actual and potential competition provides sufficient constraints on the parties. If for instance actual competitors could easily increase capacity to accommodate a significant proportion of customers currently carried by the alliance, this should in principle signal competitive constraints on the alliance's behavior on the routes concerned.

4.1.4 Remedies

If however an alliance eliminates actual or potential competition on a route, the alliance partners usually propose commitments that remedy the Commission's competition concerns. The Commission has recently taken three decisions in the aviation sector[22] that include a considerable amount of commitments that are binding to the respective airlines. In two alliance cases it granted an anti-trust exemption for six years for agreements between British Airways and Iberia and between Air France and Alitalia. In February 2004, the Commission also cleared the merger between Air France and KLM which will create the largest airline group in Europe.

With a view to address the Commission's concerns, the parties offered a large number of commitments.[23] In the case of the two alliance agreements, the main

---

[21] See Doganis (2001), p. 72; or Hanlon (1999), pp. 214-217.
[22] See Bermig (2004).
[23] Apart from the slot related commitments discussed here, depending on the case, commitments also comprise issues like frequency freeze, block-space, Frequent-Flyer, inter-lining and inter-modal agreements, as well as price regulation.

purpose of such commitments has been to strengthen already existing competitors. Due to the lack of available slots at airports, these competitors were not in a position to increase their operations and to offer a viable service in particular to time-sensitive (business) customers. The main remedy therefore has been that the parties offer additional slots to such competitors.

As some of these routes are attractive to new entrants, it cannot be excluded that demand for slots will exceed the total number of slots offered by the parties. Some priority rules therefore had to be established. The Commission considers that it is more effective to add frequencies to an existing service than to start a new service from scratch on a particular city pair. In addition, in terms of competition, it is considered that a competitor offering a package of flights per day will have more chance to compete efficiently against the parties than several competitors each offering only a limited number of flights. If demand exceeds supply of additional slots, preference is therefore given to competitors who are already established on a particular city pair. Similar considerations also apply for the slot release under the commitments of the Air France / KLM merger.[24]

With regard to long-haul services to third countries in the merger case, the Commission's entire analysis hinges on the assumption that indirect flights would offer competitive constraints for direct ones. As this is put into doubt by the governmental price regulation, a crucial condition for clearing the merger are the French and Dutch governments' declarations that they will refrain from any intervention into the price setting of indirect services on a large number of long-haul city pairs.

4.2 Predatory Pricing

Generally predatory pricing[25] is defined in economic terms as a price reduction that is profitable only because of the increased market power the predator gains from eliminating, disciplining or otherwise inhibiting the competitive conduct of a (potential) competitor. Hence, a predatory price is profitable only because of its exclusionary effects. Predation is difficult to prove since it is not easy to distinguish a price reduction with predatory intent from a reduction that

---

[24] To some extent, commitments offered under the merger exceed those offered in the alliance cases. This reflects the different market situations and takes into account that the clearance of the merger duration is given for an unlimited period of time, whereas the exemption decisions in the alliance cases were delivered for only six years.

[25] For an extensive description of predatory pricing see Motta (2004), pp. 413-454.

represents competition on merits. Competition authorities frequently use so-called 'cost-price-tests'[26] in order to determine whether a price has any legitimate business purpose or whether it is being used to squeeze out competitors and prevent entry of rivals that are at least as efficient as the incumbent firm.[27] But even when a company's price is above costs it still might have a predatory intent.[28] Thus a cost-price test is not sufficient to disclose predation and other relevant factors have to be taken into account, i.e. the market structure or the predatory intent of the company.

While the European Commission has not dealt with a predation case in the aviation sector, Bundeskartellamt, the german federal anti-trust authority, issued a decision in 2002.[29] Bundeskartellamt saw the pricing strategy of Lufthansa on the Frankfurt-Berlin/Tegel route as an attempt to drive its new competitor, Germania, out of the market and feared that emerging competition would be substantially harmed as a result. Consequently it imposed on Lufthansa a price (including passenger fees) for a one-way ticket on this route which is at least €35 above Germania's price, with a total price ceiling of €134.

In its reasoning[30] Bundeskartellamt described the structural features[31] of airline passenger services. These features increase the chances of successful predatory pricing strategies of incumbent network carriers since they are able to limit the costs of predation.[32] In addition, the hub-and-spoke systems and the implied cost advantages as well as frequent flyer programs might be sufficient incentives for incumbent network carriers to engage in predatory behavior. Bundeskartellamt used the average total cost test and assessed on the basis of additional evidence Lufthansa's overall predation strategy. Bundeskartellamt found that Lufthansa was for instance able to recoup the predation costs on the route concerned on

---

[26] Price-cost tests are comparing the price with the average total, average variable or marginal cost of the alleged predator. See for instance Baumol (1996) or Areeda (1975).
[27] See Baumol (1996), p. 50.
[28] See for instance Joskow (1979) or Tirole (1988), p. 373.
[29] See Bundeskartellamt (2002).
[30] See Ewald (2003).
[31] The features are decreasing average total costs per passenger, economies of density, entry barriers (i.e. slot shortages) and the requirement to maintain a minimum flight frequency.
[32] If a newcomer enters the market at a minimum frequency, it might be sufficient to reduce prices for a relatively small portion of total capacity. Since the option to answer this strategy with capacity adjustments is limited, the newcomer will be confronted with decreasing load factors and rising average total costs per passenger.

their whole route network and the fact that the pricing strategy was limited to the route Frankfurt-Berlin was an indication of their predatory intent.

Meanwhile, due to the emergence of low cost carriers there have been vast changes both in price levels and pricing behavior of airlines in domestic German and European markets. In its overall appraisal of the changes, Bundeskartellamt has come to the conclusion that a continuation of the legal dispute in this matter will not produce any useful effects.[33]

4.3 Loyalty Programs

Loyalty schemes[34] are part of the carrier's marketing strategy. Hence loyalty schemes such as any other marketing expenditure aim at influencing the behavior of the target group in such a way that they become more inclined to seek a specific airline's products and services. In this way loyalty programs have two fundamental purposes. Firstly, to increase the carrier's share of passenger demand and secondly to make demand less price elastic.[35] Three groups of loyalty programs can be distinguished: Frequent Flyer Programs (FFPs) targeted to individual (business) travelers, travel agents' commissions and discount schemes for corporate customers.[36]

4.3.1 Frequent Flyer Programs

The main aim of an FFP is to induce customer loyalty in order to bundle customer demand. Route network size seems to be important in the success of airlines FFPs. Firstly, an airline with an extensive range of products can offer travelers a wide selection of bonus trips, and FFPs that offer a greater number of alternatives are valued more highly by travelers. Secondly, a traveler is more likely to accumulate enough points for a bonus trip if the airline has a relatively large range of routes and departures.[37]

As regards FFPs' effects on competition three aspects are of particular interest: barriers to entry, price elasticities and switching costs. If an incumbent carrier

---

[33] Bundeskartellamt has therefore agreed with Lufthansa that it will not derive any rights from its ruling.
[34] On competitive effects of loyalty schemes see also OECD (2003) and Hanlon (1999), pp. 59-67.
[35] See Hanlon (1999), p. 50.
[36] Note that corporate discount schemes are not dealt with in this chapter.
[37] See Borenstein (1989), p. 346.

has been able to capture a large part of the potential customers (especially business passengers) into its FFP, a new entrant may find it exceedingly difficult to attain an economically viable market share. Thus, in cases where an FFP is used by an airline that is dominant in a market, the program might effectively deter entry,[38] in particular into a hub airport of the dominant carrier.[39] Alliances tend to reinforce this market dominance, which in turn strengthens the entry barriers that FFPs represent.[40]

If an airline company introduces an FFP, traveling with that company becomes more attractive than traveling with other carriers. Price elasticity is expected to decrease and, ceteris paribus, the company can charge a higher price for its flights. Competing carriers might respond to the competitive advantage of an FFP by introducing fare reductions or increasing their number of departures. The extent of the effects depends on the substitutability of services of the various carriers and the price-sensitivity of customers. If the carriers provide almost equal services, the impact on prices is likely to be considerable. On the other hand the impact on price declines the more price-sensitive customers are.

Switching costs,[41] i.e. the cost to the passenger of switching from one carrier to another, are an important factor in both empirical and theoretical analyses of FFPs' competitive effects. FFPs contribute to switching cost by providing travelers an incentive to use the same airline again, which raises the cost to the customer of switching airlines. Hence, switching costs are likely to allow FFP-carriers to raise fares above a competitive level, which means that such costs may partly be viewed as a measure of consumer welfare loss.[42]

The European Commission has dealt with FFPs in a number of cases concerning airline alliances. As part of the alliance agreements, the respective carriers allow the alliance partner's clients to collect and use accumulated points in each other's FFPs. In the SAS/Lufthansa case for instance the Commission stated that the cooperation between the two companies on FFPs was likely to be a not

---

[38] See Borenstein (1996), p. 4.
[39] See Borenstein (1989), p. 362.
[40] In the same way corporate deals might also strengthen FFPs and thereby reinforce a dominant position.
[41] On switching costs see for instance Klemperer (1995).
[42] The results of an empirical study of domestic air travel in Sweden show that the FFP of SAS has resulted in a higher ticket price for SAS travel compared with the prices of other airlines and also compared with periods during which SAS did not apply its FFP on competitive routes. Further, the results of the analysis show that the SAS's FFP has had a significant effect on switching cost for SAS travelers [see Swedish Competition Authority (2003)].

inconsiderable barrier to market entry. The Commission's condition for approval under Article 81 (3) of the EC Treaty was that any other airline which provided or wished to provide services on the respective routes, and which did not have a FFP, should be allowed to participate in the program.

In Scandinavia, the attitude towards FFPs is more critical. In a ruling, the Swedish Market Court (27 February 2001) ordered the SAS not to apply its FFP in such a way that passengers earning points were able to redeem them as bonus awards or the equivalent when using certain air travel services. The practice was deemed to be an abuse of SAS's dominant position.[43] The ruling applies to domestic air travel in Sweden between cities where SAS encounters competition through existing or newly established scheduled air passenger traffic.

The Norwegian Competition Authority ordered (18 March 2000) SAS to stop awarding FFP points on domestic Norwegian routes. Unlike the Swedish ruling, the prohibition in Norway applies on all domestic routes, competitive or not. According to the Norwegian Competition Authority the intervention appears to have been effective.[44]

4.3.2 Travel Agents' Commissions

The Commission has carried out investigations into the incentive schemes for travel agents operated by several EU airlines.[45] It had to ascertain that these incentive schemes were not used by dominant carriers to remunerate travel agents for their loyalty, thereby creating illegal barriers to entry for their competitors. In several cases the Commission's investigation has triggered an in-depth reform or even a complete replacement of existing incentive schemes with a view to bringing them into conformity with EU competition rules.

The most prominent case arose from a complaint from Virgin against British Airways' system of commissions for UK travel agents. Acting upon this complaint, the Commission investigated BA's incentive schemes. This investigation resulted in the adoption (on 14 July 1999) of a decision with fines finding that the incentive schemes BA had operated for UK travel agents were in

---

[43] In breach of Section 19 of the Swedish Competition act.
[44] Only one month after the ban on frequent flyer point collection took effect, a new entrant opened service on the four major domestic routes. As of January 2004, this carrier operates 12 domestic routes and five international ones.
[45] See Stehmann (2003).

COMPETITION POLICY IN EUROPEAN AVIATION MARKETS   35

breach of Article 82 of the Treaty that has been upheld by the Court of First Instance.[46]

The assessment made by the Commission in the said case is summarized as follows: it is well established Community law that loyalty discounts, i.e. discounts based not on cost savings but on loyalty, constitute an exclusionary abuse of a dominant position. BA schemes displayed two features: firstly, the rebates were calculated on the increase of sales realized by each travel agent during the reference period over the previous reference period. Secondly, once the threshold for receiving the rebate was exceeded, the rebate was granted for all tickets sold during the reference period, including those below the threshold.

The incentives had what the Commission referred to as "very noticeable effects on the margin". Once a travel agent is close to reaching a sales target, meaning that he will receive a higher commission rate, he is unlikely to promote air tickets of airlines other than BA as he will not only lose the higher commission over the incremental sales, but over all BA sales he made during the reference period.

Under Article 82 of the Treaty, applying dissimilar conditions to equivalent transactions with other trading parties constitutes an abuse of a dominant position. Under BA's incentive schemes, two travel agents handling the same number of BA tickets and providing the same level of service to BA received a different commission rate if their sales of BA tickets were different in the previous year. BA was consequently found to have abused its dominant position by discriminating between travel agents.

4.4 State Aid

Before liberalization of the airline sector, considerable amounts of aid (exceeding €2.5 billion in 1994 and 1995) were awarded to national airlines for restructuring in the mid-nineties.[47] Since 1997, however, aid levels to the industry have fallen dramatically though some aid continues to be authorized. In 2002, the Commission authorized state aid schemes set up by several Member States[48] to compensate airlines for the losses caused by the closure of certain parts of the airspace between 11 and 14 September 2001. Around €110 million of aid was approved in 2002. The Commission also authorized the payment of the

---

[46] Judgment by the Court of First Instance in the case T-219/99, 17 December 2003.
[47] See also Doganis (2001), pp. 201-209.
[48] France, Germany, Austria and the United Kingdom.

third tranche of €129 million, which was part of the restructuring aid package for Alitalia in 2001. Further public funding of this airline of around €1.4 billion was deemed to be no state aid.[49]

In 2004, the Ryanair/Charleroi state aid decision[50] caught a lot of attention since it was the first time a low cost carrier received state aid. The Commission, acting upon a complaint, had to determine whether the measures taken in favour of Ryanair by the Walloon Region and Brussels South Charleroi airport (BSCA), a public undertaking controlled by the Walloon Region, were compatible or not with the so-called private market investor principle. In the case of Charleroi, the Commission concluded that no private operator in the same circumstances as BSCA would have granted the same advantages. Since the private market investor principle[51] had not been adhered to in this case, the advantages granted to Ryanair constitute State aid[52] which could distort competition in favor of Ryanair.

The Commission, however, took the view that the aid granted to Ryanair at Charleroi that is intended for the launch of new routes, could be compatible with the common market in the context of transport policy.[53] The aid granted permits the development and improved use of secondary airport infrastructure which is currently underused and represents a cost to the community as a whole. However, certain forms of aid have not been authorized since they have not been awarded in a transparent and non-discriminatory manner.[54]

With this decision, the Commission gave a clear signal in favor of the extension of the low cost model. The Commission's decision does not prevent agreements

---

[49] See European Commission (2004), p. 16.

[50] Commission Decision of 12 February 2004, Official Journal, L 137, 30.4.2004.

[51] According to this principle, which has been repeatedly endorsed by the Court of Justice, aid is not considered to include investments or advantages which are granted if, at the time when an investment or a commercial contract is being considered, the public undertaking is in the same situation as a comparable private firm, guided by the same objectives of long-term profitability. This principle helps ensure equal treatment between public-sector operators, such as BSCA, and private firms.

[52] Within the meaning of Article 87(1) of the EC Treaty.

[53] By reference to Article 87(3) of the EC Treaty.

[54] In particular the discounts on airport charges such as were granted at Charleroi, which go beyond the discounts already foreseen in the Belgian legislation, the reduced ground handling fees, which are not offset by possible surpluses from other, purely commercial activities (parking, shops, etc.), one-shot incentives paid when new routes were launched, where no account was taken of the actual costs of launching such routes, and the aid provided in respect of the Dublin-Charleroi route, which is not new, because it was launched in 1997.

between regional airports and low cost airlines. Quite the contrary the Commission welcomes any arrangement which would help to solve the problems of air transport congestion, in this context secondary airports are well-placed to play an important role.

5 Conclusion and Policy Outlook

In recent years the Commission applied competition and state aid rules to the European aviation sector mainly in the fields of (European) airline alliances, travel agents' commissions and restructuring, and regional aid. Furthermore, national competition authorities also dealt with anti-trust cases as regards predatory pricing and frequent flyer programs.

In the following years these enforcement activities will be widened in their geographical scope and accompanied by a number of regulatory reform projects that concern notably the IATA tariff conferences, the slot regulation and computer reservation systems (CRS).

Since the Commission received in 2004 full power to apply the competition rules to air transport between the EU and third countries, it can now also play its full role for international airline alliances, mergers and conduct.

The current block exemptions on industry wide (i.e. IATA) consultations on passenger tariffs, slot allocation and airport scheduling ends in June 2005 and is currently under review. The essential question of the review process is whether the current IATA tariff conferences are indispensable for the provision of efficient interlining services to the benefit of consumers.

The Commission has announced to undertake a thorough review of the current regulation on allocation of slots in airports, in particular with a view to introduce potential market oriented mechanisms, such as primary and secondary trading or redistribution of grandfathered slots. In January 2004, the Transport Directorate-General published a "study to develop market oriented slot allocation schemes and to assess their feasibility".[55] From a competition standpoint, it is essential to guarantee as much market flexibility as possible for both incumbents and new entrants while at the same time providing a framework for stable air services.

---

[55] The study is published on the DG Transport web-site under the following address: http://europa.eu.int/comm/transport/air/rules/doc/2004_01_24_nera_slot_study.pdf.

The Commission proposes to deregulate CRS. While there appears to be justification for taking away the current regulatory straight jacket for CRSs, it might still be necessary to maintain some form of regulatory protection against the potential anti-competitive effects of carrier owned CRSs.

Note

The views expressed in this chapter are purely those of the writer and may not in any circumstances be regarded as stating an official position of the European Commission.

6 References

AREEDA, P.; TURNER, D. (1975): Predatory Pricing and Related Practices under Section 2 of the Sherman Act. *Harvard Law Review*, 88(4), pp. 697-733.

BAUMOL, W.J. (1996): Predation and the Logic of the Average Variable Cost Test. *Journal of Law and Economics*, 39(1), pp. 49-72.

BAUMOL, W.J.; WILLIG, R.D. (1986): Contestability: Developments since the Book. *Oxford Economic Papers*, 38(0), pp. 9-36.

BERMIG, C.; LAMALLE, M.; STEHMANN, O. (2004): New Developments in the Aviation Sector. Consolidation and Competition: Recent Competition Cases. *Competition Policy Newsletter*, 2 (Summer), pp. 65-68.

BORENSTEIN, S. (1996): Repeat-Buyer Programs in Network Industries. In: Sichel, W. (Ed.): *Networks, Infrastructure, and the New Task for Regulation*, University of Michigan Press.

BORENSTEIN, S. (1989): Hubs and High Fares: Dominance and Market Power in the US Airline Industry. *Rand Journal of Economics*, 20(3), pp. 344-366.

BUNDESKARTELLAMT (2002): Beschluss in Verwaltungsverfahren gegen Deutsche Lufthansa AG, Köln, vom 18.2.2002, Geschaeftszeichen: B9-144/01 (www.-bundeskartellamt.bund.de).

DE LA MANO, M. (2002): For the Customer's Sake: The Competitive Effects of Efficiencies in European Merger Control. *Enterprise Papers*. 11, Enterprise Directorate-General, European Commission, Brussels. http://europa.eu.int/-comm/enterprise/library/enterprise-papers/pdf/enterprise_paper_11_2002.pdf.

DOGANIS, R. (2001): *The Airline Business in the Twenty-first Century*. London: Routledge.

EUROPEAN COMMISSION (2004): Report, State Aid Scoreboard, spring 2004 update, COM (2004) 256 final, 20.4.2004, http://europa.eu.int/comm/competition/-state_aid/scoreboard/2004/en.pdf.

EUROPEAN COMMISSION (2003): Panorama of Transport. *Statistical Overview of Transport in the European Union*, Luxembourg: Eurostat.

EUROPEAN COMMISSION (1999): State Aid and the Single Market. *European Economy, Reports and Studies*, 3, Brussels.

EUROPEAN COMPETITION AUTHORITIES (ECA) (2004): Mergers and Alliances in Civil Aviation – An Overview of the Current Enforcement Practices of the ECA Concerning Market Definition, Competition Assessment and Remedies. *Report of the ECA Air Traffic Working Group*, http://europa.eu.int/comm/competition/-publications/eca/report_air_traffic.pdf.

EWALD, C. (2003): Predatory Pricing in the Airline Industry as a Challenge to Competition Law Enforcement – an assessment of the current legal practice in the US and Germany. *Study group economic concepts of Bundeskartellamt*, Bonn.

GREMMINGER, M. (2003): The Commission's Approach Towards Global Airline Alliances – Some Evolving Assessment Principles. *Competition Policy Newsletter*, 1(Spring), Brussels.

HANLON, P. (1999): *Global Airlines, Competition in a Transnational Industry*, 2nd Ed., Oxford: Butterworth-Heinemann.

JOSKOW, P.; KLEVORICK, A. (1979): A framework for analyzing predatory pricing: the debate between Chicago and Post-Chicago. *Yale Law Journal*, 89(2), pp. 213-270.

KLEMPERER, P. (1995): Competition when Consumers Have Switching Costs: An Overview with Applications to Industrial Organization, Macroeconomics and International Trade. *Review of Economic Studies*, 62(213), pp. 515-539.

MOTTA, M. (2004): *Competition Policy: Theory and Practice*, Cambridge, UK: Cambridge University Press.

NEUMANN, M. (2001): *Competition Policy – History, Theory and Practice*, Cheltenham, UK: Edward Elgar.

OECD (2003): Loyalty and Fidelity Discounts and Rebates. *Series of Roundtable Discussions*, 40, Paris.

OECD (2000): Airline Alliances and Mergers. *Series of Roundtable Discussions*. 26, Paris.

SINNAEVE, A. (1999): State Aid Control: Objectives and Procedures. In: Bilal, S. and Nicolaides, P. (Eds): *Understanding State Aid Policy in the European Community – Perspectives on Rules and Practice*, The Hague: Kluwer International, pp. 143-157.

STEHMANN, O.; TOMBOY, C. (2003): Commission Closes Probe into Major EU Airlines' Incentive Schemes for Travel Agents. *Competition Policy Newsletter,* 2(Summer), Brussels.

SWEDISH COMPETITION AUTHORITY (2003): *There is no Such Thing as Free Lounge – a Report on Frequent Flyer Programs.* http://www.kkv.se, retrieved 1 April 2003.

TIROLE, J. (1988): *The Theory of Industrial Organization,* Cambridge, Massachusetts: MIT Press.

2
BUSINESS MODELS IN THE AIRLINE SECTOR – EVOLUTION AND PERSPECTIVES

THOMAS BIEGER AND SANDRO AGOSTI

1 Business Models as a New Dimension in Strategy Theory............................42
2 Business Models and their Economies in the Airline Industry 50
3 Transformation of Airline Business Models and their Performance..............56
4 The Perspectives of Airline Business Models – Convergence or Differentiation..59
5 References..63

Summary:
As the aviation industry enters its consolidation stage, the definition of the right business model becomes crucial. Traditionally, four main business models in passenger air transport are identified: network carrier, charter carrier, regional carrier and low cost carrier. By increasing competition, many companies transform themselves and adopt elements of other business models. For example Swiss International Air Lines, a traditional network carrier, uses pricing models and service concepts of low cost carriers whereas Air Berlin, a new entering low cost carrier, provides service standards equal to average network carriers. Although most charter airlines are owned by tour operators, they more and more sell seats on an individual base. The question arises whether these are first signs of a convergence of airline business models or whether new types of models will emerge. This contribution draws on theories and various case studies. It is based on the results and data of an earlier study (Bieger et al., 2002) which have been further developed.

1 Business Models as a New Dimension in Strategy Theory

The evolution of the airline business can be considered as driven by supply side factors like technological development that laid the base for new regulations, these alone lead to a change of business rules and subsequently to strategic success factors in the industry. The evolution of the airline business can be structured in four main stages,[1] as shown in Table 2-1:

Year	Stage	Characteristics
1990	Stage 4	*Network and alliances:* consolidation stage, importance of airports, system of world wide alliances, network management, low cost carriers
1973	Stage 3	*Quality and cost:* deregulation, open sky policy, new price structures, new types of service, new entrances with new business, cost efficiency, hubbing
WWII	Stage 2	*Political:* fast progress, international standards for air transport regulation, bilateral agreements between countries, financial power, route networks
1925	Stage 1	*Technical:* adventurous form of transport, hardly any airline works profitable, supply side business

Table 2-1: Stages of Air Transport Development

This contribution focuses mainly on stage four and on questions about structure, success factors and the future of business models. As a result of stage four, the consolidation stage, a system of world wide alliances dominated air traffic. The system in some parts is being attacked more and more by low cost carriers and is still supplemented by independent regional carriers or national carriers in specific, not yet deregulated, nations. Accordingly, stage four of air transport

---

[1] For the development of air transport see also Pompl (2002).

evolution can be characterized by the evolution of different business models. Business models heavily rely on supply and demand side network effects, therefore a deeper understanding of the mechanism of the network economy is necessary.

1.1 The Network Economy and its Implications for Management Theory

The Internet as a basic innovation and other technological innovations allow completely new networks.[2] Corporate networks and customer networks allow new divisions of labor and possibilities of communication between companies, customers and suppliers. This means that network effects are becoming more and more significant. Within networks, standards for exchange platforms become very important. The network partner who is able to enforce his standard fastest will normally dominate the network in this manner and be able to generate yields.[3]

Due to the increasing significance of networks, a new analysis and action unit must be defined since companies or individual products are no longer capable of explaining economic success. The *eight trends* outlined below influence the business activities of companies in the modern network economy and, accordingly, blaze the trail towards a new strategic planning unit:[4]

1. The *globally accessible wide range of supply* and the *trend towards global sourcing and global buying* even for private individuals requires that companies define themselves to a relevant customer group and adjust the provision of their products and services to the benefit of this group. A company's system of providing products and services must be made as beneficial as possible for this group of customers to obtain a top ranking. With an ever wider range of supply, this will be the only way to preserve a prominent place in the customers' thinking and decision-making behavior.

2. In an *age of excessive stimuli* and the *economy of attention*,[5] products and services must be better anchored in the thinking of private individuals and communities. Only a company that manages to attain a positive and prominent place in the C2C (customer to customer) communication – which

---

[2] See Lyons (2002).
[3] See Bieger, Rügg-Sturm & von Rohr (2002); Shapiro & Varian (1999).
[4] See Bieger, Rügg-Sturm & von Rohr (2002).
[5] See Franck (1998).

is independent of the company itself – will be able to secure the necessary attention for a first sale and the necessary customer involvement for long-term maintenance of customer relationships.

3. In an increasing number of industries, the actual core business generates nothing or too little margins. More and *more core businesses have become actual network businesses*, not at least due to the networking of information. Owing to the network effects and the resulting ruinous fight for market shares, marginal-cost prices are often placed far below average costs in core businesses. At the same time, companies are far more exposed to the consumers' critical attention in their core business. Accordingly, prices cannot be raised at random there. Side products must therefore be commercialized consistently and new yield models must be introduced.

4. Today, *providers of capital assess a company's value on the basis of future discounted cash flow*. Corporate growth crucially determines corporate value through the free cash flow to be discounted in the future. If, for instance, corporate growth exceeds the discount rates used in the calculation of the present value, the corporate value is theoretically unlimited in perpetuity – a phenomenon that made the application of traditional assessment instruments to strongly growing areas difficult during the IT-bubble.

5. As mentioned above, *most services today are provided in corporate networks*. The background for this is the competition which increasingly forces individual companies to focus on their core businesses. Thus, they are compelled to buy complementing core competencies on the market and to market their own. Cooperation is necessary. One example is a railway company which, owing to its strength in this field, takes over marketing tasks for others and, conversely, buys in parts of its services. This raises questions as to the definition of the competencies that are necessary in the business model for the provision of the relevant services.[6]

6. Modern business models require a *great number of competencies* in companies: the competence to maintain customer relationships is as important as the competence to register changing customer requirements continuously and the competence to handle yield systems. Within the necessary competencies, a company must determine where it sees its own core commitment. It must determine its range of competence. This results in

---

[6] See also Sieber (1999).

Business Models in the Airline Sector 45

a demand for complementing competencies, which are brought in by cooperation partners or acquired on the market.[7]

7. The *selection of cooperation partners or suppliers* on the market is particularly important. Someone may offer a good solution today but may quickly fall behind in the trial and error of innovative competition. The selection of partners must therefore take its bearing from their development potentials and not from the solutions on offer at this moment.

8. A *coordination model* must be drawn for the entire business model and its constituents, which consists of coordination mechanisms between market and hierarchy as a function of the specificity of the traded assets, the risk involved, and the frequency of transactions.[8]

These trends are very strong in the airline business as well. Airlines struggle to define products for different customer segments and develop innovations especially for business customers. Through community building instruments like status programs they try to get additional attention. Sophisticated revenue models in many business models are subject to continuous innovation. The example of the low cost business shows the weight of growth in the analyst's perspective. Moreover, various forms of alliances and cooperations are common in the business.

1.2 Business Models as a New Strategic Planning Unit

According to the above-mentioned trends, we are now no longer merely concerned with the determination of a strategy and a program for a company's products and services, but rather with the establishment of a business system for a product market field within a corporate network. In this context, a clear evolution can be observed of the relevant strategic planning unit, which is particularly illustrative in the field of tourism. Table 2-2 shows the evolution of the relevant strategic planning units based on the evolution of the prevailing strategy theory and its main issues.

---

[7] See also Sieber (1999).
[8] See Williamson (1986).

Prevailing strategy theory	Main issues of strategy theory	Relevant strategic planning units	Example: tourism
Market-based view (Porter 1996) SBF (Levitt 1960)	Safeguarding of competitive advantages through optimal alignment with the demands and requirements of the relevant market, and thus safeguarding of monopolistic price ranges (and thus exploitation of market imperfection on the demand side).	Strategic business units (i.e. temporally stable product/ market combinations which can be run and possibly liquidated irrespective of others (Bieger 2002).	Business field specific strategies, e.g. hotel for children.
Resource-based view (Penrose 1959; Prahalad and Hamel 1990; Barney 1986)	Safeguarding of competitive advantages through the establishment of resources that are difficult to transfer and imitate, such as core competencies and thus efficiency yields (hence, exploitation of market imperfection on the resource side).	Company (i.e. the system which has access to these resources).	Mountain railways/ cable cars which establish core competencies in event management.
Network economy (Amit and Zott 2001; Bieger, Rügg-Stürm & von Rohr 2002)	Safeguarding of competitive edges through optimal configuration resources and products/ services in corporate networks.	Business model (i.e. network of companies which jointly run a value creating a product/ service system).	Leisure park with individual company attractions.

Table 2-2: The Evolution of Strategic Planning Units against the Background of the Evolution of Strategy Theory

On the basis of these trends, a business model system was developed based on other authors' business model structures and influenced by works on management of corporate networks.[9] This model was then tested in various case studies of network industries.[10]

According to Bieger and Lottenbach, a business model can therefore be defined in simpler terms as follows:[11]

A business model is the description of the way in which a company, a corporate system or an industry creates value on the market. This requires answers to the following questions:

- Which benefits for which customers?
- How is this benefit communicatively anchored in the relevant market?
- How are the revenues generated?
- Which growth concept is pursued?
- Which core competencies are necessary?
- What is the range of one's own company?
- Which cooperation partners are selected?
- Which coordination model is used?

The example of business models used by airlines illustrates the width of how the eight dimensions can be defined.

---

[9] See Sieber (1999).
[10] See Bieger & Lottenbach (2001).
[11] See Bieger & Lottenbach (2001), on the issue of business models see also publications by Timmers (2000), Treacy and Wiersema (1995), Nehls and Baumgartner (2000), Wölfle (2000), Amit and Zott (2001), Meinhardt (2000), among others.

	Traditional network company	**Low cost carrier with point-to-point connections**
Which benefits for which customers? *Product/service concept*	Comprehensive network of connections to the highest possible number of countries and continents for the highest possible number of customer groups and flight classes.	Selective offer of highly frequented routes; only one transport class for a wide range of leisure and business traffic.
How is this benefit communicatively anchored in the relevant market? *Communication concept*	Brand formation in complex brand systems, expensive loyalty programs.	Selective presence through classic brand awareness advertising in the relevant geographical markets, particularly also through IT.
How are the revenues generated? *Revenue concept*	Expensive revenue management system throughout the network, great dependency on sidelines in the service domain.	Simple pricing systems (e.g. Helvetic Airways with one-price fare for every flight); consistent cost optimization, also in marketing; transport often sole and sufficient source of revenues.
Which growth concept is pursued? *Growth concept*	More expensive fight about market shares above marginal costs, acquisition of new companies, buying into alliance systems, diversification, etc.	Simple multiplication model: once a route runs successfully, the next route will be opened, often – as with EasyJet – through the separation of a business unit, which results in a reduction of complexity.
Which core competencies are necessary? *Competence configuration*	Network management and marketing competence are dominant for the companies.	Market presence and cost competency dominate.

	Traditional network company	**Low cost carrier with point-to-point connections**
What is the range of one's own company? *Organizational form*	Structure pivots around the central competencies of network management and marketing, partial outsourcing of services/ internal splitting up of the company in a holding company structure.	Simple integrated flying operations with easily manageable leadership structures.
Which cooperation partners are selected? *Cooperation concept*	Great number of cooperation partners in the form of other airlines which complement the network (alliances) or service providers (e.g. ground handling), complex technological dependencies on suppliers such as aircraft manufacturers and IT developers.	No cooperation; only purchase of products and services.
Which coordination model is used? *Coordination concept*	Alliance management through possession/ interlocking capital arrangements (as in the Qualiflyer Group) or complex franchise agreements.	n.a.

Table 2-3: Business Models Used by Airlines – A First Example[12]

---

[12] See Bieger & Lottenbach (2001).

2 Business Models and their Economies in the Airline Industry

As mentioned earlier, business models have to be consistent within themselves. Revenue mechanisms have to fit to the core products and the customers benefits conveyed. Despite the various forms of business models with their huge variety of business constructions (some of them focus on new revenue mechanisms like income from non-flight services, others play heavily with brand mechanisms and try to transfer their brands to other areas/ branches), four main types, to be called generic business models, can be identified in the airline business. They are different in their form of main strategic success factors, cooperation strategy or role in networks and especially in their product strategy.

2.1 Network Model

The goal of network carriers always has been to provide global air transport networks with complete service chains, seamless customer care with a comprehensive network and lounges all around the world. Integrated loyalty program systems are important parts of these concepts as well.

Trunk or hub carriers dominate at least one hub and are leading partners in global alliances. Their main strategic success factor is the operation of a hub, a related alliance system and a sophisticated integrated cross company network management system.

A hub can be considered as a strategic resource in which heavy investments are made. A hub is an enabler of various net-economies (economies of scale, scope and density).[13] On the other hand, hubbing increases supply side complexity (risk of delays, expensive infrastructure, which lead to high costs in case the hub is given up). Most important: hubbing leads to a tendency for marginal pricing. Since passengers prefer point-to-point flight connections, transfers on a hub can only be sold at a lower price. As bigger hubs take advantage of stronger net-economies, traffic is bought with marginal prices. A hub therefore is stronger:

---

[13] See Jäggi (2000).

- the bigger the home market (and the smaller the share of cheap transfer passengers necessary to get a comparably high demand);
- the cheaper the cost level (and the bigger the flexibility for cheap transfer prices);
- the more central the geographical situation (and the shorter the transfer flights therefore);
- the more flexible the regulations (night flights) and the bigger the capacity reserves (especially the runway system).

Timetables based on waves at the hub, but also price structures, fleet use etc. have to be optimized constantly and integrated by the whole alliance. Network management is the core of the system.[14] Network management has to be developed further constantly on the base of the concept of system theories.

Network management from a system point of view is the process that controls the flow of aircraft and passengers. It comprises functions such as route network planning or scheduling, rotation planning and fleet assignment. Additionally, it involves the classical functions of revenue management such as pricing and seat inventory control. Some airlines call this bundle of function "resource planning".[15]

The better the more integrated airlines are and the more able they are to manage these processes, the better they can balance between demand and supply – and this is the key success factor of this industry: publishing the right schedule (scheduling/ route network planning) and offering the seats for the right price (pricing) to the highest paying customer (yield management/ seat inventory control). At the same time, the most efficient rotation plan (rotation planning) which best utilizes the aircraft assets (fleet assignment) has to be produced. In most airlines, this network management function is set up as an intermediary between sales/ distribution and operations. The task is highly analytical and the ability of managing the functions described above in the best balanced and most integrated manner hugely influences the success of the airline.[16]

There are different types of coordination structures in alliances. These coordination structures have to be evaluated by criteria such as:

- transaction costs involved;

---

[14] See Döring (1999).
[15] See Döring (1999).
[16] See Döring (1999); Rivera, Pompeo & Martin (1997); Farkas et al. (1997); Smith, Barlow & Vinod (1997).

- ability to guarantee effects typical to single companies like internalization of synergies and fast decisions as well as long term development of competencies.

The more an alliance is kept together by ownership, the lower the transaction costs are expected to be. On the one hand, clear ownership structures are supposed to facilitate deeper corporation structures, on the other hand they also bear the risk of additional financial exposure. In this sense, the coordination structure has to be in line with the cooperation/ integration level targeted.

According to the heterogeneous ownership structures between and within alliances, different forms of corporations evolved. Qualiflyer Group, for example, did rely on financial domination and coordination through ownership. Swissair tried to get dominating share ownership of each of the partner companies. This allowed a fast integration of, for example, network management or rules and standards. By this, it did not only consume a large amount of cash, which in the end led to the financial crisis, but also caused political and cultural shocks. Star Alliance, which is coordinated by a system of agreements, is different. Sub-alliances are kept together by financial ownership and are integrated at a fast pace, for example the Austrian Airlines Group with Austrian, Lauda and the former Tyrolean. The main partners are, on the other hand, financially quite independent, the whole formal structure is very participative.

The hub and spoke trunk carrier model relies very much on integrated network management abilities.[17] A second important success factor is the ability to build up strategic alliances on a global level and to tie together a comprehensive pattern of similarly positioned airlines. A third important success factor is the operation of a worldwide distribution system as well as a distribution organization. The driving factor for development is the goal to achieve a market share as big as possible. This guarantees the internalization of network effects.

Different "levels" can be identified among network and hub airlines. On the one side there are the dominating airlines that govern the global system. In addition to them, in there is a second group of airlines dominating secondary hubs like – in the case of Star Alliance – SAS and its hub Copenhagen all major alliances. A third group of national airlines has the character of feeder airlines. Very often, since their link to the alliance is not that strong, they have to operate a second business in form of serving national or regional traffic or even provide services for another alliance. An example of this is Croatian Airlines that operates some

---

[17] See also Döring (1999).

feeder routes together with Star Alliance companies as well as with Skyteam members.

2.2 Regional Airline Model

A second important group are the regional airlines. In a way it is not a group for itself, because very often they are linked by agreements to the major alliances and serve as a kind of feeder carrier for the main carriers of the alliance. Nevertheless, more and more regional airlines serve their own point to point routes and sometimes, like the former Crossair, even operate their own regional hub. Thereby they rely on specific strategic success factors that define their own business model.[18]

Most importantly, regional airlines serve niche markets. Very often they are the only airline serving a certain airport such as f.e. Bolzano by Tyrolean, a daughter company of AUA. They have to be flexible in their corporation, they have to watch their cost structure and they have to be very dynamic in looking for new niche markets. The ability to define, develop, dominate and protect niche markets is their key to success.

2.3 Low Cost Carriers

Low cost carriers are not just cheaper. They follow their specific and own business model, which is mainly characterized by a complete abandonment of the use of network effects. They just serve point to point routes, starting very often from one main airport. These main airports are characterized by cheap costs. Very often these are side airports of major centers like Stansted in London or Hahn-Hundsrück close to Frankfurt. The advantage is not just that the costs of making use of these airports are cheaper, but the risk of getting into delay is reduced as well, due to the fact that traffic is limited at these places. These practices allow simple processes with reduced costs. Short turnaround times (very often less than 20 minutes) and low distribution cost (e-ticketing, Internet distribution) are important elements.

---

[18] See Oum, Park & Zhang (2000).

Low cost carriers do not develop their own traffic flows in most cases. They do not open new markets. They rather concentrate and try to take over a market share as big as possible in existing markets.[19]

2.4 Charter Airlines

Traditional charter airlines seem to have died out since Canada 3000, one of the major and bigger independent charter airlines, got into trouble. More and more charter airlines are owned by big tour operators like Condor by Thomas Cook or Edelweiss Air by Kuoni. These tour operators consider charter airlines an integral part of their value chain. This integration allows them to gain a bigger share of the overall value added created by their clients. The integrated management of product development, marketing and airline operation also allows them to take advantage of a better use of resources. The big tour operators did achieve good results by developing new products to fill airplane time left. This is one reason why traditional and independent charter airlines have difficulties in competing with the integrators.

Nevertheless, tour operators can hardly cover more than 60% to 80% of the overall own production depending on the degree of seasonality of the production. In mid Europe for example the overwhelming part of the production to short and medium range destinations takes place between May and October. In the remaining part of the year, only a small number of small and medium range airplanes can be employed. Therefore, tour operators have to rely on flexible capacities they rent on the market. More and more, these capacities are rented from ordinary regional or traditional network carriers because very often these carriers have capacities left, especially on the weekends.

Charter airlines have to rely on strategic success factors like cost efficiency and integration of operation management, product development and marketing of a tour operator.

The main concepts can be described as follows:

---

[19] See Campbell & Jones (2002).

Business model concept	Network carrier	Regional carrier	Low cost carrier	Charter carrier
Product	network effects, hubs	niches, low costs, regional markets	simple processes, niche markets, cost efficiency	cost effectiveness
Communication			marketing	
Revenue				
Growth	growth and market share			
Competence				integration in tour operator value chain, capacity management
Organization		Flexibility		
Cooperation	cooperation to build global links, alliances			
Coordination				

Table 2-4: Strategic Success Factors of Business Models in Airline Business

3 Transformation of Airline Business Models and their Performance

The identified business models in the airline industry rely on distinct success factors (see Table 2-4). These success factors serve at the same time as a kind of identification criteria for prospective models. These are for *network carriers*:

- Operation of a hub, big share of transfer traffic.
- Specific customer services can form an integrated travel product like lounges, mileage programs.
- Very often integrated technical departments and catering companies.
- Complex network management.

Most network carriers are part of an alliance system. These systems can be considered as a kind of substitute for mergers since, especially in Europe and South America on account of cultural reasons and legislation, airlines still have a national character. Due to the logic of the hub structures, one airline emerges as the core airline of an alliance like Lufthansa at Star Alliance. Other partners serve as a kind of regional network carrier. They serve single airports through a regional hub. Very often a small and selected number of intercontinental relations are offered from these regional hubs, like from Copenhagen to Singapore or New York. The remaining intercontinental traffic is going through the main hub of the system. The business logic between the core and the regional network carriers is slightly different. The core airline serves as a kind of competence and operation center through a

- strong and competent network department;
- strong brand;
- strong and efficient hub; and
- big intercontinental operation.

The secondary airlines operate their secondary hubs and markets within the alliance system's market. Regional airlines in an alliance system rely on cost effectiveness because very often they have to serve a huge network of connections between medium centers and smaller cities with often only weekly and not very regular flights or even in competition with low cost carriers.

Low cost carriers rely on

- efficient and lean production processes that include short turnover times;
- cheap airports;
- incomplex network structures with point-to-point shuttle services; and
- very often low salary structures and cheap leasing rates.

On the other hand, *charter airlines* also provide low cost structures and very often they are able to match the per seat kilometer costs of low cost carriers. They serve other markets and most of all operate weekly flights because they just cannot rely on regular traffic flows and shuttle point-to-point services. By serving holiday destinations mostly on a weekly basis, the integration into a demand generating tour operating system is crucial for them.

The most important factors are:

- Lean and efficient processes, low salary structures.
- Comparably good or reasonable service.
- Reputation.
- Integration into a tour operator system.

Regional airlines serve regional airports and connect them with hubs or major business centers. Most of them are very small. Many are integrated in an alliance system. Some remain independent and operate more like a low cost carrier. The distinguishing features of such airlines are:

- Lean and efficient processes, low complexity.
- Access to regional markets.
- Technical skills or specialties in the sense of ability to serve small airports with very often difficult approach conditions and short runways.

After the decline of the airline industry following 11 September 2001, the airline sector seems to be entering into a kind of experimental state. Many companies transform themselves by experimenting with the elements of their business model (see table 2-5).

To From	Network	Charter	Low Cost	Regional
Network	Lufthansa Air France	Finnair	Swiss	
Charter			Thomas Cook Hapag Lloyd	
Low Cost	(Air Berlin)			
Regional	Emirates Gulf Air		Intersky	

Table 2-5: Transformation of Airline Business Models

In this transformation process of business models some *trends* can be identified:

1. *There seems to be a general tendency towards low cost models.* Very often, however, just some elements of the original low cost business model are applied so that for example Swiss, the national airline of Switzerland that emerged from the remains of the old Swissair and Crossair, offer comparably low prices with a no frill concept. However, they still operate like a network carrier with an intercontinental hub in Zurich. They still provide business class which means long turnover times due to catering loading and high complexity. Hubbing still means flying to expensive airports at times with congestion and networking still means expensive overheads.

2. *The strongest erosion of the traditional business model seems to take place within the charter airlines.* Charter operation gets integrated back into network carriers like in the case of Swiss or, for quite a long time, with Finnair or other regional network carriers. Some independent charter companies try to transform themselves into a kind of low cost carrier like Hapag Lloyd. Even companies that belong to strong tour operation systems start to sell single seats. The forces behind this erosion of the charter model are first of all the changing travel patterns. Tourists traveling to traditional European destinations less and less follow the one week all included tour scheme. Especially in standard destinations tourists prefer visiting for some

days. Destinations with a large amount of second home owners are even more affected by this trend. By this at least during high season the traffic flow gets more regular. Therefore, more and more of these routes are taken over by traditional low cost carriers. Examples are Nice, Northern Italy destinations etc. On the other hand, network carriers on their long haul routes very often offer limited places at very low fares to tour operators as a kind of side product. By this development traditional charter operations are more and more focusing on long haul leisure destinations which low cost carriers cannot fly to and network carriers do not fly to. Consequently, the fleet of charter airline companies consists more and more of a comparably high share of long haul planes.

3. *The business model of the regional airlines also seems to come to its limits.* Very often these airlines are increasingly integrated into network carriers like in the case of Austrian Airlines with Tyrolean. By this, network carriers are concentrating on their regional operations. They take advantage of additional economies of scale in this segment. The remaining regional carriers, mostly smaller ones, are focusing on specific regions. Very often they are also subsidized or financed by these regions like Air Dolomiti. Some of them are also business start-ups that try to introduce new business models, that means low cost, interregional operation like the Swiss Berne-based Intersky. This company operates with Dash 8-300 aircraft out of Berne and offers very low fares.

On the whole, the development of business models in the airline industry could be summarized, in a way, as get big or small, regional and low cost. This statement raises a question about limits of the dominating business models.

4 The Perspectives of Airline Business Models – Convergence or Differentiation

Traditional network companies with their huge alliance systems with complex network management and price management systems seem to come to certain limits that can be summarized as a kind of complexity trap. In search of more efficiency, more and more intercontinental flights are concentrated on big hubs. Consequently, these hubs are more and more congested. As a consequence, innovations evolve in the fields of

- hubbing: continuous hubbing, side hubbing and waving as a means of reducing complexity and risk of delays;

- disintegration of business and economy traffic combined with point-to-point connections with small long haul planes like Boeing 737-800 (compare with Lufthansa's connections operated by PrivatAir from Düsseldorf to New York).

On the other hand, with the introduction of the new Airbus 380 the function of filling flights and therefore hubbing and waving at least on certain routes will further increase. The feeder flights necessary to fill these huge planes will possibly even crowd out some regional connections which leaves a chance to increase traffic on secondary hubs.

In the sector of low cost carriers, an erosion of the original business model seems to take place. More and more, smaller airlines like Helvetic Airways, Air Berlin etc. are coming into the market. The original concept is changing for example by introducing additional service features, servicing seasonal routes or concentrating on hub type basis. As soon as they start to do so, they enter network business. Demand side net effects, like reputation, access to lounges, scope in terms of connections offered or supply side net effects like economies of density and scope and the possibility to dominate hubs, start to matter.[20] And as small companies they lose their competitive advantage compared to the big network airlines. Therefore, one limiting factor for the development of low cost carriers is the ability or discipline to stay really lean and mean and stick to the original business model. Other important limiting factors are

- reputations with regard to routes with a strong steady traffic flow. Low cost carriers avoid competition among each other. In this competition, the bigger company would have good chances to win, most probably due to economies of scale.

- limitations with regard to resources in the aftermath of 11 September 2001. Aircraft could be bought quite cheaply. Pilots were available and in lack of other attractive positions. Financial markets were ready to provide money for the growth of the low cost business model. There is a vicious circle with decreasing growth perspectives. This means decreasing attractiveness for the capital market. Even the big companies that operate new airplanes will face heavy costs with the first IT-checks booming.

Table 2-6 shows price earning ratios. It is a ratio of financial market performance of different companies structured according to their presumptive underlying business model. The low cost carrier concept seems less attractive than in previous times. This can be true due to the above-mentioned limits of the

---

[20] See also Jäggi (2000).

business model and because more and more companies deviate from the original success model. In order to avoid the above-mentioned complexity, low cost carriers always have to stick strictly to their business model. The recent results of low cost carriers (e.g. Easyjet) show that this business model also comes back to industry typical business cycles and return rates.

On the other hand, the example of Swiss shows that network carriers never can be more efficient than low cost carriers because they have huge investments into their structure and hub, because they have a different salary structure and they made these huge transaction investments into their alliance systems. Despite their size, they will never be able to match the operational costs of low cost carriers. Alliance structures will more and more be standardized to avoid complexity traps.

Buiness Models	Examples	Price-Earnings Ratio (2003)
Network Carrier	Air France	14
	British Airways	18
	Lufthansa	25
Network Regional	Austrian Airlines	6
	Iberia	8
Low Cost	Ryanair	16
	Easyjet	12

Table 2-6: Performance of Business Models[21]

What can be expected is a new division of functions. It can be expected that network carriers will leave certain routes for certain segments to their low cost peers. Within the network business model, the core network carriers perform better. The reason most probably is the bigger share of long haul traffic where, due to no competition with low costs, profitability is still higher.

---

[21] See Finanz und Wirtschaft (2003), p. 15.

The question remains of what will be the sources for strategic competitiveness in the next decade when making use of net effects as well as adaptation or experimentation of the low cost model come to their limits? First there is the question whether the customer will still have a choice whether or not the European landscape will be divided among the major alliances? At least from the main markets there will always be links to the hubs of the competing alliances and thereby a competition in the long haul traffic business will remain. Since more and more competencies in network management or cost optimization will be ubiquitous, the future strategic success factors rely on

- the customer service contact which includes individualized service packages for companies as well as individuals, individual care and attention, market access;
- the travel itself (experience, safety);
- airport infrastructure since, as mentioned above, transfer traffic will still increase.

Future research in airline management should therefore focus on customer behavior, which could include influence of brands and brand portfolios in alliances, contribution of CRM, measures to assess the customer value, contribution of different service elements to customer value or standardization and process optimization on airports.

It seems to be obvious that there is a tendency towards convergence. This can be explained by the fact that the production processes and by this also the business models are strongly determined by technical conditions, especially in the form of airplanes (e.g. all types of airlines use the same types of airplanes).

5 References

AMIT, R.; ZOTT, C. (2001): Value Creation in e-Business. *Strategic Management Journal*, 22(6/7), pp. 493-520.

BARNEY, J.B. (1986): *Organizational Economics*. San Francisco: Jossey-Bass.

BIEGER, T.; DÖRING, TH.; LAESSER, CH. (2002): Transformation of Business Models in the Airline Industry. In: Bieger, Th. and Keller, P. (Eds.): *Publication of the AIEST: Air Transport and Tourism*, 44, pp. 49-83.

BIEGER, T.; LOTTENBACH, D. (2001): *Airline-Geschäftsmodelle: Wann schaffen sie Wert?* Unveröffentlichtes Manuskript, St.Gallen: IDT-HSG.

BIEGER, T.; RÜEGG-STÜRM, J.; VON ROHR, TH. (2002): Strukturen und Ansätze einer Gestaltung von Beziehungskonfigurationen – Das Konzept Geschäftsmodell. In: Bieger, Th. et al. (Eds.): *Zukünftige Geschäftsmodelle: Konzept und Anwendung in der Netzökonomie*, Berlin, Heidelberg, New York: Springer, pp. 35-61.

CAMPBELL, P.; JONES, C. (2002): Rebel Skies – Overcapacity, Job Cuts, Bankruptcies: Amid the Collapsing Fortunes of Europe's Airlines, one Sector is Soaring. We Investigate the Region's No-frills Boom and Ask – Can it Continue? *Flight International*, 4, London.

DÖRING, T. (1999): *Airline-Netzwerkmanagement aus kybernetischer Perspektive – ein Gestaltungsmodell*. Doctoral Thesis, University of St. Gallen, Bern, Stuttgart, Wien: Haupt.

FARKAS, A. et al. (1995): Facing Low cost Competitors: Lessons from US Airlines. *The McKinsey Quarterly*, 4-1995, pp. 86-100.

FINANZ UND WIRTSCHAFT (2003): Die Airline-Industrie im "perfekten Sturm". 7. Mai 2003, 36, p. 15.

FRANCK, G. (1998): *Ökonomie der Aufmerksamkeit – Ein Entwurf*. München, Wien: Carl Hanser.

JÄGGI, F. (2000): *Gestaltungsempfehlungen für Hub-and-Spoke-Netzwerke im europäischen Luftverkehr – Ein ressourcenbasierter Ansatz*. Doctoral Thesis, University of St. Gallen, Bamberg: Difo-Druck.

LEVITT, T. (1960): Marketing Myopia. *Harvard Business Review*, 44(4), pp. 45-56.

LYONS, G. (2002): Internet: Investigating in New Technology's Evolving Role, Nature and Effects on Transport. *Transport Policy*, 9(4), pp. 335-346.

MEINHARDT, Y. (2000): *Ein Ansatz zur Systematisierung von Geschäftsmodellen*. Unveröffentlichtes Arbeitspapier, Otto-Friedrich-Universität Bamberg.

NEHLS, R.G.; BAUMGARTNER, P. (2000): Value Growth: Neue Strategieregeln für wertorientiertes Wachstum – Ein Ansatz von Mercer Management Consulting. *Management Consulting*, München, pp. 79-93.

PENROSE, E.T. (1959): *The Theory of the Growth of the Firm*. Oxford: Blackwell.

POMPL, W. (2002): *Luftverkehr – Eine ökonomische und politische Einführung*. 4th Ed., Berlin, Heidelberg, New York: Springer.

PORTER, M.E. (1996): *Wettbewerbsvorteile (Competitive Advantage): Spitzenleistungen erreichen und behaupten*. 4th Ed., Frankfurt, New York: Campus.

PRAHALAD, C.K.; HAMEL, G. (1990): The Core Competence of the Cooperation. *Harvard Business Review*, 68(3), pp. 71-91.

RIVERA, L.; POMPEO, L.; MARTIN, A. (1997): Network Agility. *Airline Business*, 8(7), pp. 56-61

SHAPIRO, C.; VARIAN, H. (1999*)*: *Information Rules: A Strategic Guide to the Network Economy*. Boston: Harvard Business School Press.

SIEBER, P. (1999): Virtualität als Kernkompetenz von Unternehmen. *Die Unternehmung*, 53(4), Bern: Haupt, pp. 243-266.

SMITH, B.C.; BARLOW, J., VINOD, B. (1997): Airline Planning and Marketing Decision Support: A Review of Current Practices and Future Trends. In: Butler, G. and Keller, M.R. (Eds.): *Handbook of Airline Marketing*. New York: McGraw Hill, pp. 117-131.

OUM, T.H.; PARK, J.-H.; ZHANG, A. (2000): *Globalization and Strategic Alliances: The Case of the Airline Industry*. New York: Elsevier.

TIMMERS, P. (2000): *Electronic Commerce: Strategic and Models for Business-to-business Trading*. Chichester: John Wiley and Sons.

TREACY, M.; WIERSEMA, F. (1995): *Marktführerschaft: Wege zur Spitze*. Frankfurt a.M.: Campus.

WILLIAMSON, O. (1986): *Economic Organization: Firms, Markets and Policy Control*. Brighton: Harvester/ Wheatsheaf.

WÖLFLE, R. (2000): Entwicklung eines E-Business-Geschäftsmodell, E-Business-Erfolgspotenziale innerhalb bestehender Geschäftsstrategien. *io management*, 9 (69), Zürich: HandelsZeitung Fachverlag AG; pp. 62-65.

3
CONSOLIDATING THE NETWORK CARRIER BUSINESS MODEL IN THE EUROPEAN AIRLINE INDUSTRY

STEFAN AUERBACH AND WERNER DELFMANN

1   Changing Environment in the European Airline Industry 66
2   Strengths and Weaknesses of the European Network Carriers' Business Model 73
3   Cornerstones of a Consolidated Business Model for European Network Carriers 83
4   Conclusions 93
5   References 93

Summary:
European Network Carriers have to face an increasingly challenging competitive environment. Market liberalization leads to new formats of competition in the European Airline industry. The low cost carrier model especially threatens the business model of traditional network carriers. Furthermore the hub and spoke structure of network carriers in combination with their embeddedness in global airline alliances has increased their vulnerability to schedule problems as well as external disturbances like 9/11, the gulf war, or SARS. In light of this the business model of European network carriers has to be evaluated and further developed to provide sustainable competitiveness.

1 Changing Environment in the European Airline Industry

This chapter analyzes how network airlines can cope with a changed environment in the European air transport markets by consolidating their business model in three main aspects. These aspects include the focus of market segmentation model, the hub operation model and the alliance strategy model. In this chapter we concentrate on issues directly related to the hub-and-spoke production model.[1]

1.1 Crisis in the European Airline Industry

After September 11 the airline industry experienced a worldwide dramatic downturn in financial performance. The financial results of the flag carriers and major airlines have largely verged to disastrous. According to industry reports the losses among the large US airlines exceeded 6 billion USD. In Europe flag carriers like Swissair and Sabena went bankrupt in the aftermath of the New York attacks. The total financial loss of the AEA airlines amounted to € 0.94 billion. In 2002 European airlines generated € 2.5 billion less revenue compared to 2001 and 13 million fewer passengers. In the same period, revenue passenger kilometers (RPK) of European major airlines went down by 4.8%, while capacity (ASK) was reduced by 8.7%.[2] While European airlines made projections for 2003 taking into consideration the impact of the anticipated second Iraq War, they were unprepared for the subsequent SARS crisis.

---

[1] See also the complementary theoretical chapter by Fritz, chapter 9 in this volume.
[2] See AEA (2003), p. 1-2.

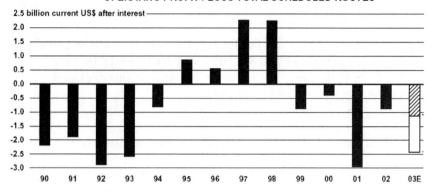

Figure 3-1: Operating Results of AEA Airlines 1990-2003[3]

Profit before taxation as a percentage of turnovers of European network carriers in the fiscal year 2002/2003 is significantly lower than the equivalents of their low cost competitors. For instance, Air France achieved 0.9%, Austrian Airlines 0.2%, British Airways 1.8%, KLM 4.2%, Lufthansa 5.6%, and SAS -0.7%. Low cost airlines Ryanair and easyJet by far outperformed the traditional network carriers with 30.4% and 13.1%, respectively.[4] The performance differential of the two segments has led to the hypothesis that the traditional network business model is no longer a competitive weapon as it was proclaimed after the deregulation of the airline industry in the US in 1978, and in Europe in 1993. Hence, many financial analysts concluded that if only point-to-point airlines earn money, point-to-point networks have to be the most efficient models of production.

The management of some of the major European airlines neglected the influence of the low cost competition, and this may have delayed necessary adjustments of their business model. The SARS crisis in 2003 and the Iraq war have relentlessly revealed the weaknesses of the traditional network carriers. But is this a sufficient indicator for the dominance of the point-to-point business model? In Europe exist approximately 75,000 origin and destination markets; only 3% of these markets have more than 75 passengers per day.[5] While virtually all of these markets have a non-stop service the remaining majority of small city-pair

---

[3] AEA (2003), p. 2-4.
[4] See Milne (2004), p. 274.
[5] Lufthansa Consulting Analysis.

markets do not allow point-to-point traffic. Although air traffic in Europe is mostly perceived as point-to-point intensive it is a fact that most traffic is routed through the hubs of the major airlines.[6]

The development of hubs was key for the financial turnaround of Lufthansa or Air France in the mid-1990s. But if hub-and-spoke networks are such powerful means to compete, what is the reason for the poor financial results of the European major carriers in the recent years?

1.2 Current Shape of the European Airline Industry

Before European air travel was deregulated, in the mid-1990s, the air transport market was clearly segmented. Scheduled carriers, focusing primarily on business travelers, controlled 75 percent of the intra-European market. The group of scheduled airlines comprises major carriers, which developed global hub-and-spoke networks (e.g. Lufthansa, Air France, British Airways, KLM, Swissair), smaller flag carriers (e.g. Olympic, CSA Czech Airlines, Malev, LOT, Aer Lingus) and smaller regional airlines (e.g. Cimber Air, Air Nostrum, Portugalia, Maersk Air, Brit Air) focusing on thin point-to-point markets and direct service. Charter airlines held the remaining 25 percent by selling seat capacity to tour operators. The typical passengers were typically budget tourists traveling between Northern and Southern Europe.

Since then the European airline industry has changed significantly. In the mid and late 1990s many carriers joined inter-airline alliances of various kinds. The number of these alliances grew from 280 in 1994 to 513 in 1999.[7] Interestingly, the number of equity investments remained rather stable while the number of non-equity alliances grew significantly.[8] The scope of these co-operations ranges from code-sharing and block-space agreements, joint frequent-flyer programs to joint ventures on certain routes or even entire traffic regions.[9] Despite intense reconfiguration relatively stable groupings around the European major carriers evolved in the second half of the 1990s, which finally have resulted in three

---

[6] See Still (2002), p. 97.
[7] See Gallacher (1997), p. 26; Gallacher, Wood (1998), p. 43; Gallacher (1999), p. 34.
[8] In 1994 79% of the total number of alliances were not based on any kind of equity participation and only 21% involved equity investments amongst the partners. In 1999 industry reports counted 87% non-equity co-operations. See Gallacher (1999), p. 34.
[9] See Doganis (2001), p. 65.

strategic alliances: Star Alliance[10] around Lufthansa, oneworld[11] led by British Airways, and Sky Team[12] led by Air France. These multilateral global groupings can be characterized by a much higher degree of integration than mere bilateral code-share agreements. Doganis describes strategic airline alliances as co-operation where the partners co-mingle assets like terminal facilities, maintenance bases, aircraft staff, traffic rights, slots, or capital resources.[13] As the ultimate strategic "alliance" he considers a full merger of airlines involved. The first real merger between airlines within Europe was the merger between Air France and KLM in 2004. Even when the airlines claim that their alliances span the globe, they all have a strong foothold within Europe. The combination of their networks in Europe creates continental multi-hub networks. From the fifteen Star Alliance members six airlines have their base in Europe. Four of eight oneworld carriers and three of six Skyteam airlines are European as well.

A second group of European airlines are smaller flag carriers, which do not belong to a strategic alliance such as Swiss, Malev, Olympic, JAT, and Croatia.

European regional carriers build the third group. These are regional carriers serving point-to-point destinations or feeding the hubs of the majors on the basis of codeshare and franchise agreements.[14]

Charter airlines traditionally concentrated on selling seat contingents to tour operators, they now are offering advanced seat reservation to individuals, too. These airlines (e.g. Air Berlin) usually have in many respects a significantly lower cost base than network carriers. Some of them have successfully entered the low cost market while others are fighting for their survival. Both scheduled and charter incumbents were shaken by the emergence of low cost carriers that targeted leisure and, to a lesser extent, business markets. Figure 3-2 illustrates that the market shares of the different carriers are expected to converge in the next 15 years.

---

[10] Star Alliance consists of 16 airlines: Air Canada, Air New Zealand, ANA, Asiana Airlines, Austrian Airlines, bmi, LOT, Lufthansa, SAS, Singapore Airlines, Spanair, TAP Portugal, Thai, United, US Airways, Varig.
[11] 8 Carriers build oneworld: Aer Lingus, American Airlines, British Airways, Cathay Pacific, Finnair, Iberia, LAN, Qantas.
[12] The Skyteam alliance has 6 members: Aeromexico, Air France, Alitalia, CSA Czech Airlines, Delta Airlines, Korean Airlines.
[13] See Doganis (2001), p. 65.
[14] See Hanlon (1996), p. 92; Alamdari and Morrel (1998).

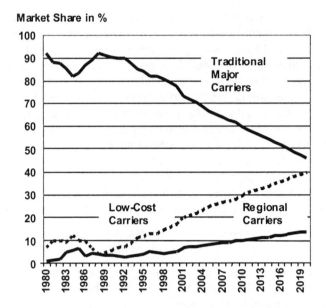

Figure 3-2: Development of Market Shares in Europe[15]

1.3 Competing Business Models: Point-to-Point vs. Hub-and-Spoke

After deregulation in the United States the major carriers developed hub-and-spoke networks as the dominant production structure in the US as well as in Europe. The underlying logistic concept aims to consolidate thin traffic flows at the hub. Airlines coordinate arrival and departure times to attract connecting traffic. Flights from the outside stations (spoke) arrive at the hub airport approximately at the same time. Passengers then have the opportunity to change aircraft while the planes are simultaneously on the ground within a short period of time. As arriving flights are bundled into so-called arrival banks, the departures to the spoke airports are scheduled in a similar way (departure bank). This structure allows the airlines to maximize the number of served city-pair markets. The theoretical optimum is reached when all flights can be concentrated within a single arrival and subsequent departure bank, respectively, because every arriving flight could be combined with $(n-1)$ departing flights to n spoke

---

[15] Source: ESG, Lufthansa Consulting Research.

airports.[16] In practice, airspace and airport infrastructure constraints as well as different flight times and sector lengths hinder airlines in achieving this. Network airlines have implemented complex decision-making processes to identify those traffic flows which might be served by connecting services and to adjust their flight schedules accordingly. This results in a set of arrival and departure banks during the day according to estimated demand and market share on the city-pair markets. With n spokes from the hub, an airline is able to serve $\frac{n(n-1)}{2}$ markets via the hub and $\frac{n(n+1)}{2}$ city-pairs per arrival and departure bank. This advantage has led to the objective in network planning to maximize the connectivity at the hub and thus to maximize the number of feasible or useful connections. To pursue this, network airlines have at least partially neglected that concentrating flights lead to lower aircraft productivity (average flight time per day) and thus to a larger fleet. A varying sector length of spoke routes and the different size of city-pair markets require a variety of aircraft types, which impact maintenance and training costs. Peaking flight operation around the banks will also lower the productivity of ground staff at the hub airport during the day. Transfer traffic requires sophisticated baggage handling systems to avoid loss or misallocation of baggage. Given the sheer size of the major network carriers in terms of fleet and number of destinations the operation of a hub-and-spoke system requires complex planning and control systems.

The traditional network carriers usually have their own sales organization co-operate with travel agencies and/or general sales agents (GSA). Bookings are made via the global computer reservation systems.[17] To compensate passengers for the burden of changing the aircraft and accepting a longer total travel time, network airlines usually offer additional services such as airport lounges, newspaper and magazines, meal service on board.

Compared to the network carriers, low cost airlines rely on a rather simple business design:[18]

---

[16] For comprehensive overviews see Doganis (1991), p. 263ff, and Hanlon (1996), p. 111ff.

[17] Although the network carriers offer online reservations and internet sales as well, the major share of tickets are still distributed via the classical channels (travel agencies, direct sales and general sales agents.

[18] It needs to be mentioned that some carriers in the market which are considered to belong to the low cost airlines offer a high level of service, like on-board TV (Jet Blue) or newspapers (Air Berlin).

- one kind of aircraft, one class of passenger, and higher seating density (approx. 15% higher than full-service network carriers);
- no frills: no airport lounges, no choice of seats, no newspapers, no meals, no frequent-flyer programs, no connecting flights, no refunds, and no possibility of rebooking to other airlines;
- no sales organization, no computer reservation systems, but intensive use of internet sales (about 90 percent of easyJet and Ryanair tickets are booked over the Internet).

As low cost airlines don't offer connecting flights, but simply shuttling between two cities and by keeping the ground logistics as simple as possible, they minimize turnaround times and maximize revenue-generating air time. In particular on short-haul routes low cost carriers' aircraft productivity is 12 – 14 h compared with 9 h for the most efficient traditional scheduled carriers. Some European low cost carriers (e.g. Ryanair) fly to and from secondary airports—located as far as 100 kilometers from city centers—thus minimizing landing and ground-handling fees. On intra-European international routes, this adds up to an operating-cost advantage per seat and kilometers flown (unit cost) of 40 to 65 percent as compared with major scheduled carriers.

Lower costs and higher seat-load factors permit no-frills carriers to offer fares 50 to 70 percent lower than those of the incumbents. The average price of the low cost carriers for a one-way ticket on international intra-European routes is €50 to €85, compared with €180 to €200 for British Airways and Lufthansa. This approach attracts price-sensitive and flexible travelers, but the lack of convenience and flexibility makes the low cost model unappealing for most passengers traveling on business.

Low cost airlines thus seem to complement both traditional and charter airlines, but there are overlaps. They compete with the charter airlines, which sell up to a third of their seats without an accompanying hotel package, for the favors of independent travelers and people who travel to holiday homes frequently. They also compete with the traditional airlines, whose weekend fares attract moderately price-sensitive leisure travelers and whose cost-conscious business passengers often may fly repeatedly on a particular route, can plan ahead, and have no need for flexible tickets.

The challenge for the low cost airlines is the selection of routes to generate enough traffic to fill the larger capacity of their aircraft. Routes should be either unserved or served by incumbents at high fares, allowing the low cost entrant to stimulate demand by lower fares. Profitable markets must also have a large leisure- and private-travel component, as this is the primary customer base of most low cost carriers.

Flying to secondary airports and low airport fees keep the carrier's costs 65 percent below those of a typical scheduled airline. For instance, Ryanair can thus offer cheap fares and still make a profit if the load factor is higher than 55 percent. Compared with Ryanair, easyJet flies to major airports such as Amsterdam, Madrid, Paris, and Zurich, which attracts in particular business passengers. According to statements of the airline 50% of its passengers are traveling on business. Due to higher airport fees the cost base of eaysJet is higher than the one of Ryanair and the break-even load factor lays around 75%. Regardless of the approach low cost carriers take, they are less vulnerable to economic cycles, since in difficult times demand for premium service tends to decline as more passengers seek less expensive travel alternatives.

2 Strengths and Weaknesses of the European Network Carriers' Business Model

2.1 Integrated Production System

As mentioned above, the hub–production model is based on the logic of bundling flights via the hub. With each additional flight to the hub the hub-airline is in the position to serve *(n-1)* additional city-pair markets. But in particular in Europe, network airlines operate intercontinental and continental flights with a variety of different sector lengths and flight times. Compared to their US counterparts, European network carriers can be characterized as an integrated production system. The success of this production system depends on how efficiently these carriers link the different parts of the system. To understand the interdependence of the different elements of this production system, it is worthwhile to describe the different traffic flows, which are transported on the same physical flights to and from the hub.

Within Europe, network carriers offer connecting services between city pairs via the hub when the passenger demand is not sufficient to justify direct flights. The same flights are booked by passengers, whose final destination is the hub city as well as by passengers originating at the hub city and with a spoke destination within Europe. Furthermore, the same physical flights to the hub may be used by passengers traveling from an intercontinental point of origin and connect to a European destination and vice versa. In the European major hubs like Frankfurt,

Munich, Paris Charles-de-Gaulle or London Heathrow passengers can connect on continental flights as well. European Airlines not only offer air services via the hub. The so-called non-hub traffic is served by direct flights bypassing the hub. Generally speaking, passengers prefer direct flights between two cities without the inconvenience of changing aircraft at the hub. For direct flights airlines are able to charge higher ticket prices than for connecting flights.

When passengers have the choice between different travel options they trade off total travel time against inconvenience of connecting flights and other factors such as frequency, price, type of aircraft according to their individual utility function and preference structure. Depending on total travel time there is not only a competition on particular routes but between different hubs.

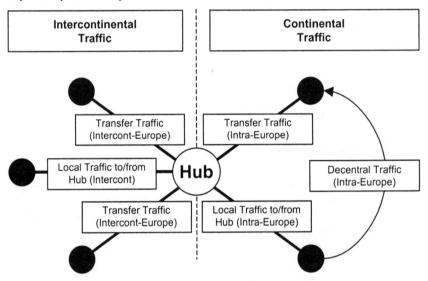

Figure 3-3: The Integrated Transport System of Network Airlines

At a first glance the logistical principle of bundling thin traffic flows at the hub seems to be most efficient by maximizing the airline's seat load factors. But low cost carriers are successfully competing against the incumbent airlines on the same routes to and from the major hubs or city-pair markets (see table 3-1).

Airline	Flight	Orig.	Dest.	Departure Time	Arrival Time	Operating Days
DBA	DI 7060	CGN	TXL	06:30	07:30	12345
Lufthansa	LH 267	CGN	TXL	06:40	07:45	12345
Hapag Lloyd Express	X3 3120	CGN	TXL	07:15	08:15	12345
Germanwings	4U 010	CGN	SXF	08:00	09:00	1234
Lufthansa	LH 269	CGN	TXL	08:10	09:15	123456
Hapag Lloyd Express	X3 3122	CGN	TXL	08:15	09:15	123456
Hapag Lloyd Express	X3 3124	CGN	TXL	09:30	10:30	12
DBA	DI 7064	CGN	TXL	09:40	10:45	12345
Lufthansa	LH 271	CGN	TXL	09:50	10:55	12345
Lufthansa	LH 273	CGN	TXL	11:20	12:25	123456
DBA	DI 7068	CGN	TXL	12:45	13:50	12345
Lufthansa	LH 275	CGN	TXL	13:00	14:05	12345
Hapag Lloyd Express	X3 3136	CGN	TXL	13:05	14:05	345
Hapag Lloyd Express	X3 3134	CGN	TXL	14:30	15:30	1345
DBA	DI 7070	CGN	TXL	15:30	16:35	12345
Hapag Lloyd Express	X3 3126	CGN	TXL	16:00	17:00	123457
Lufthansa	LH 279	CGN	TXL	16:10	17:15	123457
DBA	DI 7072	CGN	TXL	16:25	17:30	12345
Germanwings	4U 016	CGN	SXF	16:25	16:25	1234
Lufthansa	LH 281	CGN	TXL	17:25	18:30	12345
Hapag Lloyd Express	X3 3128	CGN	TXL	17:30	18:30	123457
Hapag Lloyd Express	X3 3138	CGN	TXL	18:25	19:25	12345
DBA	DI 7074	CGN	TXL	18:40	19:45	12345
Lufthansa	LH 283	CGN	TXL	19:20	20:25	123457
Hapag Lloyd Express	X3 3130	CGN	TXL	20:30	21:30	123457
Germanwings	4U 018	CGN	SXF	20:30	21:30	1234567
Lufthansa	LH 285	CGN	TXL	20:35	21:40	123457

Table 3-1: Competition on the Cologne-Berlin Route[19]

By attracting price-sensitive passengers who formerly used network airlines, the low cost competitors threaten the integrated production model of the major airlines. Low cost airlines do not only gain market share from the network carriers to fill their aircraft. According to a study conducted by Monitor Group in 2002, 43% of the passengers traveling on low cost flights are "new" customers, not flying before.[20] This is a significant increase compared to estimates of the late 1990s. For instance, Lufthansa German Airlines concluded in 1999 the market stimulation for a new low cost route would be approximately 25%.

---

[19] Source: OAG, Amadeus, Websites, calender week 03, 2005.
[20] Quoted from DGAP (2004), p. 23.

Each passenger previously traveling on a flight to and from the hub and who is now choosing a low cost direct flight is lowering the advantage of the integrated hub-production model and finally increases the cost per revenue passenger kilometer (RPK). Head-to-head competition on non-hub routes such as Cologne-Berlin is also a threat, because with any lost passenger to a low cost carrier customer loyalty decreases and limits the network airlines ability to influence this loss with their frequent flyer program. Finally this could endanger the brand premium of the major network airlines.

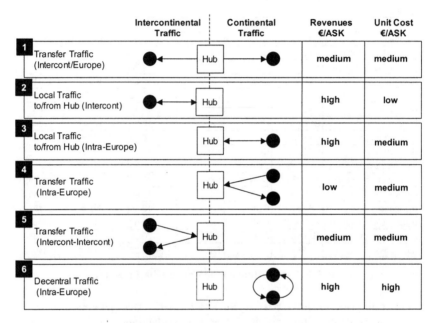

Figure 3-4: Traffic Segments and Unit Margins of Network Carriers

2.2 Cost Benefits from Hub-and-Spoke Operation: Economies of Scale, Scope and Network Density

Many empirical studies point to economies of scale, scope and density as driving forces for airlines to adopt hub-and spoke networks. Basically, hub-and-spoke networks allow airlines to achieve cost advantages through economies of scale, scope and density by adding further destinations to their networks.

Economies of scale generally mark the degression of total cost by increasing production volumes, which are measured in the airline industry by available seat kilometers (ASK) or revenue passenger kilometers (RPK). Economies of scale result from a new service to an airport, which was not yet served within the existing network. In order to isolate further network effects they define only those cost decreases as economies of scale, if the average load factor in the whole network is not affected. Economies of scale result typically from better utilization of ground staff and equipment, airport facilities, maintenance facilities, etc. Network extension with a largely harmonized fleet and engines may lead to a more efficient material management and less capital cost. The extent of cost savings per unit varies from airline to airline and the economic research has not led to a clear indication so far.[21] Hub operations with few peaks of arriving and departing aircraft do not result in high aircraft and ground staff productivity. Furthermore, hub-and-spoke networks have advantages in serving thin markets efficiently. This requires smaller aircraft with fewer seats which finally results in higher cost per ASK.

Unlike economies of scale the real network effects stem from economies of scope and traffic density. Economies of traffic density occur when the unit costs over an existing network decrease with increased traffic volume. Basically, it is less expensive to increase service on the existing network than it would be for some other carrier to provide additional service on the same routes. Economies of traffic density result from operating additional flights and/or using larger aircraft or aircraft with a higher seating capacity.[22] In order to measure the effect on unit cost researchers considered in econometric studies only those flights in the network where the number of destinations did not vary and average load factor and average stage length was constant. The studies indicated a significant unit cost advantage of 13 - 25% in networks with a high traffic density compared to those with middle or low density.[23]

---

[21] See e.g. Höfer (1993), p. 211ff.
[22] See Caves, Christensen & Tretheway (1984), p. 474; McShan & Windle (1989), p. 219ff.
[23] See Brueckner & Spiller (1994), p. 381. Formally, returns to traffic density (RTD) can be defined as the inverse of the elasticity of cost with respect to output: RTD = $1/\varepsilon$, with ε being the elasticity of total cost with respect to output. Thus the cost advantage always depends on variation of input as economies of density define the proportional increase in output made possible by a proportional increase in inputs. Economies of traffic density are statistically significant and economically large. See Fallon (2004), p. 9ff.

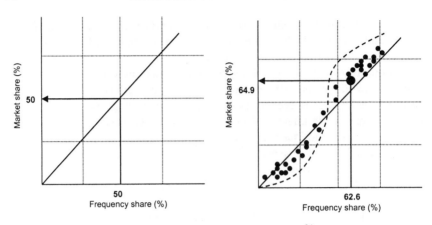

Figure 3-5: S-Curve-Effect at Alitalia[24]

The degressions in unit costs are based on the so-called s-curve effect, which describes the empirically proven phenomenon of the substantial influence of the frequency share measured in number of flights on market share.[25] The market share of an airline exceeds frequency share more than proportional, if the latter amounts higher than 50%. In the case that the capacity share is below 50% the market share is significantly lower. The traffic density results from the degree of hub dominance of the hub airline, which is influenced by the capacity share at the hub airport. The bigger the hub dominance of an airline is, the bigger the likelihood that passengers will connect to flights of the same airline. The second influencing factor is flight frequency. The higher the frequencies offered by an airline are, the higher the flexibility of passengers. For the choice of an airline or flights, respectively, this is an important criterion for the passengers.

Basically economies of scope exist when it is less costly to combine two or more product lines in one firm than to produce them separately.[26] In hub-and-spoke networks these cost benefits occur when the unit cost of service decreases as the number of markets served by an airline increases.[27] In the context of a hub-and-spoke network the airline can reduce unit cost over the existing network by

---

[24] See La Costa (1998).
[25] For a detailed description of the S-curve effect see Dennis (1998), p. 20; Kanafani & Ghobrial (1985), p. 19; Miller (1979), p. 83ff.; Tretheway & Oum (1992), p. 27ff. For an emprirical research about the S-curve effect and yield premium see Salomon Smith Barney (2003), p. 43.
[26] See Panzar & Willig (1981), p. 268.
[27] See Spitz (1998), p. 492.

serving an additional city-pair market by adjusting the flight schedule to allow for additional connection opportunities while keeping the total number of flights constant; this option is often cheaper than adding new flights. Since the airline transports passengers with a variety of different itineraries on the same flight, economies of scope increase with the number of flights, which can be concentrated within a so-called arrival or departure bank. The realization of such economies requires a critical number of aircraft and arrival and departure slots.

Both economies of traffic density and scope provide significant incentives to constantly enlarge network reach by adding new destinations. These cost benefits are accordingly one of the main drivers to enter code-share agreements and strategic alliances and increase network reach by connecting the existing networks of the alliance partners. By this means the co-operating airlines can maximize number of markets, frequency and number of passengers funneled through the network.

2.3 Hub-and-Spoke Networks and Alliances as Revenue Generators

Hubs are generally considered to be excellent means for maximizing revenue. This can be explained by

- hub dominance;
- increasing market share;
- a price premium paid for a service quality, in particular a high number of flight frequencies.

In the US post-regulation era, economic researchers examined the importance of route and airport dominance in determining the market power exercised by an airline. At their hubs, network carriers usually have a market share above 50%.[28] By hub-and-spoke networks airlines are able to link city-pair markets efficiently, even when traffic volumes on spoke routes are low. New entrants, who cannot make use of economies of scope and traffic density, face significant entry barriers. In this situation the incumbent carriers do not face any potential competition and, in theory, they can use their market power and raise fares. The

---

[28] For instance KLM has a market share (share of scheduled seat capacity) of 52.5% at Amsterdam's airport Schiphol, in Frankfurt is Lufthansa with a 58.8% market share, the dominant carrier at Paris Charles-de-Gaulle, Air France, has a share of 57.2%. For a detailed analysis see Salomon Smith Barney (2003), pp. 87ff.

empirical research in the US airline industry came to the conclusion that dominant airlines generally charge higher prices at flights arriving or departing from the hub airport. The dominant carrier's average fare per mile (yields) generated on routes to and from the hub is approx. 20% higher than the yields of competing airlines.[29] Compared to its non hub-routes the dominant carrier charges a premium of as much as 40%.[30] Whether this also applies for today's hub traffic in Europe may be doubted. In Europe the hubs are located in much higher proximity than in the US. Thus passengers have a lot more choices to fly to their final destination and the competition between the different hub-and-spoke networks prevent airlines from charging a premium. In Europe the advantage of hub dominance may impact market power and competition on thin routes. But on routes with sufficient demand, network carriers are challenged by low cost airlines with significantly lower fares and the result of heavily diluting yields.

Hub-and-spoke networks and strategic alliances are generally considered as tools to increase revenue because city-pair markets can be linked via the hub, which have not been served by an airline or its partners before. Even when the yield for connecting service is lower than for direct service, the additional passenger volumes over-compensate this and result in higher total revenues. Within an alliance the partners coordinate their flight schedules in the way that minimizes transfer times between arriving and departing flights and use the flight numbers of their partners. On the CRS screens flights are listed according to the total travel time and online connections have a higher listing priority than interline connections. The fact that 80% of all bookings are made from the first page of the CRS screens enhances the likelihood of booking an alliance connecting flight

---

[29] A 2003 study by Salomon Smith Barney revealed that the average US hub carrier enjoys a 273% yield premium versus other airlines serving the hub. The researchers even found a significant correlation between the yield premium and hub dominance measured by seat market share: for every 1% increase in seat share, yield premium rose by 0.49%. See Salomon Smith Barney (2003), pp. 5-7 and p. 43ff.

[30] See Borenstein (1989); Kahn (1993); Hendricks, Piccione and Guofu (1995). It is not only from airport capacity constraints that hub airlines derive market power. An airline with a large presence in a given city gains market power from frequent flyer programs (FFP) and travel agency commission overides (TACO). FFPs and TACOs have been found to be highly significant factors in explaining why dominant carriers can charge higher fares to and from the hub than other airlines on the same routes. FFPs and TACOs both have a strong impact on business routes and they increase market power by increasing the cost of switching from the hub airline to a competitor, lowering cross elasticity of demand and reducing incentives for price competition. See Hanlon (1996), p. 145; Taneja (2003), p. 42ff.

rather than an interline service.[31] Furthermore, the major carriers' frequent-flyer programs and their integration within an alliance are strong incentives for customer loyalty and help to keep passengers in their own alliance network and support to increase or secure passenger volumes.[32]

Hub-and-spoke networks and alliances provide a higher level of service in terms of flight frequencies. It is generally accepted that flight frequency is an important decisive factor in particular for higher yield business travelers. In some cases this may provide room for charging a service premium and to stimulate high yield demand. More important, the flight frequency has a strong impact on the market share an airline is able to gain.[33]

2.4 Cost of Network and Alliance Complexity

2.4.1 Complexity of Hub-and-Spoke Operation

The advantages for hub-and-spoke systems have been discussed above, but these networks have some features which will raise the complexity and cost of operation. There are asset-related prerequisites and process complexities, which will drive costs.

The former include airside infrastructure such as runways, taxiways and apron space of sufficient capacity to accommodate flight activities in peak times. The same accounts for the capacity of the passenger terminals. Beyond capacity, passenger terminals need to have an appropriate layout, in particular gate locations, in order to facilitate the transfer of passengers between arriving and departing flights. In Europe most airports are capacity constrained and most terminals have not been designed for efficient hub operations. Another issue is the necessity of a complex high-speed baggage handling system to allow fast transfer of baggage. Although airports usually provide this infrastructure, this will drive the airport's cost of capital and finally may impact airport charges. Hub-and-spoke operations with dense traffic during the peaks require a large fleet of aircraft associated with lower aircraft utilization. Low utilization and

---

[31] The impact of total travel time for the decision to choose a flight might be overestimated. Since there are a lot of booking engines available in the internet, which list flights according to price and availability and not according to time, in particular price-sensitive or time-flexible passengers make increasing use of low cost airlines.

[32] See Borenstein (1989); Hanlon (1996), p. 145.

[33] See explanation of the s-curve effect and Taneja (2003), p. 56.

productivity apply for cabin and cockpit crews as well as for ground staff. Different stage lengths of spoke routes require a heterogeneous aircraft fleet and related higher maintenance cost. Cockpit and cabin crews need to be licensed for the different aircraft types, which drive total number of crews and related training costs.

Process-related complexities arise from aircraft, passenger, baggage handling processes and consequences from flight delays. Peaking of flights is a serious challenge for all parties, who provide service to an airline, such as catering, fuelling, aircraft cleaning, de-icing. The likelihood of irregularities in exercising a handling process increases with the number of flights per hour. Results are extended ground time and finally departure delays. Capacity shortages during peak times and the non-availability of landing slots force airlines to stay in holding patterns in the air, which increases fuel cost.[34] Non-availability of landing slots at the hub can cause departure delays at other airports in the network, because flights to the hub are not allowed to depart, resulting in higher charges at the airport. To enable an airline to provide quick transfer of passengers between two flights, they need to use gate positions for their aircraft, which will also increase charges compared to apron positions. Baggage transfer at hubs is another critical issue. The more flights are handled at the hub per hour the higher the probability of missing luggage. AEA statistics reveal a significant larger share of missing baggage for the network carriers than for point-to-point airlines. For instance, Iceland Air missed only 7.8 bags per 1,000 passengers while the KLM missed 18.0 bags per 1,000 passengers in the period January to July 2004.[35] Each piece of luggage missed causes additional cost for tracking the bag and sending it to the destination. Furthermore, airlines usually pay compensation to passengers. Assumed that the compensation is € 100 per bag missing, the 11.9 million KLM passengers emplaned would lead to an additional cost of approx. € 21 million for the first half of 2004.

2.4.2 Complexity of Multi-Hub Operation within an Alliance

The general comments on hub-and-spoke network complexities apply also to the European network carriers' multi hub-operations and may be intensified by additional feeder flights within the alliance. But there are some additional aspects, which need to be considered. The geographical density of the European hub landscape may lead to intra-alliance traffic cannibalism and competition for feed traffic for long-haul flights. In order to avoid this intra-alliance competition

---

[34] For a more detailed analysis see Klingenberg, chapter 7 in this volume.
[35] See AEA (2004).

as far as possible, there is a need for common incentive structure for the alliance, which makes the partners contribute to the overall benefit. The coordination of flight schedules within a European multi-hub network is a tremendous effort, which requires a complex and iterative decision-making process involving planners from different companies. The fact that among many partner airlines, which build a multi-hub network in Europe, a distinct power differential is relatively weak compared to fully hub-integrated networks of regional airlines.[36]

The integration of operational systems (e.g. check-in) of the partners working with heterogeneous IT-platforms is another issue, which drives the cost of complexity. Pricing within a multi-hub network is significantly more complex than for a single airline; it requires the integration of the hub carriers yield management systems.[37]

3 Cornerstones of a Consolidated Business Model for European Network Carriers

In this chapter we will analyze the necessity for the network carriers to consolidate their market position to compete successfully with point-to-point operators. In a second step we work out how to consolidate the hub operation in order to manage the trade-off between productivity and connectivity. Finally, we will point out that alliances are not only a means to generate additional revenues in an efficient way, but also have the potential for large scale cost savings.

3.1 Consolidating the Market Position between Full-Service and No-Frills

The European major airlines do not only address service- and price-sensitive passenger segments simultaneously;[38] they also operate on routes with very different characteristics in terms of demand and stage length.

---

[36] The major airlines are using regional airlines as feeder carriers to their hubs, such as Lufthansa, British Airways and Air France. With regard to the network, the regionals transferred decision-making to the major network airline. See Steininger (1999), p. 244.
[37] For an overview on yield management techniques and systems see Auerbach and Willigens (1996).
[38] For a more detailed analysis see Flenskov chapter 4 in this volume.

3.1.1 Non-Hub Traffic on High Demand Routes

Particularly in city-pair markets with high demand, which economically justify direct services, the traditional carriers are challenged by low cost airlines. An unquestionably risen price sensitivity of travelling public both business and leisure has resulted in lower revenues per seat mile. Despite the traditional airlines' ambitious cost saving programs they have not achieved a competitive cost structure and still produce at significantly higher unit cost than their low cost competitors. To cope with this challenge, traditional airlines do not simply copy the low cost approach. On intra-European flights they continue to offer a two-class product and widely differentiate the service levels in business and coach class and corresponding ticket prices. If this strategy will be a successful one is largely dependent on the critical number of passengers' willingness to pay a higher price for better service and on the price differential between business and coach class. The remaining cost complexity of a full service operation and a lower demand in the business class market may question this approach. If they cannot generate a sustainable profit base they will finally have to leave these markets. These withdrawals from non-hub routes may be accompanied by market entries of the traditional airlines' low cost subsidiaries like in the US Delta's Song or United's Ted. In Europe, Lufthansa owns an indirect share in a low cost airline. It has a 49% stake in the regional airline Eurowings, which is majority shareholder of Germanwings. Another example is SAS, which has 100% stakes in the Swedish low cost airlines Blue1 and Snowflake. Other network carriers have withdrawn from engagements in the low cost arena: British Airways sold Go to easyJet and KLM divested its low cost arm Buzz, which was taken over by Ryanair in 2003.

3.1.2 Non-Hub Traffic on Low Demand Routes

Network carriers usually do not operate services themselves on low demand routes, which are bypassing the hub. They usually do not own the regional aircraft types with a capacity below 50 seats, which would be required for these rotes. Although many regional carriers are generating healthy profits from these thin routes it has been widely accepted that the cost of complexity to operate a regional sub-fleet and the labor cost of the majors will hinder them from reaching a positive bottom-line result. Some of the major carriers in Europe have established their own regional airlines such as Lufthansa Cityline or KLM excel, or they established franchise systems with the regionals like Team Lufthansa, and BA˙express. It may be expected that this trend will continue for the future. Routes which cannot be served on a profitable basis, will lead to a withdrawal of major airlines. They might use their subsidiaries or regional allies to participate in terms of dividends or franchise fees. At least they can leverage their brand and increase passenger loyalty.

3.1.3 Hub-Traffic

Economies of density and scope will remain the justification of the integrated production model of the majors. On any route to and from the hub the major airline will operate itself only if passenger volumes and yields are high enough to cover the total route cost. In any other case the major airline will outsource these routes to regional partners, which can build on a more competitive cost base.

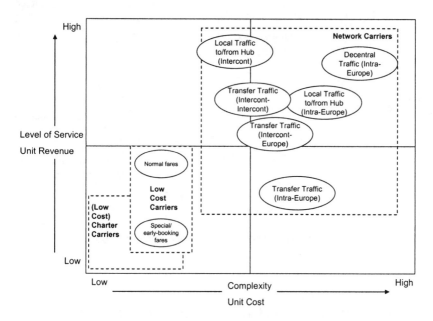

Figure 3-6: Traffic Segments

There is no way that an airline can successfully compete in any of the various market segments. However, a hub airline's integrated production model requires at least presence in all of those segments to maintain and protect their revenue generation potential and the benefits from economies of scope and density.

With regard to the future market structure we see four types of carriers surviving in the European airline industry. Regionals can defend their position against the low cost airlines if they continue to concentrate on small niche-markets. With their cost advantage compared to network airlines they could also protect their markets from a direct entry of a network airline. Furthermore, they could offer this advantage and generate additional revenues through a close cooperation with a network carrier by feeding its hub. Low cost airlines will have an increasing market share in the future as long as they are able to sustain their cost advantage driven by economies of scale. The network carriers will play the dominant role in the intercontinental markets and thin continental routes as long they can efficiently generate economies of scope and density. All other types of airlines need to move in one of these directions, otherwise they will leave the market (see figure 3-7).[39]

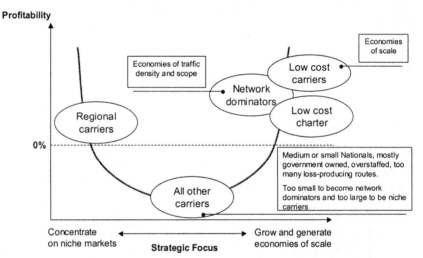

Figure 3-7: Players in the Future Airline Industry

3.2 Consolidating the Hub Operation: Managing the Trade-off between Productivity and Connectivity

The optimization of hub connectivity was the prevailing objective of network carriers in the 1990s. The hub connectivity is an indicator, which measures the number of possible connections within a defined period of time. For instance, an airline schedules flights to and from its hub in three complexes of arrival and departure banks during the day. The first complex is between 8:00 a.m. and

---

[39] For a quantitative analysis of the consolidation of the airline industry and future airline network structure in Europe see Dennis (2004) and Burghouwt and de Wit (2004).

11:00 with 30 arrivals and departures, the second one between 1:00 p.m. and 3:00 p.m. with 20 arrivals and departures and finally the third one between 6:00 p.m. and 9:00 p.m. with 30 arrivals and departures. Under the assumption that the airport has a curfew of 6 hours, the total of 160 flights would average to approximately 9 flights per hour. Actually 160 flights are operated in a time window of 8 hours and thus 20 flights per hour on average. The hub connectivity ratio here is 20/9 = 2.2. In reality, the connectivity of European hubs 1995 was between 1.0 and 2.0.[40] Figure 3-8 indicates that hub connectivity at Europe's major hubs has improved significantly.

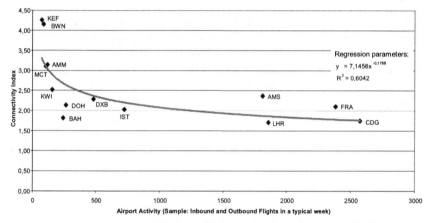

Figure 3-8: Hub-connectivity at International Airports in 2004[41]

Before the low cost airlines challenged the traditional airlines with point-to-point services the key rationale of network optimization was to maximize the number of arriving and departing flights in a tight time window, which resulted in higher hub connectivity. As mentioned, the number of city-pair markets linked via the hub is the equivalent to higher hub connectivity. In the alliances' multi-hub systems in Europe, passengers have the choice to fly via different hubs to their final destinations. At a first glance, this might appear an inefficient overlap of route networks. But recalling the s-curve effect, this is actually an increase in flight frequencies and thus – if properly managed – could lead to higher market shares and to higher revenues.

---

40   The connectivity of American Airlines's Dallas Fort Worth hub was close to 3.0. See Dennis (1998), p. 8, Doganis (1991), S. 268.

41   Source: Lufthansa Consulting Research, Compass-Analysis, OAG.

An analysis of the alliances' current European multi-hub systems it seems that each alliance member is trying to maximize hub connectivity at its own hub airport. Over the entire alliance network this finally results into low aircraft productivity. Ground time as a result of high connectivity is a costly item, not only because of opportunity cost of lost revenue, but aircraft parking fees, cost of idle crews and ground staff as well as under-utilized airport facilities in non-peak times. Nearly all hub airports in Europe are capacity constrained and thus extended flight concentration leads to further congestions. Both airport congestion and the non-availability of landing slots force airlines to fly holding patterns, resulting in higher fuel cost. Furthermore, arrival delays will cause departure delays and further decrease productivity.

When former hubbing protagonist American Airlines started to change the schedule philosophy of connectivity maximization and de-peaked the number of flights at its hubs in Chicago and Dallas Fort Worth in 2003, a debate about the risks and reward of de-peaked schedules has arisen. The announcements of American Airlines that the new scheduling paradigm has led to fleet reduction or re-assignment of 16 aircraft and related cost of capital as well as to a reduction in labor cost in the magnitude of 5%. The estimated annual cost savings add up to 1 million USD.[42] Even though there seems to be currently an intensive discussion of de-peaking, the insights are not new.[43] Basically, this is nothing else rather than managing the trade-off between productivity and connectivity: Are the cost savings from higher productivity outweighing the revenue losses due to lower connectivity?

Figure 3-9 illustrates that Delta Airlines has so many arrival and departure banks throughout the day that it could be considered to be continuous. Compared to Delta Airlines, Lufthansa has structured its operation at Frankfurt Airport in fewer banks, but the de-peaking of the schedule is already remarkable. The banks are less deep than they have been in the past.

---

[42] See Goedeking and Sala (2003), p. 93.
[43] See Ghobrial and Kanafani (1995). Delta Airlines at its Atlanta hub was always applying "continuous hubbing".

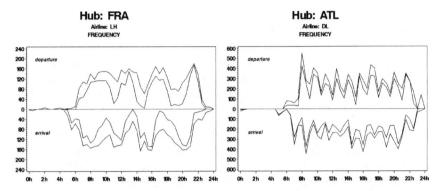

Figure 3-9: Comparison of Hub-structures at Frankfurt and Atlanta[44]

Low cost carriers' rigid point-to-point strategy is based on productivity optimization and economies of scale. If network airlines are able to manage the trade-off between productivity and connectivity successfully they may generate higher margins and finally foster their position in Europe. This counts for the connecting flights only. For the decentral traffic carried on flights bypassing the hubs, there is no advantage of high connectivity. On these flights the majors need to dramatically decrease cost/ increase productivity or differentiate the product in terms of higher quality.

As outlined, strategic airline alliances and in particular the coordination of flight schedules has led to a significant increase of connecting traffic at European major hubs. But these alliance benefits may be limited by infrastructure capacity. Most of the major European airports are slot-constrained due to physical constraints (e.g. London Heathrow), ATC constraints (e.g. Madrid Barajas or Paris Charles-de-Gaulle) and environmental limits (e.g. Amsterdam Schiphol or Dusseldorf). These infrastructure shortages in Europe limit efficient hub operation. For instance at London Heathrow, British Airways is currently limted to further increase its connecting traffic, since the airline is using Terminals 1 and 4, several miles apart. Attractiveness to connecting traffic is partly a function of infrastructure efficiency. While connecting traffic at Heathrow is only 34%, it accounts for 53% at Charles-de-Gaulle. Although most of the European major airports pursue ambitious capacity expansion programs (e.g. Terminal 5 at Heathrow), for two reasons it may be doubted that these initiatives always satisfy the airlines' and alliances' needs. First, most of the projects are long-term, but to fully exploit the advantages of a consolidated hub-

---

[44] Source: OAG, Lufthansa Consulting Research.

and-spoke operation and to cope with the low cost challenge, short-term solutions have to be identified. Second, the long-term capacity expansion of airports is in many cases focused on single parts of the airport system. For instance, at London Heathrow the new Terminal 5 may increase passenger terminal capacity and provide an efficient facility for oneworld's connecting services. But at Heathrow it is not foreseeable that an additional runway will be built.

In the short-term, additional airport capacity may be generated by a close cooperation between the home airline or home alliance, ATC and the hub airport in order to optimize airport processes. At Frankfurt and Munich, Lufthansa, the ATC provider DFS (Deutsche Flugsicherung) and the respective airport operator established joint hub-control centers (HCC) and collaborative decision-making (CDM) procedures to make the most efficient use of the existing resources.[45] So-called A-SMGCS (Advanced Surface Movement Guidance and Control Systems) are currently under development not only to increase operational safety, but also the efficiency of the resource allocation.[46]

The fact that many hub airports are restricted to develop their infrastructure according to the airlines' needs, forced the airlines to develop additional hubs, which is sub-optimal for a hub-and-spoke driven business model.

At European airports the exclusive use of terminal or gate facilities by airlines or alliances is not a widespread procedure. In the US airlines usually enter into long-term terminal or gate lease contracts, which enable them to optimize connecting services with regard to gate positions of aircraft, passenger and baggage-handling processes and intra-terminal passenger flows. In this environment the cost of the leases and the facility operation can be traded off against the additional revenues from connecting passengers. In the US infrastructure is usually considered as an important factor in hub-and-spoke operation. In Europe, a similar approach is pursued at Munich's Terminal 2. The Terminal was designed as an exclusive Lufthansa and Star Alliance hub facility. All terminal elements have strictly been planned around the processes, in particular those supporting connecting traffic. The ultimate objective is to bring down complexity and the total cost per passenger handled at Terminal 2.[47]

---

[45] A similar approach – the so-called "AAA-project" – is currently being pursued at Vienna airport. See Frank-Keyes (2004), p. 10.

[46] See ICAO (1987); ICAO (2003); EUROCAE (2001).

[47] Lufthansa German Airlines and the Munich airport operator Flughafen München GmbH (FMG) have a joint venture company, which constructed and now operates the new Terminal 2. In this company Lufthansa has a 40% stake, the remaining 60% are held by FMG.

To remain competitive as a hub-and-spoke operator the alliances and their members need to gain more control over airport capacity. Without bringing down the infrastructure-induced complexity at their hub airports or being able to manage complexity efficiently, the hub-and-spoke operation model will be vulnerable as air traffic continues to grow.

3.3 Cost Benefits from Consolidating Alliance Integration

Basically, airline alliances are considered to be a means for revenue generation. Compared to other industries with considerable alliance activity such as the computer or car manufacturing industry, airline alliances so far have achieved relatively little in the area of major cost savings.[48] According to the alliance experience in other industries, major cost savings have been produced in four areas along the value chain: bundling purchase volumes and activities, resource pooling in R&D, in production, as well as in sales and distribution.[49] While R&D plays a minor role in the airline business compared to more knowledge-intense industries, the other areas are of similar importance for airlines. A general precondition for achieving cost benefits from alliances is the coordination of decisions, which is not achieved by market-mechanisms, but by clearly defined routines and responsibilities and procedures. Although this chapter focuses on network-relevant issues, synergy potentials in sales & distribution and procurement are touched because the production model might induce them.

As outlined above, it is not that hub-and-spoke is a wrong strategy. In Europe too many hubs and widely overlapping route networks are the issue. With regard to intra-European air traffic, the alliance partners offer many routes redundantly. To consider all these overlaps and related costs as cost saving potential would be too easy. This might also be a source for further revenues as long as the alliance partners exploit the s-curve effect jointly. However, route network overlaps have to be removed as far as possible and reasonable.[50]

---

[48] Taneja (2003), p. 44; O'Toole (2002).
[49] See for instance Hamel, Doz and Prahalad (2002); Kanter (2002); Bleeke and Ernst (2002).
[50] Dennis (2004) comes to a similar conclusion. He observed that the level of integration within many of the alliances is far from perfect and it is quite possible that airlines within the same alliance will continue to compete in the way they have always done, paying little regard to the strategies of their supposed partners. See Dennis (2004), p. 7.

By now, alliances have not realized large-scale cost synergies in the area of joint procurement. Focus was merely on joint purchase of office supplies and some airport equipment (e.g. baggage trolleys). Since the member airlines are largely enjoying autonomy in their network decisions, important related decisions such as the fleet structure have not been considered on alliance level. If alliance members could agree on joint and integrated network and fleet strategy, they could bundle purchase power and negotiate more attractive conditions with the aircraft manufacturers and/or lessors. The increase in purchase power could also be vital to support re-negotiation of airport service fees.[51] As a result of a joint fleet strategy there could be a tremendous cost saving potential in the area of aircraft maintenance. First, a consolidation of maintenance bases could help to decrease cost of capital. Economies of scale due to a less heterogeneous fleet structure within the alliance and a further consolidated maintenance base structure in Europe could lead to not fully exploited synergies.

An alliance-wide approach to de-peak flight schedules at the hubs could lead to a reduction in the number of aircraft with an associated decrease in cost.

In the beginning of the alliance era, the member airlines tried to achieve cost synergies by consolidating their sales organizations. Due to the former regulation of air transport in Europe airlines naturally have a stronger presence in their home market than their partners. Thus, some airlines reduced their local staff and used the partners' sales organization as agency. Although this seems to be the most obvious synergy potential, it partially failed. The reason is the lack of a mutually agreed incentive system. As long as ticket sales for the own airline are prioritized over those of alliance partners, synergies cannot be achieved without sacrificing revenue potentials.

This chapter only touches on selected examples for cost saving potentials superficially. But all examples mentioned have in common that large scale savings are only possible if the currently pursued soft coordination of alliance activities make way for a more strictly defined set of responsibilities. There is a need to develop mechanisms to ensure that the partners' decisions will support the achievement of overriding alliance objectives and not only the individual partners' short-term ones. Certainly, this requires a mutually accepted way to re-allocate the alliance benefits to the partners.[52]

---

[51] It needs to be considered that airports should comply with "ICAO's Policies on Charges for Airports and Air Navigation Services" and according to this international standard airport charges should not be discriminatory to any airport user. See ICAO (2004).

[52] For further analysis see chapter 11 by Götsch & Albers in this volume.

4 Conclusions

A consistent business model will be the key prerequisite to achieving sustainable competitiveness in the future European airline market for all actors involved. For traditional network carriers coherent consolidation on the strategic as well as on the operative level along their entire value chain is the most important step towards regaining competitive power. This implies a clear market focus or, possibly, stringent market segmentation related to distinct business units. This fundamental restructuring will lead to a more transparent European airline industry with fewer players in all market sectors, including the low cost sector. As the consolidation potential of network carriers is higher than that of pure low cost carriers those network carriers who manage to implement the consolidated business model discussed above will not only survive but could even gain back the leadership in European airline competition.

5 References

AEA ASSOCIATION OF EUROPEAN AIRLINES (2003): *Yearbook 2003*.

AEA ASSOCIATION OF EUROPEAN AIRLINES (2004): *AEA Consumer Report for July 2004*.

ALAMDARI, F.A.; MORREL, P. (1998): Airline Franchising: Brand Extension or Outsourcing High Cost Routes?, in: Butler, Gail F. and Keller, Martin R. (Eds.): *Handbook of Airline Marketing*, New York: McGraw Hill, pp. 503-512.

ALAMDARI, F.A.; MORREL, P. (1998): Developing Effective Route Networks, in: Butler, Gail F. and Keller, Martin R. (Eds.): *Handbook of Airline Marketing*, New York: McGraw Hill, pp. 613-631.

ANTINOU, A. (1991): Economies of Scale in the Airline Industry: The Evidence Revisited, in: *Logistics and Transportation Review*, 27(2), pp. 159-184.

AUERBACH, S.; WILLIGENS, J.-M. (1999): Get IT Booked – Airline Revenue Management Systems, *Airline Fleet & Asset Management*, May-August, No. 3, pp. 34-41.

BLEEKE, J.; ERNST, D. (2002): The Way to Win in Cross-Border Alliances (Reprint), in: *Harvard Business Review On Strategic Alliances*, pp. 173-198, Boston: Harvard Business School Publishing Corporation.

BORENSTEIN, S. (1989): Hubs and High Fares: Dominance and Market Power in the US Airline Industry, in: *Rand Journal of Economics*, 20(3), Autumn, pp. 344-365.

BRUECKNER, J.K.; SPILLER, P.T. (1994): Economies of Traffic Density, in: *Journal of Law and Economics*, 37(2), pp. 379-419.

BURGHOUWT, G; DE WIT, J. (2004): Temporal Configurations of European Airline Networks, in: *Journal of Air Transport Management*, 11(3), pp. 185-198.

CARLTON, D.; LANDES, W.M.; POSNER, R.A. (1980): Benefits and Costs of Airline Mergers, in: *Bell Journal of Economics*, 11(1), pp. 65-83.

CAVES, D.W.; CHRISTENSEN, L.R.; TRETHEWAY, M.W. (1984): Economies of Density versus Economies of Scale: Why Trunk and Local Service Airline Cost Differ, in: *Rand Journal of Economics*, 15(4), pp. 471-489.

DENNIS, N. (1998): *Optimisation of Airline Schedules and Route Networks*, Working Paper, Transport Studies Group, University of Westminster, London.

DENNIS, N. (2004): Industry Consolidation and Future Airline Network Structures in Europe, in: *Journal of Air Transport Management*, 11(3), pp. 175-183.

DEPROSSE, H.; FRANKE, M. (1998): One for All and All for One: Threats and Opportunities Posed by the Partnering Principle, in: Butler, Gail F. and Keller, Martin R. (Eds.): *Handbook of Airline Marketing*, New York: McGraw Hill, pp. 503-512.

DGAP-DEUTSCHE GESELLSCHAFT FÜR AUSWÄRTIGE POLITIK (2004): *Europas Luftverkehr zwischen Marktwirtschaft und politischer Einflussnahme – Fragen zum Luftverkehrsstandort Deutschland*. Conference Proceedings, February 12, 2004, Berlin.

DOGANIS, R. (2001): *The Airline Business in the 21st Century*, London, New York: Routledge.

DOGANIS, R. (1991): *Flying Off Course. The Economics of International Airlines*, 2nd Ed., London, New York: Routledge.

EUROCAE – THE EUROPEAN ORGANIZATION FOR CIVIL AVIATION EQUIPMENT (2001): *Minimum Aviation System Performance Specification for Advanced Surface Movement Guidance and Control Systems* (A-SMGCS), ED 87A.

FALLON, J. (2004): *Market Structure, Regulation and Performance in the Airline Industry*, Conference Paper, presented at the ANU/NCEG Conference on the Performance of Air Transport Markets, June 24th & 25th.

FRANK-KEYES, J. (2004): The Way Forward. Interview with Vagn Sorensen, in: *Communiqué Airport Business*, June/July, pp. 7-17.

GALLACHER, J. (1999): Circling the Globe, in: *Airline Business*, 15(7), pp. 34-65.

GALLACHER, J.; WOOD, S. (1998): Hold Your Horses, in: *Airline Business*, 14(6), pp. 42-87.

GALLACHER, J. (1997): Partners for Now, in: *Airline Business*, 13(6), pp. 26-67.

GHOBRIAL, A.A.; KANAFANI, A. (1995): Future of Hubbing/Dehubbing in the US Airport System, in: Jenkins, D.; Ray, C.P.: *Handbook of Airline Economics*, New York: McGraw Hill, pp. 321-328.

GOEDEKING, P.; SALA, S. (2003): Breaking the Bank, in: *Airline Business*, 19(9), pp. 93-97.

HAMEL, G.; DOZ, Y.L.; PRAHALAD, C.K.; ERNST, D. (2002): Collaborate With Your Competitors – and Win (Reprint), in: *Harvard Business Review On Strategic Alliances*, Boston: Harvard Business School Publishing Corporation, pp. 1-22.

HANLON, P. (1996): *Global Airlines. Competing In A Transnational Industry*, Oxford: Butterwoth-Heinemann.

HENDRICKS, K.; PICCIONE, M.; GUOFU, T. (1995): The Economics of Hubs: The Case of Monopoly, in: *Review of Economic Studies*, 62, pp. 83-99.

HÖFER, B. (1993): *Strukturwandel im europäischen Luftverkehr. Marktstrukturelle Konsequenzen der Deregulierung*, Frankfurt: Peter Lang.

ICAO (2004): *Policies on Charges for Airports and Air Navigation Services*, Seventh Edition, Document No. 9082/7.

ICAO (2003): *Draft Manual on Advanced Surface Movement Guidance and Control Systems (A-SGMS)*, AN-Conf/11-1P/4.

ICAO (1987): *Manual of Surface Movement, Guidance and Control Systems* (SMGCS), Document No. 9476.

KAHN, A.E. (1993): The Competetive Consequences of Hub Dominance: A Case Study, in: *Review of Industrial Organization*, 8(8), pp. 381-405.

KANAFANI, A.; GHOBRIAL, A.A. (1985): Airline Hubbing: Some Implications for Airport Economics, in: *Transportation Research A*, 19(1), pp. 15-27.

KANTER, R.M. (2002): Collaborative Advantage: The Art of Alliances (Reprint), in: *Harvard Business Review On Strategic Alliances*, pp. 97-128, Boston: Harvard Business School Publishing Corporation.

LA COSTA, J. (1998): *Ensuring Fleet Strategy is Effective in Minimising Capacity Constraints and Maximizing Connection Development*, IIR Conference on network management. London: IIR Publications.

MCSHAN, S.; WINDLE, R. (1989): The Implications of Hub-and-Spoke Routing for Airline Costs and Competitiveness, in: *Logistics and Transportation Review*, 25(3), pp. 209-230.

MILNE, I.R. (2004): *Bridging the GAAP*, London: Reed Business Information.

MILLER III, J.C. (1979): Airline Market Shares versus Capacity Shares and the Possibility of Short-Run Loss Equilibria, in: *Research in Law and Economics*, 1, pp. 81-96.

O'TOOLE, K. (2002): Cost Equation, in: *Airline Business*, 18(7), pp. 42-44.

OUM, T.H.; PARK, J.H.; ZHANG, A. (1996): The Effects of Airline Codesharing Agreements on Firm Conduct and International Air Fares, in: *Journal of Transport Economics and Policy*, 30(2), pp. 187-202.

PANZAR, J.C.; WILLIG, R.D. (1981): Economies of Scope, in: *American Economic Review*, 71(2), pp. 268-272.

SALOMON, S.B. (2003): *2003 Hub Fact Book*, New York: SSMB Publications.

SONCRANT, C.U., HOPPERSTAD, C.A. (1998): Schedule Evaluation, in: Butler, Gail F. and Keller, Martin R. (Eds.): *Handbook of Airline Marketing*, New York: McGraw Hill, pp. 601-611.

SPITZ, W.H. (1998): Liberalisation of European Aviation: Analysis and Modelling of the Airline Behaviour, in: Butler, Gail F. and Keller, Martin R. (Eds.): *Handbook of Airline Marketing*, New York: McGraw Hill, pp. 489-501.

STEININGER, A. (1999): *Gestaltungsempfehlungen für Airline Allianzen*, St.Gallen/Bamberg: Difo-Druck.

STILL, S. (2002): Fortress Defense, in: *Airline Business*, 18(11), pp. 97-100.

TANEJA, N.K. (2003): *Airline Survival Kit. Breaking Out of the Zero Profit Game*, Aldershot: Ashgate.

TRETHEWAY, M.W.; OUM, T.H. (1992): *Airline Economics: Foundations for Strategy and Policy*, Vancouver: University of Bristish Columbia.

ZEA, M.; FELDMAN, D. (1998): Going Global: The Risks and Reward of Alliance-Based Network Strategies, in: Butler, Gail F. and Keller, Martin R. (Eds.): *Handbook of Airline Marketing*, New York: McGraw Hill, pp. 545-549.

4

FROM PRODUCTION-ORIENTATION TO CUSTOMER-ORIENTATION – MODULES OF A SUCCESSFUL AIRLINE DIFFERENTIATION STRATEGY

KIM FLENSKOV

1   Retrospect – the International Regulatory Framework 98
2   Changes in the Market Environment – Enabled Customers and
    a New Form of Competition 100
3   Modules of a Successful Airline Differentiation Strategy 107
4   Outlook 116
5   References 117

Summary:
The airline business is taking a major step towards maturity, fuelled by the growth and resilience in the low cost airline model. As liberalization progresses, traditional "flag carriers" and network carriers are forced to cut costs, create flexible resources, identify new ways of differentiation as well as pursuing a clearer market segmentation in order to remain competitive in liberalized markets. The two business models are likely to converge to a certain point, as the network carriers reduce costs and adapt their product offering and low cost carriers seek to pursue differentiation strategies themselves. This chapter highlights some of the reasons behind the current airline crisis and describes various elements of differentiation that in the right combination forms strategic solutions for a future market positioning.

1 Retrospect – the International Regulatory Framework

The legislation for civil aviation post World War II was established in December 1944 with the signature of the Chicago Convention. This document and its annexes set the framework for technical and commercial airline operating conditions. As regards the latter, the Convention builds on the general principle of national sovereignty of the airspace within the territorial borders of a given country, whereby one contracting state can grant traffic rights to another on certain conditions.[1] These principles still govern commercial air transport where they have not been superseded by multilateral commercial and/or technical regulations.[2]

Furthermore, air carriers have organized themselves under the auspices of the International Air Transport Association (IATA), in order to facilitate and further standardize international air transport by, for example, setting fares, defining general rules for transportation, interlining, slot-allocation and even in-flight service. The uniform classification of passengers into "segments" like economy, business and first was also harmonized within the framework of IATA.

The multilateral regulatory framework worked as a strong enabler for the development of global air transport. The downside of it was the inherent lack of efficiency: the tight control exercized by some contracting states literally abandoned every form of competition; if contracting states introduced more lax rules on pricing and market access, there were still not many incentives to enter into competition; existing business practices established by the airline business itself were hindering factors for stimulating real competition.

The US Government was the first to take the initiative to soften up the old regime: liberalization measures were introduced in the US at the beginning of the

---

[1] The general principle may e.g. include restrictions to market access in the form of single designation, limitations on traffic uplift between, within and beyond the contracting parties. Agreement between the designated carriers on capacity, traffic programmes and fares were often preconditions for approval of commercial rights.

[2] For example the European Union liberalization measures (e.g. licensing, market access, pricing, competition rules etc.) and the Joint Aviation Regulations (technical regulations).

1980s and in the EU some 6 years later establishing a more liberal framework for carrying out commercial air transport services.[3] These measures led to numerous privatizations within the air transport business, whereby former "flag carriers" have been partly or wholly privatized.

However, in spite of such efforts the air transport markets in Europe did not undergo significant structural changes, partly due to the industry's traditional bilateral structure and strong government ties, partly due to its sensitivity towards changes in the business cycle like the effects of the Gulf War in 1991 and the Asia Crisis in 1997 which coincided with the final phases of the European liberalization process.

In the US airline, liberalization led to a shakeout process in which a formerly fragmented industry developed into a grouping of major carriers as in the case of American Airlines, United, Delta, US Airways and Continental. In Europe, the former rigid regulatory regime, aided by strong government intervention,[4] largely kept the air transportation business from consolidating itself[5] and significantly developing static and dynamic efficiency[6]; there was simply too little incentive – or competitive pressure – for airlines to develop beyond the traditional airline business model.

---

[3] The European deregulation took place in 3 phases, named the three "Aviation packages", that gradually softened restrictions of capacity, prices and market access within the European Communities and later the EEA countries. The first deregulation measure was the regional directive 83/416 that enabled some degree of softening bilateral agreement constraints regarding traffic to/from secondary airports within the EC. The final package, Council Regulations 2407/92, 2408/92 and 2409/92 respectively now governs intra-EU commercial air transportation along with the application of EU competition law through Regulation 3975/87 to the air transport sector.

[4] Government intervention e.g. in the form of state aids was granted to numerous European carriers like Air France, Alitalia and Olympic, see Lawton (2002), p. 24.

[5] There were exceptions to the rule: Prior to the EU liberalization process, the British Government privatized British Airways (which was merged with British Caledonian); UTA and Air Inter were merged with Air France in the early 1990s.

[6] Through the 1990s, the airline business started to enjoy advances in the area of operations research, whereby carriers were able to access refined decision-support tools for optimizing passenger streams and revenues, crew planning and fleet assignment. Furthermore, reservations could be made worldwide with the introduction of Computer Reservation Systems and passengers could be retained with Frequent Flyer Programmes. The hub-and-spoke operating model brought significant scale-effects to airlines and enabled the traveling public to reach hundreds of destinations from remote places.

2 Changes in the Market Environment – Enabled Customers and a New Form of Competition

The past five years provided the airline industry with a host of challenges: the burst of the dot.com bubble reduced the growth of most network carriers in the late 1990s; the events of 11 September 2001 and SARS led to an estimated loss of over 25 billion US$[7] as well as the loss of 400,000 industry jobs, making several US-majors slide into bankruptcy protection (Chapter 11) and leading to the final demise of several European carriers. In 2004 most US majors are still suffering from the aftermath as well as from soaring fuel prices. Carriers like United Airlines, Air Canada, US Airways, and American Airlines are in Chapter 11 or on the brink of sliding into Chapter 11 while trying to restructure their operations.

However, structural changes in the marketplace pose the biggest challenge – or opportunity – for air carriers as changes in consumer behavior and new forms of competition hold the power to take the airline industry a further step towards maturity.

2.1 The Internet – a Transparency Provider

The Internet has proved to be a fast-growing, powerful tool to connect supply and demand in the airline business (see tables 4-1 and 4-2 below). For the airlines it means tangible advantages in the form of lower distribution and transaction costs, customer-ownership as well as a reduction in time-to-money. The savings incurred are in most cases shared with the web customers in the form of price reductions vs. other distribution channels.

In addition to this, airlines may earn further revenue from advertising banners or – even better – through selling other services in the travel value chain to their customers like car rental or hotel accommodation. In 2003, a major European low cost operator claimed that the development in such revenue could potentially lead to the situation where the carrier would be able to offer airline seats for free.

---

[7] See IATA (2003), the 25 billion US$ aggregated loss estimate are for the years 2001-02.

Low-fare carriers	Online sales in %
EasyJet	94%
Ryanair	94%
JetBlue Airways	74%
WestJet Airlines	69%
AirTran Airways	64%
Southwest Airlines	55%
Virgin Express	55%

Table 4-1: Low-fare Carrier Online Sales[8]

On the demand side, the Internet creates transparency: the relative ease of comparing product offers reduces search costs, which again limit suppliers' market power. This provides consumers, competitors as well as potential market entrants with reliable and easily accessible information on supply, which – on top of market liberalization and a barrier-free market entry – make certain routes or markets contestable.

Provider	2001	2002	2003	2002 -2003
Expedia	3,600	5,300	8,200	55%
Travelocity	3,100	3,500	3,890	15%
Orbitz	818	2,566	3,440	34%

Table 4-2: Online Majors Travel Booking Revenues[9] (in US$ million)

---

[8] See Rose (2004).
[9] See Clarke (2004).

2.2 The Low Cost Airlines

The low cost concept has its roots in the US[10] where pioneering low cost airline Southwest set up its operations in Texas in the late 1960's and remained largely unrivalled until the turn of the century. Southwest, which never made a loss in its entire history, provides its passengers with low air fares and good service, but without frills like free in-flight meals/beverages, hub connectivity, baggage transfer handling etc. Today the low cost concept has been developed further through brand and product differentiation by carriers like AirTran, AmericaWest, Frontier and JetBlue. Some US majors have also established their own low cost operators like TED (United Airlines - 2004) and SONG (Delta – 2003), just as Air Canada has done with ZIP.[11] In 2003, the low cost operators accounted for an estimated 25% of total domestic traffic in the USA.[12]

In Europe, the low cost evolution started in the UK in the mid-1980's with Ryanair, but development did not gain real momentum before the mid-1990's. The successful European forerunners were Ryanair and easyJet. In response, some of the established flag carriers established their own low cost operations (KLM – buzz,[13] Lufthansa/Eurowings – Germanwings, SAS – Snowflake, British Airways – GO[14]).

In 2001, the low cost carrier share of total number of flights carried out within the Eurocontrol Central Route Charging Area was 2.7%; in 2002, 4.2%; in spite of an overall negative growth of –2.7%, the low cost carriers' share rose by 53.9% versus 2001.[15]

The low cost model boom also reached Asia and Australia with carriers like AirAsia, TigerAir, AirDeccan and VirginBlue. Air India, as well as Kingfisher, the brewery group, is also planning to introduce their own low cost airlines in 2005.

---

[10] Lawton (2002) gives a good overview of the development of the low cost airlines in the US as well as US major's attempt to create their own low cost operations (pp. 139 ff.)

[11] Airline Business (2004): ZIP has now been integrated with Air Canada after the unions granted cost concessions

[12] See DoT (2003).

[13] buzz was sold to Ryanair in 2002.

[14] GO was spun off to a Venture Capital Company/MBO and later acquired by Easyjet.

[15] See Eurocontrol/STATFOR/Doc 43, 27 May 2003. The biggest growth was accounted for in France (82%), followed by Germany (73.1%) and the UK (34.9%).

In the Middle East, a region dominated by national carriers mostly protected through restrictive bilateral agreements, several attempts were made to incorporate low cost carriers. The first – and so far the only – low cost carrier in the region, local government-owned, Sharjah-based Air Arabia is operating low cost services between Sharjah and 12 destinations in the region.

In comparison to their full-service competitors, the low cost carriers[16] are providing value to their customers in terms of lower fares through gains in higher asset utilization (see figure 4-1) through a highly efficient point-to-point operation and in some cases the use of secondary, less congested and cheaper airports[17] that offer short turn-around times. Aircraft utilization is also improved through the offering of a standard product to all (e.g. Customer Services, catering, baggage handling and in some cases free-seating).

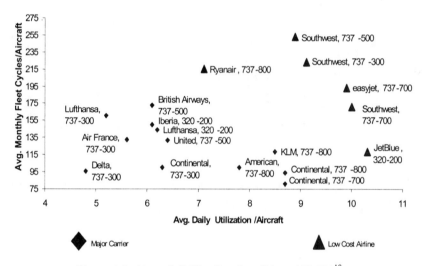

Figure 4-1: Aircraft Utilization, Seat Range 132-180[18]

Furthermore, low distribution costs through the use of Internet bookings (see table 4-1), using a single aircraft type and, perhaps more importantly, avoiding

---

[16] See chapter 5 by Klaas & Klein, and also Kaufhold & Albers' chapter 26 in this volume.

[17] The gain not only lies in the lower airport and handling fees, but also in savings in ATC fees, fuel and block-time due to simple approach and departure procedures at such airports.

[18] Source: ACAS, July 2003, Daily Utilization measured by Number of Flying Hours per Day; Fleet Cycle measured by Number of Flights per Month.

costs of complexity e.g. by not offering connecting traffic conveniences in the form of through check-in and interlining. Flexibility in the form of rebooking is non-existent or limited and, if so, a fee is normally charged for itinerary changes.

Low cost airlines do not only generate revenue by stimulating new demand through lowering prices, they also cannibalize the existing customer base of incumbent carriers.[19] They typically seek out routes that hold a certain passenger volume as well as a potential for growth that may be stimulated through aggressive pricing.

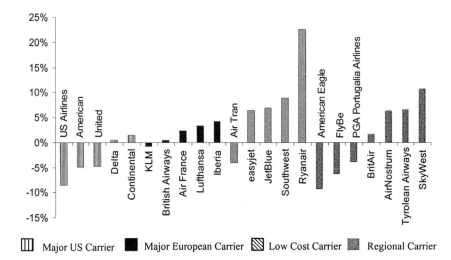

Figure 4-2: Five-Year Net Margin Average 1997-2002[20]

Figure 4-2 shows a profitability comparison between US and European major network carriers, low cost operators and regionals. Only the regional carriers seem to come close to the profitability of the low cost operators, which is not surprising since the successful regionals are very cost-conscious.

The explosive growth in the no-frills/low cost market is taking place at the expense of the incumbents, in particular major network carriers. The pressure on yield is not only threatening the profitability of directly competing short-to-

---

[19] BA's strategy programme launched in 1997 ("Future size and Shape") took account of the growth of the low cost model in the UK markets. The creation of its low-fare arm "GO" made many business-class passengers use GO instead of BA mainline services – consequently GO was spun off (Lawton (2002), pp. 120 ff.).

[20] Source: Lufthansa Consulting, ATI.

medium haul services, but it might also threaten the yield of O&D hub traffic: low cost competition to/from hub catchments might cannibalize the point-to-point customer base that is normally perceived as valuable revenue-contributors to feeder services to the extent where the network carrier is unable to deliver a better value proposition. Consequently, in order to maintain a share of traffic network carriers are forced to lower their fares to a level that is considered attractive for those passengers that might be willing to switch to low cost.

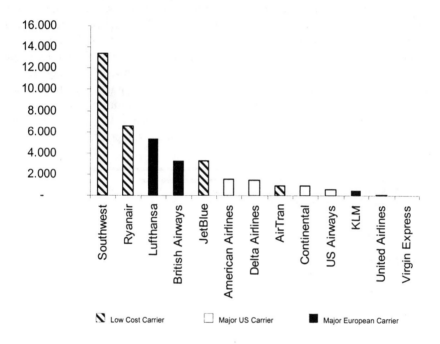

Figure 4-3: Market Capitalization Million US$ (22 August 2003)[21]

The apparent resilience and growth rates of some of the low cost/low fare carriers attracted positive attention from investors, see figure 4-3. However, the beginning of 2004 showed a stagnation of profits through aggressive price wars in Europe and a somewhat more modest growth rate among the 50 low cost operators in Europe; consequently, a consolidation in the European low cost market is likely to take place short-to-mid term.

---

21   Source: Lufthansa Consulting, Yahoo Finance/Reuters.

2.3 Fractional Ownership Programmes

Another form of air transport developed strongly during the last decade – the private jet aircraft. The cost of ownership and operation of a private or corporate jet made this form of air transport an option for the few. The development of the fractional ownership model made corporate jet travel increasingly popular in the US among small and medium-sized companies. The concept of fractional ownership implies the acquisition of a fractional share in a pool of small executive aircraft. Since the mid 1990s, the number of fractional shares has grown significantly, see figure 4-4. Companies like Netjets[22] are offering attractive packages where individual air travel needs can be met by buying a share in a pool of aircraft and individual requirements such as flexibility, range and comfort can be met. The threat from fractional ownership may not be considered to be significant to network carriers at this point, but the concept could nevertheless be an interesting alternative for customers whose time is considered very precious or for those who simply want to travel by themselves without carrying the costs and risks of operating their own private jet.

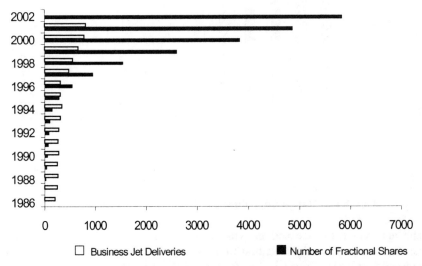

Figure 4-4: Number of Fractional Shares and Business Jet Deliveries 1986-2002[23]

---

[22] Netjets (2004): Netjets Corporate card is now being introduced into Europe. A fractional share costs 115,000€, flight hours are paid separately.

[23] Source: Lufthansa Consulting, Aviation 2003, Rolls Royce Business Jet Market Outlook.

3 Modules of a Successful Airline Differentiation Strategy

Prior to the emergence of the low cost carriers, traditional network carriers were largely competing on price and schedule attractiveness – customers only had the competing carriers with a mostly comparable product as reference. The new relation between product and price brought to the markets by the low cost carriers put the full-service carriers under pressure to lower their costs in order to bridge the price gap on competing services.

It would therefore be difficult to maintain a disproportionate price premium in contestable markets for largely identical services. Naturally, this may lead full-service carriers to seek ways of differentiation where the incremental costs will be borne by the customers.

The mushrooming growth in the number of low cost start-ups worldwide suggests also that hitherto protected markets will be put under pressure for liberalization by market demand.

Consequently, airlines need to take proactive decisions with regard to classic success parameters like keeping costs down, identifying sources of differentiation and constantly monitoring changes in the market environment that may potentially influence customer and competitor behavior in order to establish, maintain or improve their market position.

3.1 The Need to Create Resilience

The increased rate and pace of change in the air transport business demands resilience. Airlines need to create structures that enable them to swiftly adapt to changes in the market environment like a short-term crisis, a new form of competition or changes in customer behavior. The key to success lies in the airline's capability to read and understand market signals and devise the proper response to a sudden market opportunity or threat as well as the speed by which measures are taken.

The requirement for resilience is easier to meet by a "Greenfield" organization than to restructure an existing organization that has gone through an often compromise-ridden transition from government entity to privately owned or public-limited company.

Attributes that were definitive strengths and a key to success before a new form of competition entered the market are often hard to change: an organization driven by technology and production optimization may find it difficult to change the pivotal point in the organization to get more customer-orientation into product, production planning and service delivery.

Organizations that have a culture of cooperation and trust between management and employees will likely show more readiness to change than organizations characterized by a hierarchical and strongly disciplinary style of management. The success of Southwest Airlines is said to lie in its ability to build high performance relationships between management, employees, unions and suppliers.[24]

Creating resilience also means flexible production assets in terms of aircraft and fleet size as well as financing. Here airlines are challenged by long lead-times between market forecast and physical delivery as well as duopoly/oligopolistic suppliers. Thorough analysis of business cycles, production and design capabilities of manufacturers as well as competitors' anticipated fleet rollovers might open windows of opportunity to place orders at the right time to obtain low prices.

A strong focus on certain markets may entail disproportionately high business risk: in the aftermath of 11 September 2001, the Europe-North Atlantic traffic almost completely collapsed. Carriers with a disproportionately high production in this market were at a definite disadvantage to those who had a more balanced distribution of capacity between intercontinental markets.

3.2 The Need for Improving Efficiency

Independent of the business and operating model a successful airline must continuously strive to render its services at the lowest possible cost. The two major cost factors of an airline are:

Aircraft: The network carrier may have a problem in increasing short-haul aircraft utilization in a hub-and-spoke operating model, in particular if a continuous feed (depeaking) traffic is not feasible. However, a shortening of total turn-around times using the same parameters as the low cost operators will give some improvement, but may be difficult to sustain in a congested hub. Efficiency gains will be obtained through operating a uniform type of aircraft, but it could

---

[24] See Gittell (2003), pp. 18 ff.

lead to loss of bargaining power vis-a-vis aircraft suppliers. Some aircraft offer significant advantages of commonality, both technically and operationally.

Human resources: "Legacy" airlines are often characterized by having rather inflexible and expensive agreements in comparison to their low cost competitors. However, as airlines grow, personnel costs are likely to rise through demands from employee representation to get a larger share of value created.

In order to create a balance between meeting employees' salary expectations and maintaining resilience, airlines must seek to establish flexible agreements with their employees where for example the development of remuneration and working conditions is related to certain key performance indicators of the airline.

3.3 Providing Customer Value through Differentiation

From a product point of view, many airlines find it difficult to differentiate themselves from the competition, in particular on short-haul services. It was even claimed that air transport is a commodity-like service and therefore impossible to differentiate.

As competing business models were lacking, competition mainly took place on price, schedule and frequency. Consequently, management of network carriers directed a lot of attention towards measures that would optimize those parameters. On the other side, little was invested in learning about customer's wishes on product attributes and design.[25]

The strong emergence of the low cost model has given a new meaning to the concept of price competition and is likely to bring network carriers to concentrate more on delivering value to their customers beyond mere price competition: in spite of all possible efforts it is highly unlikely that a network carrier may gain cost leadership over a low cost carrier on parallel services. In order to maintain a market position or, in some cases, to survive, the only option is to differentiate their services from those of their low cost competitors and their peer competitors.

There are several ways of differentiating a seemingly uniform service like air transport and thereby delivering added value to the customers. In order to define

---

[25] See Day (1998), p. 2: "Having first-class technologies and business systems are only preconditions for success; unless a company has superior skills in understanding, satisfying and retaining customers they will not realize their full potential."

the right elements of differentiation, airlines must seek a deeper understanding of their customers' needs and expectations.

So far, customer segmentation has been limited to offering products where schedule, flexibility, seat, space and in-flight frills are working as segmentation parameters, all within the limits set by using a "standard" production platform (aircraft and airport facilities) to serve all segments. The production platform as well as monopoly/oligopoly or quasi-monopoly transport chain suppliers' (like airports) business models put certain limits on the scope for differentiation. The type of operation, for example the traditional hub-and-spoke, gives all passengers a choice of reaching multiple destinations with one or more stops within a band of fares, depending on flexibility, choice of carrier and product attributes like seat and in-flight frills, but leaves little room for serving more individual needs of flexibility, product attributes and delivery. In terms of elapsed travel time, a first or business class passenger is not better off than the economy passenger paying the cheapest fare, but he or she is often getting preferred treatment in the form of dedicated check-in, lounge access and fast track security and immigration control.

Experiments in differentiation were limited.

The Concorde enabled Air France and British Airways to offer a unique service through aircraft technology attributes: the extreme performance of the aircraft in terms of traveling speed, the narrow confines of the cabin as well as dedicated lounges with direct boarding gave the product unique attributes in terms of an unbeatable block time, adventure and exclusivity. Factors like fuel burn and rising maintenance costs put an end to Concorde operations in Summer 2003.

In order to meet the demand from a single customer segment in a more targeted manner, Gulf Air launched its all-economy, full-service division, Gulf Traveler, in 2003, operating routes to/from and within the Gulf region with its fleet of B767 widebody aircraft. Destinations outside the Gulf region include Nepal, Sri Lanka, Bangladesh, India and Pakistan. In-flight meals and entertainment are selected to match culturally to their destination countries.

Beginning in 2004, Air France launched its "Dedicate" operation where routes between Paris and "oil" destinations like Pointe Noire, Malabo, Kuwait, Doha, Kish Island, Atyrau and Tashkent will be served with A319LR aircraft in a mixed business and economy configuration. This operation was set up on request of the Petroleum industry in order to serve their specific needs.[26]

---

[26] See Pilling (2004).

Traditionally, three levers of differentiation for the air transport product can be distinguished: (a) product delivery, (b) schedule, and (c) seating and in-flight frills.

3.3.1 Product Delivery

The goal for most full-service airlines in their product delivery is to create a seamless travel experience. The reality is that airlines do not own all process steps in the travel chain and may not be aware of glitches that most likely will happen when process interphases fail. However, the increased pressure on fares as well as the commodity-like nature of the product air transport, particularly on short-haul, will enhance the importance of the delivery of a seamless travel experience as a means of differentiation.

At the airport, Customer Services may no longer be regarded as crucial with the increased use of Internet sales and automatic check-in kiosks. However, when things go wrong it is imperative that the systems are backed up by courteous and competent staff assistance that enables the carrier to recover and fulfill the brand promise. This also goes for the processes at the airport that are not within the direct control of the airlines, like security and immigration which makes it important to keep a good working relationship with the bodies responsible for such operations.

The importance of delivering the seamless travel experience goes beyond the recovery of process failures – the way the interaction with the passenger is made when things runs smoothly may be just as powerful. This goes beyond "creating" a service culture within an airline, it is more about the type and values of people airlines employ.

When JetBlue uses extra staff, including their managers, during peak travel periods like Christmas or Thanksgiving in order to help passengers with their baggage, they do not only display customer-orientation and dedication to service quality, they also help speeding up check-in procedures that enables JetBlue to operate on-time. Or, when Cathay Pacific endeavors to seat families with children in the same section of their respective traveling class, they relieve both parents and fellow travelers that may prefer to be undisturbed during a long-haul flight.

Flying low cost or low fare does not necessarily mean tradeoffs in the quality of customer service. In the annual US Airline Quality Rating 2003 JetBlue, Alaska Airlines and Southwest were ranked top 3. US Airways came in as number 5 as the first full-service carrier, down from first place in 2002.[27] A direct comparison for the European market is not available; however, according to the AEA monthly reports on missing baggage and punctuality, the smaller full-service carriers in Europe (Turkish, Air Malta, Spanair; Meridiana, Adria, Icelandair, Cyprus Airways) were performing far better than their network operator competitors.[28]

A recent survey carried out by Gallup in the US highlights the importance of people in creating a customer service experience; flight attendants and ground staff behavior was more important than schedule convenience and on-time performance when it comes to creating "...a real connection with customers."[29]

The results confirm the general quality of the product delivered by the new generation of air carriers in the US, as well as the consistency of quality that can be achieved without having to deal with the inherent complexities of hub-and-spoke operations with passenger and baggage connections. It probably also reflects the impact made by a vibrant corporate spirit of a new and successful start-up to a traveling public that are experiencing downgrades in quality from the traditional providers as they try to cut costs.

On long-haul services carriers have more interaction time with their passengers which naturally improves the chances of meeting and exceeding passenger expectations through serving meals, in-flight-entertainment and other services.

Customer interaction at all touch points must reflect the service level devised in the airline strategy and fulfill the brand promise as perceived by the customers.

3.3.2 Schedule

One of the classic differentiation factors in air travel is elapsed travel time. Time-sensitive travelers are willing to pay a certain premium for getting the shortest elapsed time, at best through a direct flight.[30] Therefore there will always be a market for services between Frankfurt a.M. (FRA) and London Heathrow (LHR) vice versa in spite of Ryanair's low cost services between Hahn airport (situated more than 100 km away from the Frankfurt City Center) and London Stansted.

---

[27] See Bowen & Headley (2004): The AQR uses criteria like on-time performance, denied boarding, mishandled baggage and customer complaints.
[28] See Association of European Airlines (AEA) (2003).
[29] See McEwen (2004).
[30] See Gillen & Morrison (2003), p. 23.

The number of frequencies offered by a certain carrier may also justify a price premium, as well as the actual departure and arrival times. Here network carriers often encounter conflicts between the convenience of point-to-point traffic and that of connecting passengers, for example on services between the Middle East and Europe: most carriers depart around or, in some cases, past midnight, mainly in order to ensure good connections to European and North American destinations. A point-to-point passenger would most likely prefer a day flight instead of a 4-5 hr night flight.

In general, price-sensitive travelers will show more willingness to accept tradeoffs in schedule convenience as well as elapsed time both on short and long haul services.

3.3.3 Seat and In-flight Frills

One of the most important differentiation factors is the quality of the seat as well as the legroom offered on a given flight. Development of the ergonomics and functionality of airline seats have mainly been triggered by carriers that wanted to offer greater comfort for their premium passengers in their Business and First Class cabins while working, sleeping or just relaxing. Seat – or rather bed – quality has become a critical product feature and airlines are investing heavily in improvements: British Airways introduced the first fully-reclinable seat in its Club World cabin in 2000; in 2003 both Air France (L'espace Affaire) and Lufthansa (Business Class) upgraded their long-haul Business Class seats to fully-reclinable standard. Emirates and Virgin Atlantic are currently working on an upgrade of their premium cabins.[31] In long-haul economy class there is movement in improving the seat comfort, in particular with airlines that wish to use the full range capability of aircraft like the A340-500/600 or the B777-200/300ER, which on certain routes entail an 18 hr non-stop flight.

On short haul, seat development is dictated move by weight and volume considerations[32] than seating comfort; the perceived or felt legroom as well as overall seating quality in short-haul cabins may to some extent be a differentiation factor if capacity is not too adversely affected.

The In-flight Entertainment offer may become a critical differentiation factor. Today most carriers offer personal video screens in premium classes on long

---

[31] See Shifrin (2004).
[32] Ryanair was considering removing the reclinable function as well as seat pockets and trays in order to incur savings on aircraft purchase and maintenance.

haul with a wide variety of films, games and news broadcasts, even on demand. Since spring 2004, Lufthansa has offered Internet access on board ("FlyNet") where travelers can access the Web through fast broadband connections with their own laptop computer.

An increasing number of carriers also offer In-flight-Entertainment on their short-haul services. JetBlue even offers Live-TV on their entire Airbus 320 fleet.

3.4 The Challenge for the Network Carriers

As competitive pressure increases from new market players, the network carrier might probably realize that it is becoming increasingly difficult to be "everything to everyone": low cost competition on short-to-medium haul services may not only threaten decentralized traffic, but also hub feeder traffic. In order to remain competitive, the network carriers need to reduce unit cost across the board to a point that enables them to offer a price that properly reflects the value delivered to the customer (in terms of schedule convenience, frequencies, service level and other product attributes) vis-a-vis its competition.

Furthermore, network carriers must find ways to increase asset utilization and productivity. Mimicking all the low cost model value drivers may prove to be a daunting task in an optimized hub structure operation that puts restrictions on aircraft utilization as well as forcing concessions from unions in terms of flexible or even increases in working hours.

The variety of services offered within a uniform business system makes it difficult to be competitive through all customer segments. Here it may be necessary for some network carriers to limit the number of geographic markets and/or the client segments they serve in such markets, in order to remain competitive and profitable. For example, a European network carrier may wish to serve certain African destinations with a business jet product, thereby only targeting high-yield passengers' need for air transport; on the other hand he may choose to serve some European markets with an economy-class product only.

Some customers may want to decide on the departure and on the level of service for a particular journey. The existing premium products offered by quality network carriers impose constraints upon the traveler, such as schedule, airport choice, aircraft size and on-board service.

A solution could be the development of different "production platforms" that are optimally structured to deliver a competitive product in each segment. Certain functions like administration and purchasing may be concentrated in a central administration unit (shared services) or outsourced, leaving the production platforms with the main task of producing and developing their respective products. Some production platforms may not be feasible within the airline's structural framework – here the services could be provided by an outside partner which again could limit risk.

Premium segment's needs and expectations may be best served by using small non-congested airports located to client's ports of origin and destination as near as possible, served by a business jet type of fleet, in order to ensure a seamless product delivery in the form of short elapsed travel time, hassle-free travel and in-flight frills according to customer specification.

The success of such a venture naturally depends on the structure of demand as well as the perceived value delivered by such services – or the willingness of customers to carry the incremental costs involved with meeting more individual needs. Here a strong brand, flexibility with regard to product choice and a bundling of services (for example outbound on executive jet, homebound on trunk or normal business class service) as well as financial exposure may render such a product an advantage over existing alternatives like complete or fractional ownership of a jet.

After devising the most cost-effective way of production, network carriers need to obtain a greater understanding of the value of their own product from an enabled consumers' perspective in order to strike the right balance between product attributes and price across segments and geographic markets.

3.5 The Challenge for the Low Cost Carriers

The main challenge for the low cost carriers is in fact to maintain their position as low cost carriers. However, as they grow and their full-service competitors reduce their costs, it will become increasingly difficult to maintain this position.

The cost leaders, like Ryanair, will strive to cut costs further in order to be able to stimulate growth with aggressive pricing. The remaining carriers are likely to try to differentiate their services as the full-service carriers cut their costs and the difference between the fares diminishes.

As some low cost operators grow bigger, it may prove difficult to keep a low cost base. Salaries are likely to increase and suppliers' price concessions may shrink. The challenge is to continuously seek new ways of reducing costs and obtain additional revenue, perhaps through developing their brands towards other industries that are linked with the low cost travel experience like hotels, nightclubs, or holiday resorts or where they may otherwise profit from owning a growing customer base.

Differentiating low cost carriers may also enter into a cooperation – or be acquired by a network carrier – with a view to provide feed-traffic to its hub, to the extent that the network carrier cannot provide it by itself at competitive rates.

4 Outlook

The coming years are likely to bring significant changes in the structure of the airline business through a higher degree of product differentiation and passenger segmentation as the industry takes another step towards maturity. In the short term, traditional network carriers are forced to reduce costs, create flexible resources, and define customer-perceived added-value to their services in order to be able to compete successfully in the future. Some of them may also be forced to pursue a clearer segmentation of the market, for example in the form of offering all-economy services in some markets and "all-business" in other.

Generally, the success of the low cost model will continue. A consolidation phase is likely to take place in Europe, as aggressive pricing will force the financially weakest players out of the market. As some "low cost" markets get mature, air transport might be offered for a symbolic amount of money if not for free, as low cost transport might fuel growth in other revenue-generating activities like "night-clubbing" or shopping inclusive tours.

The convergence between business models is likely to gain pace as low cost carriers try to profile themselves against their peers and full-service carriers manage to drive down unit costs; alliances or cooperation between low cost airlines and network carriers may evolve as product differences get marginal and network carriers are not successful in driving unit costs down on their short-haul services.

The product and price relationship brought to the market by the low cost carriers is likely to influence intercontinental services as well, where the biggest network carriers would seek to place themselves with an improved value proposition in terms of fare reductions or offering better and more individual products on their long-haul services in order to attract and retain premium customer segments, or reduce their offering to economy-class only in markets that cannot sustain a premium price.

5 References

ASSOCIATION OF EUROPEAN AIRLINES (AEA) (2003): Consumer Reports.

BAUMOL, W.; PANZAR, J.; WILLIG, R. (1982): *Contestable Markets and the Theory of Industrial Structure*. San Diego: Harcourt Brace Jovanovich.

BOWEN, B.; HEADLEY, D. (2004): *Airline Quality Ranking*. www.unomaha.edu.

CABRAL, L. (2000): *Industrial Organization*. Massachussetts: MIT Press.

CLARKE, R. (2004): Value proposition, *Airline Business*, 20(3), pp. 44-46

DAY, G. (1998): What does it mean to be market-driven? *Business Strategy Review*, 9(1), pp. 1-14.

GILLEN, D.; MORRISON, W. (2003): Bundling, Integration and the delivered price of air travel: are low cost carriers full service competitors? *Journal of Air Transport Management*, 9(1), pp. 15-23.

GITTELL, J. (2003): *The Southwest Airlines Way*. McGraw Hill.

LAWTON, T.C. (2002): *Cleared for Take-Off. Structure and Strategy in the Low Fare Airline Business*. Aldershot, Burlington: Ashgate.

MCEWEN, W.J. (2004): Skirmish in the Sky. *Gallup Management Journal Online*, 5/13/2004.

PILLING, M. (2004): Direct flights launched to offer executives non-stop travel to energy hotspots, *Flight International*, pp. 3-9.

ROSE, E. (2004): Reworking the real model, *Airline Business*, 20(3), pp. 58-60.

SHIFRIN, C. (2004): That's entertainment, *Airline Business*, 20(1), pp. 46-48.

5

STRATEGIC AIRLINE POSITIONING IN THE GERMAN LOW COST CARRIER (LCC) MARKET

THORSTEN KLAAS AND JOACHIM KLEIN

1 The Appearance of LCCs in Germany:
   Market Development and Segmentation 120
2 Fundamentals of the LCC Business Model 124
3 Strategic Positioning of a LCC:
   The Germanwings Example 132
4 Outlook into the Future:
   Further Development of the LCC Business 140
5 References 142

Summary:
The market for LCCs is a new and dynamically evolving phenomenon in Europe and specially in Germany. A glimpse at the European LCC market reveals, that not all LCCs are doing the same. In the light of Porter's strategic framework of cost leadership, differentiation and focus we define the LCC market and point out its competitive relations to the traditional market segments. We describe the fundamentals of the LCC business model and analyze the possibilities to deviate from it guided by some general rules of thumb. After these theoretical considerations we will present the strategy of Germanwings as a real life example for the strategic positioning of a LCC. Our analysis shows that the LCC market will further split up into a leisure and a business segment – demanding the pure low cost business model or allowing for specific deviations by setting a strategic focus respectively.

1 The Appearance of LCCs in Germany: Market Development and Segmentation

Airlines of the LCC-Type are a quite new phenomenon in the European, especially in the German, airline business and they just seem to fit perfectly into the present economic Zeitgeist of the German "Geiz ist geil" ("stinginess is cool") Marketing-Movement. Enabled by an ongoing globalization of the industry along with the deregulation of the European aviation markets, the emergence of LCCs brought a dynamic competition into a formerly relatively stable industry which was dominated by the established major airlines, like Lufthansa (LH), British Airways (BA) or Air France (AF). Approximately 15 years after the development of LCCs in the USA, the consequent pursuit of the low cost-strategy by Ryanair (founded in 1985 in UK), easyJet (1995 / UK) and virgin express (1992 / Belgium) marked the breakthrough of the LCC business model in Europe. The ostensible success and profitability of the LCC business led even some of the established major airlines to the foundation of low cost subsidiaries themselves, e.g. British Airways with Go (merged in August 2002 with easyJet) or KLM with Basiq Air / Transavia and Buzz (acquired in April 2003 by Ryanair).[1] The numbers are impressive: Ryanair presents a net operating margin of about 18 per cent with an average ticket price of 39 Euros in the first quarter of the actual fiscal year. For comparison, the average ticket price of easyJet is 62 Euros, of Air Berlin 93 Euros, of Deutsche BA 111 Euro and of Lufthansa 225 Euros.[2] Furthermore, the commonly accepted attractiveness of the low cost airline market segment is also assisted by multiple studies of leading management consultants who project a LCC market share of about 25 per cent in Europe until 2010 with exceptional growth rates of 20 per cent above the expected average growth of 4-5 per cent of the total European airline market.[3]

The German market has been opened up even later, first by Ryanair starting its German operations in February 2000, followed by Germanwings (a subsidiary of Eurowings / Lufthansa) in October 2002, Hapag-Lloyd-Express (a subsidiary of TUI) in December 2002 and easyJet in November 2003. And actually the young LCC-Market evolves in a highly dynamic competitive behavior with an ever

---

[1] See Pompl (2002), p. 117.
[2] See Oehler (2004), p. 12.
[3] See e.g. Mercer (2002); Hansson, Ringbeck & Francke (2002).

intensifying competition inside as well as between the traditional aviation market segments and the evolving low cost segment. EasyJet for example is going to put 25 new Airbus A 319 into operation next year – predominantly on primary German airports (e.g. 6 planes will be positioned at Berlin Schönefeld Airport from Spring 2005) – with yet incalculable consequences for the players in the German LCC market.[4] However, one consequence is unquestionable and yet clearly observable: according to the American Express Travel Index the average ticket price decreased about 20 percent in the last year and about 5 percent in the first quarter of 2004.[5]

At the present state the airline competition can be clustered into three main market segments: 1st the Network Carrier/ Alliances, 2nd the Charter Carrier and 3rd the rather new segment of the Low Cost Carrier (LCC), as shown in the following figure 5-1.

1. The segment of Network Carrier/Alliances is basically characterized by scheduled flight operations based on a dense and extensive network comprising as many of (inter-)continental destinations as economically possible. To ensure the maximum reach and density, the overall network is typically composed of the sub-networks of global leader airlines (normally traditional flag carriers) and regional co-operation partner airlines. As precondition and consequence this market segment reveals the strategic alliance as its predominant evolving organizational solution, like the Star-Alliance or the oneworld co-operation. Since for economical reasons not every origin can be linked directly to every destination of the network (all embracing point-to-point traffic), the co-ordination and synchronization of the indirect relations are ensured by hub-and-spoke-network structures which are supported by regional and 2nd tier feeder airlines. The business model is basically tailored to serve the business customer typically supplemented by extensive service offerings (e.g. quality meals), high ticket flexibility, frequent flyer programs, business lounges, and price-differentiated seat categories. But also the share of leisure travellers is continually rising, generated by special price offers for early bookings combined with a reduced service on economy seats and limited or even no ticket flexibility. Especially those price sensitive offers set the Network Carrier/ Alliances segment into direct competition with the other two market segments.

---

[4] See Feierabend (2004), p. 13.
[5] See Hagen & Hohenester (2004), p. 85.

2. The Charter Carriers segment on the other hand predominantly serves the market for price sensitive leisure travellers. Charter airlines typically maintain no complex network of multiple destinations, but offer only selective point-to-point relations mainly to centres of tourist interest. Flight schedules are often designed in close co-operation with the main professional tour operators and the distribution of the services takes place indirectly through travel agencies, selling packaged tours and sole vacation flights. However, the market is losing weight due to the low cost offers of the network carriers as well as due to the upcoming LCC market segment. Therefore some charter carriers are building some kind of hub-and-spoke network and are also selling their flights directly to the final customer by the internet (and travel agents), trying to skim the low cost market in addition to their original charter business (e.g. Air Berlin with the home hub in Nuremberg (NUE)).[6]

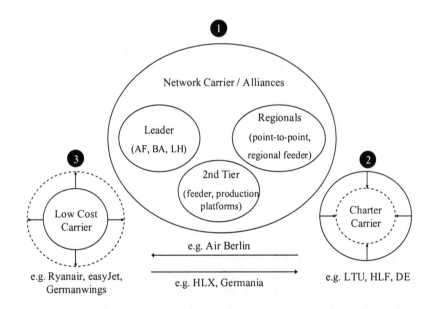

Figure 5-1: Segments of the Airline Market[7]

---

[6] See also Bieger & Agosti's chapter 2 in this volume.
[7] According to Mercer (2002).

3. As mentioned before, the LCC market is the youngest and fastest growing segment in the European airline business. In brief, a LCC-airline consequently focuses on the core air-transport service by omitting any costly service features or extras (simple product / no frills service), by optimizing the whole process chain from distribution to operation (low operating costs) and thereby consequently following the strategy of cost leadership. The widely accepted archetype LCC business model was pioneered by the US-American Southwest Airlines[8] and has been closely adopted for the European market by Ryanair. This model consists of key success drivers like short-haul routes, no frills service, minimal debt servicing costs, standardized fleet, use of cheaper secondary airports, cautious route expansion, and ticketless reservation system.[9] But the prevalent competitors in the European market show that actually not all LCCs are acting the same way, as they are following quite different strategic foci concerning their version of cost leadership. Just as Ryanair is following a pure "Southwest-Strategy" by using secondary airports, Germanwings for example tries to differentiate itself in the low cost market by targeting predominantely on the price sensitive business customer. So there seems to be no "one best way" of doing the LCC-Business and as a result the competing low cost airlines strive for their particular "differentiated" niche in the low cost segment. This is an interesting development since experts state that in the long run only two or three low cost airlines will prevail in the European market – and what is even more exciting is that it is yet unclear which version of the LCC business will prevail.[10]

In the following we will analyze the strategic airline positioning in the LCC-market. In this regard we will first shortly explain the ideal strategic options of cost advantage and differentiation. Second we will concentrate on the key success drivers of the LCC business model. On this basis we will furthermore discuss if and how an LCC can differentiate itself in a low cost environment and deviate from the ideal LCC business model. This is a relevant question, because from a theoretical point of view there is only the "either … or" option, i.e. one cannot differentiate and strive for low cost leadership at one and the same time. After these rather theoretical considerations we will present the strategy of Germanwings as an example for the strategic positioning of a LCC. Our analysis will close with fundamental considerations about the expected developments of

---

[8] See Freiberg & Freiberg (1998).
[9] See Lawton (2002), p. 39. The LCC business model will be described in more detail in the next section.
[10] See e.g. Mercer (2002).

the LCC business, which will result in some implications for the strategic positioning of airlines in the LCC-market.

2 Fundamentals of the LCC Business Model

The LCC business model can be theoretically attributed to the prominent concept of competitive strategy developed by Michael E. Porter.[11] He distinguishes three generic strategic approaches to outperform competing firms in an industry: 1st cost leadership, 2nd differentiation, and 3rd focus. Following one of these alternative strategies, a company creates a defendable position in the market against the five competitive forces originated in suppliers, potential market entrants, customers, substitute products, and competitors. Without going into details, we will give a short outline of each strategy and its main strategic implications.

1. By following a cost leadership strategy, a company on the one hand competes by comparable customary products which are lower in price relative to the competitor's offers. So as a rule cost leaders try to supply standard, no frills, high volume products at the most competitive possible price. As a consequence optimizing the whole value or supply chain with regard to low cost relative to competitors becomes the primary company goal, though quality and service must not be ignored due to the common standards of the market. Obviously this strategic approach perfectly reflects the LCC business model as low cost airlines offer a "no frills product" at a very "competitive price" in the truest sense of the word.

2. The differentiation strategy on the other hand aims at the excellence of a company to offer valuable products and/or services which are perceived marketwide as being unique, especially from the customers' point of view. For this uniqueness the firm is rewarded by higher prices and above average margins. The sources of competitive advantage by differentiation are manifold, like remarkable product design, powerful brand image, leading technology, exceptional features, or premium customer service. Accordingly much capital is invested in "being different" resulting in a broader variety of products and services in relation to the cost leadership strategy. Though costs have to be controlled carefully, they are not the primary strategic goal due to the higher contribution margin realized. While on the one hand the

---

[11] See Porter (1980).

cost leadership strategy reflected the market behavior of the LCCs, the differentiation strategy on the other hand perfectly describes the strategic orientation of the leader airlines in the Network Carrier/Alliances market segment, since they offer the broadest network of destinations combined with a great variety of integrated service options.

3. Following a focus strategy means concentration on a part of the total market, e.g. a particular buyer group, product line, or geographic market segment. The company deliberately narrows its strategic focus to this specific market niche in which it expects to achieve above average returns either by following a cost leadership or a differentiation strategy. Especially the 2nd tier and regional airlines show this kind of strategic approach by focusing e.g. on coordinated feeder services as a specific regional niche of the airline market. But setting a focus is also a viable strategic option in the LCC market, since it can be followed in line with the cost leadership strategy. We will come back to this aspect later on.

Porter explicitly points out that cost leadership and differentiation are alternative approaches and must therefore not be combined. Companies showing no clear strategic direction either of cost leadership or differentiation are "stuck in the middle", are "almost guaranteed low profitability" and consequently bear a great risk of market exit in the long run. So from this theoretical point of view two interesting questions arise regarding the strategic airline positioning in the LCC market: (1) If the LCC market is ideally reflected by the strategy of low cost leadership, what are the key success drivers that will guarantee sustainable profitability in this market segment? And (2) is it a viable option to deviate in one way or the other from the ideal type LCC business model in order to differentiate from low cost competition or to set a strategic focus? In the next two sections we will try to give some answers to these fundamental questions.

2.1 Key Success Drivers of the LCC Market

The LCC concentrates on the core air-transport service by omitting any costly service features and by optimizing the whole process chain from distribution to (ground and onboard) operations due to low costs. In other words "keep it simple" is the LCC's major guideline as it tries to achieve a *low cost market position* by offering a relatively *simple product* which is provided at the *lowest possible operating costs*. The following figure 5-2 summarizes those key success drivers in terms of a strategic triangle framework. The triangle visualizes that the three success drivers are basically of equal importance and that specific actions have to be calibrated very closely because of their reciprocity. We will elaborate on each driver's actions below (see also the summary presented in table 5-1).

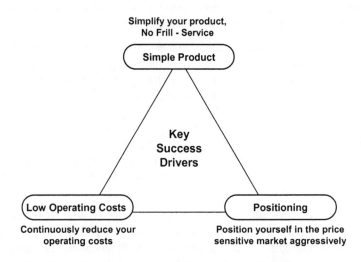

Figure 5-2: Strategic Triangle of the LCC Business

2.1.1 Simple Product

Simplifying the product and thereby offering a "no frill"-service compared to the traditional network carrier is the most obvious characteristic of the LCC business. Actions for simplifying the product are: reduced on board service by offering refreshments and snacks for charge only, free choice of seats instead of seat reservations, abandonment of frequent flyer bonus programs as well as airport lounges, clear price structures, easy to understand booking procedures, and a selective point-to-point route network to attractive destinations. All these actions supplement the other key success drivers, as a simple product on the one hand gives transparency to the price sensitive customer (positioning) and on the other hand the strive for simplicity helps to reduce costs (low operating costs).

2.1.2 Low Operating Costs

As we have learned from the generic low cost leadership strategy, offering the lowest prices does only work with achieving low operating costs at the same time. So besides simplifying their products, LCCs have to undertake any further efforts to get *and* maintain their operating costs as low as possible (of course without violating the prevailing security standards). Actions to be undertaken in this regard are: the usage of secondary, low charge airports, the standardization of the aircraft types employed, ensuring a high aircraft utilization combined with quick gate turnaround times, increasing the seating density, operating mainly short-haul point-to-point routes with high frequency, the implementation of online and direct booking systems, and the acceleration of boarding times by ticketless check in procedures.

2.1.3   Offensive Positioning

Achieving a low cost leadership position in the price sensitive airline market field, the LCC has to be offensive due to the intensive price competition and the high price elasticity of the customers. Aggressive, proactive marketing campaigns have to be tailored to attract as many flexible leisure travellers and price sensitive business travellers as possible. Most suitably those campaigns accentuate a simple, recognizable brand in combination with an omnipresent low fare image. Leading LCCs furthermore explore and set innovative low cost trends to exploit first mover advantages. Next to the struggle with other airlines, with their low prices LCCs are also in a good position to alienate additional customers from the other transportation modes such as car and rail, which also have to be taken into account at the development of the marketing toolbox.

Simple Product	- No free catering, but refreshments and snacks for charge
- No seat reservations, but free choice of seats
- Abandonment of frequent flyer / bonus programs
- No lounges at airports
- Selective point-to-point route network to well known and attractive destinations
- Simple ticket price structure and easy to understand booking procedures
Low Operation Costs	- Use of secondary, low charge airports
- Implementation of online and direct booking systems
- Ticketless check-in (simple boarding processes)
- High aircraft utilization – quick gate turnaround time
- Higher seating density (increasing the number of seats per aircraft)
- No complex interlining, no hub-and-spoke services, but short haul point-to-point services with high frequency (destinations with high passenger density)
- Standardized aircraft type / common fleet
Positioning	- Achieve and defend position of low cost leadership by attracting as many price sensitive customers as possible, i.e. flexible leisure travelers and price sensitive business travelers
- Alienation of customers from all modes of transportation (aircraft, rail, car)
- Aggressive marketing campaigns regarding simple brand and low fare image
- Exploring and setting innovative low cost trends faster than the competition, first mover advantage

Table 5-1: Action List of Key Success Drivers in the LCC Business

2.2 Deviations from the "Ideal Type" LCC Business Model

The beforehand described bundle of complementary actions concerning a simple product, low operation costs and low cost positioning is characterizing the "ideal configuration" of the typical LCC business model. The European low cost market leader Ryanair is widely considered to stick to this ideal very closely, being the low cost pioneer who transferred the American Southwest business model into the European context. And by doing this, Ryanair has become the most cost efficient scheduled airline carrier operating in Europe.[12] But instead of simply copying Ryanair's strategy the observable behavior of the subsequent contemporary market player reveals that not all LCCs do entirely adhere to all those actions proposed by the ideal type LCC business model, but deviate from it in one way or another.[13] As a consequence, regarding price-level and service-quality we can distinguish between specific strategic positions of the player in the German LCC market as shown in figure 5-3. This figure reflects our (inevitably subjective) perception of the contemporary market situation. Lufthansa as an exemplary leading network carrier is placed clearly as service quality leader with a relatively high price level. The low price segment shows differentiated market positions of the contemporary LCCs.

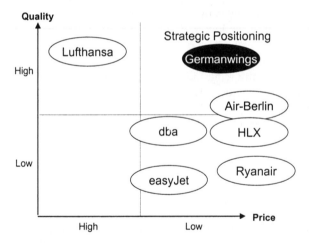

Figure 5-3: Strategic Positioning in the LCC Market

---

[12] See Lawton (2002), pp. 92-93.
[13] See e.g. Pilling (2003).

The motives for this diversity of strategic positioning in the LCC market are manifold as they depend profoundly on the idiosyncratic history of the particular company. Air-Berlin – the second largest German airline[14] – for example originates from the charter carrier market flying primarily only for professional tour operators. Having established a dense hub-and-spoke network and high flight densities, Air-Berlin is now also able to sell low fare tickets directly to end customers (mainly leisure travellers) by the internet and travel agencies. The combination of low cost and charter business reveals some interesting deviations from the ideal LCC business model, such as slots on main (e.g. CGN, MUN) as well as secondary airports (e.g. NUE, HAJ), interlining with other airlines, limited offer of free snacks and refreshments, and advance seat selection. Air-Berlin fosters an image as a potent and family friendly holiday airline with low fares. The Deutsche BA (dba) started its operations in the German charter market and radically changed its business model to low cost in 2002. Now with 130 flights per day dba concentrates foremost on German destinations with primary airports, pursuing a low cost image policy around punctuality, friendliness and reliability, tailored to address especially the price sensitive business traveler.[15]

Analyzing the variety of contemporary LCC business models in detail would go too far here, but these few examples already demonstrate that besides the ideal LCC business model applied by Ryanair/Southwest, some LCCs strive to occupy their specific niche in the low cost market by setting a strategic focus, e.g. as a low cost holiday airline or low cost business airline respectively. This is supported by selectively softening some of the ideal key success drivers in order to "re-enhance" the drastically reduced (not to say "spartanic") service quality. In this regard it is useful to divide between core and additional components of service quality (see table 5-2). On the one hand the components of *core service quality* are obligatory to every (not only low cost) airline in the business, because airline success essentially rests upon providing a safe, reliable and on schedule product with an informed and friendly customer support. On the other hand especially for a LCC the *components of additional service quality* are optional, since they inherently increase the complexity of the product related processes and therewith induce higher operation costs.

As a consequence, if an airline is looking for its niche in the low cost competition by deviating from the ideal LCC business model, it has to find the right balance between *cost*, *service* and *value to customer* for the focused market segment. Otherwise the airline will be "stuck in the middle" between cost

---

[14] Self Assessment of Air-Berlin. See the "about us" link at http://www.airberlin.com.
[15] See "dba-inside"-link at http://www.dba.de.

leadership and differentiation, which, as we have argued before, will yield low profitability and a great risk of market exit. As general rules of thumb it can be said that at first those services should be added which ideally do not induce additional operating costs. Second, the costs induced by additional services must be offset by customer value, i.e. the customer's willingness to pay (ticket price or service fee) and/or the surplus revenue generated by a higher number of passengers attracted. However, assessing this "cost-service-value" trade off is in no case an easy task to perform.

Core Service Quality Components (Obligatory)	Additional Service Quality Components (Optional)
Safety	Complimentary food and refreshments
Reliability of baggage delivery	Seat reservation service
Punctuality	Comfortable seating pitch
Informed customer support	Leather seats
Problem responsiveness	On board entertainment
Friendliness	Free newspapers
	Visible GPS navigation
	Frequent flyer program
	Primary airports
	Selective interlining
	Selected long distance hauls
	Shuttle services
	Car rental / hotel offerings

Table 5-2: Airline Service-Quality Components[16]

In the next section we will outline the strategy of Germanwings as a real life example for the strategic positioning of a LCC in the German low cost market. With regard to our preceding considerations, Germanwings reflects insofar a very interesting case, as it does not apply the LCC business model in its ideal

---

[16] According to Lawton (2002), p. 82.

specificity but deviates from it in certain ways to position itself as quality leader in the low cost segment (again see figure 5-3).

3 Strategic Positioning of a LCC: The Germanwings Example

3.1 The Company Germanwings

The Germanwings GmbH started its operations in October 2002 at the Cologne/Bonn airport (CGN) as a wholly owned subsidiary of Eurowings Luftverkehr AG and is one of the leading airlines in the German LCC market. In comparison to the observable market developments in the US and parts of the EU, the at this time rather underdeveloped low cost airline market in Germany promised a high potential for growth. Up to now this proved to be true, since in the meantime Stuttgart-Echterdingen has also become a home airport for Germanwings in August 2003. In the fiscal year 2003 Germanwings had 2.4 million bookings (buoyancy) and a total revenue of about 150 million Euros. The Company actually has 420 employees and is currently flying 14 Airbus 319/320 airplanes from both of these locations to a total of 34 destinations throughout Europe. The lowest initial fare offered is 19 Euros which comprises about 10-20 percent of the over-all tickets sold. Germanwings strives to ensure and extend its position as the leading German LCC by winning new customers at home as well as abroad and by this to explore even more European market niches.

Figure 5-4: Logotype of Germanwings

The starting point for the strategic development of Germanwings was the vision that it must be possible to offer flights designed to bring passengers directly and punctually to the heart of the European metropolises and still be economical. With its slogan "Fly high, pay low" Gemanwings highlights its core strategy, i.e. what the company brand is intended to stand for: namely to redefine and set high quality standards in the low price market segment and by this offering the best quality at the lowest price.

The necessary building blocks in order to realize this strategic goal are summarized in the Germanwings house of corporate strategy shown in figure 5-5. The roof is assembled from the building blocks of corporate identity, design, and culture. The roof symbolizes the comprehensive character as a philosophical framework covering the four ancillary pillars, represented by the classical marketing mix consisting of the dimensions Price, Product, Distribution, and Communication. In the following we will elaborate on each of these elements to illustrate the specific actions Germanwings put into practice to successfully position itself in the LCC market.

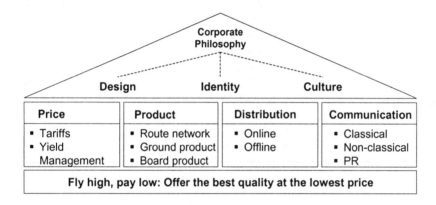

Figure 5-5: Germanwings House of Corporate Strategy

3.2 Strategic Positioning of Germanwings

3.2.1 Carving out the Direction of Strategic Impact

Being the low cost subsidiary of Eurowings, Germanwings did not start as an entirely new company, but rather as a new brand replacing the former Eurowings charter low-price product line. The decision on launching Germanwings as a new low cost brand was founded on the observation that at the beginning of 2002, the German low cost market was still clearly underdeveloped, even though Ryanair, Go, Buzz, and Germania had already started operations in the European as well selectively in the German market. Having in mind the extraordinary success and profitability of the established low cost business in other regions of Europe and

North America, the German low cost airline market promised to be a segment for sustainable growth and attractive returns.

After screening the contemporary market conditions, Germanwings identified three core factors for success in order to carve out its specific direction of strategic impact: "Offer the best quality at the lowest price" (see figure 5-5). First, the essential precondition for sustainable success turned out to be the size of the home market in conjunction with high demand intensity (i.e. potential number of attracted passengers) on the flight-relations offered. Especially the region of North Rhine-Westphalia – Germany's biggest and most densely populated federal state (population approx. 18 millions) consisting of the Ruhr basin as well as the big cities Cologne and Düsseldorf – offered a constitutive start-up potential by connecting this area to (at first 22) other major European metropolises of tourist and/or business interest.[17] Second, in contrast to other LCCs, Germanwings decided to provide a clear quality advantage concerning time and convenience in the low cost segment. Setting this strategic focus meant to deviate from the pure LCC business model by using nearby, quickly and easily accessible (primary) airports instead of secondary airports, which are typically remotely located. Furthermore, to meet the convenience of their passengers, Germanwings equipped its air fleet with "business-like" leather seats arranged in more comfortable seat pitches than in the competitors' airplanes. Third, even though offering higher quality than the competition, Germanwings strives to offer a clear price advantage through cost-leadership.

Figure 5-6: Germanwings Strategic Triangle

---

[17]   See also Garvens' chapter 15 in this volume.

The ambitious strategic challenge expressed by Germanwings promotional-promise "Fly High, Pay Low" was obvious: carefully leveling the additional services offered with the resulting operation costs. As mentioned before only those services should be added which ideally do not induce additional costs, *or* which are offset by the revenue of a higher number of attracted customers, *or* which are offset by customer value, i.e. the customer's willingness to pay. Furthermore, in order to reach the aspired position as "the quality leader in the low cost market" Germanwings first had to build a widely acknowledged corporate brand to overcome the homogeneity of the LCC market with regard to quality and by this to exploit the increased significance of the additional psychological value of a strong and established corporate branding.

3.2.2 Developing a Unique Corporate Philosophy

As the basis of the evolving corporate branding, first of all a unique corporate philosophy consisting of the elements corporate identity, culture, and design was defined and developed. Each element of this specific philosophy is in line with what company spirit and business strategy the brand is intended to represent: a unique, dynamic and young airline which strives to be the market price leader, redefining and setting quality standards in the low price market segment. In the following we will give a short overview about these elements of Germanwings' corporate philosophy.

Corporate identity

By the corporate identity a company defines and tells the market (whether actively or by default) just what its specific purpose is (i.e. why it exists) and what the corporate strategy is all about (i.e. where it wants to go). Germanwings defines its mission to prevail as a widely acknowledged low cost airline, the market leader on price with a high quality core product and innovative services. The company strives to be more than a mere low budget airline, apart from price leadership it is committed to safety, reliability and punctuality. Conformingly it redefines quality regarding convenience and time in the low price market segment.

Corporate culture

The term corporate culture denotes the behavior, habits, and rules which a group of co-workers uses to interact with each other. Corporate culture refers to both formal, written company policy concerning everything from dress code to employee relationships as well as to the informal behaviors that have become accepted by the group. The employees at Germanwings are expected to live a well balanced mixture of innovation, unconventionality, seriousness, and enterprise team spirit. In order to spread this way of behaviour throughout the whole company, every employee attends a one day cultural training. According to its corporate identity the unique Germanwings culture can be characterized as unbureaucratic, dynamic, young and entrepreneurially spirited.

Corporate design

Corporate design refers to all actions affecting the visual appearance of a company both inwards and outwards. According to its corporate image and culture Germanwings target was to create a corporate design reflecting both its simplicity and respectability. Here Germanwings developed a corporate design manual specifying all mandatory standards concerning basic design elements (e.g. corporate colors of bramble, yellow, silver; definition and use of logotype), business stationary (e.g. templates for business letters, fax, business cards), communication (standards for advertising, publications and new media), image style (illustration, photography), and signage (e.g. labelling of aircraft and vehicle fleet, corporate uniforms). This manual serves as a strict guideline with unchangeable components of corporate design which give significant expression to the Germanwings brand. They are all directly related and complement each other to a homogeneous, simple and respectable corporate design.

3.2.3 Consistent Application of the Marketing Mix

There are basically four classical variables to consider when crafting a marketing strategy and writing a marketing plan respectively.[18] They are *price*, *product*, *distribution* (also called placement), and *communication* (also called promotion) (again, see the four pillars of the house of corporate strategy in figure 5-5). The application of the marketing mix should be a consistently calibrated combination or "configuration" of these four variables in a way that will meet organizational requirements and objectives. So, when configuring the marketing mix of price, product, distribution, and communication, one has always to be aware of the corporate philosophy, the company goals as well as the specificities of the target market. In our case this is the German LCC market and – needless to say – the key success drivers of the LCC business model presented in section 2.1 (see figure 5-2 and table 5-1).

As we have mentioned before, the overall strategic goal set by Germanwings is to offer the best quality at the lowest price. With its unique corporate philosophy

---

[18] See e.g. Kotler & Armstrong (2003).

(consisting of identity, culture, and design) Germanwings consequently establishes a widely acknowledged company brand in order to differentiate itself as the quality leader in the German LCC market. This strong and established company brand is expected to be clearly associated with the perception of premium service quality by the customer (again, see the roof of the house of corporate strategy in figure 5-5). These critical aspects have to be taken into account in order to configure each variable carefully due to a consistent application of the marketing mix. In the following we will give a short overview about the specific actions of Germanwings relating to each variable of the marketing mix (see figure 5-5).

Price

Setting the right price strategy, or for short "pricing", is a very important variable of the marketing mix, because it is the only lever which directly influences the generation of sales revenue for a company. Whereas the other three variables rather induce costs, since it costs to produce and design a product as well as to distribute and communicate it. Pricing must support these remaining variables by ensuring that the generated revenues are higher than the induced costs and by reflecting the relationship between supply and demand. It is obvious that especially in the price sensitive LCC market environment with its relatively homogenous products pricing is one of the most dominant variables which has to be shaped very consciously. Due to the intense competition depending predominantly on the ticket price, pricing a specific flight connection too high or too low could mean a loss of customers or contribution margin respectively. One typical possibility to avoid this price based competition would be to reduce or "camouflage" price and product transparency e.g. by product bundling and complex price systems. But as we have explained before, the low cost business model calls for simplicity of the product as well as of the ticket price structure. Keep it simple has therefore been the main driver for the design of the Germanwings price system. The uniform lead price for all connections is actually 19 Euros including all taxes and charges which apply to about 14 percent of the seats per aircraft. Subsequently, depending on the number of tickets sold, the yield management system rases the ticket price up to about 60 percent of the corresponding tariff offered by the classical network carriers. By this rather transparent pricing structure Germanwings strongly underlines its offensive claim to be the leading low price airline in the German market.

Product

The product represents the market oriented output of a company and its attractiveness particularly determines the customer value, i.e. the customers' willingness to pay. With the product variable the characteristics of a (physical) product or (intangible) service are defined that – at its best – most suitably meets the needs of the customers. Generally those decisions affect product attributes regarding e.g. brand name, functionality, styling and design, quality, safety, supplementary accessories, and customer services, etc. The flight as the typical airline product reflects the character of an intangible service which mainly consists of the three components: route network, ground product, and on board product. In accordance with the no frill product philosophy of the LCC business model, Germanwings product decisions were also largely based on the guideline "keep it simple" – but due to its specific strategic orientation there are also some interesting exceptions to this ideal. Accordingly, the route network is constructed solely from point-to-point relations (i.e. no interlining) with attractive destinations that ensure a high passenger traffic. But in contrast to the LCC business model Germanwings rather uses primary airports to give consideration to its strategic goal to offer a clear time and convenience advantage compared to its "pure" LCC competitors (again, see the strategic triangle in figure 5-6). The ground product consists of the airline's appearance at the airports (e.g. check in desks) as well as the design of the check in procedures. In line with the LCC business model both elements are configured to reflect the simplicity of the product usage as well as the look and feel of the brand representing security and reliability. Simplicity is mainly achieved by quick and easy check in procedures via booking number and passport (ticketless check-in), free choice of seats on board and the deliberate abandonment of passenger lounges. Even though there are no free food and refreshments offered on board, the board product on the contrary is designed to support a premium "business-like" product impression. In this regard Germanwings uses comparatively extended seat pitches, leather seats, new and neat staff uniforms, cabin fragrance, and regular service trainings are conducted. All in all Germanwings uses especially the product variable to differentiate itself as the quality leader in the LCC business.

Distribution

Designing the distribution variable encompasses decisions associated with the channels that serve as a means for getting the product to the target customers. Basically the distribution system performs transactional, logistical, and informational functions. Due to the specific product character of the flight as an intangible service there are fewer logistical than transactional and informational functions to be fulfilled. Basically one can choose between two main distribution channels to sell tickets to the final customer (again, see figure 5-5). One possibility is the classical way of indirect or "offline" distribution, i.e. the distribution of the service via professional travel agents who perform the task to give competent advice and sell the product to the end customer. The second alternative is direct or "online" distribution by the particular airline itself. Especially during the rapidly evolving internet era and the inherently related diffusion of e-commerce, new information and software technologies enable the implementation of a cost efficient distribution channel via company homepages with integrated internet-shop systems. Regarding this, it is no surprise that the massive appearance of the LCC-type airlines coincides with the emergence of e-commerce. Additionally assisted by the deliberately intended simplicity (and thereby the inevitably resulting homogeneity) of the service product, it is now possible to implement quick and easy booking procedures as well as to provide additional sale and service information in an automated and cost efficient way. As a consequence of the LCC business model, next to its classical function as a generic information platform, Germanwings uses its company homepage as the pivotal distribution channel to sell tickets directly to the final customer. In the light of Germanwings' strive for quality leadership, the online distribution channel furthermore opens up the potential to enhance customer loyalty by providing additional benefits for Germanwings member passengers. Tailored especially to the price sensitive business traveler, distribution via mobile media is also possible since March 2004. To generate additional revenue, last but not least a merchandise line was developed which is available online as well as on board. All these examples show that the rather low costs induced by online distribution hold manifold potential to support and boost the LCC business. But as we have explained before, costs count in the LCC business and particularly distribution costs must therefore be controlled very carefully. This especially affects the trade-off between the induced costs on the one hand and customer value and revenues generated on the other hand.

Communication

Communication denotes all the market oriented communication activities, that is, the publication of specific information about the product (and the company) with the goal of generating a wide spread positive customer response. Marketing communication decisions include classical advertising, non classical activities like events and sales promotions as well as public relations activities (again, see figure 5-5). Regarding the Germanwings slogan "Fly high pay low" the classical advertising activities concentrate on a charming communication (on print media, placards etc.) predominantly of the price, but also other messages are transferred, such as the movement from Berlin Tegel (TXL) to Berlin Schönefeld (SXF). Due to the launch of Germanwings as a rather new brand, especially during the kick off phase, non classical activities like events and sales promotions were organized to make the customer associate the brand with certain emotions and to lay the basis for a more efficient marketing later on. Finally, public relations activities, namely 12 national and 32 international press conferences were conducted, specific background talks were given on television as well as in relevant journals. As in any other business, marketing communication is an ever ongoing activity to keep the customer up to date with the image and actual product offerings of the company. But as we have explained before, especially in the highly competitive LCC business, communication has to be intensive, proactive and innovative, to consistently stimulate the customers' interest.

4 Outlook into the Future: Further Development of the LCC Business

The presented Germanwings example highlights that there is no one best way of doing business – even in the comparatively homogeneous LCC market environment this holds all-too-true. In this regard we have shown that Germanwings deliberately deviates from the pure cost leadership ideal of the LCC business model by differentiating from its competitors as the quality leader regarding time and convenience. In the words of the concept of competitive strategy by Porter we described before, Germanwings concentrates on a specific niche of the German LCC market segment by setting a clear strategic focus. And Germanwings seems to be on the right track, since an actual customer survey spanning 1200 readers of the German business journal "Capital" voted Germanwings as the best low cost airline, impressively winning in all categories (cost/performance ratio, safety impression, service, seating comfort, catering, punctuality).[19] Building upon this solid basis, Germanwings consequently strives to ensure and extend the achieved position, by opening up further European niches and adding further service features and thereby trying to extend the price premium typical for the "business" low cost segment. In line with its business like product impression Germanwings wants to encourage the customer to buy again and increase customer value by specific measures aimed at special target groups. This will be realized by expansion and new development of differentiating elements regarding price and product (e.g. loyalty/discount programs, further products on sale, transfer shuttle services).

Just like Germanwings, the competitors such as Ryanair (even!) or Hapag-Lloyd-Express also intend to design or have already implemented additional service

---

[19] See Hagen & Hohenester (2004).

product features respectively. Such additional features comprise e.g. cooperation with car-rentals, hotel arrangements, loyalty programs, etc. (again, see the list of the optional additional service components presented in table 5-2). This general development towards service refinement strongly exhibits an increasing maturity of the LCC-business bit by bit. Since on the one hand it is a typical indicator for mature businesses to display a high product variety and diversity, there is on the other hand the great danger of losing simplicity – one of the key success drivers of the LCC business model. So in this steadily maturing business we have to severely remind of the general rules of thumb we elaborated before: as long as the product refinement does not induce (too high) extra operating costs, is valued by the customer, i.e. affecting his willingness to pay, and/or generates a surplus revenue by a higher number of passengers attracted, the competitive advantage of the particular airline will not be negatively affected.

Against the background of our strategic analysis presented in this chapter, we strongly believe that LCCs ignoring these general rules will face market exit in the long run with the utmost probability. However, it would be highly speculative and scientifically suspect to foretell how many and which LCCs will prevail, since there are many other insecure micro- and macroeconomic factors to account for, that is the financial capacity and the political will of the shareholders, state regulations due to environmental aspects, etc. But on the basis of our previous strategic considerations we can anticipate with a clear conscience that the LCC business will go on to split up further into the market segments for the leisure travelers and the price sensitive business travelers respectively. The leisure segment on the one hand is to be served at its best by airlines pursuing the LCC business model in its ideal form. The business segment on the other hand allows for well-advised and economically balanced deviations from this ideal by setting a clear strategic focus – take the successful results reflected by the "Germanwings strategy" as a clear reference!

5 References

FEIERABEND, L. (2004): Easyjet will an die Spitze. *Kölner Stadt-Anzeiger*, 181, 5th August 2004, p. 13.

FREIBERG, K.; FREIBERG, J. (1998): *Nuts! Southwest Airlines' crazy recipe for business and personal success.* New York: Broadway Books.

HAGEN, J.; HOHENESTER, A. (2004): *Fabelhafter Wettbewerb.* Capital, 18/2004, pp. 84-96.

HANSSON, T.; RINGBECK, J.; FRANKE, M. (2002): *Airlines: A new operation model.* Download: http://www.booz-allen.de/content/downloads/airlines_nom.pdf.

KOTLER, P.; ARMSTRONG, G. (2003): *Principles of Marketing.* 10th Ed., Upper Saddle River: Pearson Prentice Hall.

LAWTON, T.C. (2002): *Cleared for Take-Off. Structure and Strategy in the Low Fare Airline Business.* Aldershot, Burlington: Ashgate.

MERCER (2002): *Low Cost Airlines haben sich als Herausforderer etabliert.* Press Release, 22 May 2002, download available at http://www.mercermc.de.

OEHLER, K.D. (2004): *Ryanair erwartet weitere Preissenkungen.* Kölner Stadt-Anzeiger, 180, 4 August 2004, p. 12.

PILLING, MARK (2003): Price Promise. *Airline Business*, 19 (4) pp. 58-60.

POMPL, W. (2002): *Luftverkehr. Eine ökonomische und politische Einführung.* Berlin et al: Springer.

PORTER, M.E. (1980): *Competitive Strategy. Techniques for Analyzing Industries and Competitors.* New York et al: The Free Press.

SCHMIDT, G.H.E. (2000): *Handbuch Airlinemanagement.* München, Wien: Oldenburg.

6
LOW COST CARRIERS IN SOUTHEAST ASIA: A PRELIMINARY ANALYSIS

MARK GOH

1 Introduction 144
2 Business Model Considerations 146
3 Some Case Studies 148
4 Discussion 157
5 Conclusion 161
6 References 162

Summary:
This chapter presents an initial analysis of the state of the Low cost Carriers (LCCs) operating within Southeast Asia, so as to better understand the operating, marketing and distribution strategies of these LCCs. Next, we assess the approach of the LCCs' business model, using simple comparative analysis, drawing information from secondary sources. Preliminary evidence suggests that the LCCs in Southeast Asia are no different from those in the US or Europe when it relates pricing, in that low fares are snapped up quickly. Operationally, however, the challenges of meeting a fast turnaround time and minimizing landing charges still remain. There are also policy questions which remain unanswered.

1 Introduction

The airline industry in Asia is experiencing a somewhat belated entry of competitors in the airline business especially for the budget tourist market. This entry of Low Cost Carriers (LCCs) or airlines as it is sometimes called is, like in Europe, one of the most interesting aspects of the free market. Unlike what Pender et al.[1] describe as an LCC, LCCs in Asia are slightly different. So what really constitutes an LCC?

According to Pipa et al.,[2] an LCC (also known as a no-frills or discount carrier) is an airline that offers low fares but eliminates all unnecessary services. The typical LCC business approach is based on (http://www.wordiq.com/definition/-Low cost_carrier):

- a single passenger class configuration, usually economy
- a single type of airplane (reducing training and servicing costs – often the Boeing 737)
- a simple fare scheme (typically fares increase as the plane fills up, which rewards early reservations, practising what is commonly known as yield management)
- unreserved seating (which encourages passengers to arrive and board early)
- direct, point to point flights with no transfers
- flying to cheaper, less congested and less expensive secondary airports
- short flights and fast turnaround times (allowing better utilization of planes).

Amenities such as free in-flight catering and other complimentary services are eliminated, and instead replaced by optional paid-for in-flight food and drink.

In Europe, Pender et al.[3] have outlined the operating features that provide LCCs with a cost advantage. LCCs fly to short haul destinations of usually 1,500 km (based on the Boeing 737). Eliminating the food and other complimentary services have allowed LCCs in Europe to effectively reduce the aircraft

---

[1] See Pender & Baum (2000).
[2] See Piga & Filippi (2002).
[3] See Pender & Baum (2000).

turnaround time at the airport from the usual 50 - 65 minutes to 25 minutes as is the case with Ryanair. Also, the traditional model of booking flights with third party travel agents is replaced by a click and mortar distribution system such as the Internet and online booking.

1.1 Motivation for LCCs in Southeast Asia

There are several arguments in favour of having the LCCs in Southeast Asia. First, Asia has a population of 3.8 billion,[4] and the flying range between Asian cities is within 3 to 4 hours. For instance, using Singapore as a hub, it is possible to reach a population of 500 million (typically ASEAN, Taiwan and South China) within a 4-hour flying radius.[5] Thus, it is no surprise that aviation companies like Qantas and Virgin are interested in this market.[6] Second, as Asia continues to prosper economically, there is the attendant need to cater to a growing business audience serving a regional market. Third, as the populace becomes more educated, there should be a greater desire to travel to see the region either individually or with their families. Geographically, the reach to anywhere within Asia is far easier and cheaper than to Europe or the US. Fourth, the low cost travel market is still relatively untapped if we use Singapore Airlines' (SIA) revenue as a proxy. Saywell notes that just 7% of SIA's revenue of S$5.9 bn in the year 2003 comes from the short haul routes and leisure travel. Clearly, there is room for further expansion.

However, capturing current demographics characteristics of such travellers is still in its nascent stages as the LCC industry is still a fledging industry. Nevertheless, there are already some airlines like Tiger Airways who have targeted young adults and ValuAir who have targeted female travellers who are avid shoppers, for instance. This could suggest that the future customer basis of the LCCs is widely distributed across all demography. The advent of the LCCs will surely and fundamentally change the nature and intensity of competition in the aviation landscape.

Unfortunately, there is little work done on LCCs in Southeast Asia despite the potentially dramatic impact in the overall regional airline industry. Most studies, to date, are focused on the US market[7] and Europe.[8] Indeed, whether the

---

[4] See Sritama (2004).
[5] See Saywell (2004).
[6] See i-Sinchew (2004).
[7] See e.g., Boguslaski, Ito & Lee (2002); Ito & Lee (2003).

presence of LCCs will lead to a battle for market share or market expansion is still unclear. It is for this reason that this chapter attempts to profile the various LCCs operating in Southeast Asia and seek to understand their operating, marketing and distribution strategies.

This chapter is organized as follows. In the next section, we present some business model considerations of the current LCCs, compared with the business models employed by European LCCs. At the same time, we present some elements of the Southeast Asian LCCs' competitive strategies that have not been identified in the existing literature. Following this, we highlight data collection approach and present short descriptive case studies of key Asian LCCs in this emerging market. The next section discusses the findings of the study. Finally, we conclude the chapter with some key implications for policy making, some necessary action steps, and point to areas of future research.

2   Business Model Considerations

Basically, the LCC model, feeding on a totally new, price sensitive market, works best on two pillars, namely,

1. capitalizing on the ability to save on landing fees, and to get into and out of airports quickly
2. selling of tickets online, using a mixed distribution system.

Indeed, as Ryanair has proven in Europe, landing fees are expensive and as such LCCs want to move in and out of airports as quickly as possible to achieve that fast turnaround, and in turn maximizing operating utilization. Moreover, the destination airports are secondary airports, sometimes with little commuter traffic. Geographic differentiation is also practised by the LCCs as reported in Piga et al. (2002) in that the airlines never fly to exactly the same destination airports. Likewise, when the same destination is served by two or more airlines, the departure airport is different. Also, subsidiaries of established airlines are inclined to form LCCs as proven by BA (GO) and KLM (Buzz), providing an effective product differentiation strategy.

Later on, in our presentation and analysis of the case studies, we will show that the above practice is indeed not the case in Southeast Asia. For a start, the primary business hubs are well established and depending on the business focus

---

[8] See Pender & Baum (2000); Piga & Filippi (2002).

of the LCC, some airlines may leave from the same departure airport and arrive into the same destination airport. Industry observers in Asia feel that the new LCCs, for operating cost reasons, will probably fly tourists and weekenders to nearby resorts and leisure holiday destinations such as Phuket, Bali, Chiang Mai, Langkawi. For the business travellers and serious shoppers, popular destinations such as Taipei, Hong Kong, Jakarta and Kuala Lumpur are also strong possibilities (www.tigerairways.com/news2.html).

Table 6-1 below shows the various cost drivers involved in operating an airline both for a typical US airline and SIA, whose operating efficiency is already one of the best in Asia.[9] Indeed, labor and fuel costs appear to be the major ticket items for airlines generally. Based on Sutton,[10] the cumulative cost advantage of LCCs operating in Europe is 43%, when all operating cost factors are considered, vis-à-vis the full service carriers.

Factor	US Airlines Average Cost percentage	SIA's percentage
Labor	36.8	18.0
Fuel	12.8	19.4
Ticket commission	11.3	7.9
Airports	5.4	6.1
Food	3.2	6.7
Interest	2.0	N.A.
Advertising	1.2	N.A.
Others	23.1	41.9

Table 6-1: Dollar Value Percentage Comparison between US Airlines and SIA[11]

If we base our analysis on table 6-1, then it becomes obvious that the cost advantage for LCCs is possible only if margins are increased through either

---

[9] See Lee (2004).
[10] See Sutton (2003).
[11] Source: Air Transport Association (2001); Lee (2004).

cheaper pilots being employed or relying on more fuel efficient planes and minimizing unnecessary payloads. (In the case of SIA, pilots are the most expensive human resource on a plane as they can earn typically S$14,000 to S$16,000 a month and first officers can expect S$9,000 to S$11,000 a month. In general, the staff cost of pilots in SIA form 40% of the 15.8 to 18% wage bill.)

Although there are other Southeast Asian LCCs, we will only focus on the more visible ones such as AirAsia, Tiger Airways, ValuAir, and NokAir and present them as relevant case studies for our discussion in this paper. Another reason for doing so is the lack of publicly available information. Most of the budget carriers, like Air Paradise (Indonesia), Lion Air (Indonesia), Cebu Pacific (Philippines), Air Andaman (Thailand), First Cambodia Airlines (Cambodia) have either little public information or are written in a non-English medium.

The data collection for this chapter is based on secondary data archival research. This involved extracting the information from the relevant websites, related articles from the aviation and academic journals in the broad area of low cost carriers.

3 Some Case Studies[12]

3.1 AirAsia (www.airasia.com)

Launched originally as a subsidiary of Malaysian Airlines, AirAsia went into receivership and was re-launched as a private company in December 2001. The restructured airline then started to position itself as an aggressive cost focused airline. For a start, AirAsia uses new and innovative cost optimizing techniques such as quicker turnaround times and maximizing of flight utilization while continuing to maintain the most important elements of safety, service and schedules. All fares are quoted one way to allow passengers the flexibility to choose where and when they would like to fly. Fares are based on supply and demand, and prices usually increase as seats are sold on every flight. So, the earlier the booking was made, the cheaper the fare will generally be, which appears to be a good way to lock in passengers (especially budget travellers). AirAsia's booking system continually reviews bookings for all future flights and tries to predict how popular each flight is likely to be.

---

[12] Information is extracted from the respective websites.

AirAsia exhibits many of the characteristics that typify a no-frills carrier, operating out of 3 hubs, Kuala Lumpur, Bangkok, and Johor Baru (just North of Singapore). This strategy is consistent with reducing cost and serving large catchment areas. It uses a website to take bookings for its operations of 43 routes across 5 countries. Traditional tickets cost a lot of money to print, process and deliver to the traveller. Saving on these costs by dis-intermediating the middle man allows AirAsia (the first airline to do so although she still maintains ground offices in the respective airport terminals) to pass the savings downstream. Instead of tickets, passengers are provided with an itinerary which includes the booking number, payment and travel details upon completion of the booking. During check-in, passengers only need to mention the booking number and must present their identification card or passport for identification purposes. (However, with the heightened security as a consequence of the potential threat of terrorism, we are not sure if the airport authorities would still allow this practice to continue.)

Under the no-frills service, there are no complimentary meals or drinks. Instead, passengers are given the choice of purchasing a variety of food and drinks on board at a low price. This is made necessary as AirAsia does not permit outside food and beverage to be consumed onboard, again for reasons of down time efficiency.

Additionally, AirAsia operates a first-come-first-served boarding system that allows the individual a choice of seat. This policy encourages passengers to board the aircraft early and avoid passenger related delays at the terminal.

Similarly to the other LCCs, AirAsia concentrates on just one aircraft type, the popular Boeing 737-300 jets (see Ryanair), each with 148 seats, 2 flight crew and 3 cabin crew (see SIA 777-300 of 278 passengers and 21 crew). Clearly, AirAsia has a cabin crew to passenger ratio of 1:49.3 while SIA has a ratio of 1:14.6. This means operations and maintenance remain efficient and focused.

AirAsia does not explicitly have a frequent flyer program claiming that such customer loyalty programmes are expensive. Instead, AirAsia would rather lower the cost of travel for everybody. Although AirAsia has recently adopted the practice of promoting travel through a link-up with the Development Bank of Singapore, as a form of customer loyalty, it remains to be seen if it is successful.

AirAsia is point-to-point and does not encourage connecting flights. For passengers wishing to do so, at least 3 hours transit is required to accommodate the long distance from arrival gate to baggage collection area and then check-in again at the departure hall for the next flight. This clearly discourages business travellers even though the website of AirAsia has options for corporate customers. In summary, AirAsia is based on a simple product and service and strives for cost and time efficiency.

Figure 6-1: AirAsia's Route Network

3.2 ValuAir (www.ValuAir.com.sg)

ValuAir is positioned as the region's quality, semi-frills budget airline operating out of Singapore. Privately owned, ValuAir begun operations in May 2004 and currently flies daily on 3 routes of no longer than 4 hours to Bangkok, Jakarta and Hong Kong using 2 brand new short-to-medium range commercial passenger craft, the A320s.

ValuAir's business model is slightly different from the definition provided earlier in this chapter. It appears to fit the niche between full service and budget airlines, in that it provides free but simple meals, better seats than budget airlines (with a pitch of 32 inches, comparable to SIA rather than 28 inches for other budget airlines) and a reasonable luggage allowance of 20 kg per passenger compared to the other airlines' 15 kg baggage limit. Their philosophy is to give customers what they want and not just what they need.

ValuAir's assigned seating includes three price categories: one for children, a discounted fare with terms and conditions (Saver Fare), and a standard fare with no restrictions (Flexi Fare). Also, assigned seating is possible beforehand, unlike that of AirAsia.

Like AirAsia, ValuAir also uses one type of aircraft, the Airbus A320, with a capacity of 162 seats.

ValuAir operates on a very simple and transparent fare structure. Current on-time departure performance level is at 99.5% which is one of the highest in the industry. In another departure from AirAsia, ValuAir actually relies on the main gateways such as Singapore's Changi Airport Terminal 1, Bangkok International Airport Terminal 2 (just like SIA), Hong Kong's Chep Lap Kok Airport, and the Soekarno-Hatta Jakarta International Airport Terminal 2.

ValuAir has a sector driven frequent flyer program based on number of boarding passes accumulated and a free ticket is redeemable with 40-60 boarding pass points (each pass is worth 2-3 points depending on length of travel). Comparing this with the established airlines like SIA, passengers on SIA require 25,000 miles travelled to gain one free ticket from SGP-BKK. On ValuAir, 20 such return trips are needed. Based on SIA's Kris Flyer program, this works out to be 14 return trips. In this regard, the established flyer's loyalty is better.

With a current travel radius of 4 hours, the airline appears on target to meet its objective and it will continue to seek air rights to fly to high traffic destinations around Asia within a five-hour flying radius from Singapore. ValuAir's strategy appears similar to that of SIA's subsidiary, Silkair. The management of ValuAir believes there is a huge potential to serve both tourists (both new and revisits) and businessmen (especially the SMEs) who want a convenient and affordable way to get to and from Hong Kong, a central Asian hub and gateway to the huge China market. As for the vacation market, the airline is keen to tap in on the 75,000 Singaporeans who went on vacation to Hong Kong in Q1 2004 out of the 380,000 visitors from the South and Southeast Asia regions who did so.

The business model employed by ValuAir is basically to focus on excellent on-time performance, fast and efficient check-ins, and baggage handling.

Figure 6-2: ValuAir's Route Network

3.3 NokAir (www.nokair.co.th)

NokAir is a no-frills, low cost domestic airline based in Bangkok, Thailand, established in 2004. It is a subsidiary of Thai Airways International. Currently, it serves the major provincial cities of Chiang Mai (pop. 250,000), Hat Yai (187,920), Udon Thani (222,425), Phitsanulok (150,000), Phuket (250,000), and Khon Kaen (141,000). These cities are also popular with regional tourists.

Local Thai and international tourists seeking to explore the beauty of the different parts of Thailand, as well as family visitors or leisure and business travellers looking for value deals. The vision promoted by NokAir is to ensure that in-country air travel will no longer be expensive for Thais. NokAir will help enhance the quality of life for Thais to be comparable with those in other developed countries. Thais will now be able to explore the beauty of Thailand, families will now have more opportunities for reunion and cost conscious business travellers will now be able to pay more sensible prices for airfares. It makes use of 2 leased 737-400, with a capacity of 149 seats each. Crew and pilots are trained by the parent company, meeting the same standards as the former. Again the practice of earlier booking secures cheaper fares applies here. The focus is on safety and convenience of passengers.

Again, under the no-frills service, there are no complimentary meals or drinks. Instead, passengers are given the choice of purchasing simple soft drinks and beverages at reasonable prices, again assuming that the airline does not allow food and drinks on board. The usual weight limit of 15 kg for baggage applies.

A mixed channel approach is employed i.e. through their website (www.nokair.com.th), call centre (operating from 0800 to 2100 hrs only), and the various airport ticket offices nationally. Payment for the tickets is very convenient as it makes use of retail outlets such as cash at 7-Eleven Counter Service Plus, Master/VISA card online, and via the call center, direct transfer at the Siam Commercial Bank ATMs. NokAir intends to increase the payment reach by putting more channels in the future.

The prices of the tickets are intended to be comparable with those of the other low cost airlines in the markets (although few international airlines have access rights to the Thai domestic market). Seating can be user assigned at the point of purchase, and subject to the issuance of a booking number. There are two categories of seats, 12 business seats (Nok Plus) and the rest economy. The price differential for these categories is Bht 500 or about USD 10. NokAir claims to be the only low cost airline in Asia to offer business class seats at an affordable price.

NokAir engages in a fast turn around time of 30 minutes before returning to its hub at Bangkok. This is probably the fastest turnaround time for all the LCCs and the major airlines in the region. NokAir aims for a new mainstream audience who want to travel economically but want the same safety and reliability assurance of an established brand like Thai Airways.

Figure 6-3: NokAir's Route Network

3.4 Tiger Airways (http://www.tigerairways.com/index.html)

The latest LCC to be launched, Tiger Airways, commenced operations in September 2004, plying initially on three routes across two countries. The majority shareholder is SIA, who owns 49% of the airline. The other significant shareholder is the Ryan family, founders of Ryanair (16%). Tiger Airways is positioned as a low cost, low fare, no-frills airline to service initially the Southeast Asian region. Flights will initially operate from Changi Terminal 1 until a low cost carrier terminal is built. Two shifts are involved either morning or afternoon. There are no plans for night stops or night flights.

Currently, Tiger Airways operates on 4 leased new Airbus A320 aircraft (this is a variant of the original mid-sized craft; A320 typically can hold 100-220 passengers and can cover 5,400 km.). The seating configuration of this aircraft is

180 passengers. The present build-up plan of aircraft is 2 more A320s in 2005 and another 4 A320s in 2006. Like AirAsia, Tiger Airways operates on a single class and free seating arrangement. No personal food is allowed for consumption on-board. Snacks and refreshments are available for purchase on-board. For a start, the airline operates on 3 routes across 2 countries.

Tiger Airways' business model is similar to the others, in particular that of Ryanair, notwithstanding the fact that Ryanair is a significant minority shareholder. Using Singapore as a hub, it intends to fly to destinations that are within 4 hours flight time from Singapore, reaching a potential population of 550 million people, covering the economies of Southeast Asia, Taiwan and Hong Kong and South China. It expects to be profitable within the first year of operations and then grow 20% to 25% in each of the next five years.[13] By charging on average 40% less than the competition, it hopes to secure 200,000 customers, serving up to 10 destinations in the first year of operations.[14]

Tiger Airways seeks to adopt an aggressive pricing, operating and marketing strategy. For instance, in a recent publicity and promotion drive for its inaugural flight, Tiger Airways set a promotional price of S$1 (one-way excluding taxes) from Singapore to Bangkok, Hat Yai, and Phuket. The response from the public was overwhelming, with 4.5 million hits on the airline's website in ten hours.[15] Operationally, the airline is relying on only new A320's to provide for greater simplicity, better fuel efficiencies, and lower maintenance costs. Marketing wise, the use of new A320's will suggest to the flying audience a psychological brand value of safety for passengers.

The airline intends to sell up to 80% of its tickets through its website. Further, it wants to use the Internet as the preferred medium for marketing, sales and revenue management.

Table 6-2 contains a summary of the company profile of the above case studies.

---

[13] See Gan (2004).
[14] See Sreenivasan (2004).
[15] See Kaur (2004).

Factor	AirAsia	ValuAir	NokAir	Tiger Airways
Base (HQ)	Malaysia (Kuala Lumpur)	Singapore	Thailand (Bangkok)	Singapore
Hubs	Kuala Lumpur Bangkok Johor Baru	Singapore	Bangkok	Singapore
Date of launch	December 2001	May 2004	June 2004	September 2004
Affiliation	Private Company	Private Company	Thai Airways	Private Company
CEO	Tony Fernandes	Lim Chin Beng (SIA veteran of 25 years)	Patee Sarasin	Patrick Gan
No. of active routes	43	3	6	3
No. of destination countries covered	5	3	1	1
Type of aircraft used	Boeing 737-300	Airbus A320	Boeing 737-400	Airbus A320
Fleet size	19	2	2	4
Flight radius (in hours)	4	4	2	4

Table 6-2: Summary of Company Profile

4 Discussion

From the above case studies, it appears that there are two breeds of LCCs operating in Southeast Asia. The first breed, like NokAir, capitalizes on serving secondary markets (the smaller provincial cities and leisure locations) with little or no direct competition from the main line carriers. The second group of LCCs, like ValuAir, focuses on the primary high cost airports of Singapore and Bangkok, and emphasizes on existing cost conscious but time insensitive business and leisure markets and some new markets, accepting competition from incumbent carriers. Also, to be competitive, it appears clear that labor cost, food and ticket sales commission have to be kept to a minimum. We now take our discussion to the next level.

4.1 Target Markets

It appears that depending on where the airlines operate from or fly to, the target market may vary significantly. In all, there are 3 distinct target market groups. First, there is the up market leisure traveller who is keen to visit exotic resort locations in Southeast Asia (for example, Phuket and Langkawi). This group is catered for by LCCs such as ValuAir and AirAsia. Next, there is the group of cost conscious small business travellers who need to get on the busy industrial routes such as Singapore-Bangkok, and Singapore-Hong Kong. Finally, there is the group of domestic travellers, who are increasing in affluence and affordability, and desire to see new in-country historical or heritage locations inexpensively (e.g., Chiang Mai and Luang Prabang). Unfortunately, no study to date exists to determine the exact demographic profile of such travellers.

4.2 Branding of Product and Services

The role of creating appropriate brand values appears entrenched in the LCCs. Indeed, nearly all LCCs established both tangible and psychological brand values, and painstakingly highlighted this to their potential passengers through their websites. For instance, AirAsia sells on the basis of reliability (safety), punctuality (schedules) and service. The same applies to ValuAir and NokAir. On the service dimension, while it is clear that the LCC cannot compete with the full service carriers, the LCCs employ a pay-on-demand principle to effectively set themselves apart from the competitors.

4.3 Pricing

With advent of such LCCs into Southeast Asia, air travellers are now given greater choice in terms of choosing which airline to travel with, and at what price. Already, the lowest price possible has been set by Tiger Airways and judging from the overwhelming demand, it is apparent that the market is full of price/ budget conscious travellers for short haul trips. This has forced the pricing structure of the full service carriers to adapt to the new competitive environment.

4.4 Distribution

Putting the sale of tickets on the Net is a key cost saving factor in the supply chain of the LCC business. Based on dollar value percentages of table 6-1, it is clear that the travel agent commission forms a sizeable percentage of operating cost. Hence, going on the direct sell mode can allow LCCs to save up to 10%[16] of the operating cost of a typical full service carrier. By selling through the web, costs are kept to a minimum in terms of distribution. In fact, the distribution process for LCCs is far more simplified in that bookings are taken using established credit cards, then the booking reference number is sent to the passenger, and this number is subsequently presented at the airport check-in counter. This simplification translates to a reduction in manpower cost, paper ticket and printing cost, and other handling charges. LCCs, with their click and mortar system, may well displace the travel agents in their direct dealings with passengers.

4.5 Challenges and Implications

Low cost carriers appear to pose a threat to traditional full service airlines, since full service carriers cannot compete on price and, when given a choice, most consumers will opt for low price over other amenities. However, operating LCCs is not without its challenges and implications. We now present some of these challenges and implications

---

[16] See Lee (2004).

1. An insufficient number of good secondary airports within the commuting distance of Asia's capital cities suggest that LCCs, especially those that fly during the peak periods of full service carriers, cannot avoid the congested and more exorbitant hub or primary airports as easily as the US and European budget carriers can.[17]

2. Some of the routes flown by the LCCs are also flown by the established airlines. This can lead to a severe price war for both types of carriers, and hurt their margins in the process. Should this possibility eventuate, then it is likely that LCCs will suffer more. Most of the airlines in Asia are big state monopolies who do not like any whiff of competition. As such, they may be more prepared to defend their position and stay for the long haul with their deep pockets. Alternatively, they may provide their own LCCs to challenge the incumbents in this market. Already, Singapore Airlines, Thai Airways International, and Australia's Qantas Airways are setting up low cost subsidiaries of their own or through joint ventures, recognizing the potentially strong competition from the LCCs.

3. Unlike Europe or the US, Asian nations have been much slower to open their airports to all airlines, thus forcing the LCCs to turn to the domestic markets. For some countries, this calls for strong political will to change the status quo. The good news is that this challenge may be short lived as at the recent IATA annual conference, it was reported that the restrictions on the number of flights in the region will commence to open from 2008 and the open sky framework will be complete by 2015.[18]

4. The current boom in LCCs is being driven by the governments of Malaysia, Thailand, Indonesia, and Singapore, which are granting landing rights to the carriers in hopes of boosting tourism and business travel.[19] New LCCs will have to be sharp-clawed to compete in the region's increasingly busy budget-airline space. There is already a bunching effect on the more popular routes, for example, Singapore-Bangkok. Furthermore, the major Asian airlines are already cramming cheap seats into the back of wide body jets, creating a budget plane within the plane effect.[20] This can nullify the ability of discounters to gain real cost advantages on the busy industrial routes.

5. In Asia, consumers, especially the less technology savvy and those who do not possess credit cards, are less likely to buy online. This can limit the

---

[17] See Balfour (2004).
[18] See Sritama (2004).
[19] See Balfour (2004).
[20] See Kolesnikov-Jessop & Holland (2003).

potential growth of the LCCs who rely on cyberspace for sales growth. The restricted reach of and to the Net, and the unavailability of credit can stunt the proliferation of the LCCs' distribution system. In the case of the latter, there is a large group of Asian teens and young adults (mostly University students) who would like to travel independently and cheaply but do not own credit cards of their own.

6. The expansion of LCCs, utilizing direct point-to-point routes, will affect traditional airline hub and spoke networks. This naturally poses interesting questions for the terminal operators and policy makers. Over time, it may well be that the structure of the regional airline market will shift from a hub-and-spoke network to a point-to-point network.

4.6 Suggested Action

Despite the concern of the established, traditional airlines over the LCCs (hence the entry into the LCC market), there is really insufficient evidence to conclude that the LCCs will severely cannibalize the networks of major full service carriers like SIA, Thai Airways and MAS. The reason for this is simply due to the fact that much of the payload comes from newly generated traffic either from rural travel or exotic travel, especially on thin routes. The question that begs an answer is whether this passenger segment is lucrative enough and sustainable economically. To perhaps help the LCCs along their trajectory to becoming a viable alternative in the Southeast Asian aviation industry, the following solution steps are suggested:

1. Either the public or private sector should develop better infrastructure in the form of:

    a. Building more secondary airports in large population countries e.g. Thailand. This will serve to reduce the dependence on the busy hub airports for regional or in-country travel.

    b. Building more runways on existing airports to ease congestion and even no-frills terminals to cater for the budget conscious travellers e.g. Terminal 3 at Changi International Airport. As the estimated cost of building a runway plus the associated support structure like ancillary roads and drains can cost about USD100 million, a private-public-partnership arrangement could prove useful here. In the case of Terminal 3, the new low cost terminal is expected to be ready in 2006, and the operating and maintenance charges are expected to be 20% lower than full service terminal.[21]

2. Fine tuning the operations scheduling for LCCs by rescheduling the flights of LCCs to less busy timeslots to better manage peak period demands at existing airports. One example would be to use an auction based system for landing fee charges, which currently favour lighter planes. The system should also encourage the LCCs to fly more often during off peak times.

3. Being clear at service differentiation and redefining the expectations:

   a. Passengers taking LCCs should be perceived as taking a bus while the major airlines passengers should be seen as taxi passengers, who are willing to pay the extra for service, comfort and convenience of time and place (as in better and more convenient take-off times, and alighting directly at the air bridge rather than the tarmac).

   b. Focus on niche markets by linking smaller historical cities (of population of at least 100,000) with secondary airports. LCCs should aggressively serve the popular tourist places and world heritage preserves in Southeast Asia would include places like Luang Prabang (Laos; pop. 26,500), Danang (Vietnam; pop. 1,000,000), Pagan (Myanmar; pop. 260,000), Jinhong (Yunnan, China; pop. 363,110), and Siem Reap (Cambodia: pop. 695,485). For such locations, there appears to be a need to serve secondary airports (and hence locations/ cities) at relatively lower frequencies (once daily, say). To survive, LCCs should focus on new leisure markets with little or no current direct competition from the incumbent carriers, following the strategy of the European LCCs as mentioned in Piga et al. (2002).

5 Conclusion

While it is still early days for the aviation industry in terms of where and how the LCCs will evolve, one fact remains and that is the skies of Southeast Asia will become even more crowded in the days ahead as more airlines compete for revenue, passengers and corridors. At present, what remains clear is that the average consumer, the retail traveller, will stand to benefit from the price wars and numerous options. Indeed, the LCCs will change the shape of travel and

---

[21] See Gan (2004).

tourism in Southeast Asia. As the LCC industry matures, hopefully more data and information will be forthcoming to allow for better and rigorous academic analysis. Even as strategic plans unfold, in-depth case studies can be conducted for classroom teaching. In the meantime, we hope that this chapter will pave the way for future research on this important industry development in Asia.

6 References

AIR TRANSPORT ASSOCIATION (2001): Statement of Carol B. Hallett, President & Chief Executive Officer Air Transport Association of America Before the United States Senate Science. *Transportation and Commerce Committee Hearing on Airline Labor Relations*, accessed from http://www.airtransport.org/public/testimony/display2.asp?nid=889 on 23 August 2004.

BALFOUR, F. (2004): Will Asia's low cost airlines fly high? *Business Week* 21 accessed from http://www.keepmedia.com:/Register.do?oliID=225 on 21 August 2004.

BOGULASKI, C.; ITO, H. AND LEE, D. (2002): *Entry patterns in the Southwest Airlines route system.* Unpublished manuscript, Brown University.

GAN, A. (2004): Will surging oil prices clip their wings. *The Edge Singapore,* 30 August – 09 September.

ITO, H.; LEE, D. (2003): *Low cost carrier growth in the US airline industry: past, present, and future.* Unpublished manuscript, Brown University.

KAUR, K. (2004): Tiger's $1 fare offer causes Web gridlock. *The Straits Times*, 01 September, p. H8.

KOLESNIKOV-JESSOP, S.; HOLLAND, L. (2003): No frills' takes flight, but who will rule in Asia: the state or the entrepreneurs? *Newsweek*, Sep 08, accessed from http://www.keepmedia.com/pubs/Newsweek/2003/09/08/307765 on 20 August 2004.

LEE, R. (2004): The airline's expenses: where to cut and how? *The Straits Times*, 17 April, p. H14.

PENDER, L.; BAUM, T. (2000): Have the frills really left the European airline industry? *International Journal of Tourism Research*, 2(6), pp. 423-436.

PIGA, C.A.; FILIPPI, N. (2002): Booking and flying with low cost airlines. *International Journal of Tourism Research*, 4(3), pp. 237-249.

SAYWELL, T. (2004): Discount airlines: purrfect timing. *Far Eastern Economic Review*, 25 December 2003 – 01 January 2004.

SINCHEW, I. (2004): *Low cost carriers to drive growth* (1 August), accessed from http://e.sinchew-i.com/content.phtml?sec=5&artid=200408010000 on 16 August 2004.

SREENIVASAN, V. (2004): Tiger Airways prepares for fierce far battle. *Business Times*, 24 August., accessed from http://business-times.asia1.com.sg/story/-0,4567,12637,00.html on 31 August 2004.

SRITAMA, S. (2004): ASEAN air industry growing. *The Nation*, June 08, accessed from http://www.nationmultimedia.com on 26 August 2004.

SUTTON, O. (2003): What makes low cost carriers low cost, *Interavia Business & Technology*, March/April 2003, 58(670), pp. 14-17.

7

THE FUTURE OF CONTINENTAL TRAFFIC PROGRAM: HOW LUFTHANSA IS COUNTERING COMPETITION FROM NO-FRILLS AIRLINES

CHRISTOPH KLINGENBERG

1 Strategic Repositioning of the Traditional Airlines 166
2 The Product: Clear Distinction from the No-Frills Airlines 171
3 Production and Processes: Concentration on Value-Added Activities 175
4 Summary and Outlook 182
5 References 183

Summary:
The entry of no-frills carriers into the airline industry has caused a marked increase in the intensity of competition. The no-frills airlines are meeting the consumer trend toward extremely low-priced offers through a new business model focused on lean processes. The established airlines must therefore also face this market segment without neglecting traditional customer groups. The hub-and-spoke airlines are coming under pressure of declining yields worldwide. This has already led some American airlines to file for Chapter 11 bankruptcy protection.
From Lufthansa's perspective, this pressure should be countered not only through a significant reduction in unit costs but also with a fundamental strategic repositioning to respond to current customer demands. This challenge can be met only by introducing structural changes.

1 Strategic Repositioning of the Traditional Airlines

Since the mid-nineties, new competitors with a radically simplified business model have established themselves in Europe alongside the traditional airlines, which fly most of their flights via a hub. The significant unit-cost advantages of these no-frills airlines make it possible to offer low air fares, which, on the one hand, stimulate demand but, on the other, lead to a drop in average market revenues as a whole, thus causing the established airlines to suffer considerable losses in profits. This section describes the impact on the traditional airlines and offers a possible response to the challenge posed by no-frills airlines.

1.1 Analysis of Customer Groups

No-frills airlines entered the market in Europe more than 10 years later than in the US, but at a significantly faster pace. While Southwest, the most important American no-frills airline, registered continuous SKO[1] growth of roughly 10 percent per year since its establishment in 1971, it still took 20 years before the Texan regional carrier evolved into an airline of national importance.[2] In as late as 1998, the no-frills airlines in Europe were limited to less than 2 percent of intra-European traffic – largely concentrated in Great Britain. For the first time in Europe, Ryanair was able to stimulate the market in national traffic over the Irish Sea through considerably lower prices.[3] In 2002, the no-frills carriers achieved a market share of roughly 10 percent through 30-percent annual SKO growth, thus significantly faster than those in the United States.

The growth of the no-frills airlines is largely supported by demand stimulated by low prices. Random sampling showed that 50 percent of Ryanair customers would not have flown with another airline, but would have either selected ground transportation or not taken the trip at all.[4] The large shares of new customers have long lulled the traditional carriers into thinking the no-frills airlines were not direct competitors, especially since the product was clearly different, and the

---

[1] SKO (Seat Kilometer Offered): seat kilometer offered as standard measure for the production capacity of an airline.
[2] See Gittel (2003), p. 7ff
[3] See Binggeli & Pompeo (2002), p. 88.
[4] See Simon, Kucher & Partner (2002).

assumption became popular that business travelers would fly with low cost carriers only in exceptional cases. Since approximately 20 percent of the no-frills airline customers also have a frequent-flyer card from a traditional airline and pursue a business-travel purpose, this assumption began to waiver.[5] Finally, the entry of various no-frills carriers, such as Germanwings,[6] Hapag Lloyd Express, Air Berlin, and Germania Express, into the German domestic market in 2002/2003 and the conversion of the formerly Deutsche British Airways into a no-frills airline (dba) resulted in a massive overlapping of the no-frills route networks. Business travelers had a choice of many connections. Direct competition between no-frills carriers and traditional airlines could no longer be denied. Furthermore, it is worth mentioning that, on some US routes, the consumer is apparently prepared to pay a premium for no-frills carriers in comparison with traditional airlines (reverse premium)[7] and that the average premium paid for established airlines relative to the no-frills airlines dropped from 25 percent in 2000 to 16 percent in 2002. This is partially explained by the timeliness and reliability of the no-frills airlines.

1.2 Unit Cost Advantages of the No-Frills Airlines

The no-frills airlines limit their product mainly to what is absolutely necessary for transportation and systematically eliminate all extras ("frills"). The no-frills product is characterized by a single class on board with tight seating, provision of food and drink only for payment, the elimination of reserved seating, no offer of flexible, that is, rebookable tickets, only point-to-point flights without operation of a hub-and-spoke system specializing in connecting flights, a uniform fleet, exclusively direct sales, etc.[8] Moreover, the no-frills airlines achieve a high rate of flight hours per day.

Within these structures, the no-frills airlines can produce at unit costs that are more than 50 percent below those of traditional airlines, with sales costs and seating productivity accounting for the largest difference (see figure 7-1).[9]

---

[5] See Simon, Kucher & Partner (2002).
[6] Lufthansa has a minority stake in Eurowings and Germanwings is a 100% subsidiary of Eurowings.
[7] See Domestic Airlines Fares Consumer Report (2002), p. 43ff.
[8] For a general description of the product features of the no-frills airlines, see Doganis (2001), pp. 126ff and Lawton (2002).
[9] See Goldman Sachs (2003), pp. 21ff; Binggeli & Pompeo (2002), p. 90.

Furthermore, airports or local municipalities occasionally offer significant subsidies in lesser deveoped regions, as the Ryanair case in Charerloi shows.

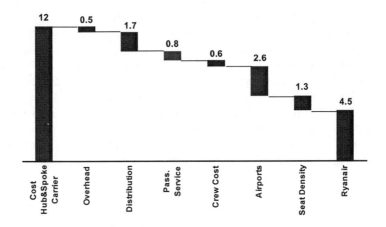

Figure 7-1: Comparing Ryanair's Average Costs per Offered Seat Kilometer in cts with Three European Hub & Spoke Carriers 2001[10]

The EU Commission deemed the financial incentives, amounting to approximately € 10 per passenger, in some cases inadmissable subsidies.[11] The average profit of Ryanair per passenger was about € 15 in 2002/2003. Furthermore, the no-frills airlines also frequently operate additional businesses at the lesser developed airports, such as joint ventures with car rental agencies. In this context, the no-frills airline acts as a regional developer, so to speak.

For the established airlines the question arises: what portion of this unit-cost difference can be eliminated without losing the network connecting short- and long-haul traffic?

1.3 Accelerated Decline in Yields in the Entire Market

Market stimulation, which is absolutely the key to the success of the no-frills airlines, can be achieved only by spectacular pricing campaigns because, ultimately, the travelers who are to be convinced to fly are those who so far have

---

[10] Source: Binggeli & Pompeo (2002), p. 90.
[11] See Baker (2004). Ryanair has appealed the decision.

not considered air travel on account of the relatively high prices, in other words, those who are not reachable through traditional channels like travel agencies with computer reservation systems. That is why, upon market entry, no-frills carriers offer and heavily advertise extremely low benchmark prices in the lower two-figure, or even one-figure, range.[12] In addition, they receive publicity, for example, when court proceedings are pending due to their unconventional business practices.

Even though only a small portion of the tickets are offered at low benchmark prices, average revenues are also significantly lower than those of traditional carriers. In the meantime, the established airlines have adapted their pricing schemes to the competition from no-frills airlines and likewise offer very low-priced tickets for early bookers. At the same time, it is interesting to note the following empirical price ceilings: Lufthansa was able to achieve considerable stimulation when it introduced its € 88 fare for a return ticket within Germany (offered since November 2002), while a € 143 ticket introduced six months earlier did not create any significant stimulation. Two-figure fares therefore send a certain signal, which no-frills carriers acknowledge. They observe a significant leveling off of demand for return fares over € 100.

One example of the drop in prices when no-frills carriers enter a market is the neighbor traffic between Germany and Great Britain (figure 7-2). Between 2002 and 2003, the no-frills carriers increased their market share from 13 percent to 36 percent, thus triggering an average 23-percent drop in the price per ticket at Lufthansa and British Midland. The market as a whole stagnated, so that the stimulation created by the no-frills carriers was merely able to compensate for the drop in business-traveler traffic. The traditional carriers (besides Lufthansa primarily British Airways) together lost 23-percent market share and additionally a quarter of their average revenues.

---

[12] Both Hapag Lloyd Express and Germanwings launched with a base price of € 19 for a one-way flight.

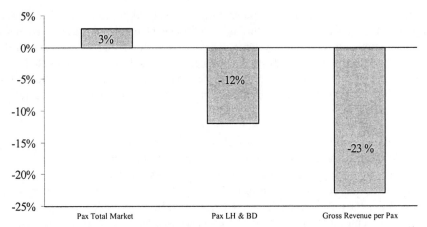

Figure 7-2: Development of Traffic between Germany and GB from 2002 to 2003[13]

Revenues fall even more steeply when various no-frills carriers compete on the same route, which, by the way, has practically never happened in the US.[14] In Europe in summer 2003, the Cologne Berlin/Tegel route was serviced by three no-frills carriers (dba, Hapag Lloyd Express, and Germanwings) in addition to the existing Lufthansa flights. The market grew by 62 percent. Lufthansa lost market share, but was able to increase the number of passengers by 50 percent by offering low-priced tickets. Average revenues across all airlines also dropped by 50 percent, though. The carrier dba lost the most, giving up its market leadership to Lufthansa. That it is difficult to achieve an operating profit in such a competitive situation is proven by the responses: Germanwings moved its services to Berlin/Schönefeld starting with the winter flight schedule and Hapag Lloyd Express selectively thinned out its connections.

One can expect a 30-percent fall in prices in the entire Lufthansa continental traffic from 2002 to 2005 due to these and similar developments – driven by changes in consumer behavior and no-frills airlines.

---

[13] BD: British Midlands, LH: Lufthansa.
[14] Here, we disregard the attempts of the established airlines to compete with the no-frills airlines by establishing their own low cost carriers.

1.4 Goals of a Strategic Repositioning

In the United States, the quality of the established airlines has significantly dropped due to the competition battle with the no-frills airlines and the triggered cost pressures.[15]

In order to counter this risk of focusing solely on costs, Lufthansa has set the following strategic goals with its Future of Continental Traffic Program:

- Lufthansa will remain a premium carrier, which sets itself apart through a high-quality product.
- Product differentiation must be strengthened.
- The Hub-and-Spoke system will remain the backbone of production, but decentralized traffic (i.e., traffic outside Frankfurt and Munich) must become profitable.
- Unit costs must be reduced by 30 percent and Lufthansa, as Europe's most punctual large airline, must expand its leadership in quality.

At the center of this strategy is the conviction that the brand premium is sufficient to profitably sell a higher level of service than that provided by the no-frills airlines. In order to achieve this goal, though, Lufthansa must first create a host of conditions regarding heavier customer focus and increased internal efficiency.

2 The Product: Clear Distinction from the No-Frills Airlines

Customer focus is expressed primarily in the question of which product to offer to the various customer groups. In past years, the airlines were loudly criticized in this connection for not offering any transparent price-service options.[16] Fully flexible tickets cost eight to 10 times more than the lowest-priced tickets. Since many customers are not aware of the value of flexibility, they explain the differing ticket prices only by the different board service. In doing so, they often

---

[15] See Taneja (2003), p. 17.
[16] See Taneja (2003), pp. 77ff.

overlook that the highest-priced ticket even within economy class is eight times more expensive than the lowest-priced ticket.

So there can be only one solution: offer price-worthy products and convey this price-worthiness to the customer in such a way that he can make his selection as an informed consumer on the basis of his specific traveling circumstances and desires, regardless of the reason for his travel. In order to increase the price-worthiness of the business-class product in particular, the board product in this class must be enhanced, especially vis-à-vis the fully flexible fares in economy class. In addition to fare conditions and the board and ground products, the choice of serviced airports is also an important product component.

2.1 Primary or Secondary Airports?

A fundamental difference between the product of traditional airlines and that of the no-frills airlines is evident in the choice of serviced airports. That is why this issue arises for a premium airline only in exceptional cases. The primary airports form the basis of business because they are located in business centers. This applies in particular to the hubs. As new market entrants, no-frills airlines are able, by contrast, to develop secondary airports and extract significant subsidies from the airport operator and local economic promotion. They thus often avoid the disadvantages of high traffic volume at the primary airports. The difference in airport fees can be significant. Ryanair pays € 4.35 per passenger[17] in Hahn, while an airline in Frankfurt has to pay roughly € 17 per passenger.[18] That notwithstanding, the choice of primary airports represents an important differentiation criterion with far-reaching consequences.[19]

2.2 Differentiation of the Product

Differentiating established airlines means not only distinguishing them from the no-frills airlines, but also differentiating within their own customer group – among those customers who are dependent on flexibility (mostly business travelers) and willing to pay and those customers who are price-conscious. Without a logical differentiation that the customer – and also airline employees –

---

[17] See Hahn airport fee regime, 1 November 2001.
[18] See Frankfurt airport fee regime, 13 November 2003.
[19] Which prices are attainable at an airport, depending on the distance to the next airport, is analyzed in Gillen & Morrison (2003), p. 21.

can understand, it will not be possible to establish an appropriate price spread on the market.

The most important product differentiation is that drawn between business and economy class and that drawn between flexible tickets, which are rebookable without a fee, and restricted tickets, which either have conditions such as a Sunday return rule or are limited in their availability, as is the case with no-frills airlines.

Lufthansa offers three product lines:

- comfort, speed on the ground, and flexibility: business class
- flexibility: fully flexible economy fares
- basic: cheap economy fares with restrictions.

In addition to this differentiation related to the specific purchased product, frequent flyers are also able to take advantage of comfort components by virtue of their status or in exchange for miles.

2.2.1 Differentiation of the On Ground Product

Comfort components on the ground include, besides lounges, above all, priority check-in with minimal waiting periods and shorter check-in and boarding pass pickup deadlines, fast lanes at the security check, priority ticketing, lounge access, and, in future, also faster separate boarding. Negotiations are underway with the authorities to introduce fast lanes also at the passport control, possibly combined with biometric passenger recognition.

The goal of differentiating the product on the ground is greater speed for customers with more valuable tickets. The flexibility of the ticket also means greater speed because it enables the ticket holder to take the next available flight.

2.2.2 Differentiation of the On Board Product

From October 1996 to March 2004, Lufthansa flew its European traffic with the following board product on the continent.

Within Germany, a two-class system, where the classes differed simply in the board service, but not in seating comfort and, on international routes, in business class, five abreast[20] with convertible seats (CVS) enabling a row to reduce from six to five seats.

Since summer 2004, Lufthansa has been flying throughout Germany and Europe with four abreast in business class, with each middle seat blocked. No other airline in Europe offers this seating comfort in business class on all continental flights. This thus clearly runs counter to the general trend of downgrading for cost reasons.[21]

Differentiating between German domestic and European routes has become dubious in the course of European integration and thus no longer is a viable product option. On European routes, the unequal treatment of passengers on both sides of the aircraft resulting from sitting five abreast (on the left, two passengers; on the right, three passengers) is being eliminated. Furthermore, and perhaps most importantly, this clears the way to introduce a new seating without convertible seats that makes tighter seating possible and thereby creates more seats per aircraft without losing comfort.

Four abreast is to be initially implemented without investments. This measure offers full flexibility in the distribution between business and economy and, at most, thereby drives out economy passengers. Yield management ensures that only the lowest-priced fares in economy will be driven out. With a share of 20-percent business class guests and one of every four flights with critical capacity (over 85 percent), two percent of all guests will be driven out, which can be compensated by adjusting business-class fares.

Thus, the new board product for business class creates the conditions for higher seating productivity and increases the attractivity and priceworthiness particularly for business-class customers.

The problem of the large price spread within economy class remains. Air France has addressed this problem by introducing a third class on international flights. The differentiation between second and third class, however, solely consists of a sandwich on shorter routes, which is not enough to explain the difference in ticket price – seating comfort is identical. This third-class product creates complexity costs and costs time in turning around the aircraft (see 3.2). Instead of differentiating on board, Lufthansa is focusing on clearly communicating the

---

[20] Five abreast: maximum of 5 passengers in a row.

[21] For example, KLM introduced a 6 abreast seating configuration in business class in October 2003, thereby discontinuing the previous 5 abreast configuration.

advantages of a flexible ticket and on transparency in terms of when these advantages make sense.[22]

3 Production and Processes: Concentration on Value-Added Activities

Apart from the goal of establishing a product assortment that is worth the price, there is also the challenge of producing this assortment profitably, thus bringing the costs of production in line as much as possible with the cost levels of the no-frills airlines. Lufthansa has set itself the goal of reducing unit costs by 30 percent. In addition, Lufthansa aims to consistently optimize productivity of its fleet, infrastructure, and personnel.[23]

3.1 Production Factor Fleet (I): Increasing Daily Capacity

A large part of the complexity involved in traditional airline processes is driven by the complexity of the aircraft fleets. This complexity has largely evolved over time as a consequence of individual decisions. Primary among them was the desire to participate in the latest technological developments and to make the airline independent on one manufacturer. Additionally, a hub-and-spoke carrier needs to cover as broad a spectrum as possible of aircraft bodies, ranging from 50 to 280 seaters.

Only after the low cost carriers introduced the uniform-fleet strategy and communal fleets (e.g. Airbus 319, 320, and 321) offering various aircraft body sizes with largely identical technology became available did the large airlines change their way of thinking. So now practically all European airlines are striving, at least as a prospect, to establish a uniform fleet. The following questions frequently arise in the discussion concerning a uniform fleet:

---

[22] It is worth mentioning that the no-frills airlines also in the meantime are considering which frills can have a differentiating effect and create competitive advantages, see Pilling (2003), p. 60.

[23] Whether this implies an entirely changed business model, as hypothesized, for example, by Price Waterhouse Coopers, or whether it represents an optimization of the existing business model remains an open question in this analysis; see PWC (2003), p. 10.

- How does one avoid reliance on one manufacturer, despite having a uniform fleet in continental traffic?

  First, apart from aircraft offering between 120 and 200 seats, the airlines still have to order long-haul types and regional aircraft. Second, airlines will bundle their aircraft demand over 10 to 15 years and call for bids in a total package request. By doing this, they achieve more favorable purchase prices rather than by traditionally playing two providers off against each other with smaller orders. So then all providers get a chance in these major calls for bids. Those providers who currently are not represented in the fleet must then assume the costs of retraining and other adjustments – a disadvantage which, in turn, is balanced out by the large volume and the advantage of the call for bids that follows.

- How do the airlines counter the risk of the entire short-haul fleet being grounded due to a technical problem?

  This has not happened with the two major manufacturers, Boeing and Airbus, in the past 40 years; however, it makes sense to hedge against this risk with the manufacturers. Since it is purely a manufacturer problem, it is also possible to do so.

An important goal in harmonizing a fleet is to standardize the training of the cockpit personnel. Since the Airbus A320 family and the Boeing 737 family each cover only a spectrum of 120 to a maximum of 200 seats, standardized training can be achieved only by doing without the 250 to 280 seaters. In the meantime, only BA, KLM, and Lufthansa operate aircraft of this magnitude (4, 1, and 10, respectively) in continental traffic. These aircraft sizes make sense during periods of peak demand at airports with limited slots. It seems hardly possible, though, to compensate for the complexity costs and higher operating costs of this type through additional revenues with only these flights. Lufthansa has therefore taken the fundamental decision to decommission the Airbus A300-600 (280 seats), to retire the Boeing 737 fleet, and thus to operate a uniform fleet of the Airbus 320 family, at the latest by the time the new runway is placed into operation at FRA.

3.2 Production Factor Fleet (II): Shorter Ground Times

A major difference between no-frills carriers and established airlines lies in the duration of a ground event. While no-frills carriers for the most part get by with 25 minutes of turnaround time, the turnaround time at a large European hub is often double. The reasons for this difference are:

- A simpler product: the elimination of preassignes seating accelerates boarding by five to 10 minutes;[24] limited catering on board and the readiness of the crews to collect trash during the flight likewise reduce the time required for cleaning on the ground by five to 10 minutes.
- Simpler airport infrastructure: due to uncertainties regarding taxi times and other delays, as well as those concerning the time it takes to taxi to the gate, large airlines plan buffer time into their ground times to stabilize operation.

It is difficult to prove to what extent better communication among ground handlers plays a role.[25] Lufthansa has introduced a system at its hubs with which the time frames for all service providers can be recorded separately and automatically. This system enables separate monitoring of all suppliers, irrespective of the prior service provided by other suppliers.

At the decentralized stations, Lufthansa introduced 25 minutes of ground time (30 minutes when the crew is changed) across-the-board for decentralized German domestic services in summer 2004, thus attaining the level of the no-frills carriers. At the Frankfurt hub, the minimum ground time was shortened to 40 minutes. Overall, these measures increase aircraft productivity by 10 percent.

Besides minimum ground times, the ground times actually planned, of course, also play an important role because only about 20 percent of all ground events can be planned into the minimum ground time due to the traffic waves at the hub. Thus, the difference between minimum ground time and planned ground time serves as a further buffer. Lufthansa has developed a simulation tool to calculate the punctuality for a given flight schedule based on the actual linking of ground events. This tool serves to plan buffers into the ground times as well as to plan reserves in such a way that maximizes their effectiveness in increasing ontime performance.

3.3 Production Factor Infrastructure (I): Hub-and-Spoke versus Point-to-Point

Hubs serve to bundle flights. Lufthansa services approximately 10,000 origin-and-destination pairs (O&Ds). One hundred and twenty of these account for 100,000 passengers per year in traffic volume and are thus sufficient for a direct

---

[24] Because passengers no longer have to look for their seats and do not get ahead of each other and the plane fills up first from the window seats.
[25] See Gittel (2003), pp. 22ff.

flight at least in double "day-trip" connections: leaving mornings and evenings in both directions with 100 – 120 seaters. With the current passenger volume, the remaining 99 percent of all O&Ds are too small to be serviced directly. This conventional way of thinking ignores the following factors:

- The market entry of a no-frills carrier can significantly stimulate the market, though with considerable price discounts.
- The data for flight planning are based on the analyses of computer reservation systems, while the data of no-frills carriers, which, of course, do not use CRS, can only be maintained manually. This leads to an underestimation of traffic volume.
- Customers are prepared to pay more for a direct flight than for a connecting flight. Ultimately, direct service represents the superior product in terms of speed, comfort, and reliability.
- Customers will drive a considerable distance for a direct flight, particularly when they represent price-induced demand. So, an airport's traditional catchment area has to be redefined for these cases.

But even when taking these factors into consideration, one still cannot conclude that the hub-and-spoke system is obsolete. Even though the doctrine originating in the US in the nineties, when the development of a fortress hub was seen as a guarantor of higher revenues, is no longer valid, the bundling function is still indispensable, especially for long-haul traffic, despite the strategic vulnerability of the hubs caused by direct flights.[26] This strategic vulnerability in hub traffic should be countered by development of a low cost business model in decentralized traffic because any withdrawal from decentralized traffic is an invitation for the no-frills carriers to service this route.

3.4 Production Factor Infrastructure (II): Depeaking

The formula for success of the hub of the nineties was to make arrival and departure waves as high as possible – in an effort to minimize the connecting times of as many connections as possible and thus to maximize revenues because the flights in the computer reservation system are sorted by total travel time. This formula must be modified for the following reasons:

---

[26] See Morrison & Winston (1995), pp. 44ff; Butler & Huston (1999).

- Minimizing connecting time reduced total travel time at the relevant O&D because it seemed the flight times themselves, for the most part, could not be influenced. The computer reservation system organizes all the connections of an O&D by total travel time, and a top spot on the screen raises the sale chances. Now, a significant portion of customers book via the Internet and use the best-buy function, which no longer organizes connections by total travel time but by price. Corporate clients also use sophisticated control mechanisms on the basis of rates contracted with the airlines and thus also do not organize the connections by total travel time. Even though there is still no scientific research on the remaining preference effect of total travel time, it seems reasonable to assume that it was significantly overestimated and plays a role less as a preference categorizing function and more as a threshold function. All connections with a total travel time below x hours are considered and those below them will be selected by price, comfort, etc.

- High arrival and departure waves are counterproductive if the infrastructure of the hub airport cannot handle these flights without delays.[27] In Frankfurt, for example, the number of maximum landings (arrival benchmark figure) was raised in three phases from 37 to 39 to 41 to 43 from 1996 to 1999. What initially was planned as a logical development of hub functionality proved, however, to be counterproductive. The summer of 1999 was the least punctual in the history of Lufthansa because air traffic control was able to handle the 43 landings in the peak period only with considerable flight delays. The following year, this shortcoming was mitigated by extending the planned block times by approximately 10 minutes for all flights to Frankfurt, thereby respectively lengthening the total travel time of all connections to Frankfurt. No serious changes in market shares could be detected, which would seem to suggest that total travel time is of secondary importance. However, these extra 10 minutes of block time considerably tie up resources.

- The wave structure not only reduces infrastructure productivity, but also produces additional problems in mastering the traffic flow due to the bunching effect, that is the statistical pile up of flights. Although during peak times, merely 43 slots are allocated for landings per hour, over 50 flights regularly enter Frankfurt airspace in the peak hour because even small statistical fluctuations in departure time, as are commonplace in air traffic, cause a pile up. These pile ups would be significantly smaller if traffic occurred in an even flow, without traffic waves.

---

[27] See Mayer & Sinai (2003), pp. 1196ff.

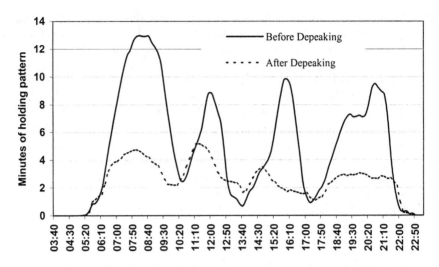

Figure 7-3: Holding Delay Before and After Depeaking in Frankfurt

Four years of various, quite successful, optimization measures by Deutsche Flugsicherung (DFS), Fraport, and Lufthansa (for example, by shortening the separation between flights on final approach) helped to mitigate, but not entirely alleviate the situation in Frankfurt. So it was decided, starting in summer 2004, to reduce the benchmark figures[28] for landings from 43 to 41 in the morning and 45 to 42 in the afternoon. At the same time, the benchmark figures for takeoffs were reduced from 48 to 43 and 44, respectively. By reducing holding pattern times, the flight time for all flights to Frankfurt falls by 7 minutes (see illustration 3) and infrastructure productivity rises by 5 to 10 percent because the peak coverage of the ground-handling service providers can be reduced by this same figure. This development thus compensates for the disadvantages of the last increase in the benchmark figure. In addition, the total number of movements could be increased from 76 at the start of 1999 to 80 and 81, respectively, as a result of the optimization measures which have since taken place. The shortening of block times compensates for the lengthening of some connecting times. From the point of view of the passenger, depeaking thus shifts flight time to ground time while simultaneously reducing total travel time.

Frankfurt is thus the first airport worldwide to reduce its benchmark figures for takeoffs and landings in order to make depeaking possible. American Airlines

---

[28] The benchmark figure gives the maximum number of permissible movements per hour.

significantly reduced its waves of arrivals and departures at Dallas Fort Worth. However, Dallas is not coordinated through slot allocation, as most European airports are, which is why Delta was able to fill the former peaks of American.[29] This defeated the effort to attain the desired even utilization of airspace and of the runway system. Nevertheless, American was thereby able to achieve a clearly more even utilization of ground infrastructure.

3.5 Production Factor Personnel

A uniform fleet and even utilization of infrastructure are important preconditions for raising employee productivity. A uniform fleet raises employee productivity in the following ways:

- Less expense for training (only 1 aircraft type instead of 3) and less expense for retraining, in particular for cockpit personnel, thus also savings in training facilities such as simulators and greater flexibility in capacity planning of the cockpit personnel.
- Simpler planning and control. This applies to rotation planning, crew planning and control, and better swapping possibilities during daily operation, for example, when an aircraft drops out of service.
- Higher technical productivity and higher productivity in preparing operations manuals. The costs of training technicians can be reduced and maintenance processes can be streamlined. It even becomes substantially easier to stock and provide replacement parts.
- The entire ground handling is simplified at the airports – from the creation of freight lists to the stowage plan for the catering.
- Planning and control processes can be considerably simplified because the final aircraft type does not have to be determined until a few days before operations.

Similar effects – albeit of a smaller magnitude – result as a consequence of depeaking. Peak coverage can be reduced by 5 to 10 percent in the entire ground handling, thereby markedly increasing capacity.

In order to further increase productivity on board, the inflight product was optimized as of summer 2004 so that all flights under 185 minutes of block time

---

[29] See Goedeking & Sala (2003).

can be serviced with the legally prescribed minimum crew in the cabin (this includes 96 percent of all Lufthansa flights in European traffic). Thus, it is no longer necessary to plan additional flight attendants who in the past were deployed in addition to the minimum crew. This step was able to lower complexity costs and thus further reduces the gap between Lufthansa and the low cost carriers.

4 Summary and Outlook

Apart from focusing on costs, the differentiation of customer groups is a central task in restructuring a network airline. This differentiation is achieved by dividing them into three fare types (basic, flexible economy, and business). Product elements on board and on the ground, each of which must be priceworthy, are assigned to each fare type.

On the basis of this clear product classification, costs must be consistently reduced by simplifying the fleet types (with the goal of a uniform fleet), by increasing seating capacity, by filling daily aircraft capacity, by accelerating ground processes, and by reducing traffic peaks at hubs (depeaking).

The described measures alone will not achieve the desired 30-percent reduction in unit costs. Parallel to working out process changes, negotiations are being conducted with the unions to reduce personnel costs. Lufthansa is likewise negotiating with all suppliers on savings that extend beyond those already outlined, e.g. through depeaking. Since air traffic still occurs within a heavily regulated environment, Lufthansa is also working together with political officials within the framework of the "Air Traffic for Germany Initiative" to lower air traffic costs in Germany.

On the sales side, travel agencies are evolving from trade representatives into service providers in Germany through the new commission model. This model eliminates the previously paid commission and replaces it with a service fee, which is paid to the travel agent. This step also eliminates a cost block which the low cost carriers, who specialize in direct sales, have not even introduced.

The coming years are expected to see further vigorous expansion by the no-frills airlines, in combination with a subsequent consolidation of the airlines. After the merger between Air France and KLM, consolidation will continue also among the traditional airlines, leaving only three or four large airlines in Europe by the end of the decade. The critical challenge will be to maintain a low cost structure when two airlines merge.

5 References

AIRPORT FEE REGIME HAHN (2001): 1 November.

AIRPORT FEE REGIME FRANKFURT (2003): 13 November.

BAKER (2004): Playing by the Rules. *Airlines Business*, 20(3), pp. 30-31.

BINGGELI, U.; POMPEO, L., (2002): Hyped Hopes for Europe's Low cost Airlines. *McKinsey Quarterly*. (4), pp. 87-97.

BUTLER, R.V.; HUSTON, J.H. (1999): The Meaning of Size: Output? Scope? Capacity? The Case of Airline Hubs. *Review of Industrial Organization*. 14(1), pp. 51-64.

DOGANIS, R. (2001): *The Airline Business in the 21st Century*. London, New York: Routledge.

DOMESTIC AIRLINE FARES CONSUMER REPORT (2002), 4.Q, Department of Transportation, at: http://ostpxweb.dot.gov/aviation/domfares/web024.pdf.

GILLEN, D.; MORRISON, W. (2003): Bundling, Integration and the Delivered Price of Air Travel: Are Low Cost Carriers Full Service Competitors? *Journal of Air Transport Management*. 9(1), pp. 15-23.

GITTELL, J.H. (2003): *The Southwest Airlines Way*. New York: McGraw Hill.

GOEDEKING, P.; SALA, S. (2003): Breaking the Bank. *Airline Business*, 19(9), pp. 93-97.

GOLDMAN SACHS (2003): *Airlines Europe: Estimate Cuts and Structural Problems*. Analyst report, April.

LAWTON, T.C. (2002): *Cleared for Take-off*. Aldershot, Burlington: Ashgate.

MAYER, CH.; SINAI, T. (2003): Network Effects, Congestion Externalities, and Air Traffic Delays: Or Why Not All Delays Are Evil. *The American Economic Review*, 93(4), September, pp. 1192-1215.

MORRISON, S.A.; WINSTON, C. (1995): *The Evolution of Airline Industry*. Washington D.C.: Brookings Institution.

PILLING, M. (2003) Price Promise. *Airline Busines*, 19(4) April, pp. 58-60.

PRICE WATERHOUSE COOPERS (2003): *Destination 2007: The One-Off Opportunity for Change in the Airline Industry.* Analyst report.

SIMON, KUCHER & PARTNER (2002): *Passagierbefragung Billigflieger in Deutschland.* Unpublished.

TANEJA, N.K. (2003): *Airline Survival Kit.* Ashgate, Burlington.

8

COMPETITIVE DYNAMICS THEORY – APPLICATION TO AND IMPLICATIONS FOR THE EUROPEAN AVIATION MARKET

CAROLINE HEUERMANN

1 Introduction 186
2 A Model of Competitive Interaction 188
3 Multimarket Contact and Intensity of Competition 192
4 Competitive Development in the European Aviation Market 199
5 Analyzing the Market from a Competitive Dynamics Perspective 206
6 Implications of Multimarket Competitive Interaction 211
7 Conclusion and Outlook 215
8 References 216

Summary:
In view of the ever increasing competition and the expected consolidation phase in the European aviation market, airlines increasingly display highly aggressive competitive moves and countermoves. The resulting dynamic interaction extends over multiple markets, since airlines vie for passengers simultaneously on similar routes or in the same regions. By analyzing the European aviation industry from a multimarket competitive dynamics perspective, this article provides first insights concerning (1) the future development of the industry and (2) develops potential implications for airline management. Two main strategic approaches to outlive the competitive battle are identified and discussed.

1 Introduction

Since the introduction of the three liberalization packages in 1987, 1990, and 1992, leading to full cabotage in 1997, the European aviation market has undergone fundamental changes.[1] The market entrance of a large number of newly founded airlines, primarily low cost carriers, and their increasing prominence during the last few years have profoundly altered the competitive landscape. The European aviation market has therefore entered a stage of intensified competition, which will inevitably lead to market consolidation throughout the next months and years.

Competitive dynamics arise from the sequence of rivals' actions and reactions in an industry. Consequently, competitive dynamics research is concerned with the analysis of strategic moves and countermoves and their impact on competitive advantage and firm performance.[2] By initiating competitive actions, firms wish to creatively strive for new competitive advantages or try to defend their competitive position against the rival's attack. Firms may make use of a wide range of competitive actions, be it more tactical ones, like new marketing activities or price variations, or more strategic ones, like new product introduction or even the entry into or withdrawal from a certain market. The pattern of behavior resulting from this range of firm actions can be quite adversarial and aggressive, but also more smooth and consensual, depending on various factors. Since the success or failure of a firm heavily depends on the responses or nonresponses of its competitors due to competitive interaction, analyzing and understanding the determinants of competitive dynamics is of essential importance for the market players involved.[3]

In the airline industry, such competitive interaction is traditionally of high relevance, which can be illustrated with cases from the US airline market as early as the beginning of the 1990s.[4] Recent developments in the European market show that the competitive game is becoming popular here as well, especially between low cost airlines: Whereas they hitherto tended to avoid direct route competition, operating from different hubs and thereby concentrating on

---

[1] See e.g. Lawton (2002), p. 19.
[2] See for an overview Smith, Ferrier & Ndofor (2001), pp. 315ff.
[3] See Baum & Korn (1999), pp. 257 and 272.
[4] See Frankfurter Allgemeine Zeitung (2003), p. 22; D'Aveni (2002), pp. 41ff.

different routes and regional markets, carriers are now beginning to encroach upon formerly untouched markets in order to fill their growing capacity.[5] Recent examples of competitive moves are numerous: While Ryanair announces its expansion at London Luton, Easyjet's home base, Easyjet is on its part opening two bases outside its home market, in Germany (Berlin, Dortmund) and thereby directly attacking Air Berlin. Moreover, Easyjet is starting operations in Cologne/Bonn, which is the home airport of Germanwings and an important base of Hapag-Lloyd Express in serving its destinations. It remains to be seen what measures the incumbents can take in order to strike back. These first examples point to the fact that competitive dynamic interaction is becoming highly relevant for European airlines' market success and thus requires a more profound examination. Furthermore, they also imply the significance of multimarket competition: Airlines not only compete in one market, but simultaneously in multiple markets in terms of certain routes and/or regions. Therefore, as illustrated by Easyjet in the above situation, competitive interactions are not necessarily limited to one market but can extend over several markets within the industry.

In view of this development, it seems useful to analyze the European aviation industry from a multimarket competitive dynamics perspective by addressing the following questions:

- *Explanation of competitive behavior*: How can the behavior of European airlines delineated above, as well as the overall increasing competition, especially among low cost airlines, be explained? Which factors influence multimarket competitive dynamics in this industry?

- *Prediction of competitive behavior:* Based on market analysis, is it possible to draw any conclusions concerning further competitive development?

- *Management implications*: To what extent does the analysis of the European airline market from the competitive dynamics perspective, provide insights for the strategic management of airline companies? How can carriers actively and deliberately influence their rivals' competitive actions and reactions?

In order to answer these questions, this chapter proceeds as follows: As a theoretical basis, a general framework of competitive interaction containing the major components involved is described. A thorough elaboration of multimarket contact and its implications for the nature of competitive moves and countermoves follows. Having set up this theoretical fundament, the current

---

[5] See Baker (2004a), p. 20.

development and the emerging competitive interactions in the European aviation market are described. Subsequently, the theoretical and empirical findings are combined to allow for a detailed market analysis from a competitive dynamics perspective and to gain first insights concerning the research questions raised above. A brief conclusion and an outlook finalize this article.

2 A Model of Competitive Interaction

In order to understand competition in a dynamic market a competitive dynamics perspective appears instructive by concentrating on those aspects or components that fundamentally shape and influence the nature of competitive interaction. Generally, several components can be identified that play an essential role in rivals' interactions: (1) The firm initiating a competitive action, (2) the characteristics of this action as well as (3) the reacting firm and (4) the attributes of its reaction; all these components are (5) embedded in their respective industry competitive environment.[6] Figure 8-1 visualizes these general components and their interplay which eventually influence organizational performance.

Industry competitive environment

- The actor
  - Awareness
  - Motivation
  - Ability

- The responder
  - Awareness
  - Motivation
  - Ability

- The action
  - Magnitude
  - Scope
  - Type
  - ...

- The response
  - Action criteria
  - Likelihood
  - Frequency, timing
  - ...

Organizational Performance

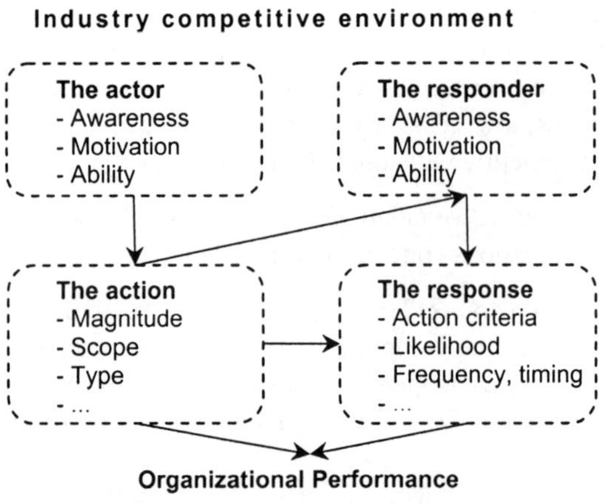

Figure 8-1: A Competitive Dynamics Model[7]

---

[6] See Smith, Ferrier & Ndofor (2001), p. 320ff.; Smith, Grimm & Gannon (1992), p. 19ff.

[7] See also Smith, Ferrier & Ndofor (2001), p. 319 and Smith, Grimm & Gannon (1992), p. 19.

The concrete result of competitive interaction will depend on a variety of different factors which it is impossible to enumerate here. Nevertheless, in order to convey a general perception of the complex conditions for interaction, some fundamental characteristics of the (re)actor, the (re)action, and the competitive environment will be considered in more detail.

2.1 The Actor and the Responder

The firm inducing a competitive action and at the same time being the beneficiary of its positive or negative outcome is the actor in competitive dynamics research.[8] There are mainly three, frequently mentioned organizational aspects which have an effect on competitive action and which analogically apply to the responder: awareness, motivation, and ability.[9]

The (re)actor's *awareness* of the competitive interdependence and the managerial opportunities arising from it influence to what extent the company realizes competitive interaction at all and whether it understands how performance could be influenced in a favorable way by engaging in competitive interaction. In this context, the organizational age and top management team's experience with similar situations in the past, affect awareness to a great extent.

The incentives and therefore the *motivation* to induce or respond to a competitive action depend on the expected future outcome of that action. The motivation to act may be influenced by the past competitive behavior of a firm and the effected gains of previous competitive action. Motivation will also be influenced by the degree of market dependence: A firm which is highly dependent on a certain market will vie for its market position more vigorously.

Finally, the *ability* to take or reply to a competitive action is dependent upon factors like the organizational slack (in terms of resources that serve as buffer, whilst giving the firm managerial flexibility to respond to competitive or environmental changes) and the organizational size (as an approximation for the ability to efficiently influence market environment and competitors as well as a measure for the amount of slack resources). Moreover, a firm's reputation regarding previous competitive behavior plays a major role as well as its overall strategic orientation.[10]

---

[8] See Smith, Ferrier & Ndofor (2001), p. 320.
[9] See Chen (1996), p. 105.
[10] See Smith, Ferrier & Ndofor (2001), pp. 320 and 327ff.

2.2 Action and Reaction Characteristics

The actor's and also the responder's organizational characteristics shape their competitive actions, the outcome of which has an effect on organizational performance. A competitive action can be defined "as a specific and detectable competitive move [...] initiated by a firm to improve or defend its relative competitive position."[11] It can be characterized by its magnitude, scope, and type. While *magnitude* is referred to as the amount of resources necessary to carry out the action, the *scope*, as the pervasiveness of a competitive action, can be measured in terms of number of competitors potentially affected.[12] Furthermore, the *type* of action gears to the fact that competitive moves can take place on distinct levels. Whereas strategic moves (e.g. the introduction of a new product line or the entry into a thoroughly new market) usually require a more significant amount of resources, involve a major redefinition of business and therefore primarily require top management attention, tactical moves (e.g. promotions or service improvements) require fewer resources due to relatively minor changes.[13] Competitive actions occurring on an operational level may be reflected by e.g. short-term price cuts or increases which mostly lie in the lower management's field of responsibility. Obviously, the competitive move induced by the actor is the condition for competitive response, but may also affect the responder's organizational characteristics depicted in the previous section.

The action criteria mentioned above equally apply to competitive responses, which actually reflect the dynamic aspect in competitive interaction. In addition, response likelihood, frequency and timing are of further interest. Firms unlikely to respond to its competitors' actions are characterized by a low response *likelihood*.[14] Moreover, the *frequency* reflects the total number of responses in a given period of time, whereas *timing* refers to the response lag or delay as the elapsed time between the competitive action and response.[15] It is assumed that firms prefer to take action against competitors whose response likelihood and frequency are low and response delay is relatively large (since quick response annunciates the willingness to fight aggressively for market positions).[16]

---

[11] Chen, Smith & Grimm (1992), p. 440.
[12] See Chen, Smith & Grimm (1992), p. 443.
[13] See Chen, Smith & Grimm (1992), p. 445.
[14] See Smith et al. (1991), p. 62.
[15] See Smith et al. (1989), p. 246.
[16] See Smith et al. (1989), p. 247 and Smith et al. (1991), p. 62.

The following example illustrates the characteristics of competitive actions and responses. The supermarket chains A and B operate in the highly competitive grocery industry, but with a different regional focus. A is now attacking B by expanding its retail network into B's home market and by establishing additional stores. This competitive action can be considered as high in *magnitude* as the building and opening of new branches requires considerable financial investments and a high amount of staff resources for planning and running the stores. Since the attack is mainly directed at a single competitor (B) the *scope* of the action can be regarded as relatively small. Furthermore, the action is strategic in *type* because A is entering a new regional market and thereby altering its long-term business portfolio. In view of this substantial attack, the *likelihood* of B's competitive response can be assessed as high. In our example, B reacts within a short period of time and initiates regular and extensive price promotions in the respective market. B's responses thus appear in a relatively high *frequency*, and the response lag (*timing*) is short. Moreover, according to the above mentioned action criteria, B's response can be regarded low in *magnitude* (as only little resources are necessary to carry out price cuts), small in *scope* (as the price cuts are primarily directed at A), and more tactical in *type* (as it entails only minor and temporary changes).

2.3 Industry Competitive Environment

Competitive actions and reactions occur in the context of the existing industry structure and characteristics. Among the factors considerably affecting competitive interaction are industry growth, market concentration, and barriers to market entry and exit. Growing demand will alleviate the necessity to vie for market positions and engage in aggressive moves and countermoves. Similarly, a higher market concentration entails a higher potential of oligopolistic collusion, thus leading to reduced competitive interaction. High barriers to entry and low barriers to exit may have a similar effect since competition does not increase due to newcomers and easy market exit.[17]

We thus conclude that the nature of competitive action and reaction and the aggressiveness that firms show when competing for market shares is significantly influenced by the characteristics of the actor and the responder as well as the industry specific competitive environment. In the competitive dynamics literature, one market attribute is considered to have major influence on the intensity of competition shaped by competitive interaction: multimarket

---

[17] See Smith, Ferrier & Ndofor (2001), pp. 324 and 334f.

contact. The fact that firms compete simultaneously in multiple markets on the one hand can be regarded as a result of their competitive behavior in terms of previous market entries; on the other hand it can be argued that multimarket contact, through the nature of the industry, arises because of the necessity to compete in various markets (e.g. due to the existence of considerable scale or scope economies). This seems to perfectly apply to the aviation industry, rendering multimarket interaction a significant issue with regard to airline competition. Accordingly, multimarket contact and its effect on the intensity of rivalry will be examined in the following section.

3 Multimarket Contact and Intensity of Competition

In order to explain the intensity of competition and the speed and scope of competitive (counter)moves of firms in an industry, the theory of multimarket competition, emanating from the realm of competitive dynamics research, refers to the degree of overlap of firms' competitive market domains. As most industries consist of more than one "market",[18] firms operating in such industries regularly "compete against each other simultaneously in several markets".[19] In literature, this is usually referred to as multimarket contact.[20] Real situations of multimarket competition can be found numerously in most industries, such as airlines,[21] banks,[22] or hotel chains.[23]

Competing on multiple markets has two major implications which are relevant for the dynamics of competitive action and reaction. Firstly, companies have to allocate available resources, e.g. technologies and competencies, to their respective markets. Secondly, the fact that multiple points of contact exist

---

[18] In this context, a market can be regarded as "the lowest geographic unit that comes under the jurisdiction of a manager who has decision-making authority over any of the competitive strategy variables [...]" (Jayachandran, Gimeno & Varadarajan (1999), p. 50). A certain sales area or distribution territory which is managed separately with regard to competitive actions and reactions could serve as an example.

[19] Karnani & Wernerfelt (1985), p. 87.

[20] See e.g. Baum & Korn (1996); Gimeno & Woo (1996); Haveman & Nonnemaker (2000); Karnani & Wernerfelt (1985); Young et al. (2000).

[21] See e.g. Evans & Kessides (1994); Baum & Korn (1996) and (1999); Gimeno (1999); Gimeno & Woo (1996).

[22] See e.g. Heggestad & Rhoades (1978) and Rhoades & Heggestad (1985).

[23] See e.g. Fernandez & Marín (1998).

between rivals, entails the potential to retaliate not only in the challenged market, but also in other markets which are possibly even more important to the rival.[24]

3.1 Concept of Multimarket Competition and the Mutual Forbearance Hypothesis

The central proposition of the theory of multimarket competition, first introduced by Edwards as the so-called *mutual forbearance hypothesis*,[25] states that with an increase in multimarket contact between rivalrous firms, the intensity of competition among them will decrease.[26] Mutual forbearance can be regarded as a situation of tacit collusion, where companies elude fierce competition by implicitly coordinating their activities in order to uniformly achieve higher performance. Consequently, firms that have only few points of contact in their market domains have a lower potential for mutual forbearance and will thus compete more aggressively with each other in these few markets.[27]

According to the literature, the rationale behind the mutual forbearance phenomenon is twofold. Based on the existence of multimarket contact between a focal firm and its rivals, the main constructs that shape the degree of mutual forbearance are *familiarity* and *deterrence*.

3.1.1 Familiarity

As the number of firms' market domain overlaps increases in multimarket competition, so do the potential starting points for a competitive or cooperative relationship. Therefore, firms with multimarket contact will probably be engaged in a larger amount of interactions. In this situation, familiarity, i.e. the extent to which rivals recognize the interdependence of their competitive behavior, is enhanced for two reasons.[28] Firstly, firms may develop a better understanding of

---

[24] See Smith, Ferrier & Ndofor (2001), p. 329. The aggressiveness of firms' behavior can thereby be deducted from relevant characteristics of the actions occurring, like the magnitude or scope of competitive moves (see section 2.2 on pp. 190f.).
[25] See Edwards (1955).
[26] See Baum & Korn (1996) and (1999); Gimeno & Woo (1996); Jayachandran, Gimeno & Varadarajan (1999); Karnani & Wernerfelt (1985).
[27] See Smith, Ferrier & Ndofor (2001), p. 329.
[28] For an extensive discussion of the role of familiarity in the interfirm context see Gulati (1995).

their interdependence caused by overlapping market domains.[29] This mutual understanding is nurtured through continuously collecting and analyzing relevant information about the individual strengths and weaknesses as well as those of the competitor. Secondly, following from the fact that firms simultaneously compete in multiple markets, the focus is automatically directed to these most important rivals. Firms pay greater attention to each other, thus exhibiting more accuracy when observing competitors' moves and collecting related information. Altogether, firms jointly pass through a range of competitive situations; they become more familiar with their competitive strategies, capabilities and interactions. This gradually emerging, common competitive history finally creates a higher degree of familiarity, entailing a better calculability of the rivals' future behavior and mutual forbearance.[30,31]

However, familiarity alone will not restrain powerful firms from engaging in aggressive competition and is not a sufficient condition for mutual forbearance to arise. In addition, firms must also have the potential to deter each other from rivalrous actions.

3.1.2 Deterrence

Simultaneous competition in multiple markets results in a larger overlap of market domains and a wider range of retaliation opportunities against competitive attacks. If challenged in one market, a firm could counterattack in a remote market where the potential damages, i.e. financial losses, to the attacker are even bigger. Competing firms will refrain from such aggressive behavior towards each other, only when they believe that the anticipated losses from such retaliation will be higher than the potential gains from aggressive competition. If firms succeed in commonly establishing such a belief, mutual deterrence is in place.[32] In this case, present behavior is linked to and influenced by expectations about future actions, denoting a 'shadow of the future'.[33]

---

[29] See Baum & Korn (1999), p. 253.
[30] See Baum & Korn (1999), pp. 253f.; Jayachandran, Gimeno & Varadarajan (1999), p. 51; Warren (1972), p. 28.
[31] Whereas according to this logic, multimarket contact is likely to encourage effective coordination among competitors, it may to some extent also be hindering, because the task of coordinating competitive behavior becomes more complex. See Evans & Kessides (1994), p. 342.
[32] See Gimeno (1999), pp. 101f.; Jayachandran, Gimeno & Varadarajan (1999), pp. 51f.
[33] See Axelrod (1984), p. 174; Parkhe (1993), pp. 799f.

As a precondition for effective deterrence, firms hold credible threats that they will retaliate against each other. This implies not only that one party is able to hurt the combatant, but also that it commits itself to compellingly do so if an attack occurs. As Schelling concisely remarks: "As a rule, one must threaten that he *will* act, not that he *may* act, if the threat fails. To say that one *may* act is to say that one *may not*, and to say this is to confess that one has kept the power of decision – that one is not committed."[34] Accordingly, firms engaged in multimarket competition will effectively deter each other from aggressive competitive behavior only when all of them are absolutely sure that severe retaliation will occur. Moreover, reciprocal deterrence involves not only a balance as regards the ability and opportunity of retaliation, but also a stability of that balance in order to keep deterrence operating.[35,36]

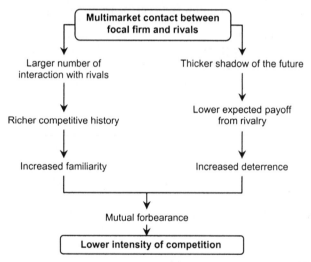

Figure 8-2: **Multimarket Contact and Competitive Intensity**[37]

---

[34] Schelling (1980), p. 187.
[35] See Schelling (1980), p. 232.
[36] In this context, it has to be mentioned that deterrence may not only be seen as a given consequence of existing multimarket contact (static view). Multimarket contact may also be actively and deliberately set up by managers in order to create a situation rendering deterrence possible. See Stephan et al. (2003), p. 405. Hence, a kind of *multimarket mentality* has to evolve. This implies that firm's endeavor to increase multimarket contact may occur "either in spite of the deterrent effect present as a function of the existing level of multimarket contact or because the deterrent is not yet strong enough to be effective". Stephan et al. (2003), p. 405.
[37] Adapted from Jayachandran, Gimeno & Varadarajan (1999), p. 53.

Although the important role of deterrence in the multimarket context is widely acknowledged, it may also be doubted as a necessary condition for the occurrence of mutual forbearance: Some authors argue that a higher risk of retaliation and severe punishment resulting from multimarket contact may, on the other hand, account also for the chance to gain a higher profit and turn out to be the winner of the competitive game, while suspending the deterrence mechanism.[38]

In summary, the literature adduces two main rationales for mutual forbearance to appear in a multimarket context, one of which relates to rival's prior interactions (familiarity), whereas the other refers to (potential) future actions retroacting on current behavior (deterrence). Figure 8-2 illustrates the fundamental coherence between multimarket contact and mutual forbearance based on the concepts of familiarity and deterrence resulting in lower intensity of competition.

Based on this theoretical framework, a range of factors facilitating or hindering the occurrence of mutual forbearance in a given market constellation and thus moderating the relationship between multimarket contact and the intensity of competition can be identified. In the next section, we will address the most important of the contingencies of multimarket competition.

3.2 Contingencies of Multimarket Competition

Drawing upon the underlying logic of the mutual forbearance hypothesis, the relationship between multimarket contact and intensity of competition can be moderated by factors either enhancing familiarity among firms or affecting their ability to deter each other.[39] In general, all factors that are relevant when analyzing competitive interactions (see section two) are expected to have implications for the multimarket contact/rivalry relationship as well. However, in connection with multimarket competition, five contingencies are mainly mentioned in the literature: spheres of influence, similarity of competing firms, their organizational structure, market concentration, and industry demand (see also figure 8-3), which are subsequently assessed.

The concept of *spheres of influence* refers to a situation in which the dominance of a firm that is competing on multiple markets varies from market to market, i.e. multimarket firms take different positions in their overlapping markets. In this case, a firm will refrain from fierce competition in the market(s) it dominates (its

---

[38] See Baum & Korn (1999), p. 253; Philips & Mason (1992), p. 396.
[39] See Jayachandran, Gimeno & Varadarajan (1999), p. 53.

so-called 'spheres of influence') because of the higher profits or market share to be at stake. In order to stabilize the market situation and prevent rivals from retaliating in its most important market(s) the firm will also abstain from attacking competitors in their key markets.[40] The existence of different spheres of influence therefore leads to a constellation "where particular firms are granted primacy in certain markets in exchange for similar status being given to their multipoint rivals in others".[41] As a consequence, mutual forbearance may be encouraged since deterrence effects intensify.[42]

In general, it can be argued that the higher the competing firms' *similarity* is concerning dimensions such as size, cost structure, or corporate culture, the easier the coordination, leading to reduced rivalry between them, will be.[43] In the multimarket context, the similarity of competing firms is usually referred to in terms of the similarity of their resource endowments, i.e. available tangible and intangible resources enabling them to successfully compete in a market.[44] Based on their individual resource pool, firms strive for competitive advantage vis-à-vis their competitors by developing competitive strengths.[45] Therefore, resource similarity of competing firms normally results in comparable competitive strengths and weaknesses and, thus, strategic similarity, too.[46] This similarity of competing firms may facilitate their ability to understand each other's strategies and competitive behavior and hence increase familiarity and the tendency to mutually forbear.[47] Moreover, resource similarity could also render the competitive threats to deter each other from aggressive behavior more credible, as firms can reciprocally assess their retaliation measures more precisely, promoting the aforementioned effect. The negative interrelation between resource similarity and rivalry has also been supported by empirical findings in the US software industry.[48]

---

[40] See Baum & Korn (1996), p. 265; Bernheim & Whinston (1990), p. 11; Gimeno (1999), p. 102; McGrath, Chen & MacMillan (1998), p. 725.
[41] Stephan et al. (2003), p. 406.
[42] It can be asked, though, whether spheres of influence are merely a motive for mutual forbearance, because firms' different market positions could also be thought of as the outcome of lower competitive intensity since firms have higher degrees of freedom in developing favorable positions in their respective key markets.
[43] See Grimm & Smith (1997), p. 166.
[44] See Jayachandran, Gimeno & Varadarajan (1999), p. 58.
[45] See Barney (1991).
[46] See Gimeno & Woo (1996), p. 324.
[47] From a resource-based view, however, the similarity of rival firms' resource endowments tends to lead to increased competition. See Barney (1991), p. 103.
[48] See Young et al. (2000).

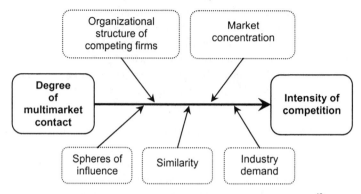

Figure 8-3: Contingencies in Multimarket Competition[49]

Concerning the efficacy of multimarket contact in lowering competition, a firm's *organizational structure* is of high relevance. An appropriate coordination of the relevant organizational units must be ensured in order to dovetail the activities in different markets. Multimarket competition will only lead to mutual forbearance if firms engaged are able to effectively coordinate their actions across geographic-product markets, and in this manner potentially enforce reciprocal threat.[50]

It is mostly argued that the degree of *market concentration* has a non-linear effect on rivalry: In markets of lower concentration firms may encounter difficulties in establishing familiarity, as well as in holding credible threats, because effective coordination is more complex in fragmented markets and no firm has enough market share or revenues at stake.[51] In conditions of high market concentration, the number of relevant coordinative links between rivals is reduced, facilitating the evolution of familiarity.[52] Moreover, firms have substantial profits at stake so they tend to refrain from retaliatory measures resulting in increased deterrence. Therefore, multimarket contact is said to have only an additional effect in concentrated markets; mutual forbearance is likely to

---

[49] Adapted from Jayachandran, Gimeno & Varadarajan (1999), p. 57.
[50] See Jayachandran, Gimeno & Varadarajan (1999), p. 58. To some extent, the right organizational structure may as well be regarded as a prerequisite or enabling factor rather than a contingency: Without a minimum of cross-market coordination of the divisions operating in multiple markets, firms would not enter into a situation of multimarket competition, rendering mutual forbearance irrelevant.
[51] See Jayachandran, Gimeno & Varadarajan (1999), p. 59.
[52] See Grimm & Smith (1997), pp. 163f.

emerge anyway, due to the smaller number of relevant competitors. Finally, in the case of moderate market concentration, multimarket contact will considerably help to advance familiarity in comparison to lower or higher concentrated markets. Hence, the alleviating effect of multimarket contact on rivalry will be strongest in moderately concentrated markets.[53]

In general, steadily growing *industry demand* facilitates forbearance among firms, because there is less need to engage in aggressive behavior in order to expand market share. The competition reducing effect of multimarket contact will therefore be promoted if there is growing industry demand. Contrary to this, stable or even decreasing industry demand, probably induces fierce competition as firms still seek to fully utilize their existing capacity and thus tend to lower prices. Consequently, mutual forbearance through multimarket contact is more difficult to achieve and rivalry increases, particularly if a reduction of capacity is very expensive.[54]

A general understanding of competitive interaction and multimarket competition has been provided in this section. We will now revert to the analysis of the European aviation market in order to gain insights concerning the currently observable development and interactions.

4 Competitive Development in the European Aviation Market

In this section, the increasingly dynamic market environment European airlines have to cope with will be briefly described. Subsequently, concrete examples of these increasing dynamics, in terms of directly observable moves and countermoves of airlines, will be given. Finally, since an emerging development towards cooperative agreements can already be observed within the European market, the set up of cooperations between low cost carriers as a recently emerging trend will be delineated.

---

[53] See Jayachandran, Gimeno & Varadarajan (1999), p. 59.
[54] See Grimm & Smith (1997), p. 166.

4.1 Increasingly Dynamic Market Environment

Since the beginning of deregulation, competition in the European aviation market has intensified considerably, and during recent years airlines have entered a fierce battle for market shares. To a great measure this is due to the market entry of numerous low cost carriers significantly contributing to the industry wide decline of fares and therefore posing severe problems to the network carriers with their complex cost structure.[55] Today, about 60 low cost carriers are operating in Europe, and their number is growing steadily as market entry is relatively easy, due to affordable prices for used aircraft and (still) extremely low charges at many secondary airports. There are diverging opinions concerning the future development of the European airline industry: On the one hand a sizeable – even if slowly decreasing – growth can still be noticed, on the other hand the problem of overcapacities is already evident and will even be aggravated as some airlines will expand their fleet on a significant scale within the course of the next years.[56]

In order to withstand the pressure to fill extant and future aircraft capacity, low cost airlines are about to aggressively expand into new markets. Whereas hitherto every low cost airline concentrated on its own market niche and tended to avoid serving those routes already offered by its competitors, the airlines are now trying to recover market shares by entering into direct competition with each other.[57] Moreover, European low cost airlines are forging ahead to operating routes which formerly were typically served by charter carriers only, above all the Palma de Mallorca destination. A fierce price competition is predicted for these routes especially.[58]

Price competition has intensified considerably since the emergence of copious low cost carriers in the European market, and it is expected to further intensify during the next months and years. The natural consequence of this intensive price competition is a profit decline that also concerns the successful and still growing airlines. For example, Easyjet recorded a significant decrease in profits as early as in 2003 and issued two profit warnings within a short period of time

---

[55] See e.g. Doganis (2001), pp. 137ff.
[56] Ryanair's open orderings with Boeing amount to not less than 125 aircraft. See Scharrer (2004). Similarly, Airbus has begun in 2003 to supply Easyjet with 120 virgin airplanes, which will more than double Easyjet's fleet over the next years. See Easyjet (2004).
[57] See Baker (2004a).
[58] See Eberle (2004).

in summer 2004.⁵⁹ Ryanair, although the most experienced low cost player and founder of the low fare concept in Europe, is already registering a significant profit loss in 2004 compared to 2003. Its stock price has reacted accordingly.⁶⁰ It can be expected that many airlines, especially newly founded companies with a weaker financial base compared to the established ones, will not survive the enduring price competition and will eventually have to retire from the market.

As the generally increasing market dynamics are ultimately reflected in competitive interaction among European airlines, recently and currently observable competitive moves and countermoves will be exemplified in the next section.

4.2 Competitive Interaction among European Airlines

The increased overall intensity of competition in the European aviation market goes along with more intense and aggressive competitive interactions of airlines. Referring to the competitive dynamics model depicted in chapter two, this corresponds with the results concerning the significance of the industry specific competitive environment in terms of growth and market concentration.

Empirical studies of multimarket competition carried out in the aviation industry, define a market as a definite single route. For this interpretation, multimarket competitive interaction could only emerge when airlines start from and land at the same airports and serve identical routes. However, it has to be taken into account that competition also and predominantly arises between airlines not only on single routes, but also within smaller or wider regions, depending on the catchment area of the respective airports served. Passengers traveling e.g. from the Rhine-Main area to Palma de Mallorca can choose between several airports (Cologne/Bonn, Dusseldorf, Frankfurt, Frankfurt-Hahn) which are reachable within a relatively short period of time. Therefore, airlines serving the Palma destination from these airports are in competition with each other, resulting in a certain market overlap, although not operating on exactly the same routes. In the following, this broader interpretation of markets and market overlap will be presumed.

The intensifying competition in the European aviation market is currently becoming especially apparent between Ryanair, which started its operations in 1985 and can therefore be considered as Europe's first "low fares / no frills"

---

⁵⁹ See Faz.net (2003) and Baker (2004b).
⁶⁰ See Faz.net (2004).

airline, and Easyjet. In July 2004, Ryanair announced a major expansion at Easyjet's home airport, London Luton, with 9 additional routes – a straight attack on Easyjet, which is also underscored by a dedicated 'declaration of war' in one of Ryanair's news releases: "Our message to Easyjet is simple – Easyjet can't match Ryanair's low fares and they can't match Ryanair's punctuality either."[61] Easyjet itself announced the launch of a number of new routes from London Luton, Gatwick and Stansted airports, primarily to Eastern European cities at reduced fares for the winter flight schedule 2004. Interestingly, Easyjet's operations in London Gatwick are most heavily intensified, and the airline intends to develop Gatwick further to become its largest European base in the future.[62] At the same time, Ryanair undertook a massive seat sale on all its routes from London Luton in October 2004 in order to cement its position there.[63]

However, Easyjet is not only Ryanair's main competitive target and does not only get attacked; it also attacks competitors on its part and appears as an aggressor in the market. In Germany, Easyjet is about to enter into a fiery competition with the incumbent German low cost and charter airlines, specifically Air Berlin.[64] In April 2004, Easyjet established an additional hub in Berlin, which is the home turf of the homonymous airline Air Berlin. Easyjet started its operations from Berlin-Schönefeld to major cities in UK and some European destinations, such as Paris and Barcelona, thereby severely attacking Air Berlin. Additionally, in May 2004 Easyjet launched the Berlin-Palma de Mallorca route, which is a traditional and particularly strong holiday route of Air Berlin and which has grown significantly within the last few years and now belongs to Air Berlin's very core business. Therefore, a vehement price battle between Easyjet and Air Berlin for the Palma de Mallorca destination can be expected. By today, Easyjet has further extended its hub in Berlin-Schönefeld by also starting operations on several Eastern European routes. In July 2004, Easyjet established an additional hub at Dortmund airport, another important base in Air Berlin's route network from which it serves a range of southern European holiday destinations. By this sustainable encroachment into the German market, Easyjet positions itself as a direct competitor with Air Berlin.

As a reaction, Air Berlin has intensified its UK destinations and has added a range of new direct routes to Manchester (from Berlin, Dusseldorf, Hamburg, Mailand/Bergamo and Paderborn) as well as the Paderborn-Southampton route in its winter timetable valid from November 2004. Furthermore, Air Berlin has

---

[61] Ryanair (2004).
[62] See Easyjet (2004).
[63] See Ryanair (2004).
[64] See Frankfurter Allgemeine Zeitung (2004a).

increased operations in London in several respects: While the London-Palma de Mallorca route has been served on a daily basis since as early as February 2004, another connection from Leipzig/Halle to London-Stansted was launched in February 2005. Additionally, Air Berlin's partner airline NIKI is offering a new connection between London and Vienna.[65] Altogether, these measures can be interpreted as a comprehensive competitive reaction to Easyjet's attack, thereby increasing the overall level of competition in the European aviation market.

Besides Air Berlin, Easyjet is also about to challenge other (low cost) airlines in the German market, namely Germanwings and Hapag-Lloyd.[66] In June 2004, Easyjet started to serve routes from Cologne-Bonn to destinations in UK and has thereby settled down as the third low cost carrier at Cologne-Bonn airport.[67] This can be interpreted as a direct attack on Cologne-based Germanwings and Hapag-Lloyd Express who have been in close competition at that airport for a while. As Germanwings is suspending some summer destinations and therefore has to face unused capacity during its winter flight schedule, it is forced to search for lucrative new routes to fill its existent capacity.[68] This adds even more fuel to the current competition, as Easyjet also has to aggressively expand into new markets in view of its increasing capacity.[69]

Moreover, upon the announcement of Easyjet to land also in Hanover, the home base of Hapag-Lloyd Flug,[70] the latter threatened its movement from that airport in order to hamper Easyjet's settlement. After extensive negotiations with Hanover airport's partners, Hapag-Lloyd finally stood against Easyjet who were refused landing rights for the time being. However, negotiations will be resumed for the summer timetable 2005.[71] Meanwhile, Hapag-Lloyd Express announced to considerably expand its capacity in Hanover concertedly with its cooperation partner Air Berlin (see following section).[72] In addition, Hapag-Lloyd Express is

---

[65] See Air Berlin (2004). The recently established alliance between Air Berlin and NIKI is briefly outlined in section 4.3 on pp. 205f.
[66] To be precise, one has to distinguish between Hapag-Lloyd Flug (HLF) and Hapag-Lloyd Express (HLX), both of which have Europe's largest tourism group TUI as the parent company. HLF as TUI's charter airline traditionally serves its holiday destinations; HLX, founded in 2002, successfully competes in the low cost market.
[67] See Easyjet (2004).
[68] See Ramm (2004).
[69] See section 4.1 on pp. 200f.
[70] More specific, Hanover is the home base of Hapag-Lloyd Flug's parent company TUI.
[71] See Die Welt (2004a) and Die Welt (2004b).
[72] See Süddeutsche Zeitung (2004).

on its part intensifying services to Central England (Manchester, Coventry) in the upcoming winter timetable in order to enforce its settlement in UK, Easyjet's home market.[73]

The examples mentioned so far depict the growing interaction among European low cost airlines. However, former flagship carriers are getting increasingly involved as well. In order to defend their market position against the steadily growing threat posed by low cost carriers on some routes, network carriers like the German Lufthansa now tend to assimilate their prices to those of their low cost competitors: As a general reaction to the very low fares offered on some routes, Lufthansa introduced 98 Euro return tickets on direct routes within Germany in 2002. The fare was even lower (€ 88) if booking took place via the Internet.[74] However, problems with this fare policy specifically arose on the route between Frankfurt and Berlin, where Lufthansa had already been involved in a fierce price battle with its low cost competitor Germania Express for some time. According to the German Antitrust authority, it was feared that Lufthansa capitalized on its dominant market position and forced Germania Express out of this market.[75] In order to ensure fair competition, Lufthansa was soon called to raise its prices again to a level that adequately reflected the differences in service offered by Lufthansa and Germania Express during their flights.[76]

The Cologne/Bonn-Hamburg route represents a second example of the intensifying competition between Lufthansa and the German low cost carriers: In response to the very low fares offered by Hapag-Lloyd Express, Lufthansa started a counterattack in 2003 by considerably increasing frequency (from 8 to 12 times a day) and decreasing prices on this route (below € 100). This aggressive reaction had not been expected by Hapag-Lloyd Express, and it initially also came to the Antitrust authority's attention.[77] Nevertheless, the competitive situation between Lufthansa and Hapag-Lloyd Express on this route has not changed significantly since 2003.

---

[73] See Hapag-Lloyd Express (2004).
[74] See also Klingenberg's chapter 7 in this volume.
[75] See Flottau (2002).
[76] See Ginten (2002).
[77] See Binder (2003).

4.3 Emerging Cooperations among Low Cost Airlines

Besides competition intensifying in such a way, the European airline market is interestingly showing trends in the opposite direction. Whereas alliance building and networking has been observable up to now with network carriers in the global context only, a number of low cost airlines are currently entering into cooperation with each other.

In this respect, Air Berlin is leading the way by establishing the first major European low cost alliance through acquiring a 24 percent stake in the Austrian NIKI in the beginning of 2004 in order to facilitate the penetration of the Austrian and Eastern European markets.[78] In February 2004, Air Berlin concluded another cooperation agreement with Germania Express, an airline it used to compete with on several routes. This second alliance is primarily concentrated on distribution, since it has the objective to partially offer the respective partner's flights on the Internet pages and thus enlarge product range.[79] Moreover, Air Berlin just announced another cooperation with HLF, TUI's charter airline. Contrary to the hitherto existing alliances, the agreement with HLF aims at matching the airlines' flight capacities in the charter market and concerns services from their joint hubs Nuremberg, Palma de Mallorca, Stuttgart and Munich with coordination of flight schedules starting November 2004.[80] Obviously, all cooperation agreements mentioned here aim, from Air Berlin's point of view, to define their market territory in view of the increasing threat from the low cost carriers, particularly Easyjet, who are expanding very aggressively within the European market.

Similarly, HLX has set up a far-reaching commercial partnership with the Italian Volareweb.com by joining their resources in the sales and distribution field in order to enhance presence in both companies' home countries. Another cooperation followed in May 2004, when HLX reached a code share agreement with Air Polonia to enter the Polish market.[81]

In the course of this chapter's description of the current developments in the European aviation market, the relevance and role of multimarket competitive

---

[78] See Air Berlin (2004).
[79] See Flottau (2004).
[80] See Financial Times Deutschland (2004) and Frankfurter Allgemeine Zeitung (2004b).
[81] See Hapag-Lloyd Express (2004). However, both of HLX' partners recently had to retire from the market: Volareweb.com has suspended all flights since 11 November 2004; shortly after, on 5 December 2004, Air Polonia announced insolvency as well.

interaction has already become implicitly noticeable. In the following, the theoretical instruments presented in chapters two and three are explicitly employed in order to analyze the market from a competitive dynamics perspective and elucidate the occurrences described above.

5 Analyzing the Market from a Competitive Dynamics Perspective

The variety of concrete examples depicted above to delineate the development in the European aviation industry, indicates that the market currently underlies competitive dynamics to a considerable extent. In this context, competitive actions and reactions which are different in magnitude, type, and scope can be observed:[82] Whereas simple price reductions on certain routes obviously require only little resource commitment and are thus characterized by a low magnitude, new market entries entail a far higher amount of (financial) resources and therefore a higher risk. Different types of actions can be identified with variations in flight frequencies and new route offers on the tactical level, or the development of new regions (e.g. Eastern Europe) and building alliances with other (low cost) airlines on a more strategic level.

Airlines like Easyjet and Ryanair continue to fiercely attack their competitors by numerous new market entries and/or aggressive price promotions. Especially Easyjet is engaged in interaction with various competitors at the same time and has struck several blows on the market, thereby exhibiting a relatively broad scope of its competitive actions. Once offended, competitors respond by retaliating (e.g. through the launch of new routes), or they engage in cooperations in order to broaden their market coverage and enhance their competitive position. It can be stated that, generally, two fundamental approaches of airlines' competitive behavior have become empirically apparent: Airlines either try to capture the market single-handedly through aggressive competitive attacks, or they strive for the development of greater networks in cooperation with other airlines. Both approaches are now analyzed in more detail.

---

[82] For an explanation of magnitude, type, and scope as relevant action characteristics see chapter 2.2 on pp. 190f.

5.1 Confrontation

By behaving aggressively and accomplishing a range of direct and clever competitive attacks, like in the case of Easyjet, a firm can try to exploit the competitive uncertainties of its competitors.[83] The most effective competitive moves are those which allow for enhancing the firm's competitive position while at the same time rendering the rival's reaction less probable or reducing its magnitude or speed. There has been some research on how competitive actions can best reduce the chances of retaliation.[84] It is argued that an ongoing series of aggressive actions is especially effective in order to stabilize and expand one's competitive position, because the rival(s) will be continuously engaged in responding and defending its/their markets. Moreover, the innovative character of actions plays a major role, as particularly those actions hitherto unknown to the market and diverging from usual competitive behavior have the potential to disrupt the competitive situation in a sustainable way. This hints to the notion that the greater the number of new competitive actions, the greater the potential for exploiting market opportunities and leaving competitors behind will be.[85]

The confrontation approach is very well exemplified by Easyjet's current behavior. Easyjet is successfully attacking several competitors at the same time by launching a variety of new routes in quick succession, thereby foraying into competing airlines' traditional home markets. By establishing its first German hub at Berlin-Schönefeld in spring 2004 and a second base in Dortmund shortly after, Easyjet attacked Air Berlin obviously severely enough to provoke an adequate reaction by the German carrier in order to defend its market position: Evidently, Air Berlin had realized to some extent that the potential to hurt Easyjet by attacking Easyjet in its home market was much greater than by exclusively concentrating on the defense of the Berlin hub. Similar behavior can be observed on the part of Hapag-Lloyd Express as a response to Easyjet's settlement in Cologne/Bonn: Instead of assembling all available resources to defend its home market, Hapag-Lloyd Express is intensifying operations in Easyjet's home territory by launching new routes to the UK. The behavior of Air Berlin and Hapag-Lloyd Express point to the fact that they have recognized the multimarket situation and used the opportunity to retaliate in those markets that are most important to their competitor Easyjet, just as multimarket theory would suggest.

---

[83] See Grimm & Smith (1997), p. 76.
[84] See Chen & Miller (1994); Ferrier, Smith & Grimm (1999).
[85] See Ferrier, Smith & Grimm (1999), pp. 373f.

Apparently, Easyjet was not able to effectively attack its rivals while at the same time hindering them from responding painfully. In these cases, the hitherto existing multimarket contact between Easyjet and its competitors had yet no significant impact in lowering competition among them: Neither a sufficient common competitive history between Easyjet and Air Berlin and, thus, a certain degree of familiarity had evolved to this day, nor was effective deterrence in place. In order to render deterrence possible, the encounter on a sufficient number of markets is a necessary condition,[86] which was obviously not given here. However, entering into new markets and subsequently developing additional multimarket contact, Easyjet, Air Berlin and Hapag-Lloyd set the stage for effective deterrence in the future.[87]

In contrast, the various market entries undertaken by Easyjet can also be regarded as successful competitive actions with respect to those airlines that likewise run an operational base in Berlin but have not (yet) responded, as in the case of dba. Provided that dba was aware of Easyjet's attack, it was obviously either not motivated or not able to respond. From a competitive dynamics perspective, reasons for this waiting or even ignoring the attacks could be due to the relevant characteristics of Easyjet's attack itself: There was a high frequency of competitive moves in terms of various market entries, which represented, at least to some extent, a new kind of competitive action since previously low cost carriers were keen not to directly compete with each other. Easyjet is the first airline to break these rules of amicable equilibrium by intensifying operations in the German market. From the fact that market entries as well as the establishment of additional bases in Germany were effected in a quite narrow period of time (less than one year), it can be concluded that moving quickly is crucial in this market, thereby slowing down the other airlines' competitive activity.[88] At the same time through these actions, the number of multimarket contacts is increased which, as mentioned above, favors mutual forbearance in a later phase of market development.

---

[86] See e.g. Stephan et al. (2003), p. 405.
[87] The increase of multimarket contact is even more likely in the network-based airline industry where firms are aware of network externalities and the network patterns of their rivals. See Korn & Rock (2001), p. 65.
[88] See also Ferrier, Smith & Grimm (1999), p. 375.

5.2 Cooperation

In order to build up extensive networks and commonly set up a favorable position in the European market, firms may strive for cooperative agreements. Whereas globally oriented cooperations and worldwide networks like the Star Alliance were established years ago, alliances between low cost airlines are a new phenomenon. As described above, the first low fare alliance was concluded in 2004 by Air Berlin and NIKI. Similarly, HLX agreed a cooperation with an Italian and a Polish low cost airline. By setting up these kind of partnerships, the airlines plan to coordinate their flight schedules and/or sales activities, which can undoubtedly be considered as an attempt to counterbalance the increasing competition and stabilize market shares.

From a competitive dynamics point of view, such alliance formation implies that aggressive competitive interaction, which can be seen as characteristic in the confrontation approach, should be eliminated here, at least regarding the respective field the alliance agreement extends over. Competitive actions and reactions are shifted beyond the alliance boundaries and are coordinated in a way that competition takes place between the alliance and one or more single competitors, or among different alliance networks. Thus in this context, we must clearly differentiate between interaction among the alliance members (i.e. within the alliance) and between the alliance and its competitors (i.e. beyond the alliance).

By setting up cooperations and building greater networks the formerly competing market players intend to reduce competition *within* the alliance and create a new market balance. The alliance is based on a cooperative agreement that explicitly postulates the ambition to jointly gain a competitive advantage. Concurrently, alliance members will refrain from aggressive competitive behavior vis-à-vis their partners. From this perspective, some kind of familiarity can play an important role in two respects: Firstly, the willingness to engage in a cooperative venture already requires a certain familiarity. Secondly, familiarity is further enhanced in the course of the cooperative agreement, as the firms gather a deeper knowledge about each other and gain a common competitive history.[89] This

---

[89] It has to be annotated, however, that alliance building tends to reduce multimarket contact since a major goal of airline alliances usually consists in widening market coverage, thereby explicitly delimiting each alliance partner's geographic activity. This then leads to the insight that the familiarity mechanism supports the cooperative behavior prescribed in the alliance agreement *in spite of* the decreasing degree of multimarket contact.

familiarity and the mutually forbearing behavior resulting from it can potentially be also transfused to markets the alliance agreement does not extend over.

Since less, or at the extreme, no single airlines but larger networks of airlines compete in the market, concentration is increased as well (which was mentioned as being influential when shaping the potential for mutual forbearance). Furthermore, it can be expected that, while cooperating, airlines become more similar to each other, at least in the areas they cooperate in, e.g. when developing common distribution and marketing channels. Finally, a cooperation agreement will help to enhance the involved firms' spheres of influence by clearly delimiting and coordinating the respective competences and responsibilities.[90] Moreover, by confining their spheres of influence, the deterrence mechanism among the members of the alliance is reinforced, thereby stabilizing the alliance constellation. In the Air Berlin-NIKI example, the market coverage of Air Berlin and NIKI is clearly defined and carefully coordinated, which ensures an amicable co-existence of both.

On a broader scope and *beyond* the alliance, deterrence is also enhanced through the wider market coverage following from the alliance agreement, resulting in an increased degree of multimarket contact with the alliance's competitors. The cooperative agreements Air Berlin concluded with other airlines then might also have a deterring effect on its competitors, in particular Easyjet: E.g. the cooperation between Air Berlin and Hapag-Lloyd Express in order to match capacities on their joint charter markets could also prevent Easyjet from further engaging in intensive rivalry with the alliance's members, because it has to fear retaliation on a greater variety of markets. Therefore, in principal, cooperations in the multimarket context also serve the purpose to bar competitors from intensifying operations in those markets covered by the alliance network. This deterrent effect, however, becomes viable only when the alliance's members succeed in credibly assuring that they will conjointly retaliate if attacked.[91] Moreover, the alliance must be organized in a way that allows for adequate coordination and an effective realization of the joint counterattack.[92,93]

---

[90] For an explanation of market concentration, similarity and spheres of influence as contingency factors of multimarket competition see section 3.2 on pp. 196ff.

[91] See Schelling (1980), p. 187. See also section 3.1.2 on pp. 194ff.

[92] For the role of the organizational structure as contingency factor in multimarket competition see section 3.2 on pp. 196ff. See also Albers (2005) for an in-depth treatise on alliance organization.

[93] In a similar sense and as an example of the mutual support within such an alliance, the case of Air Canada's impending hostile takeover by American Airlines-backed Onex Corp. in 1999 can be alleged. At that time, Air Canada's partners in the Star

To conclude, whereas within the alliance, familiarity furthers the cooperative behavior between the alliance partners in addition to the underlying alliance agreement, deterrence vis-à-vis the competitors outside the alliance helps to stabilize the market environment on a broader level. Eventually, these two effects facilitate the development of a well-balanced market situation.

In the light of the market analysis presented and the different strategic approaches identified, we come back to our three research questions raised at the beginning. In this chapter, the first question has been elaborated on and airlines' competitive behavior has been explained from a competitive dynamics point of view. In the following chapter we will discuss the remaining questions: Can the future development of the European aviation market be anticipated to some extent? What are the managerial implications following from the competitive dynamics perspective? Are there any strategic recommendations to be given concerning the competitive behavior of airlines currently vigorously vying for their market positions?

6 Implications of Multimarket Competitive Interaction

6.1 The European Aviation Market – on Its Way to a New Equilibrium?

As shown throughout this chapter, the market is at present entering a phase of intensifying competitive interaction. Whereas low cost airlines used to compete primarily with network carriers, they strive more and more to recover market shares from each other. This trend will be aggravated in view of the existing and increasing overcapacities in the market. A phase of consolidation can be expected which will lead to the insolvency of cost-inefficient airlines.[94]

Alliance, first of all United Airlines and Lufthansa, announced a financial package in combination with their unrestrained willingness to fight in order to fend off the takeover. See Chipello & Carey (1999). This strategy of deterrence finally succeeded, keeping Air Canada as a member of the Star Alliance.

[94] In the course of the last years, some airlines (like Swiss Air in 2001) had already to retire from the European market, others are fighting against insolvency as in the case of Alitalia. A recent example concerns the Dutch low cost carrier V-Bird which has suspended all flights on further notice due to insolvency since 8 October 2004. See sueddeutsche.de (2004).

Meanwhile, the competitive gap between network and low cost carriers is beginning to narrow, as exemplified by Lufthansa's price reductions and promotions, as well as increased fares for low cost flights during 2003.[95]

Experience from the US airline market also shows that steadily decreasing prices and an industry-wide yield decline, coercively result in a phase of consolidation and numerous market exits. Previously, in the 1990s, the US market could be considered as being in a kind of an overall balance, stable and profitable, even though affected by some cyclical fluctuations.[96] In this context, the competitive interplay in multiple markets and the phenomenon of mutual forbearance could be directly observed between Northwest and Continental Airlines, thereby once more supporting the assumption that multimarket competition among airlines implies some balancing effects.[97] When Northwest cut fares on various routes to West Coast cities, which represented Continental's core market, Continental Airlines turned out not to be interested at all in starting a price war with Northwest in this market. Instead, they lowered fares in Minneapolis, the most important Northwest market. Northwest recognized the competitive interdependency and raised fares on West Coast routes again onto the previous level. This reemerging equilibrium was only rendered possible as the respective airlines served the same markets, even if the focus was slightly different.

According to the development in the European market and the currently observable market entries, the overlaps of markets in terms of routes and regions are also tending to increase, favoring multimarket interaction further as a kind of self-enforcing effect. Parallel to the expected trend of consolidation, this could lead to a situation where only a few important and well-positioned airlines divide the European market among each other. In addition, a number of smaller airlines could position themselves in niches, e.g. by operating as feeder airlines on a regional basis. Therefore, a well-balanced and stable market based on some kind of tacit collusion and mutual forbearance can be expected as a result of the current development process.[98]

---

[95] See Frankfurter Allgemeine Zeitung (2004a).
[96] See D'Aveni (2002), pp. 41ff. and Doganis (2001), pp. 1ff.
[97] See Frankfurter Allgemeine Zeitung (2003); Evans & Kessides (1994), p. 341.
[98] By assuming the emergence of tacit collusion, no evaluation concerning its legal allowance is implied.

6.2 The Significance of Multimarket Competition for the Strategic Management of Airlines

It is obvious that all airlines in the market are interested in and should strive for a position that allows them to outlast the above mentioned consolidation phase. Moreover, if "multimarket structures are less competitive and capable of yielding higher, more stable performance, firms are likely to actively pursue these arrangements."[99] The above market analysis shows that, at least from an empirical perspective, there are two different or even opposed approaches to exploiting the given multimarket situation in order to outlive the current development, establish oneself as an efficient and well-positioned airline and eventually enter a stable market equilibrium founded on mutual forbearance.

An airline pursuing the *confrontation* approach mainly builds its market position upon deterrence. This airline should therefore become aware of the potential resulting from multimarket competition and strive for a market situation rendering deterrence possible, thereby carrying out those competitive attacks that are best suited to hamper the competitors' activities. Referring to the Easyjet example,[100] a quick series of competitive attacks with a broad scope, in terms of various international market entries, can set the stage for effective deterrence at a later time and ensure that Easyjet is perceived as a serious competitor. At the same time these measures also enhance multimarket contact, a sufficient amount of which is a necessary precondition for deterrence.[101] Once such a market situation is established within Europe, a credible commitment of the firms involved is essential: An airline is only able to deter if it holds credible threats and if its competitors are absolutely sure that retaliation will occur in case of attack.[102] The spreading of according information about one's ability and motivation to retaliate to set up an adequate reputation is a necessary requirement. In this context, public commitment in terms of concrete and binding announcements that reinforce the intent to retaliate may help to render deterrence more credible. At the same time, gathering deeper knowledge about which markets/routes are of genuine interest to the competitors and what is their actual (financial) ability and motivation to defend their markets can give the respective airline a valuable information advantage.

---

[99] Stephan & Boeker (2001), p. 235.
[100] See section 4.2 on pp. 201ff.
[101] See Stephan et al. (2003), p. 405.
[102] See section 3.1.2 on pp. 194ff.

Contrary to confrontation, market balance in the *cooperation* approach relies on familiarity rather than exclusively on deterrence. Familiarity plays a primary role in forming the alliance and managing it later on, since a common understanding is a prerequisite for the conclusion as well as the further existence of the alliance agreement. According to competitive dynamics research, the airlines that are especially suitable to cooperate with each other are those that make use of similar resource endowments, e.g. in terms of fleet size and composition.[103] This tends to result in strategic similarity and an increased ability to understand each other and, thus, greater familiarity. Therefore, in order to enhance familiarity and force the alliance's success, the airlines should strive to continually exchange sufficient relevant information concerning their strategic attitude and goals, at least with respect to the business field affected by the alliance agreement. In such a way, knowledge about the respective cooperation partner increases so that both/all airlines gain a certain amount of confidential and intimate information (e.g. concerning strategically important target markets/routes and intended competitive activities). This can also have a deterring effect: Airlines stick to the alliance because they fear that sensitive information could be disclosed. Moreover, the fact that the alliance members' spheres of influence are enhanced (as illustrated by the Air Berlin-NIKI example) further supports the effectiveness of the deterrence mechanism.[104] Taken together, airlines that engage in cooperation so as to outlive the expected consolidation phase should firstly focus their managerial efforts on cultivating close relationships in which active information exchange takes place and, thus, on nurturing their familiarity. Secondly, in order to ensure the reliability of the deterrence mechanism a credible degree of commitment is also necessary so that mutual forbearance is further enhanced.[105]

Even if the general tendency towards cooperative agreements should become common across the industry, the necessary consolidation and the accompanying selection process will be ensured through the careful analysis and selection of

---

[103] See also section 3.1.1 on pp. 193f.

[104] See section 3.2 on pp. 196ff.

[105] Interestingly, it seems that airlines involved in the newly concluded cooperation agreements are beginning to diverge from their initial business model, which is primarily based on reducing those activities that cause complexity and on increasing the turnover rate. Cooperating, however, adds complexity to the existing managerial processes and therefore ties a certain amount of resources, while probably decreasing turnover rates due to coordination of flight schedules. This also indicates that the business models in the aviation industry have begun to blur and that the competitive gap between network and low cost carriers has started narrowing. See Field (2004). See also Klaas & Klein chapter 5 in this volume.

potential cooperation partners. Besides the existing global airline alliances, different patterns of cooperation with a European focus may evolve: Airlines can either cooperate with similar market players which have formerly been more or less direct competitors, or possibly with regionally focused players operating as feeder airlines or serving a certain market niche.

To summarize, the two approaches discussed above imply thoroughly different forms of competitive behavior and rely on familiarity and deterrence to a different extent, while both aiming at establishing and ensuring a favorable long-term market position in order to outlast intensifying rivalry. Obviously, the solo is a riskier venture because it necessitates greater and sustainable financial volumes than the cooperative alternative; on the other hand, in case of success, the market potential can be assessed as considerable. In contrast, cooperation could probably enhance the short-term stability of the market more and would lead to a new equilibrium faster, through increased familiarity.

7 Conclusion and Outlook

It was the aim of this chapter to provide insights concerning the development of the European aviation market and the implications for the management of airlines from a competitive dynamics perspective. This analysis had to take into account the emerging and intensifying competitive interactions in the market as well as the indubitable influence of multimarket contact. From this perspective, the expected phase of consolidation and restructuring may be followed by a new market equilibrium based on mutual forbearance. Recently, airlines have already started to take measures in order to endure the competitive battle, either through cooperation, exemplified by the newly concluded cooperative agreements around Air Berlin and HLX, or through confrontation, illustrated by the very aggressive competitive attacks carried out by Easyjet. During the discussion, the different prominence of familiarity and deterrence was emphasized in each case; moreover, the generally eminent role of related information exchange became apparent.

However, the analysis presented here is only a first contribution from a competitive dynamics and multimarket perspective. Market development over the next months and years will show how long the competitive struggle will last and to what extent the balancing multimarket competitive effects will prevail in the end. Furthermore, it remains to be seen which of the empirically identified *modi operandi* will turn out to be more advantageous and what additional factors actually influence their market success or failure. Likewise, a final coexistence of airlines involved in cooperations and those going it alone seems possible by all means. In this context, the question is raised whether the effectiveness of the familiarity and deterrence mechanisms will be influenced in the future, due to the involvement of cooperative groups and networks of airlines in multimarket competition instead of single companies.

8 References

AIR BERLIN (2004): URL: www.airberlin.com (access: 05.10.2004).

ALBERS, S. (2005): *The Design of Alliance Governance Systems*. Köln: Kölner Wissenschaftsverlag.

AXELROD, R. (1984): *The Evolution of Cooperation*. New York: Basic Books.

BAKER, C. (2004a): Europe's Low cost Battle Intensifies. *Airline Business*, 20(8), p. 20.

BAKER, C. (2004b): EasyJet Shares Dive after Second Profit Warning. *Airline Business*, 20(7), p. 21.

BARNEY, J.B. (1991): Firm Resources and Sustained Competitive Advantage. *Journal of Management*, 17(1), pp. 99-120.

BAUM, J.A.C.; KORN, H.J. (1996): Competitive Dynamics of Interfirm Rivalry. *Academy of Management Journal*, 39(2), pp. 255-291.

BAUM, J.A.C.; KORN, H.J. (1999): Dynamics of Dyadic Competitive Interaction. *Strategic Management Journal*, 20(3), pp. 251-278.

BERNHEIM, B.D.; WHINSTON, M.D. (1990): Multimarket Contact and Collusive Behavior. *RAND Journal of Economics*, 21(1), pp. 1-26.

BINDER, F. (2003): Lufthansa greift Hapag-Lloyd Express an. *Die Welt* (02.05.2003), URL: http://www.welt.de/data/2003/05/02/82350.html (access: 09.11.2004).

CHEN, M.-J. (1996): Competitor Analysis and Interfirm Rivalry: Toward a Theoretical Integration. *Academy of Management Journal*, 21(1), pp. 100-134.

CHEN, M.-J.; MILLER, D. (1994): Competitive Attack, Retaliation and Performance: An Expectancy-Valence Framework. *Strategic Management Journal*, 15(2), pp. 85-102.

CHEN, M.-J.; SMITH, K.G.; GRIMM, C.M. (1992): Action Characteristics as Predictors of Competitive Responses. *Management Science*, 38(3), pp. 439-455.

CHIPELLO, C.J.; CAREY, S. (1999): Fight for Canadian Carriers Threatens Alliances. *The Wall Street Journal* (26.08.1999), Eastern Edition, New York, p. A.3.

D'AVENI, R.A. (2002): Competitive Pressure Systems. *MIT Sloan Management Review*, 44(1), pp. 39-49.

DIE WELT (2004a): Billigflieger streiten um Flughafen Hannover. *Die Welt* (03.07.2004), URL: http://www.welt.de/data/2004/07/03/299973.html (access: 05.10.2004).

DIE WELT (2004b): Easyjet würde nie ein Luftfahrt-Bündnis eingehen. *Die Welt* (19.07.2004), URL: http://www.welt.de/data/2004/07/20/307551.html (access: 05.10.2004).

DOGANIS, R. (2001): *The Airline Business in the 21st Century*. London, New York: Routledge.

EASYJET (2004): URL: www.easyjet.com (access: 05.10.2004).

EBERLE, M. (2004): Boom der Billigflieger bringt viele Verlierer – Air Berlin wächst rasant. *Handelsblatt* (23.07.2004).

EDWARDS, C.D. (1955): Conglomerate Bigness as a Source of Power. In: Edward, C.D. (Ed.): *Business Concentration and Price Policy*, National Bureau of Economic Research Conference Report, Princeton: Princeton University Press, pp. 331-352.

EVANS, W.N.; KESSIDES, I.N. (1994): Living by the "Golden Rule": Multimarket Contact in the US Airline Industry. *The Quarterly Journal of Economics*, 109(2), pp. 341-366.

FAZ.NET (2003): Gewinneinbruch bei Billigflieger Easyjet. *Faz.net* (18.11.2003), URL: http://premium-link.net/$61643$365033123$/Rub7EF1D5D213234C6989-C9039B54879372/Dox~EB1B2B9846867BA4B3631606040E7D1E0~ATpl-~Ecommon~Scontent.html#top (access: 27.12.2004).

FAZ.NET (2004): Ryanair leidet unter Preiskampf der Billigflieger. *Faz.net* (28.01.2004), URL: http://premium-link.net/$61643$365033123$/RubC8BA557-6CDEE4A05AF8DFEC92E288D64/Dox~E5B2EF96067780F0E16C1A0F2E828 610B~ATpl~Ecommon~Scontent.html (access: 27.12.2004).

FERNANDEZ, N.; MARÍN, P.L. (1998): Market Power and Multimarket Contact: Some Evidence from the Spanish Hotel Industry. *The Journal of Industrial Economics*, 46(1), pp. 301-315.

FERRIER, W.J.; SMITH, K.G.; GRIMM, C.M. (1999): The Role of Competitive Action in Market Share Erosion and Industry Dethronement: A Study of Industry Leaders and Challengers. *Academy of Management Journal*, 42(4), pp. 372-388.

FIELD, D. (2004): Cost Convergence. *Airline Business*, 20(9), pp. 32-33.

FINANCIAL TIMES DEUTSCHLAND (2004): Kooperation der Konkurrenten. *Financial Times Deutschland* (06.07.2004).

FLOTTAU, J. (2002): Druck von allen Seiten. *Die Zeit*, 14, p. 26.

FLOTTAU, J. (2004): Air Berlin und Germania bilden Allianz. *Financial Times Deutschland* (17.03.2004).

FRANKFURTER ALLGEMEINE ZEITUNG (2003): Tust du mir nichts, tue ich Dir nichts. *Frankfurter Allgemeine Zeitung*, 137 (16.06.2003), p. 22.

FRANKFURTER ALLGEMEINE ZEITUNG (2004a): Der Ausleseprozeß setzt ein. *Frankfurter Allgemeine Zeitung*, 38 (14.02.2004), p. 16.

FRANKFURTER ALLGEMEINE ZEITUNG (2004b): Air Berlin und Hapag Lloyd bieten gemeinsam Flüge an. *Frankfurter Allgemeine Zeitung* (06.07.2004).

GIMENO, J. (1999): Reciprocal Threats in Multimarket Rivalry: Staking out 'Spheres of Influence' in the US Airline Industry. *Strategic Management Journal*, 20(2), pp. 101-128.

GIMENO, J.; WOO, C. (1996): Hypercompetition in a Multimarket Environment: The Role of Strategic Similarity and Multimarket Contact in Competitive De-escalation. *Organization Science*, 7(3), pp. 322-341.

GINTEN, E.A. (2002): Die Billigflieger erobern den deutschen Markt. *Die Welt* (06.03.2002), URL: http://www.welt.de/daten/2002/03/06/0306un318516.htx (access: 09.11.2004).

GRIMM, C.M.; SMITH, K.G. (1997): *Strategy as Action*. Cincinnati: South-Western College Publishing.

GULATI, R. (1995): Does Familiarity Breed Trust? The Implications of Repeated Ties for Contractual Choice in Alliances. *Academy of Management Journal*, 38(1), pp. 85-112.

HAPAG-LLOYD EXPRESS (2004): URL: www.hlx.com (access: 05.10.2004).

HAVEMAN, H.A.; NONNEMAKER, L. (2000): Competition in Multiple Geographic Markets: The Impact on Growth and Market Entry. *Administrative Science Quarterly*, 45(2), pp. 232-267.

HEGGESTAD, A.A.; RHOADES, S.A. (1978): Multi-Market Interdependence and Local Market Competition in Banking. *Review of Economics and Statistics*, 60(4), pp. 523-532.

JAYACHANDRAN, S.; GIMENO, J.; VARADARAJAN, P.R. (1999): The Theory of Multimarket Competition: A Synthesis and Implications for Marketing Strategy. *Journal of Marketing*, 63(3), pp. 49-66.

KARNANI, A.; WERNERFELT, B. (1985): Multiple Point Competition. *Strategic Management Journal*, 6(1), pp. 87-96.

KORN, H.J.; ROCK, T.T. (2001): Beyond Multimarket Contact to Mutual Forbearance: Pursuit of Multimarket Strategy. In: Baum, J.A.C.; Greve, H.R. (Eds.): *Multiunit Organization and Multimarket Strategy*, Advances in Strategic Management, Vol. 18, Oxford: JAI Press, pp. 53-74.

LAWTON, T.C. (2002): *Cleared for Take-Off. Structure and Strategy in the Low Fare Airline Business*. Aldershot, Burlington: Ashgate.

MCGRATH, R.G.; CHEN, M.-J.; MACMILLAN, I.C. (1998): Multimarket Maneuvering in Uncertain Spheres of Influence: Resource Diversion Strategies. *Academy of Management Review*, 23(4), pp. 724-740.

PARKHE, A. (1993): Strategic Alliance Structuring: A Game Theoretic and Transaction Cost Examination of Interfirm Cooperation. *Academy of Management Journal*, 36(4), pp. 794-829.

PHILLIPS, O.R.; MASON, C.F. (1992): Mutual Forbearance in Experimental Conglomerate Markets. *RAND Journal of Economics*, 23(3), pp. 395-414.

RAMM, T. (2004): Unter Haifischen. *Touristikreport* (03.06.2004), URL: http://www.touristikreport.de/rd/archiv/5568.php (access: 05.10.2004).

RHOADES, S.A.; HEGGESTAD, A.A. (1985): Multimarket Interdependence and Performance in Banking: Two Tests. *The Antitrust Bulletin*, 30(1), pp. 975-995.

RYANAIR (2004): URL: www.ryanair.com (access: 22.09.2004).

SCHARRER, J. (2004): Easyjet versetzt Branche in Schrecken. *Touristikreport* (12.02.2004), URL: http://www.touristikreport.de/rd/archiv/6429.php (access: 31.08.2004).

SCHELLING, T.C. (1980): *The Strategy of Conflict*. Cambridge (Mass.), London: Harvard University.

SMITH, K.G.; FERRIER, W.J.; NDOFOR, H. (2001): Competitive Dynamics Research: Critique and Future Directions. In: Hitt, M.A.; Freeman, R.E.; Harrison, J.S. (Eds.): *The Blackwell Handbook of Strategic Management*, Oxford, Malden: Blackwell, pp. 315-361.

SMITH, K.G.; GRIMM, C.M.; CHEN, M.-J.; GANNON, M.J. (1989): Predictors of Response Time to Competitive Strategic Actions: Preliminary Theory and Evidence. *Journal of Business Research*, 18(3), pp. 245-258.

SMITH, K.G.; GRIMM, C.M.; GANNON, M.J. (1992): *Dynamics of Competitive Strategy*. Newbury et al.: Sage Publications.

SMITH, K.G.; GRIMM, C.M.; GANNON, M.J.; CHEN, M.-J (1991): Organizational Information Processing, Competitive Responses, and Performance in the US Domestic Airline Industry. *Academy of Management Journal*, 34(1), pp. 60-85.

STEPHAN, J.; BOEKER, W.; (2001): Getting to Multimarket Competition: How Multimarket Contact Affects Firms' Market Entry Decisions. In: Baum, J.A.C.; Greve, H.R. (Eds.): *Multiunit Organization and Multimarket Strategy*, Advances in Strategic Management, Vol. 18, Oxford: JAI Press, pp. 229-261.

STEPHAN, J.; MURMANN, J.P.; BOEKER, W.; GOODSTEIN, J. (2003): Bringing Managers into Theories of Multimarket Competition: CEOs and the Determinants of Market Entry. *Organization Science*, 14(4), pp. 403-421.

SÜDDEUTSCHE ZEITUNG (2004): Rückschlag für Easyjet. *Süddeutsche Zeitung* (17.07.2004), URL: http://www.sueddeutsche.de/wirtschaft/artikel/450/35415/ (access: 05.10.2004).

SÜEDDEUTSCHE DE (2004): Aus für Billigflieger V Bird. *Süddeutsche Zeitung* (08.10.2004), URL: http://www.sueddeutsche.de/wirtschaft/artikel/839/40799/ (access: 19.10.2004).

WARREN, R. (1972): The Concerting of Decisions as a Variable in Organizational Interaction. In: Tuite, M.; Chisholm, R.; Radnor, M. (Eds.): *Interorganizational Decision Making*, Chicago: Aldine.

YOUNG, G.; SMITH, K.G.; GRIMM, C.M.; SIMON, D. (2000): Multimarket Contact and Resource Dissimilarity: A Competitive Dynamics Perspective. *Journal of Management*, 26(6), pp. 1217-1236.

Part II

Alliances

9

COOPERATION AND INTEGRATION AS STRATEGIC OPTIONS IN THE AIRLINE INDUSTRY – A THEORETICAL ASSESSMENT

THOMAS FRITZ

1	Introduction	224
2	Theoretical Foundations	225
3	The Airline Industry	229
4	Cooperation and Integration as Strategic Options	235
5	Conclusion	248
6	References	249

Summary:
This chapter gives a theoretical comparison of Integration and Cooperation as strategic options for airlines. The analyses build on the ideas of the market-based view, the resource-based view and a transaction cost assessment. It is shown that due to the significance of economies of density and the coordinative effort in attaining these, an overlapping character of route networks tends to require integrative coordination, i.e. M&A transactions. This supports the thesis that further deregulation of international air travel will result in significant concentration efforts.

1 Introduction

Deregulation, privatization and the effects of globalization have changed the face of the international airline business in recent years.[1] In any industry, the globalization of value-creation increases the demands on the efficiency and effectiveness of the infrastructure that supports international labor division.[2] In the airline industry the specific consequences of these changes are major adjustments in network configuration and scheduling. The central instrument that has emerged for realizing such adjustments is the strategic alliance.

Major reasons for the dominance of alliance strategies among airlines are regulatory issues.[3] In the context of continuing liberalization, the airline industry has however experienced a relaxation of its strict regulatory regime.[4] This chapter therefore brackets out the issues associated with regulatory constraints and focuses instead on a theoretical analysis of the question whether today's cooperative arrangements, i.e., strategic alliances, represent a superior form of coordination or whether, in the event of complete deregulation, they will be replaced by comprehensive mergers and acquisitions (M&A). The investigation refers to scheduled airlines with extensive networks.[5] The theoretical analysis takes a comparative approach, using the market-based view, the resource-based view, and transaction cost analysis.

The chapter is divided into 5 parts. Following this introduction, section 2 defines the fundamental terms "cooperation" and "integration" and introduces the theoretical concepts used to compare these two as alternative strategies. Section 3 introduces the key success factors in the airline industry and potential effects of cooperation and integration in this context. Section 4 finally analyzes the problem of evaluating cooperation and integration as alternative strategies for airlines. The theoretical approaches are applied individually. Their specific

---

[1] See Hanlon (1996), p. 2.
[2] See Steininger (1999), pp. 105f on the effects of globalization on the airline industry.
[3] See Steininger (1999), pp. 47f; Malanik (1998), p. 12.
[4] See Walker (1999), pp. 96f.
[5] For a definition of air traffic, see Pompl (1998), p. 25. This delimitation excludes low cost carriers, as they operate in a fundamentally different business model.

explanations are then consolidated into an overall comparison of integration and cooperation. Section 5 summarizes the topic and offers an outlook.

2 Theoretical Foundations

2.1 Definition of Terms

Cooperation, i.e. strategic alliances, and integration as a result of M&A transactions are two distinct types of external strategies. In both cases, legally and financially independent companies are combined to form larger business entities.[6] The distinction between a cooperative agreement and integration is based on the criteria of legal and financial independence. Cooperative agreements are voluntary, and the participating companies retain their legal independence as separate entities and their financial independence in the business units are not affected by the cooperation.[7] In integration, on the other hand, at least one of the partners gives up its financial and legal independence. Integrations are characterized by hierarchy-based decision-making, whereas decision-making in cooperative agreements is generally consensus-based.[8] Cooperation and integration strategies can have similar motives:

> Indeed, alliances between competing firms on the one hand and horizontal acquisitions on the other are alternative moves. In order to join forces, pool assets, combine resources, and exploit synergies, firms can either choose to collaborate on well-defined areas of business while retaining their strategic autonomy, or they can completely and permanently merge their operations within a new and expanded legal entity.[9]

Still, there is a number of goals that can be realized only by integration. These include increasing revenues, taking over inefficient management and capital investment as well as satisfying the management's desire for prestige or a new challenge.[10] Since integration completely and permanently removes a competitor

---

[6] See Wöhe (1993), p. 403.
[7] See Blohm (1980), p. 1112; Straube (1975), p. 65.
[8] See Weston, Siu & Johnson (2001), p. 432; Spekman, Isabella & MacAvoy (2000), pp. 39f.
[9] See Garette & Dussauge (2000), p. 63.
[10] For a summary of M&A motives, see Copeland & Weston (1983), pp. 561ff; Weston & Weaver (2001), pp. 85f.

from the market, its competitive effect is greater than that of a cooperative agreement.

2.2 Overview of Theoretical Concepts

Comparing alternative forms of coordination, it is necessary to consider environmental factors as well as internal conditions. Among the established theories of business strategy, these two aspects are covered by the market-based view and the resource-based view.[11] Both views have been discussed extensively in academic literature, so that a sufficient theoretical grounding can be assumed. In order to evaluate the cost of coordination incurred in cooperation and integration, additionally transaction cost theory will be applied.

2.2.1 The Market-Based View (MBV)

The MBV is based on the work of Michael Porter and assumes that profitability is determined primarily by the structure of the industry in which the company operates.[12] Industry structure is analyzed in a five-forces-framework looking at bargaining power of suppliers, bargaining power of buyers, the threat of entry, the threat of substitution, and the rivalry among existing firms.[13] In the conceptual framework of the MBV, it is the central task of the manager to analyze the industry structure, to determine an ideal positioning, and to align the company's value chain to this positioning. For the positioning, Porter identified three generic strategies: differentiation, cost leadership, and concentration on selected areas ("niche strategy").[14] The main critique of the market-based view is that, as a result of its outside-in perspective, it gives too little consideration to company-internal aspects.[15]

---

[11] For an overview of various strategy theories, see Rumelt, Schendel & Teece (1994), pp. 24ff; Mintzberg, Ahlstrand & Lampel (1998).
[12] For an overview of the market-based view, see Porter (1980, 1985, and 1996). For the industrial economics underlying the MBV, see Scherer & Ross (1990); Bain (1968).
[13] For a detailed description of industry structure analysis, see Porter (1980), p. 4ff.
[14] See Porter (1999), pp. 70ff.
[15] See Rühli (1994), p. 41; Collis & Montgomery (1995, 1998).

From the MBV perspective, cooperation/integration serves primarily to improve competitive position and/or to influence the industry structure.[16] They can also be a response to changes in industry structure.[17] Cooperation/integration are superior strategies if the resulting competitive position cannot be achieved by the participating companies on their own.

2.2.2 The Resource-Based View (RBV)

The RBV assumes that a company's competitive advantages are based primarily on internal resources.[18] "Resources" are defined as the company's strategic, financial, technical, physical, people-related, and organizational assets, i.e., tangible and intangible resources. The RBV approach postulates necessary characteristics of the company's strategic resources in order to build the basis for competitive advantage. It assumes that companies are differently equipped with strategic resources (resource heterogeneity) and that these resources are at least partly inelastic in supply and difficult or impossible for competitors to copy (resource immobility).[19] It is the main criticism of the RBV that the resource concept has remained difficult to define in operational terms.[20]

From the notion of resource heterogeneity, it follows that companies are not fully equipped with all relevant resources and that access to further resources could improve competitiveness.[21] Cooperation/integration permits companies to make joint use of their resources and thus overcome resource-related constraints.[22] Unlike integration, cooperative agreements focus on selected areas, allowing companies to limit exchange processes to certain resources.[23] On the other hand,

---

[16] See Harrigan (1985), pp. 21ff; Child & Faulkner (1998), p. 18; Porter & Fuller (1989).
[17] See Hammes (1994); Hammes (1995), pp. 85ff.
[18] See Wernerfeldt (1984), Collis & Montgomery (1995), Prahalad & Hamel (1990); Barney (1996). For the academic roots of the RBV, see Penrose (1959).
[19] See Barney (1996), p. 142.
[20] A further perspective on resources relevant to competition is the resource-dependence approach, which primarily emphasizes a companies' dependence on critical resources in the assessment of exchange relationships. See Pfeffer/Salancik (1978); Child/Faulkner (1998), pp. 34ff.
[21] See Garette, Dussauge & Mitchell (2000), p. 100; Hamel (1991).
[22] See Hamel (1991); Garette, Dussauge & Mitchell (2000), pp. 99f; Spekman, Isabella & MacAvoy (2000), p. 35; Child & Faulkner (1998), p. 35.
[23] See Garette & Dussauge (2000), p. 63.

the remaining competitive relationship between the cooperation partners can be a barrier to learning and knowledge transfer.[24]

2.2.3 The Institutional Economics Perspective: Transaction Costs

Transaction cost analysis is a domain of the New Institutional Economics. It originated with Coase and was refined in particular by Williamson.[25] The focus of the analysis is not the physical transfer of goods, but the transaction costs that are incurred in the process of negotiating the physical transfer.[26] The key assumption in transaction cost theory is that the individuals involved in an exchange evaluate the transaction costs of alternative forms of coordination and take these costs into account when selecting the appropriate control structures.

The identification of transaction costs is based on two assumptions: human beings are characterized by bounded rationality[27] and they tend to exploit opportunities to maximize their own benefits (opportunistic behavior).[28] The value of transaction costs increases with the specificity of the transaction-related investments as well as the uncertainty about the partners' behavior while it decreases with transaction frequency.[29] The specificity of a transaction increases as the value of the human and material investments outside the relevant transaction decreases.

In the institutional economics perspective, market and hierarchy are the extremes of a continuum of possible control structures.[30] The existence of cooperative arrangements as intermediary forms of coordination between market and hierarchy requires them to offer an advantage in transaction costs. Advantages over market-based coordination can be the savings of costs for repeatedly negotiating contracts and improved transfer of information. Advantages over hierarchy include selective limitation of the scope of cooperation and simpler termination of cooperative relationships.[31]

---

[24] See Hamel, Doz & Prahalad (1989); Spekman, Isabella & MacAvoy (2000), pp. 231ff.
[25] See Coase (1937); Williamson (1975, 1981, 1990).
[26] See Picot (1982), p. 270.
[27] See Simon (1961), p. XXIV.
[28] See Williamson (1990), pp. 54f and 73ff.
[29] See Picot (1982), p. 272.
[30] See Jarillo (1988), p.34.
[31] See Sydow (1992), p. 143.

It is one weakness of the transaction cost approach that it does not explicitly consider the evolutionary nature of cooperation agreements. Increasing trust and stronger bonds between the partners can reduce the risk of opportunistic behavior.[32] A more general criticism is that transaction costs are difficult to operationalize, limiting the possibilities for inter-subjective validation of comparisons.[33]

Apart from these points of criticism, the transaction cost approach is a valuable analysis tool in comparing cooperation and integration strategies. The identification of transaction costs in different forms of coordination can furnish concrete evidence for comparing these alternatives.

3 The Airline Industry

The structure and specific challenges of the airline industry have been extensively discussed in economic literature.[34] As a basis for the theoretical assessment, the following section will therefore focus on key success factors for airlines as well as potential cost and revenue effects of cooperation/integration.

3.1 Key Factors for Success in the Airline Industry

3.1.1 Route Network as a Success Factor

One major success factor for airlines is to operate a flight schedule that is optimally configured to customer needs, i.e. that offers direct connections and high frequencies.[35] In designing an appropriate route network, one has to distinguish between external constraints associated with infrastructure bottlenecks and the ability to design flight schedules within these restrictions.

On the infrastructural side, airlines require slots on the target airports. In the slot-awarding process, established companies have clear advantages since they only lose a slot if they give it up. Competitors or new entrants are restricted to the

---

[32] See Child & Faulkner (1998), pp. 22f.
[33] See Schneider (1985), p. 1241; Sydow (1992), pp. 146f.
[34] See e.g. Pompl (1998); Doganis (2001); Doganis (2002).
[35] See Steininger (1999), pp.197ff.

allocation of new or unused slots.[36] The establishment of hub-and-spoke networks[37] permits airlines to realize economies of scale and density.[38] Economies of scale result from cost degression, depending on the size of the network, but independent of the network's utilization.[39] Economies of density are defined as cost advantages that arise from higher network utilization.[40] The bundling of traffic flows in hub-and-spoke networks is a significant source of such economies of density.[41]

Optimizing a hub-and-spoke network demands a trade-off between short flight times of direct flights and the service frequency of connecting flights. A larger number of one-stop-connections allows a higher frequency on these city-pairs. The challenge in network planning is not the optimization of individual routes, but the optimization of the network as a whole, considering the complete effects of any adjustment on the connection options in the entire network.

3.1.2 Yield and Cost Management as a Success Factor

The product "air travel" can neither be produced in advance nor put on stock. With the takeoff, any unsold seats become worthless.42 All other features being equal, the price is the critical feature. For price differentiation, airlines have defined sub-markets (ticket reservation classes) that reflect different types of service and reservation flexibility.[43] Therefore, in addition to the absolute price level, a sophisticated system for optimizing price and capacity availability is critical to an airline's business success.

The objective of this yield management is to maximize the total revenue from a flight by the combination of price and capacity management. The main challenges stem from the restrictions to short-term capacity adjustment and the

---

[36] See Beyhoff (1995), pp. 157ff.
[37] See Brueckner & Spiller (1994), p. 380; de Wit (1995), p. 174; McShan & Windle (1989).
[38] See Höfer (1993), pp. 214ff; Caves, Christensen & Tretheway (1984). Usually there is distinction between economies of scale and scope. In the airline industry however economies of scope occur as economies of density due to the network-based value creation. See Burton & Hanlon (1994), p. 218.
[39] Caves, Christensen & Tretheway (1984), p. 474.
[40] See Bailey, Graham & Kaplan (1985), pp. 50f; Berry, Carnal & Spiller (1996).
[41] Caves, Christensen & Tretheway conclude that a 1% increase in traffic would only lead to a 0.8% increase in total cost. Caves, Christensen & Tretheway (1984), p. 175.
[42] See Pompl (1998), p. 31; Wells (1999), p. 212.
[43] See Daudel & Vialle (1994), pp. 44ff.

fact that reservations and cancellations are not made at one point in time, but are instead distributed over time right until takeoff.[44]

The profitability effect of yield management also depends heavily on the parallel cost management. An optimization of utilization will reduce the flight-dependent unit costs. However, utilization has to be considered on different levels from individual flight to total fleet, in order to manage capacity costs and the network-wide effects of providing specific capacities.[45] Only an integrated view of the costs and revenue potential of certain capacity contingents, i.e. plane types, can guarantee that capacity is provided at an efficient level.

In the other main cost blocks - crew, maintenance, and administration - complexity is a major cost driver. For crews and maintenance, the complexity consists in the number of destinations and the different types of equipment.[46] In administration, larger networks incur higher complexity costs due to the greater amount of planning effort required. Again, complexity reduction has to be closely coordinated with network planning and yield management.

3.2 Cooperation and Integration in the Airline Industry

Following the deregulation of the airline industry, the importance of cooperation and integration has increased considerably. In the United States, the liberalization of domestic air travel was followed by a dramatic concentration process.[47] In international air travel, cooperative arrangements have so far played the leading role resulting in a number of global alliance networks. Given the scope of the alliances and the significance of the participating airlines, competition has in many areas become a race between global alliances.[48] Building on these existing alliances, potential cost and revenue effects of cooperation or integration can be identified.

---

[44] See Wells (1999), pp. 216ff; Weber (1997), pp. 101ff.
[45] Weber (1997), pp. 159ff, gives an overview of different types of utilization.
[46] See Weber (1997), pp. 169ff.
[47] For an overview of the liberalization in the US see Steininger (1999), pp. 29ff.
[48] See Gomes-Casseres (1994).

3.2.1 Potential Cost Effects of Cooperation and Integration

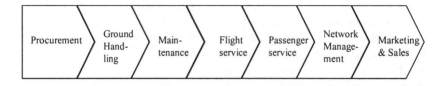

Figure 9-1: Airline Value Chain[49]

In analyzing the potential cost effects of cooperation/integration in air travel, the basic industry value chain provides a useful framework (see figure 9-1).

In the procurement of aircraft and fuel, airlines can realize economies of scale as unit cost degressions for order-specific fixed cost and an improved bargaining position through bundled demand.[50] In ground handling, they can achieve economies of density by making joint use of existing facilities.[51] In maintenance, they can also realize synergy effects through know-how pooling, centralization, and higher utilization of facilities.

Flight operations account for approximately 40% of an airline's total operating cost and comprise costs for crew and the use of the aircraft. In aircraft use, economies of density can be generated by improving utilization or increasing aircraft capacity. These economies can be realized through the consolidation or expansion of hub-and-spoke networks. Cooperative agreements are the simplest way to enlarge the network and to profit from these cost advantages. Economies of density through adaptation of network and flight schedule are thus the main driver of strategic alliances.[52] The combination of the individual networks into a multi-hub network can happen at three levels.[53]

Activities at the first level serve to optimize the existing hub-and-spoke network. Smaller airlines that operate in regional/national markets with low traffic volumes are integrated into the network as suppliers of traffic. Owing to the low marginal cost of an additional passenger, the traffic thus "supplied" is highly

---

[49] See Steininger (1999), p. 241 building on Porter (1992), pp. 59ff.
[50] See Höfer (1993), p. 214.
[51] See Steininger (1999), pp. 174ff.
[52] See Dennis (2000), pp. 75f; Burton & Hanlon (1994), p. 218.
[53] See for a similar distinction O'Toole (1999), p. 83.

profitable.[54] Cooperation arrangements are also used to outsource certain short- and medium-distance routes to regional niche players.

At the second level, the integration of networks among equal partners creates a continental multi-hub network. This helps to overcome infrastructure bottlenecks and provide ample connectivity by "de-congesting" the large hubs in a multi-hub network through redistribution and re-bundling of intercontinental traffic.[55]

Establishing partnerships across multiple continents stands on the third level.[56] Major trunk routes link the strategic gateways in a global network. The critical factor for the economic success of these trunk routes is the performance of the connecting flight network at either side of the trunk route.[57] Cooperation or integration is required because even major airlines are unable to build an effective and efficient transfer network at the ends of all trunk routes by themselves. In addition to these cost-reducing capacity effects, an extended network also offers significant revenue potential, a fact which will be discussed in the next section.

Synergy potential in the actual network management is limited to economies of scale from joint development and operation of an IT system. Similar economies of scale can also arise in marketing and sales.

3.2.2 Potential Revenue Effects of Cooperation and Integration

Larger, denser, and more extensive route networks generally satisfy customer preferences for short and hassle-free flights better.[58] In addition to seeking direct connections and high frequencies, customers also prefer single-carrier connections.[59] These preferences can be satisfied when multiple airlines form a shared network and coordinate their flight schedules.

Adding complementary destinations to a network increases the city-pair connections that can be offered. On the one hand, alliances can service a larger number of direct flights, since the partners might jointly achieve the economically necessary level of utilization. Additionally, in a hub-and-spoke

---

[54] See Tretheway & Oum (1992), p. 76.
[55] See Oum, Taylor & Zhang (1993), pp. 15f; Steininger (1999), pp. 242f.
[56] "A global airline network is defined as an airline network capable of providing service to most of the large and medium-sized cities around the world, particularly in North America, Europe and Asia." Oum, Taylor & Zhang (1993), p. 15.
[57] See Steininger (1999), p. 245; Oum, Taylor & Zhang (1993), p. 15.
[58] See Oum, Taylor & Zhang (1993), pp. 14ff.
[59] See Oster & Pickrell (1986).

network, every additional spoke served results in a disproportionate increase in the number of one-stop city-pairs. Capitalizing on this extended city-pair portfolio requires however a maximum connectivity at the hub.[60]

From a marketing perspective, partners need to form a common, seamless service network that emulates the characteristics of a single-carrier connection.[61] Central elements in all alliances are code-sharing agreements. Code-sharing agreements are marketing deals in which one airline can market flights under its own code while the partner actually operates the flight.[62] Code-sharing is currently the main tool for route network integration. It permits each partner to offer higher frequencies without expanding its own service.[63]

The realization of a seamless service also requires the harmonization of the partners' performances. This encompasses securing a uniform quality standard on the ground and in the air through similar aircraft quality and crew training levels as well as the consolidation of lounges and arrival/departure gates.[64] In addition, consumers' perceptions of flying with a single carrier are increased by consistency in branding, joint promotions[65] and reciprocal recognition of loyalty programs.[66]

While joint marketing is a prerequisite to achieving sales increases, the coordination of sales and distribution is the main driver. The closer production is integrated, the more coordinated pricing and yield management have to be. In pricing, the main task is to prevent the partners from offering different prices for similar services. Even simple code-sharing requires coordinated yield management, i.e., a division of capacity access, rules for handling over-bookings, and sharing of the marketing risk.[67]

---

[60] Connectivity expresses the number of connecting flights that can be reached in a timeframe between the minimum transfer time and the maximum waiting time of 2-3 hours. See Steininger (1999), pp. 259f.
[61] See Oum & Park (1997), p.140.
[62] See Oum, Park & Zhang (1996), p. 187; GAO (1995), p. 13.
[63] See Oum, Park & Zhang (1996), pp. 188ff.
[64] See Dennis (2000), p. 81; Steininger (1999), pp. 276ff.
[65] See Steininger (1999), pp. 282ff.
[66] See Gallacher (1997), p. 35; Gallacher (1999b).
[67] See Steininger (1999), pp. 315ff.

4 Cooperation and Integration as Strategic Options

The existing alliance constellation raises the question about its stability in the event of further M&A activities.[68] A number of authors discuss the implications of extensive liberalization for the existing culture of alliances in international air travel, a typical comment being:

> Conscious that deregulation led to increased concentration in the US domestic industry, airlines expect that liberalization on international routes will have much the same effect.[69]

In this line of thinking, strategic alliances are temporary structures that will in the long term be replaced by merged airlines.[70] Oum and Park, on the other hand, conclude that, even with the total elimination of regulatory restrictions, airlines will not attempt to build global networks through either internal growth or mergers. They base their thesis on the time disadvantages and the enormous financial challenges of such a strategy.[71] This section will address the question whether alliances are temporary structures or a superior form of coordination based on the theories introduced above.

4.1 Analysis from the Market-Based View

A comparative evaluation of cooperation and integration using the market-based view focuses on air-travel-specific approaches for improving competitive position and/or influencing industry structure. The focus of the evaluation is the extent to which strategic alliances can implement these approaches and where integration represents the superior alternative.

---

[68] "Few believe that the current shape of the global groupings is anything close to the final endgame." O'Toole (1999), p. 79.
[69] Burton & Hanlon (1994), p. 209.
[70] See Malanik (1998), p. 15.
[71] See Oum & Park (1997), p. 134.

4.1.1 Market-Based View: Improving Competitive Position

Industry structure and success factors specific to the airline industry open up various fields for improving competitive position by cooperation/integration. Table 9-1 provides a summary of the competitive forces and approaches to influence them.[72]

Competitive force	Approach to improve position
Bargaining power of suppliers	- Economies of scale (joint purchasing)
Bargaining power of buyers	- Product differentiation through FFP
Threat of entry	- Economies of density and scale - Expansion of market share at home hub
Rivalry among established airlines	- Product differentiation - Risk diversification - Limitation of competition

Table 9-1: Influencing Competitive Position through Cooperation/Integration

Four levers drive these possibilities: economies of density and scale, increase in market share and product differentiation, risk diversification, and limitation of competition.

Economies of Density and Scale

Economies of density and scale are among the most important drivers of cooperation/integration in air travel. Economies of scale, especially in purchasing, can be equally well-achieved through both cooperation and integration.[73] The realization of economies of density requires a joint optimization of the route network. In cooperative agreements this is associated with a high level of coordination effort. Especially if the optimization of the network requires restricting the activities of one partner, the conflict potential is significant. In M&A transactions, the dominant partner can simplify such coordination processes by designing and issuing suitable policies and directives.

---

[72] Due to its limited significance, the threat of substitution is not analyzed here.
[73] See Ionides (1999), p. 35.

The MBV however does not allow an evaluation of this problem. It is particularly lacking a method for assessing the cost of coordination. It is only possible to ascertain that the potential for improving competitive position exists both in cooperative agreements and integration.

Market Position at Home Hub and Product Differentiation

An airline's potential for product differentiation depends heavily on its market position at its home hub. Central differentiation features include the city-pair connections offered, which match customers' timing and geographic preferences and the value of the airline's Frequent Flyer Program (FFP) for the customer.

The fundamental prerequisite for acquiring a customer is to offer the connection that matches the customer's preferences.[74] In an alliance, this customer value can be expanded considerably through network links. At the same time, the coordination effort for the complete joint optimization of an alliance network is enormous. The expansion of an airline's position at its home hub requires an expanded slot portfolio. In M&A transactions, taking over the partner's slots creates maximum freedom for an integrated optimization of the network. In alliances, on the other hand, the slot portfolio can be expanded only by a slot swap with the partners or by code-sharing. In a swap or exchange, the partners must accept reciprocal dependencies, while code-sharing is limited to the joint use of the combined slot portfolio.

FFP is one major vehicle for differentiation and both cooperation and integration can increase the FFP's value by offering collection and redemption in a combined network. In a cooperation arrangement, the parallel exchange of customer data for marketing purposes can, however, pose a problem due to privacy regulations and the competitive importance of this information.

Risk Diversification

Demand for air travel depends heavily on regional economic conditions.[75] This regional demand is furthermore influenced by extraordinary risk factors such as terrorism. Thus, significant regional demand differences are inevitable. Risk diversification through cooperation and integration strategies is therefore focused on a geographical dimension.

---

[74] I.e. mergers have generally led to price increases on markets that were dominated by the merged players. See Borenstein (1990), pp. 400ff.
[75] See Steininger (1999), pp. 16ff.

The options for risk diversification through strategic alliances are, however, limited.[76] Geographic risk diversification is based on the notion that worse results in one region will be offset by better results in other markets. In alliances, markets are only linked to strengthen the position of the airlines in their home markets. A transfer of operating results naturally does not occur. The alliance partners profit from an increase in demand experienced by their partners only through the connecting traffic that is forwarded into their own networks.

Limitation of Competition

The competition-reducing effect of cooperation/integration needs to be considered in two dimensions. In the first dimension, such strategies limit competitive intensity between the individual airlines in jointly served markets. The limitation can be achieved through the coordination of pricing and flight schedules among the alliance partners. Integration however would fully eliminate this competition.

The second dimension is competition between entire alliance networks. One effect of the formation of stable alliance networks has been the reduction of possibilities for involving additional partners. A cooperative agreement is hardly possible when one of the target partners is already engaged in an alliance with a company that has a similar profile. An expansion of the alliance with further partners who are already involved in other alliances requires an M&A effort.

If the acquisition, however, is a hostile takeover, it will result in the usual disadvantages such as high acquisition premiums, risks to relationships with customers and suppliers, and the loss of qualified people.[77] Beyond these disadvantages, a takeover can also generate conflicts within the existing alliance. If the acquired airline has a profile similar to one of the existing alliance partners, the resulting friction can destabilize the cooperation within the original alliance. Such a strategy is therefore only relevant for airlines whose cooperation portfolio lacks specific strategic features.

4.1.2   Interim Conclusions

A MBV-based evaluation has to take into account the specific relevance of the competition-reducing approaches. For partners in a balanced alliance, the strategy of taking over competitors outside of the alliance requires careful consideration. Geographic risk diversification is also not the major factor in

---

[76] See Contractor & Lorange (1988), pp. 11f.
[77] See Copeland, Koller & Murrin (1995), pp. 435ff.

assessing cooperation vs. integration. It is primarily economies of scale and density as well as the improvement of market position in the domestic market that determine the attractiveness of joining forces among airlines.

In the realization of synergetic advantages, cooperation and integration both offer similar potential. The assessment depends on the evaluation of the respective coordination efforts. From the perspective of the MBV, it is therefore not possible to determine which alternative – cooperation or integration – is the superior choice. The value of the stand-alone MBV analysis lies primarily in identifying areas of potential synergy or improved market position.

4.2 Analysis from the Resource-Based View

4.2.1 Strategic Resources in the Airline Industry

In a resource-based analysis, the key factors for success can be synthesized into one single strategic resource: the capability to effectively and efficiently offer an extensive choice of routes. The foundation for this strategic resource is the portfolio of slots. The strategic importance of slots and the design process that builds on this portfolio should be briefly explained using the VRIO analysis.[78]

An effective and efficient network enables an airline to respond to opportunities and threats arising from external factors such as privatization, globalization, and deregulation. The resource is thus *valuable*. It is also sufficiently *rare*. While all airlines do have more or less efficient route networks, direct substitution is barred by different geographic focus. The possibility of *imitation* is very limited due to infrastructure bottlenecks. The ability of an *organization* to exploit the potential of this resource on the cost and revenue sides depends on the extent to which it can design an efficient and effective flight schedule. Effectiveness refers to superior coverage of customer preferences in combination with strong yield management. Efficiency refers to a sophisticated cost management for network and deployment planning. The ability to capture the potential of the network as a resource thus depends on high-quality network planning and yield management.

Strategic Resources and Cooperation/Integration

Cooperation and integration serve to overcome resource scarcity and/or to join complementary resources. Linkages of the route networks and extensions of frequent flyer programs alone, however, do not necessarily imply that strategic

---

[78] For a detailed discussion see Barney (1996), pp. 145ff.

resources have been linked. A combination's strategic value rather lies in using the expanded infrastructure as a basis for joint optimization. Joint optimization aimed at greater efficiency plays a particularly important role in the case of overlapping structures. For complementary structures, the key is effective linkages between the route networks.

In integration, the joint optimization can be executed by means of command-and-control directives issued in the resulting hierarchical relationship. In alliances, on the other hand, the initial step is to develop a shared perspective on decision-making that goes beyond optimization of individual alliance members.[79] In order to create and enforce such a shared decision-making perspective it is necessary to have an effective governance structure, e.g. in the form of a joint organizational unit. The tasks of this unit should include the coordination of flight schedules, pricing and yield management as well as the calculation and distribution of the value created by the alliance.[80] Taking into account the resource contributions of the partners, the joint utility of the alliance should be maximized and distributed among the partners reflecting their respective contribution.[81] This task fundamentally requires a comprehensive exchange of knowledge and data among the partners. The exchange has to encompass not only data about cost structures, future investments, and customers, but also the coordination of systems for network planning and yield management. The sharing of market forecasts and customer expectations is also vital for joint pricing and yield management.

This type of exchange includes data and capabilities that are essential to exploit the individual strategic resources. In the course of cooperation, alliance partners might acquire these capabilities, thus eroding the basis of individual competitive advantage. The exchange of knowledge therefore presupposes a high level of trust and/or the creation of safeguards. These aspects are to be discussed below in the context of institutional economics (i.e., transaction costs).

Scale or Link Alliances?

Garette and Dussauge base their distinction between scale and link alliances on the relationship of the resources contributed to the alliance. In airline alliances, the resources contributed typically have a similar profile. The main differences lie in the geographic focus of the partners' respective networks. On this basis, the focus of scale alliances in the airline industry lies in the attainment of economies resulting from joint use of resources. The levers for achieving economies of

---

[79] For the conditions of relationship-specific rents see Dyer & Singh (1998), p. 666.
[80] See Berardino & Frankel (1998a); Berardino & Frankel (1998b).
[81] See Götsch & Albers, chapter 11 in this volume.

density in the case of overlapping networks result from the bundling and redistribution of the joint traffic. Link alliances, on the other hand, aim at the combination of the partners' differing and complementary resources.[82] Applied to the airline industry, such complementarities can be observed in the linkages between geographically disjunctive route networks. In general, however, airline alliances have both overlaps and complementary routes. A categorization thus depends on whether scale aspects or link aspects dominate.

Garette and Dussauge formulate an assessment of integration versus the existing alliances like this:

> Horizontal acquisitions will always outperform scale alliances[83] [...] such scale alliances provide a way of avoiding or at least postponing mergers in industries undergoing strong concentration processes.[84]

This suggests that an integration strategy would be superior to alliances in which the scale characteristics dominate. However, this assessment is based mainly on considerations relating to transaction-cost theory.

4.2.2 Interim Conclusions

The resource-based view cannot provide an unambiguous evaluation of integration as the alternative to strategic alliances. However, it does allow alliance relationships to be classified into scale and link alliances, providing a starting point for the transaction cost analysis.

Both the market-based view and the resource-based view can furnish valuable insights for the comparison of integration and cooperation. A concrete comparative evaluation of the two requires a transaction cost analysis, which will be discussed in the following section.

4.3 Analysis from the Institutional Economics Perspective

The institutional-economics-based comparison of integration and strategic alliance follows two steps. First, the importance of specific investments for airline alliances and the possible levers for opportunistic behavior are explored. Then the risks of opportunistic behavior in the various phases of the integration

---

[82] See Garette, Dussauge & Mitchell (2000), pp. 102f.
[83] Garette & Dussauge (2000), p. 65.
[84] Garette, Dussauge & Mitchell (2000), p. 102.

process are reviewed. The assessment concludes in a comparison of the transaction costs of both alternatives, considering specificity and the uncertainty of behavior.

4.3.1 The Importance of Specific Investments in Airline Alliances

In a strict sense, an investment is defined as a payment on which later payments or income will follow.[85] In a broader sense, the sacrifice of a current benefit in the expectation of a future reward or payoff can also be understood as an investment.[86] This broader definition thus also encompasses investments whose "initial payment" takes the form of opportunity costs for not performing specific activities.

For cooperation-specific investments, the basic assumption is that the returns expected in the future without the cooperation will be much lower or non-existent. Furthermore, there is no possibility to compensate for the investments in the cooperation through investments in other activities. A high level of cooperation specificity thus means that, without the cooperation, the value expected in the future would decrease and there is only a limited possibility to offset the lack of utility in other ways.[87]

Marketing and Facilities

From a "payment"-perspective, the major alliance-specific investments are the coordination and integration of IT systems, the expansion of shared airport facilities, and expenses for joint alliance marketing.

When a partner leaves an alliance, alliance-specific marketing using a joint alliance brand loses much or all of its value for the departing partner. Only if the airline entered the alliance to enhance its own image, the image effects might last positively beyond the alliance membership.

Investments in joint airport facilities also lose their value when airlines end an alliance and have to resume offering separate facilities. Investments in airport facilities and marketing are partner-specific, but it is possible to reverse such investments without future disadvantages. Therefore, they are only a small barrier to ending an alliance engagement.

---

[85] See Hax (1993), pp. 1f.
[86] See Eisenführ (1997), p. 1.
[87] Williamson also bases his analysis on a broad concept of investments; see Williamson (1996), p. 13.

The coordination of the IT systems or the introduction of a new alliance-wide IT system can generate high costs for specific investments in human resources and equipment that retain their value only if the alliance continues. It is this high investment specificity that has delayed airlines in reaching agreement on a joint IT strategy.[88]

Re-configuration of Route Networks and Schedule

From a "utility"-perspective, the key alliance-specific investment is the coordination of the timing and geographic aspects of flight schedules. The coordination of supply demands the cancellation of services to certain city pairs at certain times by at least one of the partner airlines in order to achieve a better alliance-wide distribution of services.

Geographic coordination for example would require the partners in an alliance to focus on specific areas and limit the extent of overlapping services. Such efforts can go as far as requiring a partner to abandon certain city pairs and is thus also associated with the loss of revenue. These opportunity costs require compensation, i.e., a payoff from the alliance.[89] This payoff can be achieved on the routes in the remaining geographic focus area. The exclusion of parallel offerings by the alliance partners have to be associated with higher frequencies and greater utilization rates for the incumbent partner.

Changes in flight schedules usually require changes in the airline-specific slot portfolio. Such a change can be completed by a swap with alliance partners. However, once a slot has been given up, it is hardly possible to get it back.[90] Thus it is hardly possible to reverse an investment in a jointly optimized flight schedule after the end of the alliance.[91] In strategic alliances at the second level, in particular, with their focus on increasing efficiency, very extensive coordination is required. At the first level, by contrast, smaller markets are integrated into a larger network. Synergies arise primarily from the additional traffic fed into the larger network, without requiring a major overhaul of the flight schedules. Likewise, the linkages of intercontinental markets at the third level require a focus on connecting the hubs, which does not impose great restrictions on the partners' individual flight schedule designs.

The adjustment of the own schedule in order to optimize the alliance is thus an alliance-specific investment. Following the joint optimization, the productive

---

[88] See Pommes (1998), p. 28.
[89] See Berardino & Frankel (1998a), p. 83.
[90] See Beyhoff (1995), pp. 157ff.
[91] See Berardino & Frankel (1998a).

value of the individual network becomes partner-specific and its utility outside the alliance decreases sharply. As well as the integration of IT systems, these investments effectively "lock in" the partners to the alliance.[92]

4.3.2 Opportunistic Behavior in Strategic Alliances

Opportunism and Joint Value Creation

In the course of a strategic alliance, the risk of opportunistic behavior exists primarily in the exploitation of alliance relationships in order to maximize individual utility at the alliance's expense. Incentives for such behavior occur because the partners remain legally independent companies accountable to individual stakeholders.

The basis for coordination of joint production are mechanisms for measuring and distributing the value created in the alliance.[93] After the "alliance value" has been quantified, it must be fairly distributed to the partners, i.e., consistent with their contributions to the alliance.[94] Given the complexity of the production relationships in networks, agreement on a generally accepted allocation key to offset such asymmetries is a major challenge. The perfect transfer mechanism would create an incentive structure in which individual utility maximization would lead to actions that maximize the utility of the entire alliance. Bounded rationality of human decision-making, however, makes the attainment of such a perfect system nearly impossible. Latitude remains for opportunistic behavior, for example, as a result of poor data transfer in flight schedule planning or the rejection of certain adjustments to the flight schedule.

Opportunistic Behavior and Specific Investments

Preventing opportunistic behavior requires an effective sanction mechanism. One sanction could be a credible threat to quit the alliance. However, the partners' pre-existing specific investments reduce the credibility of this type of sanction. In alliances with largely integrated IT systems or based on far-reaching coordination of flight schedules the partners are heavily dependent on the continuation of the alliance. Furthermore, most potential substitute partners are already engaged in alliances of their own. After exiting an alliance, an airline

---

[92] See Gomes-Casseres (1994); Shapiro & Varian (1999).

[93] Berardino & Frankel state: "[...] if the alliance is to hold together, a formal way of compensation must be devised so that the alliance can overcome the very real individual interests of the partners." Berardino & Frankel (1998a), p. 86.

[94] See Albers (2000), pp. 58f.

does not necessarily face open doors to other alliances, particularly when the other alliances already include a partner with a similar profile. The credibility of an individual partner's threat to quit the alliance is therefore limited.

Given this constellation, another possible sanction arises in the threat of exclusion from the alliance. This would typically require the assent of all of the alliance partners and is thus only realistic if the opportunistic behavior hurts all members of the alliance. When a partner is excluded from an alliance, the damage incurred from the investment specificity might however outweigh the problems caused by the opportunistic behavior. The power of the threat to eject a partner from an alliance is thus also limited.

The specificity of investments and the risk of opportunism are central obstacles to more intensive alliance cooperation. At the same time, the dynamics of cooperation and integration in the airline industry create incentives to avoid excessive dependency on certain partners.[95]

4.3.3 Opportunistic Behavior in the Integration Process

In an M&A process, the risk of opportunistic behavior exists in all phases of the integration. These risks represent sources of increased transaction costs.[96]

The costs of evaluation, organizational integration, the consequences of a drop in motivation and a loss of flexible responsiveness all represent bureaucratic costs and are considered the typical transaction-cost disadvantages of hierarchical integration.[97] Financing costs and a possible strategic price premium are not bureaucratic costs in the strict sense, but can still be understood as integration-specific transaction costs.

The risk of a drop in motivation plays a role in hostile takeovers as well as in friendly mergers. Changes in the corporate environment usually result in uncertainty among employees throughout the whole post-merger process. This often leads to the loss of highly skilled individuals.[98] In addition, the danger of cultural incompatibility between the two companies further raises the risk of failure for the integration.[99] This risk is significant in the airline industry where

---

[95] See O'Toole (1999), p. 83.
[96] See Weston, Siu & Johnson (2001).
[97] See Williamson (1996), pp. 81ff.
[98] See Copeland, Koller & Murrin (1995), p. 439.
[99] For the concept of "corporate culture" and its significance for economic success see Besanko, Dranove, Shanley & Schaefer (2004), pp. 579ff.

company cultures tend to reflect national cultures.[100] In this context, a track-record of cooperation in existing alliances may have a risk-reducing effect.

A loss of flexibility results from the limits placed on the companies' options for action. On completion of an integration, the option of switching partners is significantly reduced. A divestment, for example, would generate its own set of transaction costs on top of the original integration costs.

4.3.4 Interim Conclusion

In transaction cost theory, the superior institutional arrangement is the one associated with lower transaction costs. Hierarchical integration is advantageous in situations in which the risk of opportunistic behavior in a cooperative arrangement leads to transaction costs that are greater than those of the M&A transaction. This is primarily the case when the utility of a cooperative agreement depends heavily on cooperation-specific investments and the level of uncertainty about the partners' behavior is high.

The more the partners' networks overlap, the greater the need for adjustments. But the further schedules and route networks have to be adjusted to specific partners, the higher the specificity of the alliance investment becomes. The least specificity would be found in alliances in which the overlaps are limited to connections between two otherwise disjunctive hub-and-spoke networks.

Increasing partner-specific coordination of one's own network increases the susceptibility of the alliance to opportunistic behavior. Consequently, the transaction costs for flight schedule coordination in alliances are higher the more the participating networks overlap. At the same time, extensive coordination and elimination of such overlaps is the largest source of synergy potential in alliances, especially on the cost side.

Berardino and Frankel propose a catalog of organizational principles to support such extensive coordination, including long-term contracts and high financial penalties.[101] Regardless of its completeness, this catalog conveys an impression of the design and scope of the transaction costs associated with coordination. In order to realize the additional efficiency potential, the airlines would need to turn to hierarchical integration. This raises the question of whether the potential benefit exceeds the transaction costs associated with an M&A transaction.

---

[100] For the importance of national cultures in corporate culture see Mead (1998), pp. 123f.

[101] See Berardino & Frankel (1998b), p. 71.

A weakness of transaction cost theory remains the problematic operationalization of transaction costs. A quantification of the costs of opportunistic behavior can, for example, only be estimated based on evaluation of the probability and extent of the risk of such behavior. The analysis above can, however, lead to a qualitative statement: the more the partners' flight schedules overlap, the more likely it is that an integration strategy is superior to the existing airline alliances.

4.4 Results of the Theoretical Analysis

In isolation, neither the MBV nor the RBV support the derivation of a concrete comparative assessment. Nevertheless, the market-based view offers valuable analysis tools for understanding the competitive significance of strategic alliances. The resource-based view provides additional insights into the design and potential linkages of strategic resources in the airline industry (table 9-2). On the basis of these insights, transaction cost theory helps to derive a concrete comparative assessment. The result largely coincides with the conjectures supported by the market-based view and the resource-based view.

	MBV	**RBV**	**Transaction Cost**
Main Insight	Analyzing main drivers in cooperation and integration - Economies of scale/ density - Reducing competetive pressure	Classifying alliances in scale and link alliances depending on the relationship of the partners' route networks	Identifying/ evaluating coordination cost in cooperation /integration, depending on the classification from the resource based perspective
Short-coming	Evaluation of coordination cost not possible	Evaluation of coordination cost not possible	Full operationalization of coordination cost remains difficult

Table 9-2: Comparison of Results from the Theoretical Analyses

In combination, the two strategy theories, the MBV as well as the RBV, and the transaction cost approach can all make a valuable contribution to the comparison of integration and cooperation as alternative strategic options.

In summary, mergers or acquisitions are most likely to be superior to strategic alliances when the networks of the future partners have many overlaps, i.e., when the synergy potential outweighs the transaction costs of integration.

5 Conclusion

This evaluation focused on the importance of joint flight schedule configuration between partner airlines. Superior coordination is the most important source of synergy potential for cooperative agreements or integration. At the same time, capturing this potential is a central problem within strategic alliances.

The theoretical investigation supports the view that deregulation will unleash an industry concentration process. Given the significance of geographic overlaps, it can be assumed that this concentration process will start at a regional level. A likely scenario is the emergence of continental mega-airlines with large market shares and regionally optimized networks on the major international markets. These strong regional players will continue to form strategic alliances in order to secure global market coverage. The starting point for such an industry consolidation could be the existing airline alliances in which concentration processes would occur. The long-term collaborative experience embedded in these alliances would considerably reduce the transaction costs in partner selection, partner evaluation, and coordination of an integration process. On the other hand, the concentration process could also be kicked off by airlines that are excluded from alliances or whose alliances fail, as a way of overcoming competitive disadvantages.

Such considerations are becoming increasingly prominent in the industry. It is therefore necessary to develop the analyses presented here into concrete, quantifiable models for the decision for or against an M&A strategy. An optimal decision model should additionally incorporate criteria such as financial variables, intercultural aspects, and regulatory issues.

6 References

ALBERS, S. (2000): *Nutzenallokation in strategischen Allianzen von Linienluftfrachtgesellschaften*. Arbeitsbericht Nr. 101 des Seminars für Allg. BWL, Betriebswirtschaftliche Planung & Logistik, Köln.

BAILEY, E.E.; GRAHAM, D.R.; KAPLAN, D.P. (1985): *Deregulating the Airlines*. Cambridge, Mass. et al.: MIT Press.

BAIN, J.S. (1968): *Industrial Organization*. New York: Jai Press.

BARNEY, J.B. (1996): *Gaining and Sustaining Competitive Advantage*. Reading, et al.: Addison-Wesley.

BERARDINO, F.; FRANKEL, C. (1998a): Keep the Score. *Airline Business*, 14(9), pp. 82-87.

BERARDINO, F.; FRANKEL, C. (1998b): Alliances: The next Step. *Airline Business*, 14(9), pp. 68-71.

BERRY, S.; CARNALL, M.; SPILLER, P. (1996): *Airline Hubs: Costs, Markups and the Implications for Customer Heterogenity*. Cambridge, Mass.

BESANKO, D.; DRENOVE, D.; SHANLEY, M.; SCHAEFER, S. (2004): *Economics of strategy*, 3rd Ed., New York: Wiley.

BEYHOFF, S. (1995): *Die Determinanten der Marktstruktur von Luftverkehrsmärkten*. Köln.

BLOHM, H. (1980): Kooperation. In: Grochla, E. (Ed.): *Handwörterbuch der Organisation (HWO)*, 2nd Ed., Stuttgart: Poeschel, pp. 1112-1117.

BORENSTEIN, S. (1990): Airline Mergers, Airport Dominance and Market Power. *American Economic Review, Papers and Proceedings*. 80(2), pp. 400-404.

BRUECKNER, J.K.; SPILLER, P.T. (1994): Economies of traffic density in the deregulated airline industry. *Journal of Law and Economics*, 37(2), pp. 379-419.

BURTON, J.; HANLON, P. (1994): Airline Alliances: Cooperating to Compete? *Journal of Air Transport Management*, 1(4), pp. 209-227.

CAVES, D.W.; CHRISTENSEN, L.; TRETHEWAY, M. (1984): Economies of Density versus Economies of Scale: Why trunk and local Airline cost differ. *RAND Journal of Economics*, 15(4), pp. 471-489.

CHILD, J.; FAULKNER, D. (1998): *Strategies of Cooperation: Managing Alliances, Networks and Joint Ventures*. Oxford et al.: Oxford University Press.

COASE, R. (1937): The Nature of the firm. *Economia*, November 1937, pp. 386-405.

COLLIS, D.J.; MONTGOMERY, C.A. (1995): Competing on Resources: Strategy in the 1990s. *Harvard Business Review*, 76(4), pp. 118-128.

COLLIS, D.J.; MONTGOMERY, C.A. (1998): *Corporate Strategy - A Resource Based Approach*. Boston: Irwin McGraw-Hill.

CONTRACTOR, F.J.; LORANGE, P. (1998): Why Should Firms Cooperate? The Strategy and Economic Basis for Cooperative Ventures. In: Contractor, F. J. / Lorange, P. (Eds): *Cooperative Strategies in International Business*, Lexington: Lexington Book.

COPELAND, T.E.; KOLLER, T.; MURRIN, J. (1995): *Valuation – Measuring and Managing the Value of Companies*. 2nd Ed., New York et al.: Wiley.

COPELAND, T.E.; WESTON, F.J. (1983): *Financial Theory and Corporate Policy*. 2nd Ed., Reading, Mass. et al.: Addison-Wesley.

DAUDEL, S.; VIALLE, G. (1994): *Yield management – applications to air transport and other service industries*. Paris et. al: Presses ITA.

DENNIS, N. (2000): Scheduling issues and network strategies for international airline alliances. *Journal of air transport management*, 7(6), pp. 75-85.

DEWIT, J. (1995): An urge to merge? *Journal of Air Transport Management*, 2(3/4), pp. 173-180.

DOGANIS, R. (2002): *The airline business in the 21st century*. London: Routledge.

DOGANIS, R. (2002): *Flying off course - The economics of international airlines*. London: Routledge.

DYER, J.H.; SINGH, H. (1998): The Relational View: Cooperative Strategy and sources of interorganizational competitive advantage. *Academy of management Journal*, 23(4), pp. 660-679.

EISENFÜHR, F. (1997): *Investitionsrechnung*, 11th Ed., Aachen: Verlag der Augustinus-Buchhandlung.

GALLACHER, J. (1997): Power to the plans. *Airline Business*, 13(8), pp. 34-37.

GALLACHER, J. (1999): Playing your Cards right. *Airline Business*, 15(8), pp. 46-50.

GARETTE, B.; DUSSAUGE, P. (2000): Alliances versus Acquisitions: Choosing the right option. *European Management Journal*, 18(1), pp. 63-69.

GARETTE, B; DUSSAUGE, P.; MITCHELL, W. (2000): Learning from competing partners: Outcomes and durations of scale and link alliances in Europe, North America and Asia. *Strategic Management Journal*, 21(2), pp. 99-126.

GENERAL ACCOUNTING OFFICE (1995): *International Aviation: Airline Alliances Produce Benefits, But Effect on Competition is Uncertain*. GAO/RCED-95/99.

GOMES-CASSERES, B. (1994): Group vs. Group: How Alliance Networks Compete. *Harvard Business Review*, 72(4), pp. 62-74.

HAMEL, G. (1991): Competing for competence and inter-partner learning within international strategic alliances. *Strategic Management Journal*, 12, Special Issue-Summer, pp. 83-103.

HAMEL, G.; DOZ, Y.; PRAHALAD, C K. (1989): Collaborate with your competitors and win. *Harvard Business Review*, 68(1), pp. 133-139.

HAMMES, W. (1994): *Strategische Allianzen als Instrument der strategischen Unternehmensführung*. Wiesbaden: DUV.

HAMMES, W. (1995): Der Zusammenhang zwischen strategischen Allianzen und Industriestrukturen. In: Schertler, W. (Ed.): *Management von Unternehmenskooperationen*. Wien: Überreuther, pp. 55-114.

HANLON, P. (1996): *Global Airlines. Competition in a Transnational Industry*. Oxford et al.: Butterworth-Heinemann.

HARRIGAN, K.R. (1985): *Strategies for Joint Ventures*. Lexington, Mass.: Lexington Books.

HAX, H. (1995): *Investitionstheorie*. 5th Ed., Heidelberg: Physica.

HÖFER, B. (1993): *Strukturwandel im europäischen Luftverkehr – Marktstrukturelle Konsequenzen der Deregulierung*. Frankfurt am Main et al.: Lang.

IONIDES, N. (2000): Two years old and still growing. *Airline Business*, 14(6), pp. 34-35.

JARILLO, J.C. (1998): On strategic networks. *Strategic Management Journal*, 9(1), pp. 31-41.

MALANIK, P. (1998): Strategische Allianzen statt Fusion. Die "sanfte" Variante des Strukturveränderungsprozesses oder einfach nur das bessere Konzept? In: Deutsche Verkehrswissenschaftliche Gesellschaft e.V. (DVWG) (Eds.): *Strategische Allianzen im Bereich Transport Verkehr Logistik*, Schriftenreihen der DVWG, Reihe B, pp. 1-15.

McShan, S.; Windle, R. (1989): The Implications of Hub-and-spoke Routing on Airline Costs and Competitiveness. *The Logistics and Transportation Review*, 25(3), pp. 209-230.

Mead, R. (1998): *International Management – Cross-Cultural Dimensions*. 2nd Ed., Oxford: Blackwell.

Mintzberg, H.; Ahlstrand, B.; Lampel, J. (1998): *Strategy Safari - A guided tour through the wilds of strategic management*. New York: Free Press.

O'Toole, K. (1999): Reworking the model. *Airline Business*, 15(11), pp. 78-83.

Oster, C.V.; Pickrell, D.H. (1986): Marketing alliances and competitive strategy in the airline industry. *Logistics and Transportation Review*, 23(4), pp. 371-387.

Oum, T.; Park, J. (1997): Airline alliances: current status, policy issues, and future directions. *Journal of Air Transport Management*, 4(3), pp. 133-144.

Oum, T.; Park, J.; Zhang, A. (1996): The Effects of Codesharing Agreements on Firm Conduct and International Air Fares. *Journal of Transport Economics and Policy*, (2), May, pp. 187-202.

Oum, T.; Taylor, A.J.; Zhang, A. (1993): Strategic airline policy in the globalizing airline networks. *Transportation Journal*, 32(3), pp. 14-30.

Penrose, E.T. (1959): *The Theory of Growth of the firm*. Oxford: Blackwell.

Pfeffer, J.; Salancik, G. (1978): *The external control of organizations*. New York et al.: Harper & Row.

Picot, A. (1982): Transaktionskostenansatz in der Organisationstheorie: Stand der Diskussion und Aussagewert. *Die Betriebswirtschaft*, 42(2), pp. 267-284.

Pommes, C. de (1998): Are you IT-compatible? *Airline Business*, 14(7), pp. 26-27.

Pompl, W. (1998): *Luftverkehr – Eine ökonomische und politische Einführung*. 3rd Ed. Berlin et al.: Springer.

Porter, M.E. (1980): *Competitive Strategy – Techniques for Analyzing Industries and Competitors*. New York: Free Press.

Porter, M.E. (1985): *Competitive Advantage*. New York: Free Press.

Porter, M.E. (1996): What is Strategy? *Harvard Business Review*, 74(6), pp. 61-78.

Porter, M.E. (1990): *Wettbewerbsstrategie - Methoden zur Analyse von Branchen und Konkurrenten*. 10th Ed., Frankfurt: Campus.

PORTER, M.E. / FULLER, B. (1989): Coalitions and global strategy. In: Porter, M. E. (Ed.): *Competition in global industries*, Boston: Harvard Business School Press, pp. 315-343.

PRAHALAD, C.K.; HAMEL, G. (1990): The Core Competence of the Corporation. *Harvard Business Review*, 69(3), pp. 79-91.

RÜHLI, E. (1994): The Resource-Based-View of Strategy. In: Gomez, P.; Hahn, D.; Müller-Stewens, G. and Wunderer, R.: *Unternehmerischer Wandel: Konzepte zur organisatorischen Erneuerung: Knut Bleicher zum 65. Geburtstag*. Wiesbaden: Gabler, pp. 107-134.

RUMELT, R.; SCHENDEL D.E.; TEECE, D.J. (1994): *Fundamental Issues in Strategy: A Research Agenda*. Boston: Harvard Business School Press.

SCHERER, F.M.; ROSS, D. (1990): *Industrial market structure and economic performance*. 3rd Ed., Boston: Houghton Mifflin.

SCHNEIDER, D. (1985): Die Unhaltbarkeit des Transaktionskostenansatzes für die "Markt oder Unternehmung"- Diskussion. *Zeitschrift für Betriebswirtschaft*, 55(1985), pp. 1237-1254.

SHAPIRO, C.; VARIAN, H.R. (1999): *Information Rules: A Strategic Guide to the Network Economy*. Boston: Harvard Business School Press.

SIMON, H.A. (1961): *Administrative Behavior*. 2nd Ed., New York: Macmillan.

SPEKMAN, R.E.; ISABELLA, L.A.; MACAVOY, T.C. (2000): *Alliance Competence*. New York et al.: Wiley.

STEINIGER, A. (1999): *Gestaltungsempfehlungen für Airline-Allianzen*. Dissertation an der HSG St. Gallen: Bamberg.

STRAUBE, M. (1972): *Zwischenbetriebliche Kooperation*. Wiesbaden: Gabler.

SYDOW, J. (1992): *Strategische Netzwerke. Evolution und Organisation*. Wiesbaden: Gabler.

TRETHEWAY, M.W.; OUM, T.H. (1992): *Airline economics*. Vancouver: Centre for Transportation Studies.

WALKER, K. (1999): The great global debate. *Airline Business*, 15(9), pp. 96-98.

WEBER, G. (1997): *Erfolgsfaktoren im Kerngeschäft von europäischen Linienfluggesellschaften*. Dissertation an der HSG St. Gallen: Bamberg.

WELLS, A. (1999): *Air Transportation – A Management Perspective*. 4th Ed., Belmont: Wadsworth.

WERNERFELDT, B. (1984): A resource-based view of the firm. *Strategic Management Journal*, 5(2), pp. 171-180.

WESTON, F.J.; SIU, J.A.; JOHNSON, B.A. (2001): *Takeovers, restructuring and corporate governance*. 3rd Ed., New Jersey: Prentice-Hall.

WESTON, F.J.; WEAVER, S. (2001): *Mergers and Acquisitions*. New York et al.: McGraw-Hill.

WILLIAMSON, O.E. (1975): *Markets and Hierarchies*. New York et al.: Free Press.

WILLIAMSON, O.E. (1981): The modern corporation: origins, evolution, attributes. *Journal of Economic Literature*, 19(4), pp. 1537-1568.

WILLIAMSON, O.E. (1990): *Die ökonomischen Institutionen des Kapitalismus*. Tübingen: Mohr.

WILLIAMSON, O.E. (1996): *Transaktionskostenökonomik*. 2nd Ed., Hamburg: Lit.

WÖHE, G. (1993): *Einführung in die Allgemeine Betriebswirtschaftslehre*. 18th Ed., München: Vahlen.

10
INTERFACE MANAGEMENT IN STRATEGIC ALLIANCES

MATTHIAS GRAUMANN AND MARCUS NIEDERMEYER

1 Introduction 256
2 Strategic Alliances Between Airlines 256
3 The Star Alliance 258
4 WOW Alliance 261
5 Organization Design Concept 263
6 Conclusion 270
7 References 271

Summary:
This chapter analyzes the design of interface management in strategic alliances forged by airlines. Interface management hereby encompasses all measures that contribute towards reducing the dysfunctional effects of interdependencies between alliance members. On the basis of two case studies, a theoretical concept for interface management is presented and efficiency hypotheses of alternative measures are formulated.

1 Introduction

The term "strategic alliance" denotes a formalized, long-term relationship between one company and another. The objective of their cooperation is to improve each other's competitive position by compensating the weaknesses of the one partner with the strengths of the other.[1] It is obvious that efficient organizational measures have to be taken to make full use of the potential advantages arising out of strategic alliances. One area which deserves special attention is *interface management* between the alliance partners. These partners stand on an equal footing and that consequently rules out the issuance of hierarchical directives as a coordination tool within the alliance.[2] Strategic alliances are thus characterized by an interplay between autonomy and interdependence. On the one hand, the alliance partners are free to decide whether or not to enter an alliance and maintain their legal independence; on the other hand, they knowingly accept mutual dependencies.[3]

The aim of the chapter is to provide support for decision makers who deal with interface management in practice. It is evident that such a support must be based on a theoretical concept. After a brief discussion of the advantages of strategic alliances in section 2, two case studies are presented in section 3 and 4. The phenomena described are integrated into an organization design concept presented in section 5 which is then used for an efficiency assessment of alternative interface management measures. The result and the limitations of the selected approach are described in section 6.

2 Strategic Alliances Between Airlines

Liberalization of the regulatory restraints in the aviation industry in recent years profoundly changed the competitive environment in which airlines operate.[4] Strategic alliances are seen as a means of retaining existing success potential and

---

[1] See Sydow (1992), p. 63 and Backhaus & Piltz (1990), p. 2.
[2] See Steininger (1999), p. 13.
[3] See Lutz (1993), p. 36 and Tröndle (1987), p. 16.
[4] For an overview see Doganis (2002).

INTERFACE MANAGEMENT IN STRATEGIC ALLIANCES   257

opening up new opportunities for further achievement in the future.[5] The envisioned advantages are seen both on the cost front as well as in the earnings dimension resulting from cooperating in specific areas. Such cooperation areas can emerge from systematizing competition-critical resources[6] just as they can from the basic structure of competition in the air traffic business[7] or the usual value chain.[8] The following cooperation areas are conceivable for both passenger and cargo airlines:[9]

- Through joint procurement alliance partners gain advantages from higher-volume purchasing. Their increased leverage in the market from bigger volumes helps them to obtain better terms (i.e. delivery service and purchase prices).

- Cooperation on ground handling permits more economical use of existing ground facilities. This results mainly from an increase in flight frequencies arising from integration of airport ground stations.[10]

- Joint maintenance generates benefits if maintenance operations are coordinated or if individual partners specialize in servicing specific aircraft types.[11]

- Alliance partners cooperating on flight operations can exchange crew members beyond their own corporate borders. As long as working hours are not exceeded, this helps them to save hotel expenses and reduce downtime.

- Network and flight-schedule cooperation offer additional advantages. Alliances allow airlines to expand existing hub-and-spoke networks as well as to build up new continental multi-hub networks and global networks.[12] Traffic feed between partners gives them higher load factors and better utilization of aircraft capacities. This – and possible deployment of bigger aircraft – leads to lower costs.[13]

- On the marketing front joint advertising and common branding enables alliance partners to realize cost benefits and enhance their image. The larger

---

[5] See Pompl (1998), pp. 395 ff.
[6] See Graumann & Jeronimo (2002).
[7] See Althen, Graumann & Niedermeyer (2001), pp. 422 ff.
[8] See Diegruber (1991), pp. 241 ff.
[9] See Oum, Park & Zhang (2000), pp. 32 ff. and Steininger (1999), pp. 228 ff.
[10] See Steininger (1999), p. 187.
[11] See Doganis (2001), p. 64.
[12] See Delfmann (1998), pp. 177 f. and Steininger (1999), pp. 228 ff.
[13] See Oum, Park & Zhang (2000), p. 13.

the number of destinations served, the greater the coverage of individual advertisements.[14] By branding a joint alliance image airline partners can highlight such attributes as uniform quality or greater reach.[15]

- Another cooperation area is codesharing. This is a tool for offering customers seemingly more flight connections.[16] The customer gains the impression that only a single carrier is involved even though an onward flight is operated by an alliance partner. For the customer this makes an airline's portfolio of services more attractive.

- Cooperation also allows frequent flyer programs to be pooled and extended to a greater number of routes. This results in benefits for customers. It cements their loyalty to an airline and thereby boosts sales.[17]

If the advantages outlined in the individual cooperation areas are to be utilized, this inevitably involves the organizational dimension of a strategic alliance. The following considerations focus on the so-called interface management as a specific constituent of organizational measures.[18] The analysis starts by looking at two strategic alliances and the measures they use to foster cooperation. The first example is the Star Alliance.

3 The Star Alliance

3.1 Motives for Founding the Alliance

The idea behind the founding of the Star Alliance originated at a meeting of the executive boards of United Airlines and Deutsche Lufthansa AG. The airlines were not content with operating a multitude of point-to-point connections which made it impossible for them to provide customers with a seamless travel experience from one end of their journey to the other. The establishment of hub-and-spoke networks may have allowed single airlines to lay on extensive air services to international destinations, but if passengers wished to journey on

---

[14] See Steininger (1999), pp. 307 ff.
[15] See Steininger (1999), pp. 282 ff.
[16] See Hanlon (1996), pp. 101 ff. and Pompl (1998), pp. 108 ff.
[17] See Oum, Park & Zhang (2000), pp. 12 ff.
[18] See Brockhoff & Hauschildt (1993), pp. 396 ff.

beyond that point there were no connecting flights from a single source. In order to enhance customer benefits, United Airlines and Deutsche Lufthansa decided to inter-connect their hubs. This allowed them to serve a worldwide market which neither could have served on its own. From the customer's viewpoint, their cooperation additionally allowed them – through codesharing – to extend each other's route network. It also offered alliance partners an opportunity to give their own flights greater weight than those of competitors in computer reservations systems.

Those advantages won over other airlines. Starting with five founding members in May 1997, Star Alliance now numbers 17 partner airlines. The alliance roll call includes Air Canada, Air New Zealand, ANA (All Nippon Airways), Asiana Airlines, Austrian Airlines, bmi, LOT, Lufthansa, Scandinavian Airlines, Singapore Airlines, Spanair, Thai Airways, United Airways, US Airways and VARIG. In its global network, the alliance now serves about 700 destinations in 128 countries. In the competition between alliance systems the Star Alliance is regarded to rank as the leading airline grouping in the air transport industry.

3.2 Interface Management

In order to ensure coordination of interdependencies, the Star Alliance has set up a wide-ranging managerial regime. The fundamental strategy for cooperation is decided by a so-called Chief Executive Board (CEB) twice a year. The CEB consists of the Chief Executive Officers (CEOs) of the partner airlines. The decisions taken at the CEB meetings are communicated to the Alliance Management Board (AMB) as a basis for the overall planning and steering of the alliance. The AMB comprises of the alliance managers of the individual airlines (at Lufthansa they are also called "interface managers") who are appointed solely for the purpose of managing the alliance. The alliance management at Lufthansa AG is a staff unit which reports to the Executive Board. Different organizational solutions are in place at the other Star Alliance airlines.

The alliance managers maintain close contact with the line units at their own company and their managerial equals at partner airlines. Their job is to transport the alliance's objectives into their company. They assume principally a steering function and support line units – when necessary – in their coordinating activities. Should the fronts between negotiating line units harden, the alliance managers contact one another and attempt to steer the line units from a neutral perspective into accepting a compromise. It is not, however, possible for alliance managers to nursemaid all the individual coordinating activities. It is, therefore, often left to the interdependent units to reach an agreement on their own. Only when problems prove intractable, the alliance managers intervene.

Additionally to those measures the airline partners founded a dedicated company: the Star Alliance Service GmbH based in Frankfurt am Main. This company has about 60 employees who act independently of the alliance's member-airlines. In order to underpin that independence, external consultants are also recruited in pursuance of the company's aim of implementing specific alliance projects. A project manager is appointed to oversee each project; he/she is authorized to get together a team of staff from appropriate company units in the individual alliance carriers. Since the airlines in the Star Alliance do not surrender their decision-making authority, there are times when projects are undertaken by only some of the partner airlines. Figure 10-1 illustrates the interface management in the Star Alliance.

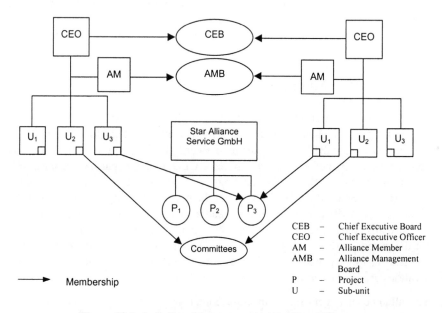

Figure 10-1: Interface Management in the Star Alliance

4 WOW Alliance

4.1 Motives for Founding the Alliance

Prompted by the success of the Star Alliance Lufthansa Cargo, Singapore Airlines and Scandinavian Airlines agreed to engage in closer cooperation in 1999. The three cargo carriers were attracted by potential benefits from economies of scale and the opportunity of utilizing the advantages to improve their market position.[19] About a year later the three airlines jointly founded the "New Global Cargo" alliance which in 2002 was renamed "WOW".[20] The decision to enter into an alliance was based on two considerations:

- The first was the perceptible trend towards consolidation in the airfreight business; in view of that, the three carriers deemed it sensible to add greater weight to their own position by entering into closer cooperation.

- The second was the forecasted market growth of between five and six per cent in the industry in the years ahead. The three considered it difficult to accommodate that projected additional demand on their own, since capacity increases in the air traffic business require substantial capital expenditure.

In July 2002, Japan Airlines joined the WOW alliance as its fourth member. The alliance now commands a fleet of 43 freighters and the belly-hold capacities of 760 passenger aircraft. Those resources are utilized in a network of more than 500 destinations in 103 countries around the globe.

4.2 Interface Management

In order to exploit potential alliance advantages efficiently an interface management is required to coordinate interdependencies in the six cooperation areas in the WOW alliance: Sales, Product/Marketing/Branding, Handling, IT, Network Access and Purchasing.[21] In the WOW alliance the chairmen of the

---

[19] See Althen, Graumann & Niedermeyer (2001), p. 438.
[20] This is not an abbreviation. WOW was deliberately chosen as an exclamation of amazement and enthusiasm.
[21] These are the usual terms in WOW parlance. In fact, the cooperation areas listed above are otherwise termed joint purchasing, ground handling, network and flight

partner airlines stake out the fundamental framework of cooperation. The chairmen meet every two to three months to determine the strategic direction of the alliance. Six virtual teams are formed to implement the objectives of the six cooperation areas, to assign individual tasks to line units in their cooperation area and to oversee coordination in the execution of those tasks. These virtual teams are staffed by employees from each of the involved airlines. They communicate, if necessary, several times a week by phone, email or videoconferencing. If coordination is of greater importance, they meet personally. The teams function in cooperation with line units principally as moderators and mediators. Figure 10-2 illustrates the interface management in the WOW alliance.

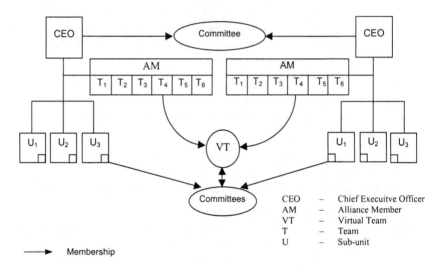

Figure 10-2: Interface Management in the WOW Alliance

---

schedules, marketing, code-sharing, coordination of customer discounts and IT resources.

5 Organization Design Concept

5.1 Interface Management

The specific measures taken by the Star Alliance and by the WOW to reduce the dysfunctional effects of interdependencies shall now be integrated into a generally applicable concept. Once such a concept is established, it can be used to provide support for solving concrete interface management problems in other strategic alliances.

The integration starts with the assumption that interface management is part of the organizational structure of an enterprise. From a very formal viewpoint, organizational structures consist of abstract *units* like "position", "department" or "division" under which concrete staff are subsumed. Organizational structures further consist of *rules* which apply to the different units i.e. to the staff after subsumption. The purpose of the rules is to direct the actions of staff towards the enterprise's aims.

Since units and rules can be regarded as variables, the question arises whether the utility-maximizing combination of units and rules can be found in an organizational decision model. Such an attempt, however, suffers from the problem of immense complexity. Thus, finding the optimal combination is regarded to be unrealistic. In order to cope with the problem of complexity it has proven useful to work with two heuristic simplifications when looking for a (satisfying) organizational structure:

- The first simplification is that intended rationality and its decision-making aspects should be modelled separately from problems like shirking or moral hazard. The terms that are normally used in the pertinent literature to denote these two dimensions are "cordination" and "motivation" respectively.[22] The following considerations apply exclusively to the *coordination of decisions* taken by employees; motivational questions are neglected. The prime assumption hereby is that the quality of coordination is influenced by interdependencies between organizational units.[23] Sequential or reciprocal interdependencies can affect the timing and the quality of production

---

[22] See Milgrom & Roberts (1992), p. 25 and Frese (2000), pp. 7 ff.
[23] See Frese (2000), pp. 258 ff. and Graumann (1999), pp. 79 ff.

processes; pooled interdependencies can lead to competition between organizational units and can prevent them from making use of synergies.[24] The coordination benefits ensuing from an organizational structure depend on how well it succeeds in mitigating or defusing these three kinds of interdependencies – they cannot be totally eliminated.

- The second simplification is that the task of shaping an organization can be divided into the configuration of a framework structure and the subsequent interface management style.[25] An interface exists when a potential need for coordination arises between organizational units because of interdependencies between them. Consequently, "interface management" encompasses all measures that can contribute towards reducing the dysfunctional effects of interdependencies.

Measures taken in interface management can be distinguished according to whether the affected units handle coordination themselves or in consultation with an additional integration unit. In addition, it should be clarified whether coordination begins and ends with the passing-on of information (communication), whether it takes the form of consulting or whether a specific decision-making mandate for coordination is conferred (see table 10-1).[26]

	Self-coordination	External coordination
Communication	Immediate information exchange or information committee	Integration unit for supporting information exchange
Consultation	Consulting committee	Integration unit for consultation
Decision-making	Decision-making committee	Integration unit with decision-making authority or decision-making committee with integration unit as member

Table 10-1: Measures of Interface Management

---

[24] See Thompson (1967), pp. 54 f.
[25] See Frese (2000), pp. 14 ff.
[26] See Frese (2000), p. 404.

5.2 Interface Management in Strategic Alliances

The interface concept described above can be applied not only to a company's internal issues but also to overlapping issues involving one or other of its partners.[27] A look at Star Alliance's managerial organization shows that the Chief Executive Board corresponds to a top-level decision-making committee (see again figure 10-1); the Alliance Management Board and the Star Alliance Service GmbH can be understood as integration units which play a part in the interdependency coordination process and selected members from line units come together in decision-making, communications or consulting committees. Additionally, direct communications are permitted between interdependent units. Special importance is hereby attached to personal contacts. Employees from the different airlines are supposed to familiarize themselves with other corporate cultures and build up mutual trust so as to attain a common understanding of alliance problems.

In the WOW alliance the regular meetings of the CEOs on strategic priorities correspond to a meeting of a top-level, decision-making committee. The virtual teams are equivalent to integration units which fulfil their purposes accordingly in the interdependency coordination process. Selected employees from the line units meet in information, consulting or decision-making committees whereby most of the meetings are initiated by the virtual team. By and large, since an integration unit is normally involved in the coordination process, most of the measures taken by the interface management in the WOW alliance can be regarded as external coordination and not as self-coordination, because an integration unit is normally involved in the coordination process.

In addition to the interdependencies within a partner airline interdependencies may also exist *between the partners*:[28]

- Resource interdependencies are created when alliance partners jointly utilize scarce resources e.g. ground handling capacities, maintenance machinery or crews. Competition or a lack of synergies are potential problems arising out of this situation.

- Market interdependencies come about when alliance partners are active in the same market sector. Again, competition between the airlines or a lack of synergies may be the consequence and need coordination.

---

[27] See Galbraith (2002), p. 149 ff and Brockhoff (1994), pp. 8 ff.
[28] See Graumann (1999), p. 157.

- Process interdependencies occur when a company requires output from partners in the alliance as input for its own production. Deutsche Lufthansa decides, for example, to offer connections on a specific US American route. As a prerequisite it needs to use United Airways' network to provide for connection flights. An interdependency is created because Lufthansa restricts the decision-making scope of United Airways since the latter must adapt its flight schedules to fit in with its partner's plans. A lack of coordination would affect Deutsche Lufthansa's product quality and customer satisfaction.

Although interface management in a strategic alliance can be described with the usual terminology of organization theory, there is a pivotal difference which needs to be underlined from interface management within a single company. Measures in a company are taken within a hierarchical structure so that in the final instance the authority to issue directives always rests with management. Such hierarchical dicta or pronouncements are not possible in an alliance. In an alliance each partner decides autonomously the measures that are confined within its own corporate contours. As a consequence, the measures enacted by *inter*organizational interface management can only apply to the relevant *intra*organizational structure. In that light, all the measures listed in table 10-1 are applicable to interface management within an alliance. In individual instances that means:

- In the process of self-coordination an immediate exchange of information can take place directly between the alliance units affected by the interdependencies. Alternatively, committees (information committee, consulting committee, decision-making committee) can be formed from authorized staff of the affected companies.

- In the case of external coordination it is not possible for one partner to assume the coordinating function for all other alliance partners in the absence of a hierarchical regime within an alliance. The integration unit must, therefore, be an internal company unit which brings the alliance perspective to bear within that company. The decisive difference from intra-company interface management is that in the inter-company instance an interface remains in existence but is transferred to the integration unit. Such an integration unit can be harnessed to exchange information, be a member of a consulting committee or be invested with decision-making authority.

5.3 Efficiency Review

In order to appraise the benefits generated by alternative interface management measures, criteria are sought which must meet two requirements: they must (1) stand in a means/end relationship to long-term alliance aims and there must (2) be a comprehensible cause/effect relationship between possible measures and the criteria.[29] Since the considerations here apply only to the coordination dimension, appropriate criteria must focus on the different types of interdependencies.[30] The criterion of efficient interface management is, thus, the *quality of interdependency coordination*. The extent to which this aim is reached can be appraised only indirectly through proxy attributes.[31] They are defined:

- as the quality of the decisions of interdependent units; and
- as the duration of the decision process.[32]

The proxy attributes are inversely correlated: improving the quality of decisions through improved information processing prolongs the decision-making process (as long as the technology remains unchanged). Since decision quality from the point of view of the benefits ought to be rated more highly than the duration of the decision process, the trade-off is assumed here to be in favour of quality.

Two factors principally influence the decision quality and the duration of the decision process:

- the communication channel; and
- the conflict handling.[33]

Communications can, on the one hand, run through a central unit (i.e. an integration unit) which evaluates and distributes information (two-stage and indirect communications). Alternatively, communications can be channelled directly (one-stage) from unit to unit.[34] Central (i.e. integration) units have an advantage over line units in that they specialize in handling coordination tasks. It may also be assumed that they – unlike line units – are not in any way inclined to indulge in departmental egoism or office politics. However, indirect communications from the time angle are disadvantageous in that it takes longer

---

[29] See Keeney (1996), p. 82.
[30] See Frese (2000), pp. 264 ff.
[31] For the concept of proxy attributes see Keeney (1996), pp. 110.
[32] See Laßmann (1992), p. 146.
[33] See Laßmann (1992), pp. 149 ff. and pp. 157 ff.
[34] See Laßmann (1992), pp. 149 ff.

for the information to reach the recipient than is the case with direct communications. At times accuracy can also suffer in transmission. Conversely from both aspects direct communications are conceivably better than the indirect mode. In turn, the disadvantage of direct communications is that coordination-relevant information and the destination where it is needed first have to be ascertained. It can be plausibly assumed that individual units frequently lack sufficient know-how and, hence, judge the coordination relevance of information less accurately than professional integrators do.

Conflicts arise in situations of perceived competition for scarce resources and are manifested in disagreement or disputes.[35] Contrary to the traditional view that conflicts are inherently disruptive the pertinent literature points out that a conflict can also induce positive effects. As examples the sources cite the development of new ideas or the mobilization of capacity reserves.[36] From the point of view of the underlying decision-oriented organization theory the effects on the information levels of those involved should be kept in mind when diverse strategies of conflict handling are considered. In this respect open conflict and pseudo-harmonization (conflict suppression) are to be distinguished.[37] From the point of view of a decision-oriented or, in a narrower sense, coordination-related perspective preference should be given to the open waging of conflicts between the interdependent units of different airlines. The reason is that only in that way can the problem-related knowledge of those involved be fully harnessed and the quality of decisions enhanced more effectively than is the case when conflicts are suppressed through pseudo-harmonization.[38] It should, however, be noted that improving decision quality inevitably prolongs the decision-making process.

Bearing the theoretical propositions in mind and taking the analyses of the experiences in the Star Alliance and in the WOW alliance into consideration, the interdependency-related efficiency of alternative measures of interface management can be cautiously assessed. The formulation of efficiency hypotheses is based on three assumptions:

- Communication measures possess merely an informative character. They can only serve to reduce the uncertainty resulting from interdependencies about the actions of other units; they do not eliminate an interdependency nor do they resolve conflicts.[39] Consulting measures with which the

---

[35] See Coleman (1994), pp. 869 ff.
[36] See Berkel (1992), col. 1092.
[37] See Frese & Heppner (1995), pp. 63 ff.
[38] See Laßmann (1992), p. 164.
[39] See Laßmann (1992), p. 284.

involved units have ready solutions for a decision (the ultimate decision rests with other units) are more effective. It is, however, decision measures that strip interdependencies of their benefit-reducing effect which exercise the greatest impact on coordination.

- Problems arise with self-coordination when the employees of line units are not adequately equipped with qualifications to overcome communication barriers (keyword: corporate cultures of alliance members) or are threatened with being over-burdened by coordination-related communication tasks. The potential benefits of direct communications can then boomerang. In such a case, the benefits of direct communications can be better utilized by forming a committee, since this reduces the number of communication relationships.

- External coordination by an integration unit has advantages over self-coordination in resolving conflicts if it is assumed that integration units are better qualified. This assumption is ostensibly justified because the employees in integration units are exclusively responsible for coordinating interdependencies whereas representatives on committees are additionally burdened with other tasks. At the same time, there is a risk of inaccurate transmission of information since the integration units have only an indirect insight into the functions of line units. This can adversely affect the time taken in decision-making and decision quality.

	Self-coordination			
	Direct communication	Information committee	Decision-making committee	Consulting committee
Decision quality	-	+/--	++/-	+/-
Decision duration	+/-	++/-	++/-	++/-
Interdependency coordination	+/--	+/-	++/-	+/-

+: good, ++/-: quite good, +/-: average, +/--: middling-to-poor, -: poor

Figure 10-3: Efficiency Hypotheses concerning Self-coordination Measures

Figures 10-3 and 10-4 show the efficiency hypotheses formulated here. It is assumed that "efficiency" is specified as the coordination of interdependencies and that the proxy attribute "decision quality" carries more weight than the proxy attribute "decision duration". The result of the hypotheses is that external coordination with integration units has slight advantages over self-coordination with interdependent units in strategic alliances.

	External coordination			
	Communication through integration unit	Integration unit with decision-making authority	Decision making committee with integration unit	Consulting through integration unit
Decision quality	+/--	+/-	+	+/-
Decision duration	+/--	++/-	+	++/-
Interdependency coordination	+/--	+/-	+	+/-

+: good, ++/-: quite good, +/-: average, +/--: middling-to-poor, -: poor

Figure 10-4: Efficiency Hypotheses concerning External Coordination Measures

6 Conclusion

Organization design does not have to be intuitive. The selection of organizational measures can be based on a concept which distinguishes between objectives, actions and disruptive interdependencies influenced by the actions while influencing the objectives. The application of the concept to the problem of interface management in strategic airline alliances led to the result that external coordination with integration units has slight advantages over self-coordination with interdependent units. To fully appreciate this result three limiting aspects need, however, to be borne in mind. For one, trend assessments are always tainted with the usual problems of qualitative reasoning and are, therefore, open to question.[40] Future research must strive for metrical relations. Moreover, the considerations only apply to the dimension of coordination and disregard motivational aspects. Although this selectivity can be justified with the heuristic organizational procedure introduced in section 5.1 the result is to be considered only as a first step. A comprehensive analysis of interface management needs to include motivational aspects as well.

Acknowledgements

The authors are indebted to Dipl.-Kff. Nina Brandt for preparatory research.

7 References

ALTHEN W.; GRAUMANN M.; NIEDERMEYER M. (2001): Alternative Wettbewerbsstrategien von Fluggesellschaften in der Luftfrachtbranche. *Zeitschrift für betriebswirtschaftliche Forschung*, 53(6), pp. 420-441.

BACKHAUS, K.; PILTZ, K. (1990): Strategische Allianzen – eine neue Form kooperativen Wettbewerbs? In: Backhaus, K. and Piltz, K. (Eds.): *Zfbf Sonderheft No. 27, Strategische Allianzen*, pp. 1-10.

BERKEL, K. (1992): Interpersonelle Konflikte In: Gaugler, E. and Weber, W. (Eds.): *Handwörterbuch des Personalwesens*, Stuttgart: Poeschel, pp. 1085-1094.

BROCKHOFF, K. (1994): *Management organisatorischer Schnittstellen unter besonderer Berücksichtigung der Koordination von Marketingbereichen mit Forschung und Entwicklung*. Göttingen: Vandenhoeck und Ruprecht.

BROCKHOFF, K.; HAUSCHILDT, J. (1993): Schnittstellen-Management – Koordination ohne Hierarchie. *Zeitschrift Führung und Organisation*, 62(6), pp. 396-403.

COLEMAN, J.S. (1994): *Foundations of Social Theory*. Cambridge MA., London: The Belknap Press.

DELFMANN, W. (1998): Hub-and-Spoke-Systeme. In: Klaus, P. and Krieger, W. (Eds.): *Gabler Lexikon Logistik*, Wiesbaden: Gabler, pp. 177-178.

DIEGRUBER, J. (1991): *Erfolgsfaktoren nationaler europäischer Linienluftverkehrsgesellschaften im Markt der 90er Jahre*. Konstanz: Universitätsverlag Konstanz.

---

[40] See Parsons/Fox (1991).

DOGANIS, R. (2001): *The Airline Business in the Twenty-first Century*. London, New York: Routledge.

DOGANIS, R. (2002): *Flying off Course. The Economics of International Airlines*. London, New York: Routledge.

FRESE, E. (2000): *Grundlagen der Organisation. Konzept – Prinzipien – Strukturen*. Wiesbaden: Gabler.

FRESE, E.; HEPPNER, K. (1995): *Ersatzteilversorgung. Strategie und Organisation*. München: TCW Transfer-Centrum.

GALBRAITH, JAY R. (2002): *Designing Organizations. An Executive Guide to Strategy, Structure, and Process*. San Francisco: Jossey-Bass.

GRAUMANN, M. (1999): *Organisationstheoretische Untersuchung der Rückversicherungsunternehmung. Ein entscheidungslogisch orientierter Ansatz*. Berlin: Duncker & Humblot.

GRAUMANN, M.; CLEMENTE, J. (2002): Wettbewerbskritische Ressourcen von Regionalfluggesellschaften – Eine VRIO-Analyse auf der Grundlage des Ressource-based View. *Zeitschrift für Verkehrswissenschaft*, 73(1), pp. 39-71.

HANLON, P. (1996): *Global Airlines*. Oxford et al.: Butterworth-Heinemann.

KEENEY, R.L. (1996): *Value-Focused Thinking. A Path to Creative Decisionmaking*. Cambridge/MA. et al.: Harvard University Press.

LABMANN, A. (1992): *Organisatorische Koordination*. Wiesbaden: Gabler.

LUTZ, V. (1992): *Horizontale strategische Allianzen – Ansatzpunkte zu ihrer Institutionalisierung*. Hamburg: Steuer- und Wirtschaftsverlag.

MILGROM, P.; ROBERTS, J. (1992): *Economics, Organization and Management*. Englewood Cliffs: Prentice-Hall.

OUM, T.H.; PARK, J.-H.; ZHANG, A. (2000): *Globalization and Strategic Alliances : The Case of the Airline Industry*. Amsterdam et al.: Pergamon.

PARSONS, S.; FOX, J. (1991): Qualitative and Interval Algebras for Robust Decision Making Under Uncertainity. In: Sing, M. G. and Travé-Massuyès, L. (Eds.): *Decision Support Systems and Qualitative Reasoning*, Amsterdam etc: North-Holland, pp. 163-168.

POMPL, W. (1998): *Luftverkehr. Eine ökonomische und politische Einführung*. Berlin et al.: Springer.

STEININGER, A. (1999): *Gestaltungsempfehlungen für Airline-Allianzen*. Bamberg: Difo-Druck.

SYDOW, J. (1992): *Strategische Netzwerke. Evolution und Organisation*. Wiesbaden: Gabler.

THOMPSON, J.D. (1967): *Organizations in Action*. New York: McGraw-Hill.

TRÖNDLE, D. (1987): *Kooperationsmanagement. Steuerung interaktioneller Prozesse bei Unternehmenskooperationen*. Bergisch Gladbach, Köln: Eul.

11

SYNERGY ALLOCATION IN STRATEGIC AIRLINE ALLIANCES

BJÖRN GÖTSCH AND SASCHA ALBERS

1 Introduction 276
2 Synergies, Alliances, and their Governance 277
3 A Typology of Airline-Alliance Specific Synergies 279
4 An AHP-based Approach to Synergy Allocation among Airlines ...286
5 Limitations, Conclusion, and Outlook 298
6 References 299

Summary:
Multi-partner alliance networks are now common practice in the airline industry. The aim of the participating firms is, in a general sense, the generation of synergies providing them with a competitive advantage vis-à-vis their competitors. Synergies, however, are by definition only generated by joint efforts and materialize in a way that does not allow to directly relate individual inputs to the total outcome. Thus, the problem of attributing fair shares of the synergistic effect among the alliance partners needs to be resolved, as it is the raison d'être of the alliance for its member firms. This paper (a) conceptualizes the problem of synergy assessment and its relevance for alliances in general, (b) specifies concept and problem for the airline sector, and (c) proposes a general model for synergy assessment and distribution in airline alliances.

1 Introduction

For its fiscal year 1998, Lufthansa reported that its strategic alliances contributed some €250 million to its "success" of that year.[1] In 2003, US Airways announced an expected $75 million increase in revenues per year by joining Star Alliance.[2] Alliance membership apparently pays off. However, even though prominently placed in annual reports and executive statements, a conceptually founded and, thus, satisfactory method for quantifying and distributing the benefits (synergies) for individual airlines is still missing – even more, the problem is widely neglected in both strategic management and airline specific research.[3] The question, however, bears relevance, as alliances are contractual arrangements exclusively serving the strategic aims of their still autonomous partner firms.[4] Therefore, alliance membership needs to yield benefits for each individual partner firm – otherwise members will tend to leave the alliance. Depending on size and importance of the partner, this can put the whole alliance at stake and therefore threaten the competitive position of the remaining members.

In this chapter we systematically address the problem of synergy allocation within strategic airline alliances and provide a framework aiming to support airline representatives in their efforts to practically resolve this difficult question. Setting the stage by delineating a comprehensive synergy understanding as well as the essential nature of the synergy allocation rule within alliance governance systems, we propose a four stage model for synergy assessment and allocation, specified for the airline alliance context: (1) identification of the realized synergy effects by a brain trust, (2) identification of each partner's contribution, (3) deduction of "fair shares" per partner based on Saaty's Analytic Hierarchy Process (AHP),[5] and (4) computation of required monetary transfers among the partners. Limitations and a critical review and outlook finalize the chapter.

---

[1] See Lufthansa (1999), p. 51.
[2] See Baker & Field (2003), p. 42.
[3] See Albers (2000); Gulati & Wang (2003); Jap (1999); (2001).
[4] See Das & Teng (2000b), Zajac (1998), p. 321.
[5] See Saaty (1980).

2 Synergies, Alliances, and their Governance

Strategic alliances are inter-firm cooperative arrangements aiming at the achievement of advantages for the participating firms.[6] These advantages, however, come in various forms and need to be allocated among the alliance members. It is generally accepted that a failure to adequately link inputs of and benefits for the individual alliance member firm leads to instability and potential alliance failure. For the alliance's balance and internal consistency it is important that each of the partner firms receives an output that is related to its efforts for the alliance, that is, to its respective input. In this context, the term *appropriation concern* has been coined. It denotes "the firm's concern about its ability to capture a fair share of the rents from the alliance".[7] These concerns arise from behavioral uncertainty (potential opportunism) of the alliance members and their individual utility calculi.[8] Mechanisms are needed to limit the opportunism and thus counter appropriation concerns in strategic alliances.[9] This task is one of the original purposes and raisons d'être of the *alliance governance system*.[10]

The alliance governance system can be regarded as the organizational setup which is used to govern, or manage, an alliance. Alliances are interpreted as organizations of a second order which are used to coordinate and control cooperative processes and actions among firms as first order organizations. Alliance governance systems consist of (a) structural and (b) instrumental design components and vary in shape significantly.

The instrumental components – of primary interest here – are also called alliance governance mechanisms and can be grouped into coordination, control, and incentive mechanisms.[11] Coordination mechanisms are required to align management and value adding processes among the alliance partners, control mechanisms ensure that the partners behave as intended and conform to the terms of the alliance agreement which has been closed at the outset of the alliance. The third group of mechanisms, the alliance incentive mechanisms bear

---

[6] See e.g. Das & Teng (1998), p. 491; Parkhe (1993b), p. 795.
[7] Gulati & Singh (1998).
[8] See Das & Teng (2001), p. 6; Oxley (1997), pp. 392ff.; Pisano et al. (1988).
[9] See Gulati (1998).
[10] See Albers (forthcoming).
[11] See Albers (forthcoming).

importance as the alliance member firms are autonomous entities which are only hardly subject to fiat or direct supervision as usually employed in hierarchical settings. As these authoritative means of coordination are not as widely applicable as in the intra-organizational context, negotiation and bargaining as mechanisms which achieve behavioral effects on a voluntary basis increase in importance. As Parkhe remarks: "[Alliances] are voluntary cooperative relationships in which participating firms are exposed to the risk of opportunism. This problem suggests a need for [...] negotiating a partnership structure that provides *incentives* to forbear and discourages opportunistic tendencies [...] both ex post [...] and ex ante."[12] Inter alia, these incentive effects can be achieved by adequate synergy allocation mechanisms which distribute rewards in the form of a share of the jointly generated synergies as the output of the alliance among the partner firms.[13]

Even though perceived benefits might be immaterial, a potential compensation of those immaterial benefits which exceed potential material returns of the partner will be performed in monetary terms. In general, three basic types of synergy allocation rules for alliances can be identified.[14] A typology based on the dimensions of (1) *number of payment(s)*, (2) *compensation basis*, and (3) *relation to the actual performance* is proposed (see table 11-1).[15] Along the first dimension, a one-time payment is distinguished from a stream of payments over a period of time, which may, however, vary with regard to the actual amount of the payment as well as the time periods in which the payments occur. The second dimension covers the fact that payments can be agreed upon in absolute amounts, be based on individual exchanges (transactions) between alliance partners or as a fraction of the results of the activities after a certain time period. Finally, the third dimension refers to the fact that the compensation can be wholly based on forecasts (*ex ante*), and is thus independent of the actual performance of the cooperative venture, or be assessed and determined relative to the business performance and is thus specified only *ex post*.[16]

---

[12] Parkhe (1993a), p. 233 [emphasis added].

[13] In addition to the synergy allocation rules, a second class of alliance incentive mechanisms are so-called safeguards. See Albers (forthcoming).

[14] See Albers (forthcoming), but also Contractor & Lorange (2002), Contractor & Ra (2000).

[15] See Albers (forthcoming).

[16] Two especially simple rules stand out of the variety of conceivable methods. The *equality rule* states that each participating partner receives an equal share of the results of the joint activity, thereby neglecting a detailed input assessment. However, the rule has been heavily disputed from the theoretical viewpoint as the correlation between provided inputs and outputs is seen as arbitrary at best. From a practical

Even though addressed separately here, the actual compensation rule within an alliance will most likely consist of a combination of the types presented here, allowing to combine different advantageous aspects of alternative types to a certain extent.

Type	Number of payment(s)	Compensation basis	Relation to actual performance	Example
I	One-time	Absolute sum	Independent	Lumpsum payment
II	Continuous	Transaction	Dependent	Royalties, markups
III	Continuous	Result of activity	Dependent	Dividends

Table 11-1: Typology of Alliance Compensation Rules[17]

3 A Typology of Airline-Alliance Specific Synergies

The empirical relevance of airline alliances has been undisputed for a long time and is still growing. In addition to the formation of large, multi-lateral alliance networks like Star, Oneworld, and Skyteam which also attract the majority of public attention, bilateral agreements among carriers are still by far the major cooperation arrangements among airlines. Airline Business' Alliance Survey indicates that over the last 5 years the total number of alliances remains rather constant at around 500 alliances. However, fluctuation among the partners appears to be rather high since the number of agreements put into operation over the last 5 years oscillates around 60 to 90.[18]

A variety of drivers for the formation of these agreements are frequently advanced which ultimately relate to the generation of synergies in order to

---

viewpoint, the advantages of transparency and practicability are praised, submitting as well that the equal division is the adequate answer to the impossibility of linking inputs to outputs. Underlying the *non-redistribution rule* is the assumption that inputs and effects of the alliance activities accumulate in a fair way among the partners as the respective tasks for the alliance are performed (provided they behave as specified), rendering a separate assessment and distribution of synergy effects needless.

[17] Adapted from Albers (forthcoming).
[18] See AirlineBusiness (2004).

achieve competitive advantages among the partner firms.[19] A specification of these effects which are at the center of airline alliances is useful for our purposes.

Synergy can be interpreted as the delta between the state of isolated activity and the state of the joint performance of an activity.[20] *Positive* synergy is achieved if the joint performance of activities by actors A and B results in a value V which is comparably better than the isolated performance of these tasks (i.e. $V(A) + V(B) < V(A+B)$).[21] Accordingly, negative synergies are achieved if $V(A) + V(B) > V(A+B)$. Furthermore, it appears useful to distinguish *potential* synergies from actually *realized* synergies. The latter represent the actual *synergy effect* whereas the former are the fraction of synergies that can be realized only under optimal conditions.

Our subsequent analysis focuses on the impact of cooperative behavior, that is, on synergies achieved by cooperative actions. This impact can be classified along three dimensions: Whether it is mainly (a) positive or negative, (b) related predominantly to changes in cost-, revenue-, or risk positions, and (c) if it materializes only once at the outset of the alliance or is enduring and recurrent.[22] The following airline-alliance specific overview of potential synergies focuses on the discussion of the first two dimensions. We thereby concentrate on the enduring and recurrent types of synergy, i.e. the existence of continuous synergy effects, as we assume that only those require a professionally installed synergy allocation system.

3.1 Potential Synergies of Airline Alliances

Positive potential synergies can be categorized into revenue increasing, cost decreasing, and finally risk reducing synergies, and vice versa for *negative* potential synergies (see figure 11-1).[23] In the following paragraphs we provide a short overview of potential synergies along these three categories.

---

[19] See e.g. Oum et al. (2000), pp. 7ff.
[20] See Albers (forthcoming); Weber (1991), p. 100.
[21] See Copeland & Weston (1988), p. 684.
[22] See Rodermann (1999), pp. 126ff.
[23] See Hax (1992), p. 969; Weber (1991), pp. 104ff.

3.1.1 Revenues

Airline alliances are hitherto mainly associated with revenue increasing effects rather than cost reductions.[24] Revenue increasing effects especially result from the *connection of the partners' route networks*. Airlines significantly increase their destination portfolio and strengthen their market position, especially if their networks are configured in a hub and spoke layout.[25] The higher number of destinations enhances not only the customers' flexibility but also the number of flight connections customers can choose from. Moreover, a coordinated global destination network facilitates the incorporation of the seamless travel idea.[26] It seems reasonable that this will positively influence the booking habits of customers in favor of the alliance members and, thus, will enlarge revenues due to a significant increase in customer utility.[27]

An important prerequisite to increase market share is market access. The access to foreign markets is yet still regulated by bilateral air service agreements in most regions on the globe. The connection of route networks represents a viable option to access markets and hereby customer groups which would have been out of reach without the alliance membership. The most popular form of connecting route networks occurs by crafting code share agreements among the partners.[28] If antitrust immunity is granted, the alliance members are even allowed to coordinate their pricing schemes which may also lead to revenue increases.[29]

Another source of revenue augmentation is the coordination of flight schedules which leads to an increase in total available connecting markets (due to the increased availability of connections within an acceptable timeframe) as well as, in most cases, to shorter total travel times. The *connection of alliance partners' frequent flyer programs* (FFP) biases the customer towards flying within the alliance network and thus increases the overall revenue for the alliance members.[30] Revenue increasing tendencies also arise from a more efficient utilization of the hub and spoke structure of the route networks. By concentrating all available slots from all alliance members at an airport the respective alliance

---

[24] E.g. statement of Jürgen Weber (former CEO of Lufthansa AG) in 2000.
[25] See Delfmann (2004), pp. 193ff.; Pompl (2002), p. 141; and Auerbach & Delfmann chapter 3 in this volume.
[26] See Sterzenbach & Conrady (2003), p. 212; Pompl (2002), p. 142.
[27] See Steininger (1999), p. 197.
[28] See Steininger (1999), pp. 296-306; Sterzenbach & Conrady (2003), pp. 199ff.
[29] See Sterzenbach & Conrady (2003), p. 227.
[30] See Sterzenbach & Conrady (2003), p. 208.

will put itself in a powerful position due to the limited slot availability at major hub airports.[31]

A major concern resulting in *decreased* revenues may arise from negative reactions of customers in order to avoid dependence on a single supplier.[32] Due to the structure and importance of FFP or other alliance-specific discount programs, this issue seems not very relevant in the airline context. In contrast, the reduction of autonomy of each single alliance member and, closely connected, the limited flexibility to respond to environmental changes is extremely relevant for airline alliances.[33] Another source of revenue disadvantages is the possible loss of individual market relationships due to alliance membership.[34]

3.1.2 Costs

Cost decreasing effects of cooperation occur to a significant extent in the form of increased market power. Joint purchasing activities can substantially reduce costs for services, components, or entire aircrafts due to volume discounts.[35] Moreover, cooperation offers the opportunity to exploit comparative cost advantages among the alliance members. This is particularly relevant on the operational (i.e. joint operation of terminal facilities, lounges, sales activities, etc.) and research and development level (i.e. CRS, IT, etc.).[36] Code sharing also has cost-related effects. The coordination of flight schedules, for example, usually increases the seat load factor which then leads to decreasing unit costs.[37] Lower unit costs may also arise from a creation of a multiple hub environment as mentioned above. By adapting the flight schedules to the main waves at the mega hubs, the ground times of aircrafts decrease significantly.[38]

Despite many positive effects, code sharing does have some cost disadvantages as well. All coordination activities such as those of connected route networks or

---

[31] See Dennis (2000), p. 82.
[32] For a more intensive discussion see Ossadnik (1995a), p. 21; Sandler (1991), p. 167; Weber (1991), p. 112.
[33] See Sterzenbach & Conrady (2003), p. 235; Pompl (2002), p. 146; Porter (1985), pp. 331ff.; Provan (1983), pp. 79ff. and pp. 85ff.
[34] See Pompl (2002), p. 146.
[35] See Jäckel (1991), pp. 100ff.; Pompl (2002), p. 148.
[36] See Jäckel (1991), pp. 109ff.
[37] See Sterzenbach & Conrady (2003), p. 202; Pompl (2002), p. 142.
[38] See Dennis (2000), p. 82; Hanlon (1999), pp. 133ff.

harmonized flight schedules cause additional transaction costs.[39] These arise, inter alia, from the need for additional communication, from cultural or strategic differences among the cooperating airlines and the necessary integration costs.[40]

In addition, introducing a new alliance brand name (like Star Alliance or Oneworld) also requires financial investments. Finally, cooperation involves opportunity costs on the individual firm level. Resources that are utilized to fulfill alliance tasks cannot be assigned to individual tasks. Apart from that, individual goals cannot be targeted without regard to alliance-specific requirements.[41]

3.1.3 Risks

The last dimension of positive potential synergies is the risk reducing influence. This effect results primarily from the reduction of behavioral uncertainty among the alliance partners.[42] Chen and Chen show that parallel code sharing agreements lead to higher seat load factors and lower risk at the same time.[43] Moreover, Dennis argues that the connection of route networks creates efficient market entry barriers for new entrants like low cost carriers and thus strengthens the market position of the alliance members.[44] Finally, joint and consequently more powerful lobbyism may also reduce the overall market risk for the alliance members.[45]

In the risk increasing dimension, especially the loss of individual autonomy plays a major role. Under certain circumstances, alliance membership can create such dependence on the alliance that an autarkic survival of the alliance members is impossible. This issue becomes primarily relevant when an alliance consists of members of different size and power. Thus, alliance membership can also increase the individual risk of an airline company.[46] Another aspect is the

---

[39] It can be shown, for example, that the adaptation of several flight schedules to the very few major waves at hub airports causes significant additional coordination tasks (see Dennis (2000), pp. 75ff.) and at the same time increases the vulnerability of the entire network (see Hanlon (1999), pp. 134ff.).
[40] See Sterzenbach & Conrady (2003), p. 235; Pompl (2002), p. 146; Jäckel (1991), pp. 148ff.
[41] See Jäckel (1991), pp. 148ff.
[42] See Steininger (1999), pp. 139ff; Jäckel (1991), pp. 97ff. and Albers (forthcoming).
[43] See Chen & Chen (2003), p. 31.
[44] See Dennis (2000), p. 84.
[45] See Jäckel (1991), pp. 101ff.
[46] See Pompl (2002), p. 147.

intensive exchange of sensible business data among the alliance partners. Furthermore, the probability of intervention of antitrust bodies gets higher the larger the alliance is.[47]

The following figure summarizes a catalog of positive and negative potential synergies. (The enumeration is of course incomplete, but provides a reasonable starting point for all subsequent steps of the analysis.)

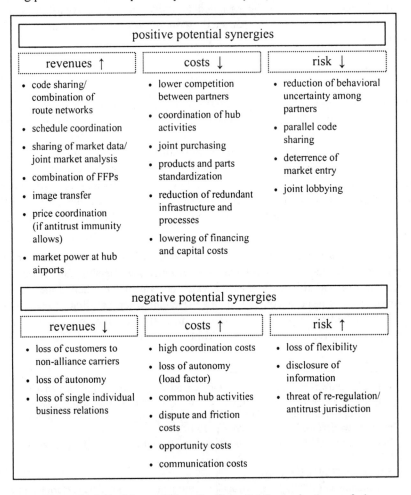

Figure 11-1: Positive and Negative Potential Synergies (examples)

---

[47] See Steininger (1999), pp. 162ff.; Jäckel (1991), pp. 152ff.

3.2 Synergy Allocation as Critical Success Factor for Strategic Airline Alliances

Airline alliances are not associations which serve charitable purposes. Airlines join in alliances because, in their utility calculi, the benefits of alliance membership offset the risks and reductions of their autonomy.[48] This, however, implies that as soon as the benefits appear to decline and hence appear to fall short of the associated cost positions for one specific member firm, this alliance partner will face an increasing incentive to exit the alliance. The loss of one member firm can, however, have considerable consequences for the remaining alliance partners, as by definition the "pie" of benefits which is generated over all alliance members is reduced and, thus, trigger further partners to leave the alliance. This domino effect can ultimately threaten the existence of the whole cooperative venture.

This scenario is obviously the worst case for an alliance and has not yet materialized. However, if an alliance member A voluntarily engages into actions which temporarily result in a worsening of its own position but result in benefits for partner B, an adequate compensation of A by the privileged member B is a rational action if B's benefits exceed A's losses.[49] In such a case the alliance synergy allocation rule should encourage and favor A's actions. It seems obvious that these kinds of agreements are more welcome in boom periods than in recessions – although it can be stated that their importance is significantly higher in the latter one. Especially in recessions alliances show their stability in terms of efficiency and loyalty as the issue of interdependency versus autonomy of the partners becomes even more relevant compared to periods of economic growth.[50]

It has been found that an alliance tends to be more stable when the member firms consider their outcome as being fair in relation to their input.[51] Thus, the effect of a synergy allocation system is twofold. First, it helps to ameliorate the atmosphere within an alliance due to the incorporation of the win-win principle and second, it increases its overall stability by introducing a mechanism to secure that each member firm will be compensated on the basis of its particular contribution to the overall performance of the alliance.

---

[48] See Netzer (1999), p. 34.
[49] If B's benefits are lower than A's losses, the action should be avoided as in this case B cannot compensate A and would hence benefit on A's expense.
[50] See Das & Teng (2000a), p. 85.
[51] See Backhaus & Piltz (1989), p. 9; Perlmutter & Heenan (1986).

Despite its undisputed importance and the considerable scholarly attention airline alliances have received, the issue of synergy allocation has yet been widely neglected.[52] Especially, a theoretically founded but also practical method which allows airline managers to address this complex topic has, to our knowledge, not yet been advanced. We therefore attempt to initiate a first step into this direction and propose an approach based on the so called Analytical Hierarchy Process (AHP).[53]

4 An AHP-based Approach to Synergy Allocation among Airlines

4.1 Introduction to the Synergy Allocation Model

A theoretically founded method which ensures an *objectively fair* allocation of synergies is yet to be put forward.[54] However, philosophers have taken a stance with objectivity – whenever human beings are involved, pure objectivity is basically not achievable.[55] It thus becomes necessary to identify allocation rules which adequately approximate an objectively fair distribution.[56] Our goal is to develop an airline-specific synergy allocation process that adheres as closely as possible to the stated objective and requirements based upon the AHP-based synergy allocation rule introduced by Ossadnik with regard to the allocation of synergy effects in mergers.[57]

---

[52] See Albers (2000); Gulati & Wang (2003); Jap (1999); (2001).
[53] For a detailed description of the method itself and its contribution to the allocation model refer to section 4.4 of this chapter.
[54] See Ossadnik (1995b), p. 75; Thomas (1969), p. 49; Thomas (1974), pp. 128ff.
[55] As Rescher put it, "objectivity calls for not allowing the indications of reason, reasonableness, and good common sense to be deflected by 'purely subjective' whims, biases, prejudices, etc. Accordingly, objectivity always strives for the sensible resolution while subjectivity gives rein to temper and lets personal inclination have its way. This does not require excluding values (how humans ever achieve that?), but rather insists on not being deflected from the path of reason by rationally inappropriate prejudicial influences". Rescher (1997), p. 4.
[56] Ossadnik proposes the partner-specific contribution to the creation of the particular synergy effect for these purposes. See Ossadnik (1996), pp. 45ff.
[57] See Ossadnik (1995a); (1996).

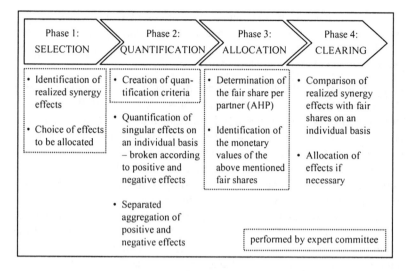

Figure 11-2: Four-phase Synergy Allocation Model

The suggested allocation process consists of four phases: Selection, quantification, allocation and clearing. The first phase (*selection*) functions as filter stage in order to reduce the complexity for the subsequent phases. Here, realized synergy effects are identified. It is then necessary to examine which effects should, need, or can be allocated. Within the second phase (*quantification*) the selected effects are quantified and aggregated in order to show the overall positive and negative synergy effect. The synergy allocation takes place in the third phase (*allocation*) in which the AHP is applied to determine fair shares per partner. At the end of this phase, the respective monetary values of these fair shares are calculated. Finally, the *clearing* phase closes the process. Here, the required money transfers among the alliance members are computed. Certain activities are performed by a committee[58] of airline experts to ensure inter-subjectively plausible propositions and claims. The phases will be depicted in detail within the following sections.

4.2 Identification of Realized Synergy Effects

Complexity reduction is the major task of the first phase of our model. To these ends, we propose the installation of an inter-organizational expert committee

---

[58] Detailed information on the expert committee issue will be provided in the following chapter.

(EC). Each alliance member sends one representative to the EC. These representatives act as advocates for their firms and represent the firms' interests. The delegate can be an employee of the respective firm or any other airline expert acting on behalf of the particular carrier.

An expert as part of the EC is considered to be a "highly trained, competent professional specialist who judiciously collects data, carefully weights their information content, and makes an informed judgment".[59] Whenever a complex decision is to be made for which no definitive evidence is available, it is reasonable to use experts finding a sound and well-founded decision.[60] This is clearly the case in our context: Synergies, as has been depicted above, come in various forms and are ambiguous to interpret. Another reason for using the expert committee is the large number of firms involved in the process. In order to find reasonable and practical solutions, a decision body is needed which (a) has the knowledge to find reasonable solutions, which (b) has the support of the top decision makers, and finally which (c) is manageable in terms of size. An expert committee consisting of one delegate per alliance member firm meets all mentioned criteria (given that the number of partner firms is manageable).[61]

However, it cannot be overseen that some problems are associated with using expert decision making. In their study, Spence and Brucks found that experts usually need significantly more time to come to a decision than novices do.[62] Apart from that it is argued that whenever there are experts involved in decision making, there is for various reasons some sort of disagreement between these experts as well.[63]

---

[59] Macintosh (1985), p. 43. Spence & Brucks define "an expert as someone who has aquired domain-specific knowledge through experience and training" (Spence & Brucks (1997), p. 233).

[60] See Mumpower & Stewart (1996), p. 191.

[61] Within a study Spence & Brucks compare the judgment abilities of experts to those of novices. They found that inter alia experts display greater consistency in their decisions. Moreover, it becomes obvious that experts usually make fewer larger errors than novices do. In addition, it turns out that they tend to be more effective in solving problems which have no initial structure – which is clearly the case in our context. Finally, the authors show that experts are more confident in their judgment which facilitates the communication of the decision (see Spence & Brucks (1997), pp. 240ff.)

[62] See Spence & Brucks (1997), p. 240.

[63] See Mumpower & Stewart (1996).

We are convinced that the use of the AHP as an evaluation tool reduces the sources of disagreement because the experts are not required to evaluate all synergies at once, but in pairwise[64] comparisons only.

However, we feel that using experts performing the necessary judgments within our synergy allocation model is the best possible compromise concerning decision accuracy on the one hand and the necessary reduction of problem complexity on the other hand.

Also, which functions are to be transferred to the EC in this phase of the process? At first, the EC needs to identify the realized synergies effects (i.e. the synergy categories depicted in section 3.1). Thereafter, the EC determines which realized synergy effects are to be allocated based on the underlying criteria *necessity* and *strategic relevance*.

The big global airline groupings (e.g. StarAlliance or Oneworld) involve a significant number of synergy potentials. *Necessity* refers to the goal to reduce complexity by eliminating those synergy effects that are (a) present at all alliance members at similar levels or (b) only side effects of the cooperation and thus do not need to be considered during the allocation process. In this case, the concerned synergy effects will not be considered and drop out of the process at this early stage. For example, the costs of performing the synergy allocation process exist for any participating firm and probably have limited influence on the overall process of synergy allocation. They should therefore not be included in the process.

The decision on *strategic relevance* of a synergy effect is rather political – based on the willingness of the partners to include a particular synergy effect into the allocation system. The idea behind this criterion is twofold. First, it allows the EC to actively influence the allocation process and second, it allows the members to exclude those effects from the allocation process which could disequilibrate the overall alliance or disturb the atmosphere within the alliance. An example is the effect of image transfers through alliance membership. In this case, it would be necessary to categorize the alliance brands into more and less valuable brands which may lead to disturbances within the alliance as it divides the alliance in "good" and "better" member firms. In order to avoid this problem beforehand it seems reasonable to exclude effects arising from image transfer from the entire allocation process already at this level.

---

[64] *Pairwise* refers to the fact that only pairs of synergy effects are compared and evaluated. When all possible pairs have been evaluated, the AHP provides a technique to determine the relative importance of each effect (see section 4.4).

4.3 Assessment of Individual Partner's Contribution

In the course of this stage, three tasks need to be completed. The first step includes the determination of quantification criteria by the EC. Then, the alliance members (i.e. not the EC) need to quantify their realized synergies on an individual level using the quantification criteria developed beforehand. This quantification process needs to be performed for negative and positive synergy effects separately. Finally, the last task is to aggregate these quantified synergy effects in order to show the total negative and the total positive synergy effect generated by the respective alliance.

Before the quantification process can start, every single synergy effect needs to be examined on its quantification feasibility, i.e. is quantification possible considering both an arguable assignment of resources to the process of quantification and a sufficient accuracy of quantification. We admit that there is space for interpretation, but this again reflects the various options of the alliance members to actively influence the allocation process. One should admittedly note that the degree of accuracy of synergy quantification is technically limited.[65]

Although the problem structure of quantifying synergies is clear, a generally accepted method for quantifying synergy effects is missing.[66] Quantification of synergies means measuring the difference between the status quo (i.e. autarkic operation) and the joint performance of activities. The major problem in this stage is that the quantification heavily relies on the quality of the prognosis on the outcome of the joint alliance activities and/or the performance in autarkic operation. In other words, depending on the perspective, at least one of both figures is based on estimations. Thus, it is necessary to agree on practical and sufficiently accurate criteria. In addition, the quantification process for every synergy effect is different due to diverse underlying problem structures. Therefore, the EC needs to develop a unique quantification process for every single effect that should be included into the allocation process as e.g. for measuring synergies from *decreased flexibility* definitively requires different quantification instruments than those created through *joint purchasing*. This implies that the estimations may significantly vary in terms of accuracy among the various synergy effects. However, this very important step on the way towards a fair synergy allocation is also on the task list of the EC.

---

[65] See Hax (1992), p. 971; Gälweiler (1989), column 1937.
[66] See Weber (1991), p. 110. Gupta & Gerchak propose a model for quantifying *operational* synergies in a merger/acquisition, but they admit as well that "suitable modifications" are needed to use the model in different contexts (Gupta & Gerchak (2002), p. 519).

As we cannot cover every effect in this chapter, we use the example of quantification of synergies resulting from *code sharing* to provide an idea of the quantification process.

Example: In this case we only consider synergy effects which evolve from code share agreements among the alliance members. Another assumption is the application of the so called *free sale model*.[67] We can observe different synergy effects in terms of synergy categories on busy and not busy routes. In the former case, both concerned airline companies would perform the particular flight and thus the frequency of flights between the involved destinations will increase.[68] Synergy effects caused by *increased revenues* can be deducted from increased ticket sales compared to the pre-alliance status. In the latter case only one partner would perform the flight and the alliance would significantly *decrease its operating costs* on this particular city pair. In this case the saved costs could be denoted as synergy effect.

At the end of this phase the extent of the total negative and positive synergy effect and the extent to which every alliance member realized synergy effects have been assessed. In the following phase the alliance member's individual contribution to the total negative and positive synergy effect is to be determined.

4.4 Deduction of Fair Shares per Partner

The primary task of this phase – the deduction of fair shares per partner – also belongs to scope of functions of the EC. Ossadnik proposes a synergy allocation model based on the Analytic Hierarchy Process (AHP)[69] in order to evaluate the firm's individual contribution to the emergence of a respective synergy effect.[70] The underlying assumption is that the experts are able to deduct a fair share per partner based on the importance of a particular resource for the creation of the respective synergy effect.[71] Thus, alliance members who bring relatively more valuable resources into the alliance receive a larger share of the synergy effects and vice versa.

---

[67] A *free sale agreement* is a distinctive form of code sharing among airlines. Each carrier is allowed to sell tickets on its own discretion, but the operating carrier receives a previously specified transfer price for each passenger (see Pompl (2002), p. 140).
[68] See Oum et al. (1996), p. 188.
[69] See Saaty (1980); (1996); (2001b); Saaty & Vargas (2000).
[70] See Ossadnik (1995a); (1995b); (1996).
[71] See Ossadnik (1996), p. 45.

The AHP basically involves four steps.[72] In the *first* step, a decision hierarchy needs to be determined by "breaking down the decision problem into a hierarchy of interrelated decision elements".[73] In our case a potential hierarchy scheme is illustrated in figure 11-3.[74]

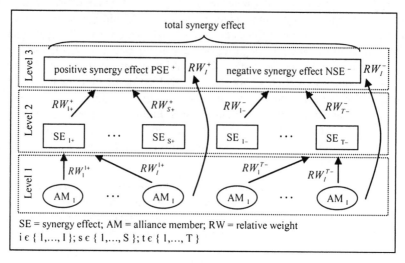

Figure 11-3: Deduction of Fair Shares – Overview

The individual contribution either of the alliance partner (level 1 and 3) or the respective synergy effect (level 2) will be expressed by so-called *relative weights* (RW) which show to what extent the respective decision element contributes to the achievement of the objectives of the next higher level. These levels symbolize the hierarchic structure of the deduction process as the results of the first two levels are needed as input data to calculate the relative weight of each alliance partner for the creation of the overall positive (PSE$^+$) and negative (NSE$^-$) synergy effect which is the target figure of the last level (level 3).

---

[72] See Zahedi (1986), p. 96.
[73] Zahedi (1986), p. 96.
[74] One may argue that it is not always possible to decompose a complex decision problem into a well-structured hierarchy scheme, but this seems the only way to solve the synergy allocation problem at reasonable time and costs.

1st round of evaluation (level 1)	
RW_i^{s+}	= relative weight of alliance partner i on positive synergy effect s
RW_i^{t-}	= relative weight of alliance partner i on negative synergy effect t
2nd round of evaluation (level 2)	
RW_{s+}^{+}	= relative weight of positive synergy effect s on total positive synergy effect PSE^+
RW_{t-}^{-}	= relative weight of negative synergy effect t on total negative synergy effect NSE^-
3rd round of evaluation (level 3)	
RW_i^{+}	= relative weight of alliance partner i on total positive synergy effect PSE^+
RW_i^{-}	= relative weight of alliance partner i on total negative synergy effect NSE^-

Table 11-2: Relative Weights on the Different Levels (level 1-3)

Within the *second* step, the input data for the subsequent calculations is generated. The input data "consists of matrices of pairwise comparisons of elements of one level that contribute to achieving the objectives of the next higher level".[75] These comparisons are performed by the EC and require the experts to rank the respective decision elements in pairwise comparisons using an ordinal rating scale (see table 11-3) which expresses the importance of one decision element in comparison to another one.

Intensity of Importance	Definition	Explanation
1	Equal importance	Two activities contribute equally to the objective
3	Moderate importance	Experience and judgment slightly favor one activity over another
5	Strong importance	Experience and judgment strongly favor one activity over another
7	Very strong importance	An activity is favored very strongly over another; its dominance is demonstrated in practice
9	Extreme importance	The evidence favoring one activity over another is of the highest possible order of affirmation
2, 4, 6, 8	[Intermediate values]	

Table 11-3: Ordinal Rating Scale[76]

---

[75] Zahedi (1986), p. 98.
[76] Adapted from Saaty (1994), p. 26.

The result of this process is a matrix containing evaluations for each possible combination. As in a comparison of an element with itself each element is equally important, the diagonal of the matrix always shows the value one. In fact, only half of the matrix needs to be filled by the experts because the lower triangle elements of the matrix are the reciprocal of the upper triangle elements.[77] It has been empirically shown that preference statements can be inconsistent. Thus, based on the finding that for fully consistent matrices their largest eigenvalue λ_{max} equals the dimension of the matrix, Saaty developed two formulas in order to determine the consistency of a matrix.[78] The consistency index (CI) as $CI = (\lambda_{max} - n)/(n - 1)$ and the consistency ratio (CR) as $CR = CI/RI$ with RI being the so called random index which can be interpreted as average consistency index.[79]

The following example illustrates the course of the evaluation process. Given an alliance with three alliance member firms (AM_i with i = 1, 2, 3) and three identified positive (SE_{s+} with s=1, 2, 3) and negative (SE_{t-} with t=1, 2, 3) synergy effects, we can derive a decision hierarchy as depicted in figure 11-3. The appropriate matrix of pairwise comparisons regarding the relative weights of the alliance member firms in terms of the creation of a particular positive synergy effect – e.g. synergies from code sharing – can read as shown in the following figure (level 1). The same logic applies to all subsequent higher levels – except the last one (see step 4 on page 296).

	AM_1	AM_2	AM_3	RW_j^{1+}	
AM_1	1	1/2	2	0.311	$\lambda_{max} = 3.0536$
AM_2	2	1	2	0.493	C.I. = 0.025
AM_3	1/2	1/2	1	0.196	C.R. = 0.04

Figure 11-4: Matrix of Pairwise Comparisons (level 1)

In our example the importance of the input of AM_2 regarding the synergy effect realized through code sharing agreements is classified to be moderately higher than those of AM_1. The following matrix e.g. reveals that the importance of SE_{1+}

---

[77] See Zahedi (1986), p. 98.
[78] See Saaty (1994), p. 41; Saaty (2001a), pp. 80ff.
[79] See Saaty (2001a), p. 81.

(e.g. code sharing) is considered to be higher than those of SE_{3+} (e.g. joint sourcing activities) in terms of creating a positive synergy effect.

	CS	FFP	Sourcing		RW_{s+}^+	
CS	1	2	3		0.540	$\lambda_{max} = 3.0092$
FFP	1/2	1	2	→	0.297	C.I. = 0.01
Sourcing	1/3	1/2	1		0.163	C.R. = 0.02

Figure 11-5: Matrix of Pairwise Comparisons (level 2)

In the course of the *third* step, the "eigenvalue" method[80] is used to calculate the relative weights[81] of the decision elements in the various matrices; here the alliance member firms (AM_i) or the particular positive (SE_{i+}) and negative (SE_{j-}) synergy effects, respectively. The calculations in this step are rather complex and cannot be performed without the aid of special software tools.[82]

The relative weights from n rows of a matrix M using the "eigenvalue" method can be obtained by the following equitation: $M \cdot E = n \cdot E$ with E being the vector of relative weights. In this case, n is called the eigenvalue and E is considered to be the right eigenvector of matrix M. As the evaluator does not know E and it is thus impossible to determine the relative weights accurately, the matrix contains inconsistencies as mentioned above. It has been shown that following the afore-mentioned equation the estimation of E can be derived from $M^* \cdot E^* = \lambda_{max} \cdot E^*$ with M^* being the observed matrix M, λ_{max} being the largest "eigenvalue" of M^*, and E^* being the right eigenvector of matrix M^*.[83]

---

[80] For a detailed description of the "eigenvalue" method refer to Zahedi (1986); Schneeweiss (1991).
[81] The *relative weight* (RW) shows to what extent the respective decision element contributes to the achievement of the objectives of the next higher level. The values of the RWs are normalized on a scale from "0" (means "no contribution") to "1" (means "sole responsibility").
[82] At this point we do not want to hide that the "eigenvalue" method used in AHP is only one method to derive RWs from pairwise comparison matrices. See Saaty & Vargas (2000), p. 8.; Saaty (2001b), p. 55; Zahedi (1986), p. 99.
[83] See Zahedi (1986), p. 99.

In our case we can see that the relative weight of AM_2 ($RW_2^{1+} = 0.493$) is significantly higher than those of AM_1 ($RW_1^{1+} = 0.311$) and AM_3 ($RW_3^{1+} = 0.196$). In other words, the contribution of AM_2 to the creation of SE_{1+} is higher and, thus, this will also lead to a higher portion of the realized positive synergy effect. On the second level we can see that SE_{1+} contributes most to the overall positive synergy effect PSE^+, because its relative weight ($RW_{1+}^+ = 0.54$) is obviously the highest among the represented synergy effects. We use this procedure on every level of the evaluation process, i.e. all RW-values indicated in table 11-2 are calculated by means of the same scheme.

The *fourth* step concludes the calculation scheme by aggregating the relative weights in order to express the contribution of the respective decision element (in our case a particular alliance member firm) to the overall positive or negative synergy effect. We call this aggregated value the relative weight of the particular alliance member firm regarding the creation of the overall positive (PSE^+) or negative (NSE^-) synergy effect. The aggregated relative weight of a particular alliance member firm AM_i can be computed from

$$RW_i^+ = \sum_{s=1}^{S}\left[RW_i^{s+} \times RW_{s+}^+\right] \text{ for positive and from } RW_i^- = \sum_{t=1}^{T}\left[RW_i^{t-} \times RW_{t-}^-\right]$$

for negative synergy effects. In our example a possible solution is illustrated in figure 11-6.

AM_1 : $0.311 \cdot 0.54 + 0.053 \cdot 0.3 + 0.4 \cdot 0.16 = 0.2478$	$(= RW_1^+)$
AM_2 : $0.493 \cdot 0.54 + 0.253 \cdot 0.3 + 0.2 \cdot 0.16 = 0.3741$	$(= RW_2^+)$
AM_3 : $0.196 \cdot 0.54 + 0.694 \cdot 0.3 + 0.4 \cdot 0.16 = 0.3781$	$(= RW_3^+)$

Figure 11-6: Aggregated Calculation of RWs (level 3)

The last task that has to be completed is the deduction of fair shares in monetary units. Assuming that the overall positive synergy effect realized and quantified within the second stage of our synergy allocation process (see section 4.3) is €500 million, according to the model the fair share of AM_1 accounts for €123.9 million ($RW_1^+ \times PSE^+ = 0.2478 \times 500 = 123.9$) and so on.

4.5 Computation of Required Money Transfers

The task that needs to be completed in the last stage is comparably easy – at least on a theoretical level. During the clearing phase the realized synergy effects are being compared to the calculated fair share of each alliance member firm. It is not very probable that these two values are of equal size. Therefore, an allocation of synergy effects based on the partner's individual contribution is necessary. All positive balances go to a specific balancing account from which all negative balances are settled. Thus, the balancing account is only used for this particular transaction and shows an account balance of zero at the end of the allocation process. The following figure illustrates the process using another fictive example.

positive synergy effects (in million Euro)				negative synergy effects (in million Euro)			
AM_1	100	124	+24	AM_1	-100	-150	-50
AM_2	150	187	+37	AM_2	-50	-20	+30
AM_3	250	189	-61	AM_3	-50	-30	+20
total	500	500	0	total	-200	-200	0
	realized phase II	share phase III	balance phase IV		realized phase II	share phase III	balance phase IV

clearing

AM_1	+24 + (-50) = -26		AM_1 **transfers** 26 million Euro to clearing account
AM_2	+37 + 30 = +67		AM_2 **receives** 67 million Euro from clearing account
AM_3	-61 + 20 = -41		AM_3 **transfers** 41 million Euro to clearing account
total	0		

Figure 11-7: Required Money Transfers

In sum, after having identified the realized synergy effects, the fair shares per partner, and required money transfers, finally, all realized synergies have been balanced out at the end of the term.

As we argued before, a fair synergy allocation system is not only a tool to improve coordination within the alliance, but also does have positive effects on the alliance members' motivation to support the alliance idea, i.e. a cooperational instead of a self-serving approach. Thus, we feel that this procedure may help the alliance to enhance the stability and by doing so, improve the performance of multi firm alliances.

5 Limitations, Conclusion, and Outlook

This chapter dealt with a core problem in alliance management: the assessment and distribution of jointly generated synergies in alliances. We have embedded the synergy allocation problem in the context of the alliance governance task, proposed a typology of airline alliance-specific synergies, and suggested an approach to assessing and distributing the synergy effects among airline alliance partners. However, and as always, there is ample room for further refinement and research.

A rather obvious and central issue is the heavy reliance on the subjective judgment of the expert committee. We agree with this weakness and admit that the expert committee is only a second best solution. However, in the general absence of a theoretically founded synergy allocation rule it is necessary to employ an alternative allocation method. The AHP is a rather accepted tool to evaluate the importance of different alternatives. Furthermore, practicability and acceptance of the employed method was given a high priority in assessing the problem. The inauguration of an expert committee which is able to accumulate trust and confidence by all alliance partners appears to be the most practical solution which is at present conceivable. The employment of experts is an adequate possibility to receive viable results with an adequate use of limited resources. This is especially true if we are dealing with complex decision problems. It has been empirically shown that under these circumstances experts perform significantly better than every other sort of decision making body.[84]

It was stated here that it only hardly possible to find a theoretically founded basis for the quantification of every single synergy effect. In these cases the model necessarily relies on sufficiently accurate estimated values. A related aspect is the model's reliance on the assumption that the alliance members indicate their individual realized synergy effect correctly and without any opportunistic motivation – an admittedly optimistic prerequisite. However, we feel that without the necessary amount of trust a long-term mode of cooperation (what

---

[84] See Mumpower & Stewart (1996), p. 191.

alliances without doubt are) will not be successful in the long run – neither in the accomplishment of the alliance goals nor in the fair distribution of jointly gained profits.

The AHP is a well-known method in theory as well as in practice and a considerable amount of research focuses on applying the AHP to different fields and contexts.[85] Considering the increasing emergence of strategic alliances in the past and the high relevance of the synergy allocation issue, we hope that the proposed model can serve as a sound basis for future research.

Airline alliances have long been suspected to be temporary phenomena. The large groupings have only emerged in the late 1990s, but since then exhibited significant growth. Star Alliance and Oneworld as well as the late-starter but currently rapidly growing Skyteam will continue to shape the global airline industry for the next decade. The struggle to attract and bind valuable partners to one particular grouping will grow stronger and stronger – currently impressively exemplified by the battle around Air China by Star and Oneworld. As long as several alliances vie for a certain partner, the alternative benefits to be gained from membership will be an important issue to consider by the prospective, but also the long-standing, partners in an alliance network. A transparent and consistent synergy allocation regime is therefore likely to become an important asset in the future. One option for its configuration was presented here.

6 References

AIRLINEBUSINESS (2004): Airline Alliance Survey 2004, in: *Airline Business*, 20(9), pp. 53-80.

ALBERS, S. (forthcoming): *The Design of Alliance Governance Systems*. Köln: Kölner Wissenschaftsverlag.

BACKHAUS, K. & PILTZ, K. (1989): Strategische Allianzen - eine neue Form kooperativen Wettbewerbs?, in: K. Backhaus & K. Piltz: *Strategische Allianzen*. Düsseldorf, Frankfurt am Main: Verlagsgruppe Handelsblatt, pp. 1-10.

BAKER, C. & FIELD, D. (2003): Where are they now?, in: *Airline Business*, 19(7), pp. 42-46.

---

[85] To the authors' knowledge, the latest application of the AHP e.g. deals with the allocation of intangible assets (see Saaty et al. (2003)).

CHEN, F.C.-Y. & CHEN, C. (2003): The Effects of Strategic Alliances and Risk Pooling on the Load Factors of international Airline Operations, in: *Transportation Research (Part E)*, 39(1), pp. 19-34.

CONTRACTOR, F.J. & LORANGE, P. (2002): Why Should Firms Cooperate? The Strategy and Economics Basis for Cooperative Ventures, in: F.J. Contractor & P. Lorange: *Cooperative Strategies in International Business: Joint Ventures and Technology Partnerships between Firms*. Amsterdam et al.: Pergamon, pp. 3-28.

CONTRACTOR, F.J. & RA, W. (2000): Negotiating alliance contracts Strategy and behavioral effects of alternative compensation arrangements, in: *International Business Review*, 9(3), pp. 271-301.

COPELAND, T.E. & WESTON, J.F. (1988): *Financial theory and corporate policy*. Reading, Mass.: Addison-Wesley.

DAS, T.K. & TENG, B.S. (2001): A risk perception model of alliance structuring, in: *Journal of International Management*, 7(1), pp. 1-29.

DAS, T.K. & TENG, B.-S. (1998): Between trust and control: Developing confidence in partner cooperation in alliances, in: *Academy of Management Review*, 23(3), pp. 491-512.

DAS, T.K. & TENG, B.-S. (2000a): Instabilities of strategic alliances. An internal tensions perspective, in: *Organization Science*, 11(1), pp. 77-101.

DAS, T.K. & TENG, B.-S. (2000b): A resource-based theory of strategic alliances, in: *Journal of Management*, 26(1), pp. 31-61.

DELFMANN, W. (2004): Hub-and-Spoke-System, in: P. Klaus & W. Krieger: *Gabler Lexikon Logistik: Management logistischer Netzwerke und Flüsse*. Wiesbaden: Gabler, pp. 193-194.

DENNIS, N. (2000): Scheduling Issues and Network Strategies for International Airline Alliances, in: *Journal of Air Transport Management*, 6(2), pp. 75-85.

GÄLWEILER, A. (1989): Synergiepotentiale, in: N. Szyperski: *Handwörterbuch der Planung*. Stuttgart: Schäffer-Poeschel, column 1935-1943

GULATI, R. (1998): Alliances and networks, in: *Strategic Management Journal*, 19(4), pp. 293-317.

GULATI, R. & SINGH, H. (1998): The architecture of cooperation: Managing coordination costs and appropriation concerns in strategic alliances, in: *Administrative Science Quarterly*, 43(4), pp. 781-814.

GULATI, R. & WANG, L.O. (2003): Size of the Pie and Share of the Pie: Implications of Network Embeddedness and Business Relatedness for Value Creation and Value Appropriation in Joint Ventures, in: V. Buskens, W. Raub & C. Snijders: *The Governance of Relations in Markets and Organizations*. Oxford: Elsevier Science, pp. 209-242.

GUPTA, D. & GERCHAK, Y. (2002): Quantifying operational synergies in a merger/acquisition, in: *Management Science*, 48(4), pp. 517-533.

HANLON, J.P. (1999): *Global Airlines: Competition in a Transnational Industry*. Oxford; Boston: Butterworth-Heinemann.

HAX, H. (1992): Synergie als Bestimmungsfaktor des Tätigkeitsbereiches (Geschäftsfelder und Funktionen) von Unternehmungen, in: *Zeitschrift für betriebswirtschaftliche Forschung (ZfbF)*, 44(11), pp. 963-973.

JÄCKEL, K. (1991): *Kooperationsstrategien im Linienluftverkehr vor dem Hintergrund zunehmender Integrationsentwicklung in Europa*. Bergisch Gladbach, Köln: Josef Eul.

JAP, S.D. (1999): Pie-expansion efforts: Collaboration processes in buyer-supplier relationships, in: *Journal of Marketing Research*, 36(4), pp. 461-475.

JAP, S.D. (2001): "Pie sharing" in complex collaboration contexts, in: *Journal of Marketing Research*, 38(1), pp. 86-99.

MACINTOSH, N.B. (1985): *The social software of accounting and information systems*. Chichester; New York: Wiley.

MUMPOWER, J.L. & STEWART, T.R. (1996): Expert Judgement and Expert Disagreement, in: *Thinking and Reasoning*, 2(2/3), pp. 191-211.

NETZER, F. (1999): *Strategische Allianzen im Luftverkehr: Nachfrageorientierte Problemfelder ihrer Gestaltung*. Frankfurt am Main: Lang.

OSSADNIK, W. (1995a): *Die Aufteilung von Synergieeffekten bei Fusionen*. Stuttgart: Schäffer-Poeschel.

OSSADNIK, W. (1995b): Die Aufteilung von Synergieeffekten bei Verschmelzungen, in: *Zeitschrift für Betriebswirtschaft (ZfB)*, 65(1), pp. 69-88.

OSSADNIK, W. (1996): AHP-based synergy allocation to the partners in a merger, in: *European Journal of Operational Research*, 88(1), pp. 42-49.

OUM, T.H.; PARK, J.H. & ZHANG, A. (1996): The effects of airline codesharing agreements on firm conduct and international air fares, in: *Journal of Transport Economics and Policy*, 30(2), pp. 187-202.

OUM, T.H.; PARK, J.H. & ZHANG, A. (2000): *Globalization and strategic alliances : the case of the airline industry*. New York: Pergamon.

OXLEY, J.E. (1997): Appropriability hazards and governance in strategic alliances: A transaction cost approach, in: *Journal of Law Economics & Organization*, 13(2), pp. 387-409.

PARKHE, A. (1993a): Messy Research, Methodological Predispositions, and Theory Development in International Joint Ventures, in: *Academy of Management Review*, 18(2), pp. 227-268.

PARKHE, A. (1993b): Strategic Alliance Structuring: A Game Theoretic and Transaction Cost Examination of Interfirm Cooperation, in: *Academy of Management Journal*, 36(4), pp. 794-829.

PERLMUTTER, H.V. & HEENAN, D.A. (1986): Cooperate to Compete Globally, in: *Harvard Business Review*, 64(2), pp. 136-152.

PISANO, G.P.; RUSSO, M.V. & TEECE, D.J. (1988): Joint ventures and collaborative arrangements in the telecommunications equipment industry, in: D.C. Mowery: *International collaborative ventures in US manufacturing*. Cambridge, Mass.: Ballinger Pub. Co., pp. 23-70.

POMPL, W. (2002): *Luftverkehr. Eine politische und ökonomische Einführung*. Berlin: Springer.

PORTER, M.E. (1985): *Competitive Advantage: Creating and Sustaining Superior Performance*. New York: Free Press.

PROVAN, K.G. (1983): The Federation as an International Linkage Network, in: *Academy of Management Review*, 8(1), pp. 79-89.

RESCHER, N. (1997): *Objectivity: The Obligations of Impersonal Reason*. Notre Dame, Ind.: University of Notre Dame Press.

RODERMANN, M. (1999): *Strategisches Synergiemanagement*. Wiesbaden: Gabler.

SAATY, T.L. (1980): *The Analytic Hierarchy Process: Planning, Priority Setting, Resource Allocation*. New York; London: McGraw-Hill International Book Co.

SAATY, T.L. (1994): How to Make a Decision: The Analytic Hierarchy Process, in: *Interfaces*, 24(6), pp. 19-43.

SAATY, T.L. (1996): *Decision making with dependence and feedback: the analytic network process: the organization and prioritization of complexity*. Pittsburgh, PA: RWS Publications.

SAATY, T.L. (2001a): *Decision Making for Leaders*. Pittsburgh, PA: RWS Publications.

SAATY, T.L. (2001b): *Decision making with dependence and feedback: the analytic network process: the organization and prioritization of complexity*. 2nd Edition, Pittsburgh, PA: RWS Publications.

SAATY, T.L. & VARGAS, L.G. (2000): *Models, Methods, Concepts & Applications of the Analytic Hierarchy Process*. Boston: Kluwer Academic Publishers.

SAATY, T.L.; VARGAS, L.G. & DELLMANN, K. (2003): The Allocation of Intangible Resources: The Analytic Hierarchy Process and Linear Programming, in: *Socio-Economic Planning Sciences*, 37(3), pp. 169-184.

SANDLER, G.G.R. (1991): *Synergie: Konzept, Messung und Realisation - Verdeutlicht am Beispiel der horizontalen Diversifikation durch Akquisition*. Bamberg: Difo-Druck.

SCHNEEWEISS, C. (1991): Der Analytic Hierarchy Process als spezielle Nutzwertanalyse, in: G. Fandel & H. Gehring: *Operations Research*. Berlin et al.: Springer-Verlag, pp. 183-195.

SPENCE, M.T. & BRUCKS, M. (1997): The Moderating Effects of Problem Characteristics on Experts' and Novices' Jugdements, in: *Journal of Marketing Research*, XXXIV(May 1997), pp. 233-247.

STEININGER, A. (1999): *Gestaltungsempfehlungen für Airline Allianzen*. Bamberg: Difo-Druck.

STERZENBACH, R. & CONRADY, R. (2003): *Luftverkehr: Betriebswirtschaftliches Lehr- und Handbuch*. München, Wien: Oldenbourg.

THOMAS, A.L. (1969): *The Allocation Problem in Financial Accounting Theory*. Sarasota: American Accounting Association.

THOMAS, A.L. (1974): *The Allocation Problem: Part Two*. Sarasota: American Accounting Association.

WEBER, E. (1991): Berücksichtigung von Synergieeffekten bei der Unternehmensbewertung, in: J. Baetge: *Akquisition und Unternehmensbewertung*. Düsseldorf: IDW-Verlag, pp. 99-115.

ZAHEDI, F. (1986): The Analytic Hierarchy Process - a Survey of the Method and Its Applications, in: *Interfaces*, 16(4), pp. 96-108.

ZAJAC, E.J. (1998): Commentary on 'alliances and networks' by R. Gulati, in: *Strategic Management Journal*, 19(4), pp. 319-321.

12

THE EFFECT OF HIGH-LEVEL ALLIANCE FORMATION ON THE PROFITABILITY OF PARTICIPATING AIRLINE CARRIERS

JENS RÜHLE

1 Introduction 306
2 Empirical Analysis 307
3 Findings 320
4 References 322

Summary:
This chapter addresses the measurable, financial benefits for airline carriers, resulting from a high-level alliance formation. Such a formation is defined as being a market-oriented and efficiency-seeking partnership at the same time. A profitability index is constructed for 12 chosen carriers by dividing operating revenue by operating costs and further adjusting for macroeconomic influences. The index is used for comparing profitability gains or losses for aligned carriers vs. a benchmark. Results indicate that aligned carriers transform their new strategic options into financial benefits, measured by an increase of the constructed profitability index compared to their unaligned peers.

1 Introduction

The airline industry has been a major focus of managerial literature in recent years for a number of reasons. Despite the fact that worldwide deregulation efforts have occurred during the last two decades, national interests and restrictions still keep airlines from engaging in a truly liberal competition.[1] As a result, international mergers have not been viable options for airlines until the beginning of the new millennium.[2]

To overcome these constraints, airlines have been forming strategic alliances which have helped the carriers to ensure worldwide seamless service and try to reduce costs. While the first airline alliances have been very simple in nature and scope, the trend in the industry has been to form more complex arrangements that involve several partners and co-operation in different business segments. This trend has been visible since the beginning of the 90s, but only from 1997 involved several partners with a stated intent of creating a tight collaboration with the ultimate goal of jointly increasing revenue and saving costs. In the future it is believed that competition will evolve from competition among carriers to competition between these alliance networks, as each carrier will serve a specific role in the alliance[3] and an overall strategy of the alliance dictates carriers' future options.

This chapter deals with profitability increases through alliance formation in the airline industry. Although many studies have dealt with airline alliances (e.g. Rhoades/Lush (1997) on typology and stability; Park/Cho (1997) on market share; Kleymann/Seristo (2001) on levels of integration; Gudmundsson/Rhoades (2001) on survival analysis), few studies have addressed the profitability issue of airlines (compare to Oum/Park/Kim/Yu (1999); Oum/Yu (1998)). The data material used in this chapter covers a broader and more recent time horizon than the aforementioned studies and also focuses exclusively on airlines pursuing a so-called high-level alliance strategy. A high-level alliance formation is defined as one that focuses both on market-oriented and efficiency-seeking goals at the same time.

---

[1] See Doganis (2001), p. 19; Hanlon (1996), p. 2.
[2] Mergers have only recently been seen among large carriers (i.e. Air France / KLM in 2004).
[3] See Oum & Zhang (2001), p. 290.

An empirical analysis of member carriers of the high-level alliances *oneworld*, *Star Alliance*, and *Wings* is conducted. The chapter assesses the short- to medium-term financial benefits of a high-level alliance formation in this analysis.

2 Empirical Analysis

The analysis takes its point of departure in 1989 with the announcement of the high-level co-operation between KLM and Northwest Airlines. Data material is being collected for 12 participating carriers throughout the entire time span including the year 2000. The year 2001 has been excluded as the terrorist attacks of September 11th are an exceptional event in airline history with an artificial demand downswing, resulting in data not suitable for this analysis.

2.1 Selection of Carriers and Alliance Groups

2.1.1 Sample of Chosen Carriers

For the analysis a sample of airline carriers was chosen meaning to represent the traditional flag carriers and reflecting tendencies of profitability increases or losses through alliance formation in the industry. The sample airlines had to fulfill two main criteria: (1) they had to be relatively equal in strategy and size, and (2) a clear distinction between aligned and unaligned carriers in the sense of the definition in a given time period had to be possible.

The first criterion was chosen due to possible side effects of a wide variety of carriers with different parameters that would influence the analysis. Choosing carriers with a truly global perspective only, partly rules out regional economic influences and effects of airline size. Being a global carrier with passengers from all over the world will result in the fact that an economic downturn starting in one region and possibly spreading over to the rest of the world with a certain time lag will not have a marked influence on the analysis, compared to relatively large but only regional players (e.g. Ryanair, Southwest Airlines).

Choosing carriers with a certain minimum size ensures the airlines are sufficiently similar, both in terms of resources and having the same array of strategic decisions available. Again, relatively small but international carriers (e.g. Lauda Air) might not have the same alternatives as larger flag carriers.

The second criterion provides the different groupings for the analysis. If an increase or even a decrease in profitability can be observed after alliance formation, the results have to be benchmarked against the unaligned industry trend at that point of time. The outcome could just as well occur because of increased competitive forces, due to further liberalization processes in that time period, technological advancement reducing input costs, etc. A comparison is therefore necessary to calculate for advantages/disadvantages resulting through a partnership with other carriers and showing an overall time trend for the whole industry. Additionally, when analyzing alliance outcome, a benchmark of unaligned carriers is always chosen. The outcome of the alliance is getting compared to the chosen benchmark for each year. Through that procedure not only the overall trend for all carriers is being ruled out, but also further short term yearly fluctuations that could not be eliminated through the adjustment procedures mentioned.

Twelve carriers were chosen for the sample. All of them are global carriers, are relatively large in size and can be classified as industry leaders, meaning that they do not appear to be followers in the sense of lacking strategic direction. All of the members of the alliances are core members, thus they have been part of the initial members and are therefore influencing further strategic development of the alliance substantially. The unaligned carriers on the other hand are so large that they could have possibly urged on an alliance formation by themselves, but refused to do so at that given time period. Their decision to remain outside of a high-level alliance can also be seen as a mutual decision during the time span of this study.

According to the findings of Oum et al.,[4] high-level alliances in terms of more collaboration aspects are the ones that contribute the most to profitability increases. The definition of a high-level alliance in this chapter has been defined as narrower. Compared to the mentioned study in which alliances are mainly market-oriented, the focus of this study contains only those alliances that intend to collaborate on a market-oriented and efficiency-seeking-oriented basis, meaning that the goal of the partnership is to increase revenue while

---

[4] See Oum, Park, Kim & Yu (1999).

simultaneously trying to decrease costs.[5] These formations are expected to evolve increasingly between global carriers.[6]

	Alliances		
Grouping	Wings	Star Alliance	oneworld
Total members	2	15	8
Member carriers chosen	Northwest Airlines, KLM	Lufthansa, United Airlines, Scandinavian Airlines, Thai Airways	American Airlines, British Airways, Qantas, Cathay Pacific
Founding year	1989	1997	1999
Analyzed from[7]	1990	1997	1999
Share of RPK[8] (2000)	7.4 %	21.4 %	16.2 %

Table 12-1: Alliance Groupings Taken for the Analysis

Table 12-1 shows the different alliance groupings. Northwest Airlines and KLM have been the first ones to establish a far-reaching alliance agreement, named *Wings*. Founded in 1989 and still in place today, this partnership can serve as an example for a successful, long-lasting partnership. Nevertheless, the alliance had to endure tensions as well, such as when KLM worked on a merger deal with British Airways in 1999, which was finally turned down due to government

---

[5] Alliances being market-oriented are focusing on an increase of their respective revenue (i.e. code-sharing arrangements, joint frequent-flyer-programs), an efficiency-seeking-oriented partnership is trying to decrease costs while offering the same service or increase output with same costs (i.e. joint procurement, staff exchange).

[6] See Oum & Zhang (2001) on airlines; for an overview of the evolution of alliances in general: Doz & Hamel (1998).

[7] *Wings* effectively operated from the 2nd half of 1989, *Star Alliance* from the 1st half of 1997, *oneworld* from the end of 1998.

[8] RPK – Round passenger kilometres.

concerns in both countries about monopoly power on served routes. Expansion of the alliance was urged with Continental and Alitalia at the end of the 1990s, but a successful outcome did not occur.

The founding members of *Star Alliance* agreed on a far reaching strategic alliance in May of 1997. *Star Alliance* has been the first multi-partner alliance and was able to attract new strong partners in the consecutive years. *Star Alliance*'s strategy has been to rely on anchor carriers (LH for Europe, UA for North America, Thai for Asia, although more anchor carriers were needed in Asia due to its sheer size, and therefore Singapore Airlines was included as an additional anchor carrier) for each continent and feeder carriers (e.g. Spanair, Braathens) that feed the international hubs of these large anchor carriers with traffic.[9]

American Airlines, British Airways, Qantas, and Cathay Pacific founded the global alliance grouping *oneworld* in February of 1999 in response to the earlier formation of *Star Alliance* that was sensed to be a direct threat to the competitive position of each participating member.

The aforementioned problems resulting through merger talks between KLM and British Airways for *Wings* also resulted in problems for *oneworld*.[10] The merger talks between these two airlines and resulting reconsideration of the other members future options possibly slowed down the process of pursuing an integrated approach for *oneworld*. Although the alliance fits the criteria of a high-level alliance, most of its activities, at least until 2000, were mainly market-oriented and therefore mainly in the first phase of a multi-partner alliance.[11]

The chosen benchmark carriers were the ones that did not match the chosen criteria for a high-level alliance as stated above. Two independent carriers with a global perspective and a strong background have been chosen for comparison as well as the alliance carriers being unaligned before their alliance agreements. The independent airlines are Continental and Japan Airlines. Both formed agreements with other carriers.[12] Co-operation efforts of these carriers involve

---

[9] See Oum & Zhang (2001), p. 290.
[10] See Baker (2001), pp. 42–45.
[11] See Doganis (2001), p. 88.
[12] Continental has code-sharing agreements with Air France (April 1997) and America West Airlines (June 1994) as well as more recent ones with Northwest and KLM; Japan Airlines has developed code-sharing relations to a large number of larger carriers (e.g. Swissair, Thai, Air France, Alitalia, Cathay Pacific; American Airlines).

mainly code-share and block seated agreements.[13] These agreements are easily dissolvable and do not involve a large commitment to the other partners. No alliance agreements are known that are directed at cost savings only. Thus both airlines can be treated as unaligned carriers in the sense of our definition.

2.2 Introduction of the Profitability Index

The chosen profitability index is constructed by dividing a carrier's aggregated operating revenue by its aggregated operating costs. An index value above 1 expresses a positive operating result, while a result below 1 shows an operating loss for the airline at that particular year.

Adjustments have to be made since the analysis is a time series analysis and influences not accountable by managerial competence want to be excluded as far as possible. Adjustment of the index to economic influences is discussed in section 2.3.

The adjusted index can be used in different ways. By standardizing it to the base year of 1989 for each carrier, the average profitability increase from 1989 for participating carriers for each year can be constructed by using a linear regression model.

Note that by adjusting the index to a base year, it does not show if the carrier is profitable in the sense of revenue exceeding the costs in that particular year. Therefore one should refrain from making inferences on operating profitability of that particular year for the carriers after the adjustment has been made.

2.3 Adjustments to Financial Data

2.3.1 Discussion of GDP Influences over Time

In order to compare alliance influences over time, adjustments to the data have to be made resulting from macroeconomic influences as these influences are not under managerial control.

1. *Testing for business cycles*: The initial step of the adjustment procedure was to test for business cycles. The constructed profitability index for each individual carrier was analyzed in terms of yearly fluctuations. These

---

[13] See Ombelet (2001), pp. 49–70.

fluctuations were found and seemed to appear relatively similar in shape across airlines. Therefore the data was tested for correlation between GDP growth and profitability.

2. *Correlation analysis*: A correlation analysis was performed to see if an adjustment of the data to GDP growth appeared reasonable. The analysis was performed by correlating the overall operating profit of the airlines to US-GDP growth over the entire time span. A correlation analysis with the sample data shows that profitability is highly correlated with economic growth (0.80).[14] The idea of taking US-GDP was derived from the thought that the US has been the driving force of the world economy in the 90s and because most of the participating carriers' demand was affected by passengers wanting to travel to or from the United States.

After testing the correlation of US-GDP-growth and overall profitability, this test was repeated for each carrier separately. This analysis revealed that for Cathay Pacific as well as British Airways this correlation did not hold true. British Airways only showed a correlation of 0.08, while Cathay Pacific had a negative correlation of -0.23. Therefore all the carriers' profits were tested against their respective countries of origin GDP. British Airways showed a correlation of 0.46, while Cathay had a significant correlation of 0.81. The other carriers' correlation to their respective home country's GDP was either lower or insignificantly different compared to the US. Therefore it was decided to adjust Cathay Pacific and British Airways with their respective home GDPs, while the other carriers were adjusted with US-GDP.

An explanation for the different outcomes for these two carriers could be the following: British Airways has a very important national market, which has been the first one to be challenged by no-frills providers with Ryanair appearing at the beginning of the 90s. Downward pressure on the national economy will hit national carriers even harder as competition in the home market is higher and competitors will fight more fiercely for market share than in markets with less national competition.

The reason for adjusting Cathay Pacific with its home GDP can mainly be explained with the fact that it has been hit hard by the Asian financial crisis at the end of the nineties and is only focused exclusively on overseas travel in the absence of any significant domestic market. Demand from passengers traveling to Asia from Europe and the US has declined substantially during that time period. Direct competitors in the Asian region have been hit hard as

---

[14] Values with a range of -1 (total negative correlation) to +1 (total positive correlation).

well, but these have larger national markets that are believed to be more resistant to the crisis as business relations in the home environment remain relatively stable, while international hesitance for establishing new business relations due to high future uncertainty led to less inter-continental travel.

The different adjustment procedure is therefore believed to reflect the special business environment of these two carriers and will be more reasonable than using US-GDP as for the other carriers.

3. *Data collection*: Financial data has been gathered from Air Transport World's (ATW) yearly publishing of the major airlines' results. Additionally, some of the companies' annual reports or web-published results had to be consulted, due to some missing or implausible data.[15] ATW provided the inputs of operating results, operating costs and operating revenue, stated in US-Dollars. In cases of input provided by the annual reports, the financial data had to be converted from the carrier's reported currency into US-Dollars.[16] Exchange rates of the year's ending were taken for conversion.

In order to compare different years, information about economic upswings or recessions had to be taken into account. Data about GDP increases were taken from the International Monetary Fund's (IMF) 2001 publication.[17]

4. *Adjustment procedure*: Although initial thought suggested that revenue and costs would be differently affected by an economic up- or downturn, as of fixed costs and time lags of wage increases, etc., only a slightly different correlation between GDP and revenue and GDP and costs could be found. A sound method for different adjustment to these positions could not be implemented.

Thus a different approach was taken, suggesting that the profitability index should be adjusted directly. In order to compare different years and exclude macroeconomic noise through up- or downturns, an average growth of GDP for the analyzed time period was calculated. The difference between average and actual growth of each year was then taken to adjust the constructed profitability index.

---

[15] ATW reports financial data of the recent and previously reported fiscal year. Unfortunately the reported numbers were sometimes not congruent. The input for the model was therefore crosschecked in cases of larger differences between years with companies' annual reports.
[16] Historic exchange rates from Pacific database (University of British Columbia).
[17] See International Financial Statistics Yearbook (2001).

For example, for the year of 1994 United Airlines has a profitability index value of 1.0384. Average growth of US-GDP for the time period of 1989-2000 is 3.22 % and the actual growth in 1994 is 4.00 %, resulting in a difference of 0.78 %. Accordingly the profitability index for United Airlines is downward-adjusted by 0.78 % to 1.0303 to reflect this above-average economic year.

2.4 Analysis

2.4.1 Calculation of the Profitability Index for Chosen Carriers

During the time of the study, the airlines increased their operating revenue substantially. The compound annual growth rate (CAGR) for all carriers during the time of 1989-2000 is 5.94 %. Taken into account that real prices for airline travel have dropped throughout this time period,[18] a larger increase in transported passengers had to occur.[19]

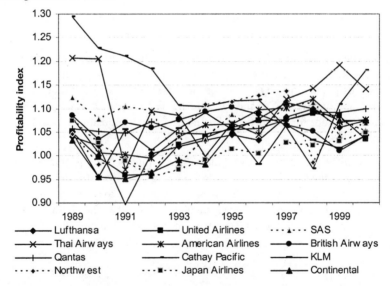

Figure 12-1: Unadjusted Profitability Index: For All Carriers

---

[18] See Findlay/Snyder (1998), p. 20.
[19] Such an increase could be measured by RPKs of transported passengers and is being published yearly in ATW.

The initial calculation of the profitability index is neither adjusted for GDP influences nor in relation to a certain base year. Hence this profitability index allows observing trends across the time period as well as the profitability situation for each carrier in each particular year.

Any index value above 1 reveals an operating profit for the particular carrier. Fluctuations can be observed in the graph.

Interestingly, Thai Airways and Cathay Pacific start off with the highest observable profitability index in 1989. Both Asian carriers were relatively small in size in 1989. It is believed that the Asian carriers with the exception of Japan Airlines have a cost advantage through cheap labor.[20] While revenue can be generated in a normal way, due to direct competition with other carriers, the cost advantages lead to a higher profitability.

Adjustments to size and cost advantages resulting through different home markets have not been modeled in the analysis. This is due to simplicity reasons as well as the approach of the analysis, which focuses on increases and decreases of profitability and not on the absolute value of the index. Therefore, in a next step the index is adjusted to the base year of 1989 for each carrier. The constructed profitability index is shown in Figure 12-2. An index value of 0.95 for a carrier in 1995 states a 5 % drop in profitability compared to the value of 1989. At that point, a direct proposition about the absolute profitability of a carrier for that year is no longer possible.

---

[20] See Oum & Yu (1998), pp. 127-134.

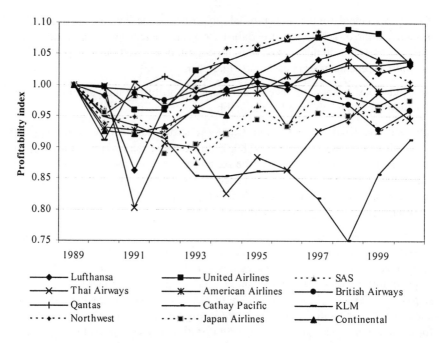

Figure 12-2: Profitability Index – Base Year 1989 for All Carriers

2.4.2 Adjusting the Years with GDP

As discussed before, an adjustment for macroeconomic influences has to be made. The correlation between US-GDP and operating profit for all carriers is 0.803. Figure 12-3 shows the relation between these two. As mentioned above, a different adjustment was chosen for Cathay Pacific and British Airways, as their home country's GDP proved to be more closely related to their operating profit.

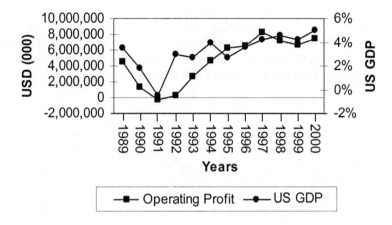

Figure 12-3: Correlation between US-GDP and Overall Operating Profit

Figure 12-4 shows the GDP volume graphs used in the analysis. Hong Kong's GDP is especially affected by regional influences in 1998-2000. The strong growth during the beginning of the 90s can also be observed.

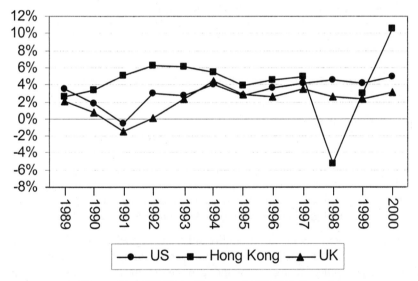

Figure 12-4: US, Hong Kong, and UK GDP Growth

Through the GDP adjustments, seasonal fluctuations have been eliminated or smoothed out. With this data material, which has been graphed in Figure 12-5, it is possible to see if – for the further analysis – an adjustment in terms of industry trend had to be made. A regression analysis was performed to check for an overall trend. Such a trend would result in a bias in the comparison of alliances and their benchmarks over time. Such a trend could not be observed for the analyzed time horizon and therefore no further adjustment was necessary.

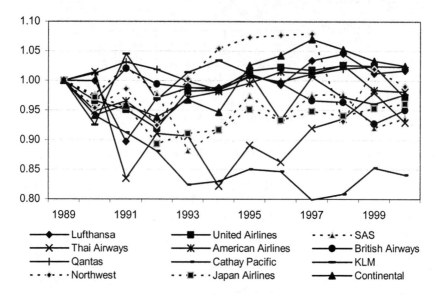

Figure 12-5: GDP-Adjusted Index: Base Year 1989 for All Carriers

2.4.3 Comparison of Alliance Groups

To compare the outcome of alliance formation, the year prior to the stated first year of the alliance agreement has been taken as the base year. Any change of the profitability index is therefore calculated in respect to a change to that particular year. The aligned carriers involved are adjusted to that base year, as well as those carriers being unaligned at that point of time in order to serve as a benchmark.

After that, an average of the profitability index for aligned and unaligned carriers is being constructed. As the development over the lifetime of an alliance is of interest, a percentage change of the index to the previous year is calculated. In a final step, the changes in profitability of aligned and unaligned carriers are being compared.

Note that the later the alliance has been established, the fewer are the carriers of the benchmark for the initial years. Carriers that establish an alliance agreement drop out of the benchmark at that particular formation year. *Wings*-members will therefore never serve as a benchmark as they have been the first ones to establish a partnership in the year of 1990.[21] *Star Alliance* serves as a benchmark until the year of 1997, *oneworld*-members until the year of 1999.

2.4.4 Alliance Outcome

When comparing the alliance outcomes (Table 12-2), both *Wings* and *Star Alliance* have been very successful partnerships in terms of profitability increases. Both alliances have on average outperformed their benchmark by almost 3.3 % per year. *oneworld* has only been observed for two years of cooperation. The results do not show a real difference to its benchmark. An overall trend of profitability increase is not visible for any partnership in particular.

	Wings	Star Alliance	oneworld
Aligned years observed	11	4	2
Av. Difference to benchmark	3.28 %	3.29 %	-0.18 %
Trend	No	No	No

Table 12-2: Performance of Selected Alliances

The performance of individual carriers within the alliances has been relatively equal over time. Within *Wings* KLM showed some weakness in 1996, compared to NW, but was able to recover soon afterwards. In *Star Alliance* Thai Airways was the only carrier performing against the negative industry trend of its peers. Within the last two years of cooperation *Star Alliance* expanded its member circle to a total of 15. The full effects of expansion with a larger network and greater potential savings are expected to accrue over time. *oneworld*'s outcome has to be seen in respect to merger talks that took place between KLM and BA in 1999. As BA, a strong core member of the alliance, was trying to merge with KLM the members of the alliance could have been unsure about the further

---

[21] Alliance formation was set to 1990 as discussed above.

development of *oneworld*, and were consequently revising their future options and holding off with further cost-saving investments.

2.5 Shortcomings of the Model

Due to the limited availability of data and method of analysis certain considerations have to be taken into account. The shortcomings of the model are believed to be the following:

1. *Blending of effects*: The introduced profitability measure blends several effects together and takes only a look at the final outcome. Although this approach is quite reasonable, deeper insight knowledge could be gained, when analyzing revenue and cost positions individually.[22] Alliance formation is likely to affect these positions differently. Unfortunately these positions are not available yet for the desired recent years in a standardized format as published from ICAO.

2. *Sample size and time length*: The sample period for testing profitability increases though alliance formation is too short for making a final statement. Also the chosen benchmark is small due to the fact that most major global carriers are involved in some sort of alliance. Although covering twelve years with the analysis in total, only four years of *Star Alliance* and two years of *oneworld* could be implemented.

3 Findings

Three alliances have been studied during a time period of 12 years, all of them are in different stages of integration and size of alliance to which they belong, but are equal in terms of their overall goal, which is to increase market coverage and reduce input costs through alliance formation.

The difference in profitability changes has been part of the analysis, meaning that the initial absolute value in the profitability index of the individual carrier does not play an important role. Therefore it does not matter how well or bad a carrier was performing at the chosen base year, which allowed us to focus on alliance relevant increases over time.

---

[22] E.g. Revenue (scheduled, non-scheduled, incidental) and costs (fuel, labor, capital, materials).

The analysis was able to show that high-level alliance formation allowed participating carriers to increase their profitability compared to their unaligned counterparts. This finding holds true for both *Wings* and *Star Alliance* members, which show a surprisingly similar fluctuation in profitability compared to their fellow alliance members. The alliance members performed on average at around 3.3 % better each year than unaligned carriers. This difference shows that partnership efforts of these carriers paid off for them.

Similar findings for *oneworld*, the shortest observed alliance, however, did not exist. It is assumed that this could be due to the fact that the alliance was experiencing some uncertainty through one of their core members pursuing merger talks with a major competitor, which possibly slowed down integration processes and therefore the chance to target cost savings intensively through joint operations. It is also still too early to make final conclusions for the grouping that has only been implemented with two years of cooperation into the analysis.

With the exception of *oneworld*, participating alliance members were benefiting from the alliance in a similar pattern. That means that no single carrier boosted the results for the entire alliance in a positive or negative way and therefore the statements about profitability increases for an alliance as such can be upheld without further explanation.

An increase of profitability could *not* be observed during the years of co-operation (learning effects, amortization of initial investment, etc.).

It should be taken into account that the airline industry has always been a very difficult environment to work in – which especially holds true for the 90s – in which not only global liberalization processes have occurred, but also a technical revolution has taken place. In such a time of turbulence, a large volatility in profitability can always be observed with some companies making right or wrong decisions.

Nevertheless, the aligned carriers coped with all these changes more successfully than the unaligned carriers. As a conclusion it can be figured that alliance formation is the mode of competition in the airline industry and unaligned carriers will have a very hard standing in the times to come, as a tighter integration in the future will possibly result in a greater cost saving potential.

The introduced empirical model was able to analyze carriers' financial performance with few chosen input factors, as data availability was still limited. Resulting shortcomings of the model had to be accepted. Thus, this model can be seen as the first attempt to analyze high-level alliances compared to their unaligned counterparts in respect to profitability increases. Therefore a follow-up analysis of such high-level alliance formations with data available for 1) at least five years of collaboration; 2) a larger sample; and 3) a more detailed analysis in respect to revenue and cost positions checking for different value creation potential for these positions would be the next logical step.

4 References

BAKER, C. (2001): The global groupings. *Airline Business*, 7(7), pp. 42-45.

DOGANIS, R. (2001): *The Airline Business in the 21st Century*. London, New York: Routledge.

DOZ, Y.L.; HAMEL, G. (1998): *Alliance Advantage: The Art of Creating Value through Partnering*. Boston: Harvard Business School Press.

FINDLAY, C.; SNYDER, R. (1998): *Productivity Commission 1998, International Air Services*. Report No. 2, AusInfo, Canberra.

GUDMUNDSSON, S.V.; RHOADES, D.L. (2001): Airline alliance survival analysis: typology, strategy and duration. *Transport Policy*, 8(3), pp. 209-218.

HANLON, P. (1996): *Global Airlines – Competition in a transnational industry*. Oxford, Boston, Johannesburg et al.: Butterworth Heinemann.

INTERNATIONAL FINANCIAL STATISTICS YEARBOOK (2001): *International Financial Statistics Yearbook*. Washington: International Monetary Fund.

KLEYMANN, B.; SERISTO, H. (2001): Levels of airlines alliances membership: balancing risks and benefits. *Journal of Air Transport Management*, 7(5), pp. 303-310.

LEVINE, M.E. (1987): Airline Competition in Deregulated Markets: Theory, Firm Strategy, and Public Policy. *Yale Journal on Regulation*, 4(2), pp. 393-494.

MORRISON, S.A.; WINSTON, C. (1995): *The Evolution of the Airline Industry*. Washington, DC: The Brookings Institution.

OMBELET, H. (2001): Airline alliance survey 2001. *Airline Business*, 7(7), pp. 42-45.

OUM, T.H.; PARK, J.-H.; KIM, K.; YU, C. (1999): *The effect of strategic alliances on firm productivity and profitability: evidence from the global airline industry*. Working paper, Faculty of Commerce, University of British Columbia.

OUM, T.H.; YU, C. (1998): *Winning Airlines: Productivity and Cost Competitiveness of the World's Major Airlines*. Boston/Dordrecht/London: Kluwer Academic Publishers.

PARK, N.; CHO, D.-S. (1997): The effect of strategic alliance on performance. *Journal of Air Transport Management*, 3(3), pp. 155-164.

RHOADES, D.L.; LUSH, H. (1997): A typology of strategic alliances in the airline industry: Proposition for stability and duration. *Journal of Air Transport Management*, 3(3), pp. 109-114.

13

MARRIAGES AND DIVORCES: STRATEGIC ALLIANCES IN THE NETWORKED ECONOMY – THE CASE OF AIR NEW ZEALAND

KATHRYN PAVLOVICH

1 Introduction 326
2 Air New Zealand's Alliance History 330
3 Discussion 337
4 Conclusion 340
5 References 341

Summary:
Airlines are vital for a nation's competitiveness in the global economy, yet the airline industry is undergoing significant restructuring through consolidation making it difficult for regional airlines to survive. This case study illustrates the engagement of airline partnerships necessary to maintain global connections. Through a relationship metaphor, the case illustrates the ebb and flow of strategic partnerships that create networked patterns of interconnectedness This chapter suggests that the dominant motivation is to achieve scale economies which is off-set by the instability these alliances cause. There is a need for longer term alliances based upon trust and quality of relationships that now characterize the networked economy.

1 Introduction

1.1 Background

At the edge of the world lies the small nation of New Zealand; its land mass equal to that of Japan, yet with a population of only 4 million people. The rich fertile landscape means agriculture has historically been the mainstay of its export economy, stemming from colonial ties as an isolated farm for the United Kingdom. New Zealand's isolation from the rest of the world has enabled the country to develop a slower pace of life, attractive to many overpopulated and corporate environments. So today, the economic dependency on agriculture has diversified to include tourism, as the low population base and rich green landscape is seen by the rest of the world both idyllic and nostalgic of how the world once was. Indeed, New Zealand is Middle-earth, home of the Lord of the Rings trilogy, which depicted the varying scenic displays this country has to offer. In 2003, the tourism industry directly contributed $5.9billion to the economy from its two million international visitors.[1] To achieve this, the airline infrastructure is essential to the national economy, and the national carrier, Air New Zealand (AirNZ), plays an active role in the promotion of international tourism, as evidenced in its Boeing 747s currently painted with Lord of the Rings theme (see figure 13-1).

Figure 13-1: Lord of the Rings Themed 747

---

[1] See Tourism New Zealand (2004).

1.2 Case Study Purpose

The airline industry has undergone significant changes in the last two decades. On one hand there has been regulatory loosening of the open-skies agreements which has increased competitive rivalry as international competitors enter previously protected airspace. On the other hand, the regulatory environment continues to have constraints on market entry, capacity and pricing because of the need to protect national airlines. These limitations are conducive to airline partnerships as they encourage bi-lateral agreements which increase the survival of national airlines. The formation of global alliance networks depend on national airlines as feeders and carriers into the global transport system, which highlights issues of national sovereignty versus global competitive sustainability as core to these alliance endeavours. Indeed, it was predicted in 1993[2] that national airline policies would typically encounter strategic scenarios of a) consolidating into a global alliance network, and b) deconsolidating an airline's position in order to increase competition. These tensions become particularly poignant to peripheral nations such as New Zealand, where the costs associated with flying, and the time involved (26 hours flying time from Europe to New Zealand) have a direct impact on the nation's economy. This case study examines these tensions of network consolidation and deconsolidation from a peripheral airline perspective, as AirNZ struggles for survival in the global arena.

The strategic alliance literature is dominated by motivations and management of partnerships.[3] When examining alliance form, the literature tends to fall into two primary categories: exploiting of existing capabilities and exploring new opportunities for capability building.[4] These two aspects draw on earlier work[5] where in 'X' coalitions firms divide the activities among themselves, and 'Y' coalitions firms share identical activities. This was expanded[6] to include 'scale' and 'link' alliances. Scale alliances are those where partners contribute similar resources within the same stage of the value chain in order to achieve economies of scale from the collaborative activity. These partnerships are more appropriate for resource pooling and often include rivals for market positioning postures. This perspective is based on a premise that firms that proactively organize their internal structures and market relationships have the highest chance of survival.[7] It is these scale alliances that are most evident in the airline industry, where these

---

[2] See Oum, Taylor & Zhang (1993).
[3] See e.g. Arino & de la Torre (1998); Doz (1996); Gulati (1998); Parkhe (1993).
[4] See March (1991).
[5] See Porter & Fuller (1986).
[6] See Garrette & Dussauge (2002).
[7] See Uzzi (1997).

coalitions form for sharing of similar activities to leverage scale economies. Inter-airline collaboration, whether through co-marketing agreements, code-sharing of interconnected flights, integrative computer reservation systems (CRS), or cross-equity arrangements are seen as a vital component of airline survival.[8] In contrast, link alliances combine complementary skills and resources to build longer term customer-supplier relationships. These link alliances tend to engage more in exploring and learning activities, which result in a more rapid evolution of the alliance structure.

Underlying this case study is a thesis that these 'scale' alliances endorse the characteristics of flexibility and fluidity of network relationships. It illustrates how airlines fit into the scale alliances motivated by global deregulation in the industry and resulting in consolidation and rationalization as airlines need economies of scale and global reach for survival. Thus, the purpose of this case is two fold: first, the case of AirNZ illustrates the consolidation/deconsolidation argument whereby firms need 'scale' alliances in order to transact efficiently through both internal capabilities and external market access. Second, the case discussion posits that the very nature of these consolidation-deconsolidation transactions encourage a transient fluidity characteristic of network relationships;[9] and that a focus on 'scale' alliances becomes short-term profit orientation rather than the more stable longer-term learning 'link' alliances.

Finally, the framing of this involves references to partner relationships, where motivations, partner selection, courtship, marriage contracts, divorce, and carving up the estate are all aspects of relationship management which apply appropriately to the organizational context. Underlying this discussion is a thesis of pattern and interconnection, through a series of successes and failures. Indeed, while AirNZ has received consistent commendations as one of Asia-Pacific's top airlines,[10] it is also recovering from the brink of bankruptcy following the disastrous acquisition of an Australian domestic airline, Ansett. The case begins with the historical development of the airline. It then turns to discuss its alliance partnerships. Issues surrounding the use of partnerships to enhance economies of scale for survival are considered, and, as changing partnership allegiances occur,

---

[8] See Evans (2001).
[9] This fluidity is a central characteristic of network theory, with seminal work by Granovetter (1985, 1992).
[10] 1982-1990 – Best Carrier to the Pacific (Executive Travel, UK); 1994, 1996, 1999, 2000, 2001 – Globe Award as "Best Airline to the Pacific" (UK Travel Weekly Globe Awards); 1998, 1999, 2000, 2001 – Best Pacific Airline (Travel Trade Gazette); 2001 – "Best Airline based in Australasia/Pacific (Official Airline Guide Awards); 1998 – survey of 3500 business-class passengers ranked AirNZ top of 65 airlines (Inflight Research Services) – source:http://www.airnewzealand.co.nz.

what is the future of small scale airlines, facing an uncertain future. Finally, discussion regarding connectedness is posited as it relates to participation in the networked economy. The data related to AirNZ has been sourced from secondary sources, notably the company's website: http://www.airnewzealand.co.nz and newspaper sources.

1.3   History of Air New Zealand

AirNZ began its life as Tasman Empire Airways Ltd (TEAL) in 1940, with flying boats being used across the Tasman Sea between New Zealand and Australia. In 1953, the New Zealand and Australian governments took ownership of this airline, which by this time had developed an international network to include Asia, US, UK and Europe. The New Zealand government took full ownership for $1,622,800 in 1961. Included in the transfer of shares was a bi-lateral agreement between Australia and New Zealand for trans-Tasman services between Qantas and Air New Zealand. TEAL was renamed Air New Zealand in 1965. Concurrently, the domestic market was serviced by a national airline formed in 1947, the New Zealand National Airways Corporation (NAC), which merged with Air New Zealand in 1978 to service both the national and international markets.

1984 was a pivotal time in New Zealand's history with the election of a Labor Government which radically changed the country's protected infrastructure. AirNZ did not escape, and it was privatized in 1989 to a consortium of Brierley Investments (65%, with 30% to be sold to the public – this included a Kiwi Share which ensured that majority shareholding remained in New Zealand), Qantas (19.9%), Japan Airlines (7.5%) and American Airlines (7.5%).

Thus, after fifty years of state ownership whereby the monopolistic airline was both protected and stable, AirNZ entered the commercial market at a time when the industry was becoming increasingly consolidated and unstable. The regulatory environment had changed, and an increasing number of open skies policies allowed the introduction of large international and low cost carriers. To survive in the turbulence of the 1990s, AirNZ needed global partnerships to access both routes and passenger loadings.

Today, AirNZ flies 22 million passengers to 20 countries. It achieved revenues of $4.3billion, and gross profit of $39million in 2002.[11] The airline is the 17th largest in the world, and employs 8,000 people. Currently it owns 42 aircraft, and

---

[11]   See AirNZ Annual Report (2003).

leases a further 44. However, on 2 June 2004, an announcement was made to upgrade the long haul fleet with the lease and purchase of 10 Boeing 777 jets, at a cost of $1.4 billion.[12] These long haul flights have a differentiated service, while low cost services are used on domestic flights and across the Tasman to compete directly with the new arrival, Virgin Blue. At a regulatory level, in line with global trends, the New Zealand government deregulated the air space in 1989 to allow foreign airlines to compete in the New Zealand skies. By 1998, there were 5 agreements. AirNZ became a member of the Star Alliance in 1999, which now includes Air Canada, Air New Zealand, ANA, Asiana Airlines, Austrian, BMI, LOT Polish Airlines, Lufthansa, Scandinavian Airlines, Singapore Airlines, Spanair, Thai Airways International, US Airways, United and VARIG. Three more airlines were also given approved on 5 June 2004, to join the Alliance: South African Airways, TAP Air Portugal and Blue 1, the Finnish airline which is the first regional airline to join the alliance consortia. Participation in the world's largest airline consortia gives AirNZ passengers access to 755 airports in 132 countries, and over 575 lounges.[13] While increased aircraft loadings, routing access, sharing of airport thoroughfares and computer reservation systems are immediate benefits for the airlines, for passengers' benefits include more seamless travel, with shorter transfer times between connections, automatic baggage transfer, reduced ticket handling, use of consortia airline advantages and sharing of frequent flyer programmes.

2 Air New Zealand's Alliance History

2.1 Courtship

The following section illustrates the changing partnerships AirNZ has been engaged in. The close relationship between Qantas and AirNZ begins; followed by a historical review of other global partnerships. Finally, the unfortunate Ansett acquisition is discussed before then turning to the current context. This section uses a partnership metaphor to illustrate the relational aspects to alliance dynamics, which is also reflected in the many newspaper quotations.

---

[12] See New Zealand Herald (2004a), p.1
[13] See http://www.staralliance.com/star_alliance/star/frame/main_10.html.

2.1.1 The AirNZ - Qantas Relationship

Close cousins, AirNZ and Qantas have been involved in partnerships since the inception of TEAL in 1940. During this time, there was significant involvement from both Australian and New Zealand governments. With the sale of TEAL to New Zealand in 1961, a bi-lateral agreement was signed between the two airlines for the trans-Tasman services. By 1989, following a joint service agreement, 80% of trans-Tasman crossings were shared flights between Qantas and AirNZ. In 1992, the relationship deepened with the announcement of a "Beyond Rights" contract which enabled AirNZ to begin services to Taipei and Bangkok via Brisbane. This gave AirNZ significant access to the growing Asian markets and enabled the airline to leverage from the on-travel flights. However, the open skies agreement AirNZ wanted in Australia never eventuated as the Australian government unilaterally blocked it during the final negotiations in 1994. This forced the AirNZ airline management into a buy-in situation. The acquisition of Ansett Holdings in 1996 (discussed in detail in a later section), soured the relationship between AirNZ and Qantas, and their flight sharing, begun in 1961, ceased. Further, Qantas announced its 19.9% sale of AirNZ to US institutions in 1997.

2.1.2 Other Code-Sharing Agreements

A key aspect of global competitiveness has been attaining access to markets, which have occurred through code-sharing agreements. Code-sharing involves the use of one aircraft while retaining partner airline flight codes to respond to customer preferences.[14] These alliances enable airlines to rationalize their operations and reduce costs. With the significant growth of tourism to New Zealand, opportunities existed for the national carrier to bring these travelers direct to New Zealand. Thus, AirNZ engaged in multiple code-sharing arrangements at a global level, in the Pacific, Europe and Asia and the Americas. Prefacing this commentary is figure 13-2, which illustrates the extensive network of alliances AirNZ has developed to ensure global access.

---

[14] See Hannegan and Mulvey (1995).

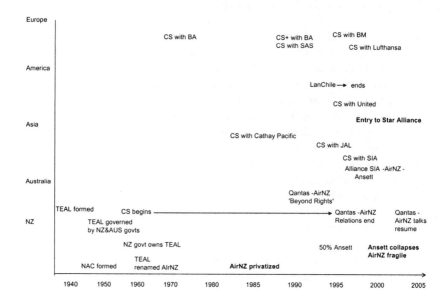

Figure 13-2: Timeline of AirNZ Alliances

2.1.3 Europe

Courting in Europe began in 1974 with a code sharing agreement with British Airways (BA), using an AirNZ flight from Auckland to Los Angeles and a British Airways flight from Los Angeles to London. This was expanded in 1978 to include Miami and Montreal as connecting points during the Northern Hemisphere summer. This relationship with BA was extended in 1990, with an agreed round-the-world service. However, after the formation of the BA-American Airlines-Qantas alliance, AirNZ needed to find another partner in the UK. It did so with a formal agreement with British Midland in 1996. Other strategic partners were also sought in Europe, and AirNZ signed an alliance with Scandanavian Airlines System (SAS) in 1990 to promote each other as preferred onward carriers. This was a significant alliance in that both airlines had connecting points in Los Angeles, Singapore, Bangkok and Tokyo. The alliance with Lufthansa too in 1998, was important, as together these partnerships gave

AirNZ direct access to the United Kingdom, Scandanavian and German markets which are key tourism markets for this country.[15]

2.1.4 Asia

Code-sharing in Asia began in 1986 with Cathay Pacific, as the parties introduced a bi-partite service between Auckland and Hong Kong. In 1995, JAL sold its shareholding in AirNZ to Brierley Investments, and the following year AirNZ and JAL formed a code-sharing agreement.

In 1997, under a new open skies agreement, a significant alliance was being formed between AirNZ, Ansett and Singapore Airlines (SIA) which would enable the group access to a combined 200 destinations in 47 countries. The partnership would give SIA access to other New Zealand cities. Because of the significant tourism markets coming from Asia, this agreement enabled SIA to use the South Island city of Christchurch as an exit point. In the commercial agreement, AirNZ managed the Christchurch-Singapore route using its 767s, while SIA operated the Auckland-Singapore route using its 747s. Each airline purchased seats on the others' flights, with each airline charging the same price for fares under the Escapade system.[16] This tri-partite alliance was claimed to be a response to the British Airlines[17]– American Airlines – Qantas alliance recently formed, which pre-empted the development of the oneworld global alliance.

At the same time, SIA was trying to increase its shareholding in AirNZ. It was widely known that SIA wanted a shareholding in Australasia, but was outbid by British Airways for a stake in Qantas in 1992. Having recently acquired a share in AirNZ through its recent purchase of Brierley Investments, the New Zealand airline was an obvious second choice. Negotiations continued, with SIA proposing a 49% interest in AirNZ. This proposal was prevented by the New Zealand government restrictions of foreign ownership. Talks ceased after AirNZ's purchase of Ansett.

2.1.5 The Americas

In the Americas, a code sharing alliance was announced in April 1991 with American Airlines, using Honolulu as the connecting point. However, this alliance was terminated in November the same year. The South American market

---

[15] See New Zealand Tourism Board (1997).

[16] See *The Independent* (1997), p. 6.

[17] British Airways had a 25% shareholding in Qantas.

was penetrated with a marketing alliance in 1994 with Lan Chile, but this too was terminated in 1998, when Lan Chile chose Qantas as its preferred partner, and joined the fledgling oneworld Alliance. This left AirNZ without a South American link. While Varig was part of the Star Alliance, it did not have direct flights to Australasia. A second connecting airline to Australasia, Aerolineas Argentinas, had also signed with Qantas. This restricted New Zealand's ability to directly access South America.

Code sharing services across the Tasman and Pacific began with United Airlines in 1996, which initiated AirNZ's entry into the Star Alliance. This agreement gave AirNZ access not only into the North American market, but also to preferred supplier privileges with the world's largest airline. This code sharing agreement was expanded in 2000 to cover 18 US cities, while United was given code sharing access into the New Zealand domestic market. By this time, United and AirNZ shared 130 flights each week. The sudden invocation of Chapter 11 by United Airlines in 2002 precipitated its withdrawal from the New Zealand market, with AirNZ taking over as preferred airline of choice.

Thus, AirNZ had access to Europe, Asia and North America through significant code sharing agreements. Through these loose and mobile courtships, it can be seen that these three groupings ultimately contributed to the global Star Alliance when it was finally formed as a global consortia. The strategic courting had culminated in a more committed de-facto relationship, while each airline retained individual independence absent from a marriage.

2.2 Marriage – The Ansett Alliance

AirNZ's foray into equity partnerships began in 1996 with the 50% purchase of Ansett Holdings for $475million. As AirNZ was never given the open skies agreement in Australia it greatly needed, it pushed the airline into buying its way into the Australian marketplace. Ansett Holdings owned 100% of Ansett Australia, the domestic airline, and 49% of Ansett International. Part of this purchase agreement included AirNZ having a pre-emptive right to buy the remaining 50% if it was available. In 1999, News Corporation Limited released this 50%, and in February 2000, AirNZ completed its 100% ownership for $630million.

The purchase was complex, as there was "another marriage partner keen to take up the matrimonial bed".[18] Rumours abound that the Australian government was

---

[18] See *New Zealand Herald* (1999a), p. D4.

courting Singapore Airlines as preferred shareholder, with such headlines in New Zealand newspapers: "Hell hath no fury as an airline scorned".[19] Certainly SIA was a stronger partner to do the recapitalization the Ansett fleet urgently needed. SIA indicated strong interest in the purchase, but was in no hurry to commit. While market analysts noted that it was "now a matter of SIA and Ansett coming to the table and agreeing on a shareholding structure for a merger that was agreeable to AirNZ",[20] the relationship between the AirNZ and SIA management teams appeared strained. Nevertheless, SIAs interest gave AirNZ 30 working days to decide whether to exercise their right to purchase. On one level, the purchase appeared a sound strategic decision for AirNZ, as it gave the airline the much needed market access and landing rights into the large Asian markets through Australia. However, in retrospect, the purchase price was too high for the aging fleet and its decreasing market share.[21] The full picture of why AirNZ chose to have full ownership may never be disclosed, but one market analyst claimed that part of the decision to buy was because AirNZ was treated as a junior partner by Newscorp, despite having a 50% ownership stake. This experience made AirNZ fearful of a similar fate with SIA.[22]

An event early in 2001 took AirNZ by surprise and was a catalyst for bringing the Ansett airline to its knees. In April 2001, just before peak travel of Easter Weekend, pylon cracks were found in Ansett's Boeing 767 fleet. The grounding of the entire fleet for further inspection by the Australian Airline Authority occurred at a time when airlines enjoy full loadings. Indeed, there were widespread rumours of influence from the Australian government on the Australian Airline Authority due to the unpopular purchase of Ansett by the Kiwi cousins. During the four-day Easter holiday, the airline had to pay out $500,000 to fly 145,000 ticketed passengers in hired planes or rival airlines. The first of the grounded B767-200s was only cleared for service on April 20. During this time, the Australian domestic airline industry was under extreme pressure, as cut-price Virgin Blue entered the market. Indeed, airfares between Australian cities reached an all-time low, with a 900 kilometre flight costing an unsustainable $66 one way.

Shortly before the collapse of Ansett, AirNZ was approached by Qantas wishing to purchase 25% of the airline from Brierley Investments. However, AirNZ's preferred partner remained SIA, and in June 2001, AirNZ Board endorsed the recapitalization of the company through lifting SIAs shareholding to 49%. However, this was declined by the government. Talks continued between Qantas

---

[19] See ibid.
[20] See *New Zealand Herald* (1999b), p. C3.
[21] See *New Zealand Herald* (1999c), p. E2.
[22] See *The Independent* (1999), p.1.

and Singapore Airlines, but the impending Ansett collapse was soon to overtake these negotiations.

2.3 Divorce

Losses by Ansett Group were being documented at $1.3million per day, and on 13 September (the day following the closure of US airports after New York hijacks), AirNZ announced a $1.425billion loss. Ansett Australia was placed into voluntary administration. The Australian public reacted with hostility, with particular sympathy for the 16,000 retrenched employees. An example of the depth of feeling was evident in the Australian trade unions blocking AirNZ aircraft from leaving Australia, one of which was to return the New Zealand Prime Minister home. An airforce jet had to be flown to an Australian airbase to return her to New Zealand. Interestingly, at the same time as the grounding of the Ansett fleet, Qantas New Zealand (operating in the New Zealand domestic market) also went into liquidation with a loss of 1100 jobs and $20million in debts, but it did not receive the same hostility in New Zealand as the Australians gave the Ansett collapse.

The collapse of Ansettt placed AirNZ on the brink of bankruptcy, and state intervention was needed for it to survive. Existing shareholders of Brierley Investments and SIA were not prepared to invest in the airline, and the New Zealand government was forced to recapitalize it for $885million, which occurred in January 2002. This gave the government an 82% interest in the airline, and returned it to the near-State control situation of 20 years ago. Other shareholdings fell to Brierley Investments (5.5%), SIA (4.5%) and other investors (8%).

2.4 Carving Up the Estate

Since AirNZ's near bankruptcy, there have been on-going negotiations for an alliance with Qantas, and both airlines have been keen to renew their partnership together. The proposal for Qantas to buy a 22.5% share in AirNZ at a purchase price of $550million was supported by the incumbent major shareholder, the New Zealand Government. The intended alliance, however, was not endorsed by either the Australian or the New Zealand Commerce Commissions. The Australian Commerce Commission claimed that placing the New Zealand services under one management structure would then give the alliance overwhelming dominance in the domestic and trans-Tasman market. The New

Zealand Commerce Commission argued that a merger would be anti-competitive, and not be in the public's best interest. Indeed, it argued that rather than increasing visitor flows, it would decrease benefits for New Zealanders, believing that 172,000 tourists would be deterred by higher fares, compared to an increase of 39,000 for domestic tourism if the alliance went ahead.[23] On 20 October 2003, AirNZ announced it would appeal the decision against its proposed alliance with Qantas,[24] with a decision from the Tribunal expected in July 2004. However, questions arise regarding involvement in the global alliance network. If a closer alliance structure is approved, what are the repercussions of AirNZ's participation in the Star Alliance as code and capability sharing is likely to impact on the transparency of the global alliance structure. Will AirNZ be able to remain in the world's largest airline alliance structure? Will competition be affected with only one global alliance in Australasia?

Meanwhile, the trans-Tasman environment remains highly competitive, as foreign airlines see market entry opportunities as the weakened airlines struggle to survive. The arrival of Emirates in 2003 has created price wars within the trans-Tasman sector, while the imminent arrival of Virgin Blue in the domestic context has seen the introduction of budget fares by the existing competitors. Nevertheless, the courtships continue. On 22 January 2004, Qantas announced that it was breaking ties with its New Zealand domestic carrier, Origin Pacific, preferring to go it alone. A spokesperson for Origin Pacific stated that they wanted a long term relationship, "but unfortunately after a few months, Qantas fell in lust with the girl next door. That has made things a bit difficult".[25] The following day, Origin Pacific announced its agreement for domestic travel with KLM Royal Dutch Airlines. On the same day, the formation of a new charter company was announced to compete against AirNZ in the Pacific. Thus, the fluidity of the network-based partnerships continues, with this evolution of partnership formation exhibited in a timeline in figure 13-3.

3 Discussion

This case study has illustrated the evolution of alliances that AirNZ has enacted to compete within the global context. Two theoretical themes require further discussion. First, consolidation of the industry is clearly evident, as AirNZ has

---

[23] See *New Zealand Herald*, (2003a), p.1
[24] See *New Zealand Herald* (2003b), p. E1.
[25] See *New Zealand Herald* (2004b), p. E3.

engaged in an increasing complex array of alliance activities as portrayed in figure 13-2. This consolidation is apparent in the emergence of the two major global networks, with the increasing rate of partnerships occurring during the 1990s as these coalitions began to form. This strategic posturing enabled airlines to increase their capacity and lower their cost structures while also enhancing service benefits to international markets. Thus, accessing and maintaining global reach through comparative cost structures was the primary motivator for the formation of these coalition blocks.

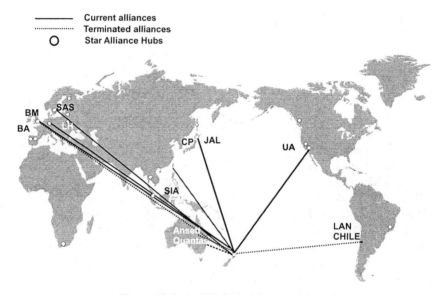

Figure 13-3: AirNZ Global Partnerships

Yet during this time, these alliances were fickle and tenuous, with deconsolidation clearly evident in the case data which confirmed the temporal state of the AirNZ coalition agreements. A code sharing partnership with American Airlines using Honolulu as a connecting point began and ended in 1991; the British Airways code sharing agreement begun in 1974 was concluded in 1995 with the formation of the British Airways-American Airlines-Qantas alliance, a precursor to the emergent oneworld network; a South American connection with Lan Chile in 1994 was also terminated in 1998 after Lan Chile preferred to slumber with the oneworld Australasian Qantas partner. The long-lasting Qantas code sharing arrangement also ended in divorce after AirNZ's ill-fated purchase of Ansett Australia.

This process of consolidation into two major global alliance networks implies elements of duopolistic behaviour, as the desire for efficient internal cost structures encourages potential collusion between the two networks. However, the deconsolidation that occurs through the fickle coalition agreements assists efficient competitive rivalry to continue through the instability of the internal alliance relationships. Thus, both external drivers of economic restructuring and global competition and internal forces of risk sharing, economies of scale and access to specialized resources have continued to shape the overall industry structure.

The second area of discussion concerns the scale and link alliances. While the Trans-Tasman alliance was durable in that it lasted for 35 years, it was somewhat unreliable and erratic as the Qantas partner became increasingly testy. Qantas demonstrated a self-centredness that left AirNZ disadvantaged in the marketplace on several occasions, e.g. the wooing of Lan Chile which gave Qantas the only direct routes into South Amercia, and the withdrawal of the Beyond Rights agreement that would have enabled AirNZ feeder access into the Asian markets via Australia. This response to direct competition in Qantas's own territory suggests that the potential forthcoming matrimony could be based on less than equal terms, with the larger partner more concerned with its own interests than in the health of the joint relationship.

The focus on alliances for economies of scale has not encouraged loyalty for deeper capability sharing typifying longer term learning alliances. These more intense relationships are characterized by a frequency of past ties, the identity of partner routines and systems, and compatible management philosophies that embed the relationship into one based on trust and fine-grained information exchange that build mutual value.[26] Indeed, the function of trust, through partner chemistry, competence of organizational action, and responsibilities towards the relationship are a vital component of the link alliances for the development of new products, new processes and new technologies that embody new knowledge. Indeed, at a macrolevel, one could suggest that the airline industry's focus on efficiency and rationalization has not created a positive strategic future, as witnessed by the increasing number of airline insolvencies. While it is claimed that, "Alliances are complex organisational arrangements that require multiple levels of internal approval, significant search in identifying partners, detailed assessments for ratifying contracts, and considerable attentions to sustain the

---

[26] See Uzzi (1997).

partnership",[27] it is through deeper and richer collaborative endeavours that true strategic value is created.

4 Conclusion

In summary, this case study offers understanding of how collaborative partnerships within networks are a fluid and dynamic process of ebb and flow. Embedded within this is a competitive and cooperative dynamic of knowledge and resource transfer through the global community. Both scale and link alliances were discussed, and the case illustrated the dominance of scale alliances caused by external environmental conditions. Deregulation and rationalization have pushed airlines towards consolidating into alliance networks which now dominate the competitive landscape as a mechanism to assist regional airline survival. Yet the fluid and shifting state of the alliances also destabilize the performance of regional players, as they can be exploited, expropriated and squeezed out from the marketplace.

Yet national airlines are essential infrastructure in the global context, as they play an important role in national identity. One example of this is the use of the Lord of the Rings theme on AirNZ's internationally-bound Boeing 747s. The making of this movie in New Zealand has played a vital role in the international promotion of New Zealand tourism over the last two years, and AirNZ has been an important part of this marketing strategy. Thus, issues of the national airline survival are paramount and the natural tensions between national sovereignty and identity versus necessitating scale alliances for survival become apparent. Yet, as noted in this case, through engaging in partnerships, boundaries between firms become nebulous and key aspects of partnering involve quality of the relationship. This suggests that in the future world of the networked economy, longer term partnerships need to be based upon partnering quality and complementarity of function to develop learning relationships, rather than solely on more transient economies of scale.

---

[27] See Gulati (1999), p. 402.

5 References

AIR NEW ZEALAND (2003): Annual Report.

ARINO, A.; DE LA TORRE, J. (1998): Learning from Failure: Toward an Evolutionary Model of Collaborative Ventures. *Organizational Science*, 9(3), pp. 306-325.

DOZ, Y.L. (1996): The Evolution of Cooperation in Strategic Alliances: Initial Conditions or Learning Processes? *Strategic Management Journal*, 17(7), pp. 55-83.

EVANS, N. (2001): Collaborative Strategy: An Analysis of the Changing World of International Airline Alliances. *Tourism Management*, 22(3), pp. 229-243.

GARRETTE, B.; DUSSAUGE, P. (2002): Alliances with Competitors: How to Combine and Protect Key Resources? *Creativity and Innovation Management*, 11(3), pp. 203-223.

GRANOVETTER, M.S. (1985): Economic Action and Social Structure: The Problem of Embeddedness. *American Journal of Sociology*, 91(11), pp. 481-510.

GRANOVETTER, M.S. (1992): Problems of Explanation in Economic Sociology. In Nohria, N.; Eccles, R.G. (Eds.): *Networks and Organizations: Structure, Form and Action*. Boston: Harvard Business School Press, pp. 25-56.

GULATI, R. (1998): Alliances and Networks. *Strategic Management Journal*, 19(4), pp. 293-317.

GULATI, R. (1999): Network Location and Learning: The Influence of Network Resources and Firm Capabilities on Alliance Formation. *Strategic Management Journal*, 20(5), pp. 397-420.

HANNEGAN, T.; MULVEY, F. (1995): International Airline Alliances: An Analysis of Code-Sharing's Impact on Airlines and Consumers. *Journal of Air Transport Management*, 2(2), pp. 131-137.

MARCH, J.G. (1991): Exploration and Exploitation in Organizational Learning. *Organization Science*, 2(1), pp. 71-87.

NEW ZEALAND HERALD (1999a): AirNZ irked at being overlooked in wedding plans. 29 March, p. D4.

NEW ZEALAND HERALD (1999b): AirNZ Says Talk of Ansett Buy Just Speculation. 6 May, p. C3.

NEW ZEALAND HERALD (1999c): *Win-Win for AirNZ.* 24 April, p. E2.

NEW ZEALAND HERALD (2003a): *Commerce Commission Rejects AirNZ-Qantas Deal.* 10 October, p. 1.

NEW ZEALAND HERALD (2003b): *AirNZ to Appeal Commerce Commission Ruling Against Alliance.* 20 October, p. E1.

NEW ZEALAND HERALD (2004a): *AirNZ to Spend $1.35b on New Boeing Long Haul Fleet.* 2 June, p. 1.

NEW ZEALAND HERALD (2004b): *Qantas Drops Origin Pacific;* 22 January, p. E3.

NEW ZEALAND TOURISM BOARD (1997): *Inbound Tourism Statistics;* New Zealand Tourism Board: Wellington.

OUM, T.H.; TAYLOR, A.; ZHANG, A. (1993): Strategic Airline Policy in the Globalising Airline Networks. *Transportation Journal*, 32(3), pp. 15-30.

PARKHE, A. (1993): Strategic Alliance Structuring: A Game Theory and Transaction Cost Examination of Interfirm Cooperation. *Academy of Management Journal,* 36(4), pp. 794-829.

PORTER, K.; FULLER, M. (1986): Coalitions and Global Strategy. In: Porter, M. (Ed.): *Competition in Global Industries.* Cambridge, MA: Harvard University Press, pp. 315-344.

THE INDEPENDENT (1997): *Singapore Open Skies Pact Will Bolster Air NZ, SIA, Ansett Alliance.* 27 June, p. 6.

THE INDEPENDENT (1999): *Dirty Tricks Dog AirNZ/SIA Deal.* 26 May, p. 1.

TOURISM NEW ZEALAND (2004): *International Visitors Survey.* Wellington: Tourism New Zealand.

UZZI, B. (1997): Towards a Network Perspective on Organizational Decline. *International Journal of Sociology and Social Policy*, 17(7/8), pp. 111-155.

Part III

Airports

14

FORCES DRIVING INDUSTRY CHANGE – IMPACTS FOR AIRPORTS' STRATEGIC SCOPE

ROBERT J. AARONSON

1 Introduction 346
2 External Shocks 347
3 Changes to the Structure of the Airline Industry 350
4 Community Integration Factors 355
5 Strategic Factors 357
6 Conclusion 359

Summary:
The airports industry currently is undergoing an unprecedented transformation on a global scale. The external events and developments within the air transport sector that have impacted the industry over the past few years have challenged every facet of airport management. Airport operators today face a formidable array of competing management priorities. Reflection and order is needed. In this chapter the major industry change drivers for airport companies are identified, systematized and assessed, aiming to provide a basis for airport managers' strategic orientation in a turbulent environment.

1 Introduction

The airports industry of 2004 is different from the airports industry of even five years ago, let alone the industry of 10 or 20 years ago. While it is difficult to generalize across a global industry, airports in 2004 can be described as tougher, more street-wise and battle-hardened, and infinitely more commercial than they used to be.

This new industry profile replaces a more stable and reactive airport business, driven by the optimism induced by more than 50 years of virtually continuous growth. During these earlier days, traffic growth seemingly had no end and the biggest challenge in managing an airport was to build facilities fast enough to keep up. Managing costs at busy airports was not a top priority.

The end result of this transformation is that airports have changed both the way they do business and the very nature of their business. Most importantly, the industry has met and confronted the forces driving the transformation with pragmatic, streamlined, business-driven decision making processes more capable of meeting the challenges and risks of the future.

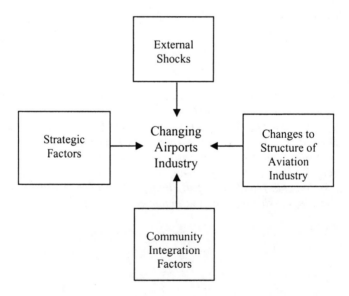

Figure 14-1: Factors Impacting the Airport Industry

The purpose of this chaüpter is to describe the forces driving the transformation of the industry. These forces come from four sources, as set out in Figure 14-1. These sources include: external shocks; shifts in the structure of the aviation industry; community integration factors and strategic/management factors. It is the convergence of these factors which has driven the extent of change the airports industry has undergone. These factors are examined in more detail in the following sections.

2 External Shocks

The past few years has seen the airports industry impacted by more external shocks than ever before. These external shocks have had a major impact on both traffic levels and the way in which airports are managed.

The industry has experienced external shocks before – particularly the 1991 Gulf War and the oil crisis of the early 1970s. There are, however, three remarkable features of recent external shocks as follows:

- *their nature* – the shocks have torn at the very fabric of the aviation industry and eroded, at least temporarily, public confidence in air transport;
- *their scale* – the shocks have impacted on the aviation industry on a global scale; and
- *their timing* – a number of unrelated major external shocks have hit the industry within a short time span.

The external shocks that have impacted on the airports industry in recent years have been well documented and generally fall into two categories: terrorism/war; and health risks. These are discussed in more detail below.

2.1 Terrorism/Wars

The aviation based terrorist attacks on New York and Washington DC on 11 September 2001, subsequent attacks on Bali and Madrid and military actions in Afghanistan and Iraq have had a major impact on the aviation industry, including airports. While traffic levels are now returning to pre-2001 levels, the primary impact of the terrorist attacks and wars has been to change the psychology of traffic growth.

Prior to the terrorist attacks and wars, the aviation industry held the perception that strong traffic growth was a permanent trend. This perception was held with good reason. For more than 50 years, strong growth had been the norm. Moreover, as long as the post-war economic boom continued and more countries in Asia joined the second and first worlds, there was every prospect of this strong growth continuing, led by these emerging markets, but bolstered by steady growth even in "mature" aviation markets such as North America and Europe.

The terrorist attacks and wars changed this perception radically. For a sustained period of time, traffic levels slumped. More importantly, however, the attacks/wars brought home to airports a reality that they may have forgotten over the good years – that traffic levels were lifestyle driven and the lifestyle itself could be the very target of terrorists. This is a frightening reality for an industry based on the investment of billions of dollars into highly specialized and sunk assets (ie assets with little alternate uses) such as runways and terminals. For the first time in 50 years, large numbers of airports saw their credit ratings tumble as markets absorbed the fact that airport investment had downside risk. Compounding the gloomy financial outlook was the insurance market, which quickly raised premiums across the air transport sector.

This change in the perception of traffic growth has many ramifications for the airports industry. These include changes to:

- *airport operating costs* – a renewed focus on aviation security has resulted in shifting the industry into a higher cost curve as additional measures are put in place to screen carry-on and checked baggage, and to secure airport perimeters. Further, the perception of greater risk is necessitating that greater financial reserves be maintained;

- *passenger travel habits* – the addition of extra security measures to the normal passenger processing procedures created the "hassle factor", with passengers having to arrive at airports earlier and both passengers and airlines having to incorporate greater flexibility into hubbing/transfer arrangements;

- *terminal flow patterns* – additional security arrangements have changed the timing and pattern of the flow of passengers through terminals. Limitations on the movements of meeters and farewellers at some airports has restricted the access of these persons to airport shopping;

- *retail* – reductions in traffic levels and changes to passenger travel habits and terminal flow patterns have had an impact on retail spend and hence income to airports. Many airports found their prime retail space "devalued" by new, security-driven design changes;

In addition, the change in the perception of traffic growth has changed the way many stakeholders assess risk as follows:

- *investors* – investors have changed the way in which they assess and price patronage risk. It is unlikely the premium paid for future traffic growth in some airport privatizations will be repeated;
- *lenders* – lenders have also changed the way in which they assess repayment risk; and
- *concessionaires* – concessionaires have changed the way they assess patronage risks through contract structures such as minimum guarantees. The industry has seen a major push by airport concessionaires for airports to share traffic associated risks. Some airport operators have structured risk-sharing agreements which have actually increased the airport's share of retail income, but this option is not attractive for airports with low traffic volumes.

For airport managers, the impacts include:

- *capital expenditure* – airport managers have changed the way they assess large scale, traffic-driven capital expenditure and payback period risks. Many airports have deferred substantial capital expenditure projects until traffic returns to historical growth rates; and
- *diversification* – airport managers have realized the importance of diversifying their income sources away from sole reliance on traffic driven revenue in order to be able to survive external shocks.

While traffic levels are returning to pre-2001 levels, the industry is only too conscious of the fragility of the recovery in the event of additional terrorist attacks involving aircraft or aviation infrastructure, including airports. Airport economics, and indeed the individual passenger's airport experience, can be characterized by one word: "uncertainty".

2.2 Health Risks

Simultaneous with terrorism and war related shocks, the airports industry also experienced major external shocks related to health risks. The emergence of the SARS virus and the Avian Flu dramatically reduced traffic levels at a large number of airports, particularly those in Asia or those with large volumes of traffic from and to Asia.

While the traffic effects of health risks such as SARS and Avian Flu are not expected to be as long term or as far reaching as the impact of terrorism/wars (assuming they are now under control), the occurrence of these external shocks at the same time as the terrorism/war shocks served to heighten traffic declines and reinforce new perceptions of traffic risks associated with airports and air travel in general. The notion that SARS could spread throughout Asia and to Canada highlighted, in a starkly negative way, that the same efficient global air transport system that underpins a "just in time" global economy can indeed hasten the spread of disease across the globe.

3 Changes to the Structure of the Airline Industry

Concurrent with the external shocks, airports have also had to deal with changes to industry structure experienced by their primary customer, the airlines.

Three of the most profound changes are:

- the commoditization of air travel;
- technological change; and
- geopolitics.

These are examined in more detail below.

3.1 The Commoditization of Air Travel

One of the most profound changes of recent years has been the advent of low cost carriers and the consequent commoditization of air travel. The changes have resulted in a change to the way in which the aviation industry is structured. Consequently, the knock-on effects have had major implications for airports, both in terms of the way that airports are managed and to the structure of the airports industry itself.

3.1.1 Impact on Airport Management

In most mature economies, air travel has now grown to the extent that the average person takes about four flights per annum. In these economies, air travel stopped being "special" some time ago.

Moreover, the advent of low cost carriers and reasonably priced charter flights have reduced the cost of travel, in some cases dramatically. This has made travel more available to a greater proportion of the population. As a result, air travel is not the preserve of society's higher socio-economic groups, but in demographic terms, has broadened its customer base to include middle income families.

This "commoditization" of air travel is also having a profound change on the way in which airports are built and managed. These include:

- *traffic patterns* – the advent of low cost carriers has resulted in a greater distribution of traffic. While major airports and hubs will continue to be important, growth is also being experienced by smaller, regional and secondary airports that can act as substitutes for major airports in major population centers;

- *growth* – low cost carriers have been one of the few sources of traffic growth over the past few years;

- *terminal design & cost* – the commoditization of air travel has put cost pressures on airports on terminal designs. When air travel was "special" and elitist, terminals were designed as architectural statements with liberal use of marble, glass and stainless steel. This emphasis is however changing towards functionality and cost effectiveness to serve the broadening passenger base, a large percentage of which does not expect or require elaborate facilities;

- *reduced meeter & fareweller ratios* – as travel becomes less special in mature economies, we are witnessing a reduction of meeter and farewellers ratios. While meeter and farewellers ratios are higher is less mature economies where travel is still special (e.g. China), in many countries these ratios are declining; and

- *changes in passenger processing and retail* – the aircraft gate driven approach to passenger processing by low cost carriers is changing the flow patterns within terminals and changing the location requirements for retail and food and beverage outlets.

The changes listed above are fundamental changes to the airports industry. A recent example in Malaysia highlighted the magnitude of this change. At Kuala Lumpur International Airport, the fast growing Asian low cost carrier – Air Asia – has put pressure on the airport to build a new, low cost carrier specific passenger terminal at Kuala Lumpur International Airport. This is despite the existing terminal facilities (having been built only recently) having more than sufficient capacity to accommodate Air Asia's requirements. Air Asia, however, has stated that the existing terminal facilities do not appear to meet its requirements. While there are likely to be many issues involved, the example highlights the changing pressures on airports arising from the advent of low cost carriers.

Other airports are feeling this pressure too. In Asia, Singapore Changi has also announced it is reviewing the need for purpose built low cost carrier terminals, while a number of airports in Australia, including Sydney and Melbourne, built very low cost terminals (relative to the quality of other terminals in Australia) to accommodate the entrance of low cost carrier Virgin Blue into the Australian market. In Europe, Geneva International Airport is opening a new terminal with services for low cost carriers or, indeed, for "legacy" carriers which want to cut costs. This development, which involves no cross-subsidies to the low cost carriers, has been welcomed by IATA.

These examples highlight the way in which the advent of low cost carriers has moved the goal-posts for many airports, who need to react quickly to capture this more cost conscious segment of the aviation industry.

3.1.2 Impact on Structure of the Airports Industry

The commoditization of air travel and the emergence of the low cost carrier has brought with it two other phenomena – the rise of the secondary airport and the re-emergence of the importance of point-to-point traffic. These phenomena have also impacted on the structure of the airports industry and are described below:

The rise of the secondary airport – Many large capital city or hub airports have a cost structure that reflect the days when air travel was special and fares and yields were relatively high. Low cost carriers tend to avoid these airports and seek airports with much lower cost structures. Consequently, many airports previously regarded as the "poor cousins" of the airports industry – regional airports; second airports in large cities and general aviation airports – have experienced a traffic growth explosion. Examples include Birmingham, Bristol, Liverpool, Luton and Stansted in the UK, Rome-Ciampino and Treviso in Italy, Hahn in Germany, Charleroi in Belgium, and many airports serviced by Southwest Airlines in the USA (the airline which pioneered low cost domestic services) as well as Avalon and the Gold Coast in Australia (and many more). In Europe over the past few years, regional/second airports have consistently experienced faster growth than their larger counterparts.

> The emergence of these regional/secondary airports has not taken much traffic pressure off the major hubs and capital city airports because low cost carriers have grown the market by reducing prices. Consequently, much of

the traffic to these regional/secondary airports does not erode traffic at the larger airports – rather it is new traffic, built around latent demand for travel to new destinations at lower prices.

- *The re-emergence of the importance of point-to-point traffic* – for the past decade or two, hubbing strategies have dominated the way airlines thought about servicing smaller markets with not enough traffic to justify point-to-point services. This had major implications for many airports, who invested substantial amounts in aviation and terminal infrastructure to accommodate large pulses of fast turnaround traffic – Dallas Fort Worth and Atlanta are prime examples.

Low cost carriers however have dramatically changed the definition of a market with not enough traffic to justify point-to-point services. Today, more cities and towns are receiving direct services as lower travel costs inducing a greater percentage of the population to travel. Moreover, the vast increase by the airlines in using regional jets for short haul traffic has made many new city pairs viable and made it practical to overfly hubs. These changes are driving the way many airlines think about strategies such as hubbing, with knock-on effects for airports.

These two phenomena, the rise of the secondary airport and the re-emergence of point-to-point traffic highlight some of the high level risks borne by the airports industry due to its large scale investment in sunk infrastructure which makes it hard to be nimble and responsive to major changes.

3.2 Technological Change

The other factor changing the structure of the aviation industry, with consequent knock-on effects for airports is changing aircraft technology.

The strongest growth in demand for new aircraft appears to be at either end of the aircraft size spectrum.

At one end of the spectrum, Boeing and Airbus appear to have no trouble selling aircraft in their B737/A319/320 aircraft categories, partly driven by the fact that these aircraft appear to be the aircraft of choice for low cost carriers. This is obviously shaping the way in which airports are being planned and developed.

At the other end of the spectrum, Airbus are forging ahead with the development of the 500-600 seater A380. Again, this is having a major effect on the way in which airports are being planned and developed. A number of airports are busily spending hundreds of millions of dollars preparing for the A380 – for example by widening runways and taxiways, by building new aircraft parking positions, by installing additional aerobridges etc.

The A380 however, may have even more fundamental effects on travel, traffic patterns and consequently on airports. This could be similar to the last time a major change in aircraft technology was introduced into the market – i.e. the B747. In hindsight, the advent of the B747 dramatically changed the world of air travel and, of course, airports.

Airports that were historically mandatory technical stops for other aircraft were suddenly being over-flown by the B747. The most vivid example that comes to mind was Nadi Airport in Fiji. Nadi was a mandatory technical stop for most passenger aircraft traveling between Australia and the USA until the longer range B747 came along and was able to service the route directly. This technological change had a major impact on Nadi and highlights the risks that can be inherent in investing in large scale sunk assets.

There are other examples across Europe, Asia and the USA: airports such as Bahrain, Dubai and Anchorage in Alaska were all in the technical stop category: now all three have "reinvented" themselves either as destinations in their own right or in the case of Anchorage, as a base for cargo operations where aircraft reconfigure their loads for trans-Pacific transport. The A380 might have as significant an effect on travel patterns and airports as the B747 had a few decades ago. One result of the A380's introduction could be the lessening of airside and airspace congestion on heavily traveled routes, as more passengers can be accommodated in each aircraft movement.

3.3 Geopolitics

Changes in the geopolitics of regions has the potential to substantially change the game for airlines and airports. Such geopolitical changes could include open skies arrangements between Europe and the USA and the addition of more countries to the European Union.

The biggest geopolitical change on the horizon, and the one with the potential to most impact on the airline and airports industries – in the long term – is China. The gradual opening of China to global aviation, together with its billion plus population and phenomenal rate of economic growth, will have a significant impact on airlines and airports in the region.

For years, the airports of Hong Kong, Singapore, Kuala Lumpur and Bangkok have acted as gateways into China and Asia. However, the gradual liberalization of aviation in China means that the populations of the massive, fast growing cities of China are becoming accessible to aviation. And the Chinese airports are already gearing up for the change, with massive investment in airport infrastructure in Shanghai, Guangzhou and Beijing. These Chinese airports are likely to become their own gateways into China and will also gather momentum as major hubbing points in Asia. China's sustained high rate of economic growth also means that the domestic aviation market is likely to grow with the burgeoning number of middle class passengers.

This may have massive consequences for other airports in Asia, which could see a reduction in their hub airport status and a greater focus on their origin and destination traffic.

4 Community Integration Factors

The third factor driving the transformation of the airports industry is the need for greater integration with the communities they serve.

The flip side of decades of travel growth and the commodization of air travel is that airports need to become more integrated with the communities they serve. This need is driving airports to change they way they operate and manage their business to ensure they retain their franchise from the community for continued operations.

There are two key community integration factors:

- minimizing the environmental impacts of airports; and
- integrating airport planning with their communities.

These two community integration factors are examined in further detail below.

4.1 Minimizing Environmental Impacts

There is little question that airports and their users have an impact on the environment. In recent years however, airports have been under significant pressure to minimize that impact. This pressure is driven by changing community standards. Mature economies around the world have increasingly been demanding higher standards of environmental sustainability from their business communities and airports have been no exception. Whether in regard to aircraft noise, aircraft related emissions or the operations of the airport itself, minimizing environmental impacts is a major goal in everyday airport decisionmaking.

A related and perhaps more fundamental question related to the impact of changing community standards on gaining approvals for expansion and growth of airport infrastructure. Although the traffic reductions experienced in recent years have temporarily taken the pressure off airport expansion, particularly from a runway perspective, gaining approvals for additional aviation infrastructure to meet growth requirements has never been more difficult and is a major issue for many airports around the world.

Lead times for achieving the requisite approvals for new runway developments are commonly 5-10 years, with many airports adopting even longer term strategies to achieve expansion aims. Even terminal developments approval lead times are growing, with BAA's experience with Terminal 5 at London's Heathrow Airport being a case in point.

In addition to the impact that long lead times for new capacity has on the airports industry, the knock-on impacts include growth in congestion and costs for airline customers and the traveling public. Airports unfortunately take much of the blame for environmental degradation, when in reality, other players in the industry, particularly airlines and airframe and engine manufacturers have not been very responsive to community pressure. The industry needs to agree to tougher standards on noise and on emissions: new technologies make such standards achievable, but are not always adopted for economic reasons.

4.2 Integrating Airports and Their Communities

As airports have grown in terms of traffic levels and their importance to regional and national economies, the way they integrate with local planning regimes has also changed significantly, putting further pressure on airports to adapt in order to survive.

A key issue relates to property development. As airports increasingly seek to take advantage of their property assets, turning land surplus for aviation requirements over to commercial, industrial and retail uses, they are increasingly finding opposition from off-airport, competing developments and coming under pressure to integrate into local and regional planning regimes. Typically however, such local and regional planning regimes do not adequately take into consideration the non-aviation aspects of the airports business. Consequently, these regimes often seek to limit non-aviation development on land surplus to aviation requirements.

With non-aviation activities now accounting for, on average, over 50 per cent of all airport revenue, and with the growing contribution of property and development revenues, integration with off-airport non-aviation development and planning regimes is becoming an increasingly complex issue, potentially stifling the revenue growth of the airports.

5 Strategic Factors

Simultaneous to all of the changes and pressures set out above, the airports industry has also been undergoing an ownership revolution. The key impacts of this ownership revolution are described below.

The frontier of this revolution of course has been airport privatization. Since the pioneering privatization of BAA plc in 1987, it is estimated that literally hundreds of airports have experienced privatisation in some form or another. Typical forms of privatization have included Initial Public Offerings (IPOs - eg BAA plc, Copenhagen, Vienna, Auckland, Beijing etc), trade sales (as have occured in places such as Australia, New Zealand, Mexico, Argentina etc) and private financing of particular developments such as terminals.

Privatizations have typically been driven by governments seeking to avoid funding ever-growing airport capital requirements and airports realizing an opportunity to divest high value assets to facilitate debt reduction or investment in other areas.

These privatizations have seen three interesting trends emerge as desribed below.

5.1 Multinational Airport Operating & Ownership Companies

Apart from airports privatized by IPO, the vast majority of privatized airports have been sold by trade sale, a mechanism which typically maximizes price. The majority of airports that have been sold by trade sale have been purchased by other airports. A number of airport companies are now multi-national airport operators. Such operators, mainly European based airports, include BAA plc (which operates airports in the UK, the USA and Australia), Amsterdam Schiphol, Copenhagen, ADP, Frankfurt, Vienna, Zurich and Vancouver.

These multi-national airport operators appear to have been highly successful in transferring airport operating and commercial skills across borders to generate revenue and captial growth.

5.2   Institutional Airport Ownership

Another outcome of the airport ownership revolution has been the emergence of institutional airport owners. These institutions have raised significant amounts of capital to purchase stakes, and often controlling interests in airports. The harnessing of institutional investment funds to raise capital to purchase airports has been strongly adopted in Australia.

Investment funds established and managed by Australia's Macquarie Bank (and related entities) for example, now own major airports such as Sydney (the world's most expensive single airport to date), Rome, Bristol and Birmingham. Other investment funds have also been established in Australia to facilitate investment in airports.

These investment funds have the potential to revolutionize airport ownership and the way in which the heavy capital requirements of the airports industry are met.

5.3   Growth in Corporatization

Even airports that have not undergone privatization are faced with an ownership revolution. In the USA, one of the final bastions of the public sector airport, more airports are being corporatized and established as autonomous self-funding authorities. The trend towards corporatization has been driven by similar factors to privatization, including the need to remove governments from funding airport capital requirements. However, there also appears to be growing recognition that separation of airport management from the political process improves efficiency and productivity as well as introduces a commercial focus otherwise unavailable in goverment ownership.

6 Conclusion

The foregoing describes many of the complex phenomena which are transforming the airports industry. The rules of the game and the nature of key relationships among aviation stakeholders are evolving rapidly, and old business models are becoming obsolete. Most airport operators are successfully meeting the challenges of this new environment. What remains constant is the commitment of the airport operator to a safe and secure global aviation network, and to providing sufficient capacity to underpin trade, investment and tourism flows.

15

THE AIRPORT OF THE FUTURE AGAINST THE BACKDROP OF DRAMATIC CHANGES IN THE AVIATION SECTOR

MICHAEL GARVENS

1 The Quick are Swallowing up the Slow 362
2 Low Cost is Stimulating the Market 363
3 Aggressive Price Policy 365
4 Cologne Bonn is Market Leader in Europe 367
5 Catchment Area of Almost 20 Million People 369
6 Corporate Design Highlights Quality and Emotion 371
7 The Airport Experience 372
8 Future Risks and Opportunities 374

Summary:
This chapter outlines the dynamic processes and radical changes that are currently shaking up the international aviation market. The changes are affecting both airlines and airports to the same extent. An airport like Cologne Bonn must face up to new requirements and conditions. Two years ago, Cologne Bonn grasped the exceptional economic opportunity presented by the advent of low cost air travel. Today, the airport is Germany's leading low cost airport. It is now faced with the challenge of consolidating this market position. Despite the risks that the market generally holds for airports, the future is looking bright for Cologne Bonn.

1 The Quick are Swallowing up the Slow

The European aviation market is currently in the middle of a period of radical change; possibly the most drastic since the launch of charter flights in the 1950s. These changes are highly dynamic and are taking place at breathtaking speed. Without exception, all players on the aviation market are affected: the airlines and the airports; the major and the minor; the famous and the obscure. Only one thing appears to be certain: those who do not move with the winds of change will very quickly find themselves lagging behind. More than ever, it is true to say that the fast are swallowing up the slow.

Airport operators that do not want to end up on the bottom of the pile are well advised to keep a close eye on developments in the airline sector. This is the source of the impetus that will determine the future of the airports. It is vitally important for airports to anticipate market developments at the earliest possible stage. Long gone are the days when airports were able to act like monopolists, virtually untouched by competition and blessed with all the benefits that such a status brings. While in the past airports were like mountains that could patiently sit and wait for Mohammad, times have changed and the roles are reversed: the mountains must now go to Mohammad.

The traditional airline hierarchy is currently being turned upside down. The arrival of the low cost carriers unleashed another dramatic decline in yield, which is placing traditional flag carriers under considerable cost and competitive pressure. Inner-European point-to-point traffic is increasingly coming under the control of the low cost carriers as a result of the high discrepancy in unit costs. The only possible exceptions to this rule are routes that are largely travelled by business travellers and the feeder routes that bring passengers to the large hub airports. Even the intercontinental hubs – the flag carriers' natural habitat – will in the medium term be dominated by a small number of large alliance carriers. Smaller national carriers – let's call them secondary carriers – can only survive in clearly defined niches or alliances. It is likely that other airlines, well-known companies included, will exit the market in the future. In short, other big names will share the fate of Pan Am, Sabena, and Swissair.

The low cost airlines, on the other hand, are not only poaching passengers from the traditional airlines, they are also increasingly attacking the tourist market. They have begun to fly to the so-called warm water destinations that have filled holiday catalogues for many years. But the integrated tourism companies and independent charter airlines have not thrown in the towel or resigned. The TUI-Group, for example, has joined forces with Hapag-Lloyd Express and established its own low cost carrier, which will use the group's charter subsidiary, Hapag-Lloyd Flug, to offer an increased number of single seats to warm water destinations. TUI has also opened the first 'low cost travel agency' in Hamburg and in so doing is now selling low cost tickets on the High Street that were previously only available via Internet.

2   Low Cost is Stimulating the Market

The charter airline Air Berlin has transformed its sales platform, which previously sold mainly single tourist seats on its Mallorca shuttle, into a low cost portal: the so-called City-Shuttle. This was naturally accompanied by a massive increase in Air Berlin's aircraft capacity. This development will eventually lead to a greater mix of low cost passengers and package holidaymakers on the flights offered by these carriers. The medium-term result of this development will be the merger of the low cost and charter sectors.

Many of the critics, especially those who voiced scepticism in the early days of the low cost era, have been forced to eat their words. They doubted the durability of the business and even compared it with the stock exchange hype that surrounded Germany's 'Neuer Markt'. The theory that low cost would only lead to a cannibalization effect can now safely be said to be disproved. Today we know that low cost has in fact stimulated the entire commercial aviation market because approximately 60 per cent of all passengers say that they would either not have travelled at all or would have opted for a different mode of transport were it not for the low cost carriers. This is reflected in passenger volumes on a few selected routes from Cologne Bonn Airport, which have significantly increased not only for the low cost carriers, but also for their competitors, e.g. Lufthansa. The daily seat capacity on the Cologne Bonn–Rome route – currently one of the airport's most popular routes – has increased from 50 to 300.

The regional carriers are destined to waste away into niche carriers. They are left the risky task of trying out routes (warming them up) only to lose them to a competitor if the route turns out to be a success. The sector's decline in yield is hitting regional carriers hardest because the small aircraft they use do not allow for a significant reduction in unit costs. Unless they are linked to alliances, where regional carriers render feeder services from tertiary markets, business will become increasingly difficult for them.

The current market dynamic is characterised by the fact that all of these processes are taking place simultaneously and at high speed. Not a day passes without the announcement of a piece of news changing the outlook for the sector. At the same time, there is a clear discrepancy between airlines on the one hand and airports on the other. In terms of speed, airports cannot keep up with the incredible pace of developments. It would also be correct to say that the deregulated airline market is colliding head on with the regulated airport market.

The fact of the matter is that the liberalization of the European aviation market is the reason that all these changes are taking place. 'Open Sky' has replaced the bilateral contracts, which previously fixed both capacities and prices, and were based not on market demand, but on each country's national interests. The national carriers were part of a country's national identity. They were a country's flagship and brand, and consequently enjoyed the solicitous care lavished on them by their national government. In that era, it would have been unthinkable for a national carrier like Swissair to fall prey to bankruptcy.

It was only the disappearance of regulations that paved the way for the establishment of no-frills carriers; a development that has set so many other changes in motion. The causality of liberalization and low cost business is also evident on the Asian market, which is still heavily regulated, and where low cost business is developing relatively quickly, initially only within national borders, but increasingly also on cross-border routes. [1]

Without exception, the excessive decline in yield that followed the appearance on the market of low cost carriers has exerted enormous cost pressure on all airlines. They are being forced to adapt their costs to the falling average yield and to pass on the pressure to external service providers. At Lufthansa, for example – and this is likely to be the same for other airlines too – expenditure on 'duties and fees' is second only to expenditure on 'personnel'. As a result, the airline's focus has shifted not only to air traffic control, but also to the airports. It must also be said in this regard that the additional cost pressure resulting from the liberalization of ground handling services in Europe varies from country to country. Some airports like Amsterdam and Copenhagen have already outsourced these divisions. Others, like BAA (British Airports Authority), never had them in the first place. German airports on the other hand are generally more affected by this because they provide more of these and other services than other airports.

---

[1] See also Goh chapter 6 in this volume.

The concentration trend in the airline sector has also led to an increase in purchasing power, i.e. the concentration of power on the side of the purchaser from the airport's perspective. In future, airport services will be bought by alliances, not individual airlines. This will lead to a further decrease in duties and fees, a change that had already been set in motion by the upswing in the low cost carrier business. The first casualties can already be named: airports like Brussels and Zurich, and soon Amsterdam too, which are home to weak national carriers that are noticing a downturn in their intercontinental business. Airports would be well advised to pursue a policy of building up a low-risk customer portfolio that does not give any one customer the strength of a monopolist.

3 Aggressive Price Policy

The trend among airlines is towards an increasingly close link between network management and procurement. This means that the size of airport duties and fees, which only played a secondary role in the past, will become increasingly influential in terms of the choice of routes. The rigid duty models at German airports are therefore outdated. Airports need flexible models that are based on more than just the weight of the aircraft and the number of passengers on board. Criteria such as the time of the flight in peak or off-peak times, the number of new routes for an airline, or the type and extent of infrastructure use are much more useful in terms of setting duties.

Ryanair has consistently applied this new approach to duties and, in doing so, has developed a completely new business model. While in the past it was the passengers who financed the airline service, Ryanair now gets other bodies who benefit from the transportation service (airports, tourism associations, or regional administrative bodies) to finance part of it. The size of the duties and fees is therefore a decisive factor when it comes to deciding whether or not to fly to a particular airport. Issues of market potential or infrastructure links are less important. This new approach naturally requires a highly aggressive price policy that increases with the willingness of passengers to travel several hundred kilometres to the airport of departure, as was the case in the early days of the low cost boom with Hahn Airport, which was safely able to call the entire Federal Republic of Germany its catchment area.

The so-called Charleroi ruling has, however, set certain limits to this aggressive price policy. While about 75 per cent of the subsidies paid to Ryanair in Charleroi were declared admissible, 25 per cent were declared inadmissible. However, this ruling actually reinforces the low cost business. It has created more legal security because it defines what sort of discounts and price flexibility are permitted and what are not. De facto, German airports have been getting around their rigid duty model with airline-related subsidies for the development of routes. In other words, a practice that has been on the agenda for quite some time has now been legalized. So, how can an airport deal with this cost pressure?

First of all, there are shifts on the income side of things. Airports are now being forced to develop the 'non-aviation' sector – i.e. the income from gastronomy and retail trade – which was considered of little importance in the past. No less than 60 per cent of London-Stansted's income comes from this sector. Fraport even claims that non-aviation business is responsible for generating 80 per cent of its EBITDA. The airport as an infrastructure supplier is a thing of the past. There are no limits to the amount of creativity that can be used to try and open up new market segments and to make arrival and departure a unique experience for passengers. Does this mean that the airport of the future will be an exciting leisure experience? Why not?

The second core task is undoubtedly the reduction of costs to a competitive level. The situation where costs could dictate prices was only tenable as long as the market was characterized by regulated airport monopolies and regulated airlines; in other words, as long as the market was subject to planned rather than free market economy regulations.

There is potential for optimizing the interface between the airline and the airport in terms of ground processes, check-in, and security checks. In these areas, a higher degree of automation would ensure a reduction in personnel costs. However, in this regard, we are witnessing the collision of incompatible systems. Uniform automated check-in terminals, for example, or biometric procedures for security and immigration checks would make processes much less expensive, which in turn would have a positive influence on the overall cost of travelling and consequently the competitive factor.

The example of the airports in Dusseldorf and Cologne Bonn, which are situated at a distance of about 60 km from each other, illustrates the new competitive situation perfectly. Since both airports share a market environment, it is becoming increasingly justifiable to ask whether an airport is competitive in terms of its unit costs. Only those that are will be able to hold on to their market share in the long term.

The question of a cooperation between airports, especially between Dusseldorf and Cologne Bonn – an idea born out of the airline alliance concept – has been flogged to death. However, the sales synergies that can be achieved in the airline sector cannot be transferred to airports.[2] Moreover, there is hardly any identifiable potential for synergy on the cost side. Finally, one simple rule applies: it is the airline that decides where it wants to fly. Anything else would be dirigism and economically rather unwise.

4   Cologne Bonn is Market Leader in Europe

The dynamic changes in what used to be a homogenous aviation market mentioned at the start of this chapter are naturally of long-term significance for Cologne Bonn Airport. In order to keep up with the competition, airports now have to react quickly and flexibly to the latest developments and make rapid business decisions. Two years ago, Cologne Bonn Airport realized that the fact that the airline world in Europe will never again be the same as it was just a few years ago was an exceptional opportunity. Cologne Bonn made the two new airlines Germanwings and Hapag-Lloyd Express its new partners. The airport's decision in favur of the low cost carriers, which at the time was a brave one, was greeted with many critical statements from both inside and outside the airport. However, since the arrival of Germanwings and Hapag-Lloyd Express in autumn/winter 2002, business development has been extremely positive. None of the people involved could have predicted just what a success story low cost would turn out to be. The critics soon held their tongues.

For Cologne Bonn Airport, the strategic reorientation was a survival strategy because at the turn of the millennium, the airport found itself in the most difficult economic crisis of its history. While about 6.4 million passengers used Cologne Bonn in the year 2000, this figure had slipped to 5.47 million in 2002. While all German commercial airports were experiencing significant downturns in passenger numbers, nowhere was the decline as dramatic as in Cologne Bonn, where figures were trapped in a downward spiral. There can be no doubt that 2001 was the most difficult year in the history of international aviation since the Second World War. The consequences of the terrorist attacks of 11 September, SARS, and the overall economic climate created massive problems for both airlines and airports. However, the economic turbulence that Cologne Bonn Airport was experiencing was not only the result of these negative global factors; it was also to a great extent due to the airport's own mismanagement.

---

[2]   See also Pal & Weil, chapter 17 in this volume.

The best way of describing the airport at the time would be to call it a 'sleeping beauty'. Cologne Bonn had for the most part resigned itself to the rapidly falling passenger figures and the departure of many airlines. Most people at the airport wrongly believed that this period of economic drought could be bridged with the help of its second pillar: cargo. Then and now, Cologne Bonn is one of the leading cargo airports in the business. It is the second most important cargo airport in Germany after Frankfurt and the fifth most important in Europe. In terms of commercial air travel, however, there was not enough imagination and courage at the airport to really get to grips with its crisis. No-one at Cologne Bonn was willing to take any entrepreneurial risks even though the airport boasted three runways, free slots, a new terminal, high-rise car parks, a train station at the airport (which only opened in June 2004), and state-of-the-art infrastructure. The fact that the airport has a densely-populated catchment area rounded off the list of valuable aces that Cologne Bonn had up its sleeve.

A new era dawned at Cologne Bonn Airport in autumn/winter 2002 with a comprehensive management reshuffle and the arrival of Germanwings and Hapag-Lloyd Express. With their low cost, price breaker concept, they really stood out from the conventional airlines that had previously dominated the market. Both airlines chose Cologne Bonn Airport as their first hub. They quickly increased the number of destinations on offer from 17 to 41, thereby creating a dense European network of flights to the continent's most important conurbations, economic centres, and tourist regions. The low cost airlines' summer timetable for 2004 included over 55 destinations. Easyjet, one of two European big players in the low cost sector, joined Cologne Bonn in June 2004. The long-term nature of the low cost airlines' operations is proof of the fact that the low cost business is not a flash in the pan. The 2004/05 winter timetable includes 21 more destinations than the timetable from the previous winter season.

In 2003, Cologne Bonn became Germany's fastest-growing commercial airport, not only in terms of percentage growth, but also in terms of absolute figures. Thanks to its strategic re-orientation, Cologne Bonn went from being bottom of the class to being Germany's undisputed trailblazer in terms of airport growth. Seen over the entire year, Cologne Bonn posted passenger growth of about 43 per cent. In comparison with the 5.46 million passengers that passed through the airport in 2002, 7.84 million passengers arrived at and departed from Cologne Bonn Airport in 2003. Within the space of only a few months, Cologne Bonn not only became the number one low cost airport in Germany, it also became the market leader in continental Europe. To date, London-Stansted is the only other airport in Europe to boast more low cost passengers.

The successful low cost concept has transformed Cologne Bonn into a trendsetter and standard-setter among German airports. While it previously brought up the rear, Cologne Bonn is now a serious competitor. The best proof of this is the fact that Cologne Bonn poached approximately 800,000 passengers from Dusseldorf airport alone in 2003. However, while Cologne Bonn was ahead of the field in the early days of low cost travel, other airports are now seeking to close the gap by copying its successful concept. It is now clear that only those airports that have a large share of low cost travel in their programme have the potential for growth. Forecasts for Germany predict that in the coming three to five years about 20 to 30 million passengers will opt to travel with low cost or no-frills airlines. Cologne Bonn Airport not only wants to defend its leading market position in this sector in the long run, it also intends to increase its lead. Cologne Bonn aims to break the 10 million passenger barrier for the first time with the help of the low cost carriers in 2006.

In order to ensure that this happens Cologne Bonn Airport must ensure that it really stands out from the competition, and must present itself to its customers – i.e. passengers and airlines – as an attractive product.

5 Catchment Area of Almost 20 Million People

Three factors play an important role for airlines when deciding on an airport: infrastructure, geographic situation, and catchment area. Cologne Bonn is well placed in all three categories. The infrastructure in and around the airport allows for further growth of the airlines. At the turn of the millennium, the airport invested over € 500 million in its development. The design was generous and was based on a comprehensive concept: a new terminal (T2) was built, two large high-rise car parks with a total capacity of almost 12,000 parking spaces were constructed, and the airport's train station was opened in the summer of 2004. There are only few airports that have implemented the concept of intermodality – the linking of a variety of transport modes such as airplane, train, and car – as close to its ideal as Cologne Bonn.[3]

But on the air travel side of operations too, Cologne Bonn is ready for future passenger growth. Its three runways are some of the best in Germany. Unlike other German airports, there are at present no capacity limits here. The largest of the runways is 3,815 metres long and the only runway in the state of North

---

[3] See also Fakiner, chapter 18 in this volume.

Rhine-Westphalia capable of coping with any size and weight of aircraft for intercontinental flights without restrictions. Furthermore, Cologne Bonn is open for business 24 hours a day. Unlike airports with capacity bottlenecks, Cologne Bonn still has attractive daytime airline slots to offer.

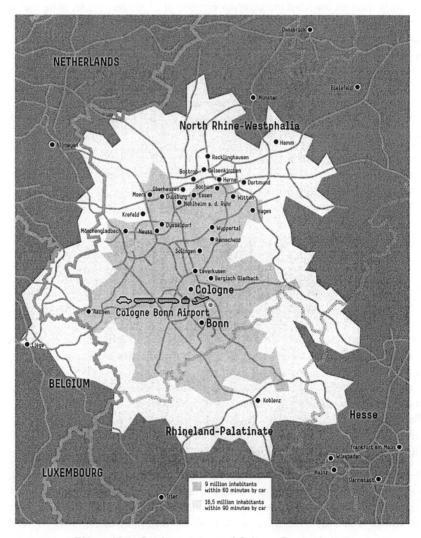

Figure 15-1: Catchment Area of Cologne Bonn Airport

The airport's situation and catchment area are also of significance (see figure 15-1). Cologne Bonn Airport is not only situated in the heart of Europe, it is also located in one of the most densely populated and most powerful economic areas in Europe. North Rhine-Westphalia is the economically most important state in Germany. In other words, it is still 'the' centre for science, industry, trade, media, trade fairs, and conferences in the Federal Republic. About a quarter of Germany's total exports originate in North Rhine-Westphalia.

At the same time, Cologne Bonn Airport also boasts a catchment area with a significant passenger potential. In other words, its market environment is very good. Approximately 15 million people live within a radius of 100 kilometres. This has increased to almost 20 million people with the opening of the new rail link in the summer of 2004. No less than 174 high-speed ICE trains, regional express trains, and suburban trains stop at the airport train station every day. This figure is set to increase with the publication of the winter 2004 rail timetable. This will improve the links between Cologne Bonn Airport and some German regions, like the Rhine/Main region: trains will leave for Stuttgart via Rhine/Main once every two hours. The existing good links to the Ruhr region will be further enhanced by six additional daily connections to and from Dortmund and Berlin. This means that a total of 35 trains per day will travel this route. Generally speaking, Cologne Bonn Airport's rail link is an important building block in the strategic positioning of the airport in terms of its battle for more passengers with Frankfurt, Hahn, or Dusseldorf.

In short, Cologne Bonn Airport offers airlines outstanding working conditions. It reflects a respectable blend of infrastructure, situation, and catchment area. There are airports, like Hanover or Leipzig, where the infrastructure is excellent, but the passenger potential of their catchment areas is modest. Exactly the opposite is the case with Hamburg and Dusseldorf. Remote airports like Hahn or Weeze-Laarbruch, on the other hand, cannot seriously be considered to be high-quality airports; after all, passengers using them can only fly from the middle of nowhere to the middle of nowhere.

6 Corporate Design Highlights Quality and Emotion

Even if passengers fly low cost, that does not mean they want to do without comfort and the convenience of a large airport. This includes ease and speed of access to the airport and a broad and comprehensive selection of shopping facilities, cafés, restaurants, and bars. Moreover, travellers enjoy starting their journey in an airport that has its own identity and an unmistakeable, contemporary image. The decision to strategically orientate Cologne Bonn Airport around low cost carriers created an urgent need for a new airport corporate design. A variety of logos and brands was already in use and the colour coding was not binding. This proliferation of designs, which could probably be more accurately described as rank growth, began in the 1980s. The corporate design no longer had anything to do with the airport's new orientation. Wedged between two major airports, Dusseldorf and Frankfurt, it was also important for Cologne Bonn to develop a clear image profile that really stood out from the images of the other two airports.

The Paris-based agency Intégral Ruedi Baur et associés came up with the new corporate design, unusual with respect to traditional airport corporate design. The complex and previously complicated airport world is now presented as being pleasant, colourful, and straightforward; a status that is neatly summed up by the slogan 'so simple'. Traditional airport logos project an image of airports as being serious places, from which people want to escape as quickly as possible. Cologne Bonn Airport's design now reflects freshness, joy, and imagination. An image is conveyed that airports and air travel have nothing to do with stiff exclusivity; travel has become more democratic and affordable for all sections of the population. In doing so, it has made the airport very popular. Passenger surveys conducted at the airport back up this claim. Well over 70 percent of those surveyed said they found the new corporate design 'good' or 'very good'.

Psychological components like renown, support, trust, acceptance, or uniqueness, which people associate with Cologne Bonn Airport, can only be considered one of many stepping stones on the road to strengthening the airport even further. Once passengers are inside the airport, the next step is to offer them even more incentives and attractions. After all, airports have the potential to be much more than straightforward infrastructure facilities with runways and waiting rooms. In other words, they are more than just transport stations from which passengers want to escape; they can be places where passengers are happy to spend more time.

7 The Airport Experience

Airports like London-Stansted already derive over half their turnover from the income from car parks, restaurants, cafés, and retailers. At Cologne Bonn Airport too, the non-aviation sector is becoming an increasingly vital source of income in the value-added chain. Nevertheless, Cologne Bonn is still highly dependent on the turnover from its aviation business. While proceeds from the non-aviation sector now account for 20 per cent of total turnover, the aim is to increase this to over 30 per cent by the year 2006.

The start of low cost air travel from Cologne Bonn also resulted in major gains for the airport's shops, bars, and restaurants. The growth in proceeds from non-aviation activities in both terminals is increasing faster than the growth in passengers. In the gastronomy sector alone, the airport has succeeded in doubling its turnover: in short, low fares and deli fare make the perfect pair. Cologne Bonn has taken a lead among the airports of North Rhine Westphalia in the gastronomy stakes. It was the first airport to trump with well-known brands such as Leysieffer and Burger King. There is still a substantial amount of as yet unexploited potential for growth in this business sector. This is why the airport company is increasingly investing in the development of more shops, bars, and restaurants. In addition to the two brands mentioned above, other well-known international brands like Esprit, Kamps, or Gosch also opened shops at the airport. This selection of shops will gradually be complemented by other popular brands such as Starbucks or Subway.

In order to develop the non-aviation business, the airport is using more than just available space in the two terminals. The airport is reacting to the considerable increase in passenger numbers by investing heavily in the expansion of Terminal 1. Construction of the Central Departure Hall began at the gable end of Terminal 1 in November 2003. The building will be a steel and glass construction and was designed by architect Helmut Jahn, who was also responsible for Terminal 2, the two high-rise car parks, and the airport railway station. The new departure hall will measure 5,000 square metres, about half of which will be used for shops, bars, cafés, and restaurants (i.e. a market place). The new building will be open for business by mid December 2004.

The most important rule is that 'business follows the passengers'. In other words, it is important to arouse their interest wherever they are. In this regard, it is also vital to achieve the right blend of products and sectors. When planning the non-aviation sector, an airport must always take into account that today's passengers do not spend money in airports with a view to meeting their basic needs. They more often buy things on an impulse. Ideally speaking, they will buy whatever takes their fancy. The airport must make it as easy as possible for customers – and this applies equally to passengers and visitors – to make such impulse purchases. Restaurants and shops must be open; there should be no thresholds or barriers.

Attractive shops and cafés help passengers in the airport while waiting for their flights. The more comfortable they feel, the more willing they are to spend money. Statistics show that low cost passengers, who have paid little for their air tickets, spend even more money than conventional passengers, because many of them have an above-average household income. Moreover, just under 40 per cent of all low cost travellers in Cologne Bonn are business passengers: an economically powerful and therefore attractive customer target group. They often use their time at the airport for optimum time management. Instead of sacrificing their precious leisure time on shopping in city centres, they go shopping at the airport before their flight departs. This is another service that should not be underestimated.

Cologne Bonn Airport is in the early stages of creating an airport experience; a world in which visitors and passengers will enjoy shopping and relaxing. Smart flying can be cleverly combined with smart shopping, smart drinking, and smart eating. In order to further enhance the attractiveness of the airport, it is however important to take steps to prevent the airport resembling the monotone, forgettable facelessness of German shopping streets. The ambience in which passengers and visitors shop and eat must be original and stand out from the norm. It is important to ensure an exclusive selection that is not readily available in the immediate vicinity of the airport. It is essential to arouse the curiosity of passengers and customers for a colorful holiday world that begins before take-off in the terminals at Cologne Bonn Airport. As is the case in the aviation sector, where Cologne Bonn was among the first on the low cost bandwagon, the airport can also be considered as ranking among the trendsetters in the non-aviation sector.

This means that airports must constantly keep up to date in their non-aviation business and must continually react to changing conditions. They must always have their eyes firmly fixed on the future. New divisions must be opened up, even if the obstacles initially seem insurmountable. Every new idea must be taken seriously and included in calculations. When viewed in this light, the idea of flying to Cologne Bonn specifically to bag a few bargains at the airport's factory outlet centre does not seem unrealistic.

8 Future Risks and Opportunities

Even if other divisions become increasingly important, aviation naturally remains the core business at Cologne Bonn Airport. Both in the low cost and cargo sectors, the airport aims not only to consolidate, but also to increase its lead on the market. Other airlines and new routes will be acquired. The tight European route network from Cologne Bonn offers airlines an exceptional foundation on which to build up long-haul business. Cologne Bonn is in this regard particularly attractive for airlines that are not part of any alliance.

In spring 2005, Cologne Bonn Airport intends to become the first airport to implement a concept that has never before been attempted in the aviation business: long-haul flights to the USA starting at € 99. What makes this offer even more attractive is the fact that the airlines would link the low cost hub Cologne Bonn with other hubs in the United States. This would make Cologne Bonn the first European airport to offer both low cost long-haul flights within an intercontinental network and low cost point-to-point flights. This project may, however, be endangered by a further-increase in the price of kerosene on the world market. The reason for this is that the price of kerosene accounts for about a third of the cost of long-haul tickets.

In view of the dynamic processes on the aviation market, an airport like Cologne Bonn is exposed to medium- and long-term risks that must be taken seriously. At the same time, it must be said that it has many strengths and faces a variety of opportunities. As recent years have demonstrated, unexpected catastrophes like the terrorist attacks of 11 September or epidemics like SARS can result in a nosedive in international air travel. However, political decisions can also have a (negative) effect on business. It is essential not to overlook the fact that airports also depend to a large extent on legal or administrative regulations at both national and international level. An amendment of the Airport Noise Pollution Act could, for example, result in restrictions on night flights. The consequences of such an amendment for Cologne Bonn Airport and for the freight carriers that are based there (UPS and DHL) would be fatal. The lion's share of the cargo at Cologne Bonn – approximately 600,000 tons are expected to be moved in 2004 – is shifted at night.

There are of course other risks of a more financial nature: e.g. the Federal Government is demanding the payment of interest on ground rent from Cologne Bonn Airport. Instead of the current annual payment of € 500,000, the Federal Government is now demanding € 15 million. It is vital that a mutual agreement is reached by the Ministry of Finance and the airport in the spirit of economic efficiency. Other economically problematic factors include, for example, Cologne Bonn Airport's continuing dependence on turnover from the aviation sector, which is characterized by declining proceed margins.

As far as the airlines are concerned, it can already be said with certainty that there will be a comprehensive consolidation of the low cost market over the coming years. It is likely that several airlines will exit the market. The developments that have already hit the classic carriers – and in particular, the flagship airlines – will also hit this sector. Who would have thought a short time ago that Sabena would ground flights permanently or that Swissair or Alitalia would end up in such dire economic straits. One does not need to be a prophet to predict that some players on the low cost stage will not be able to withstand the massive competitive pressure and consolidation trend and will eventually disappear from the market. This is why it is vital for the survival of any airport to ensure that its portfolio of customers is wide and varied in all sectors: low cost travel, cargo, and non-aviation. It would be very unhealthy to become economically dependent on one airline. The best example of the disastrous effects of such a policy is the economic collapse recently experienced by the airports in Brussels (Sabena) and Zurich (Swiss).

In my opinion, however, Cologne Bonn Airport can look with confidence into the future. It is already Germany's market leader in the low cost and express cargo sectors. It boasts an outstanding infrastructure and will not be troubled by worries about capacity bottlenecks in its terminals and runway system in the coming years. Its intermodal links, situation, and catchment area are second to none. It is only with innovative business ideas and a simultaneous outstanding selection of shops and services for passengers and airlines, that an airport like Cologne Bonn Airport can consolidate its position on this fiercely competitive market in the long term. In this regard, every airport will have to fend for itself, at least in Germany anyway. The aviation market will set the pace and fix the conditions. The airlines will only fly to those airports that offer them the best chances of success. The idea of a cooperation – e.g. between Cologne Bonn and Dusseldorf – which was raised by certain politicians, can no longer be realized in the current market environment. Dirigistic meddling in airport affairs must be rejected! After all, the old rule applies here: where there is demand there will also be supply.

16

INTERNATIONALIZATION STRATEGIES FOR AIRPORT COMPANIES

BENJAMIN KOCH AND SVEN BUDDE

1 Airports' Going Global	378
2 Airports and their Role in the Air Transport Chain	379
3 Internationalization of Airport Companies	381
4 Conclusion and Closing Remarks	405
5 References	406

Summary:
Today's airport management is not only a matter of maintaining and enhancing the performance of a single airport. It more and more targets at developing the airport operator's overall business from a local infrastructure and service provider towards an internationally active enterprise. This forces airports to intensively evaluate the opportunities and threats involved in any international project. This chapter is based on the results of an explorative study of several international airport operators. The research results are presented and discussed to derive a set of success factors for airport internationalization projects.

1 Airports' Going Global

Ongoing liberalization and privatization developments have caused dramatic changes in the worldwide aviation industry over the past years.[1] These changes equally affected the competitive positions of airlines and airports. After the first privatizations appeared in the market with airport companies such as BAA plc., Vienna Airport, Fraport AG or Unique Zurich Airport being sold – completely or at least significant shares of them – to private investors, a new market of privately operated, competitively acting and profit-oriented airport operating companies has come to life.

Part of this development has been a trend, especially of these privatized companies, to start expanding their business portfolio. This has not only been realized by increasing the commercial activities[2] at their own airports, but by acquiring shares of other airports as well. Those airports have been both within the investors' home countries and abroad. Rapidly, airport internationalization has become a new area of competitive activity within the aviation industry. New, internationally active airport groups have emerged, even "global mega airport companies"[3] are expected to emerge.

An airport's strategic decision to internationalize its business as well as the successful design of the implementation and realization of such development projects are subject to the influence of a multitude of different and very specific aspects. Thus, for every single project in the field of airport internationalization the unique and individual framework of risks and opportunities has to be taken into account as its crucial development background.

So far, the analysis of the phenomenon of airport internationalization – the participation in global projects as well as the underlying decision criteria – has hardly been dealt with in scholarly literature. This chapter intends to start closing this gap.

---

[1] On the general developments in the aviation industry see for example: Doganis (2001) or Graham (2001).

[2] For the topic of the commercialization of airports please refer to Graham (2001), Freathy & O'Connell (1998) or Doganis (1992).

[3] Doganis (1999), p. 112.

To reach that goal, we examine a set of representative case studies on various airport operators who have been involved in internationalization projects over the past years. Based on the analysis of selected projects, we discuss the problems and shortcomings experienced and finally derive several crucial success factors, which can be classified into company and project specific success factors.

The empirical analyses can serve as a means of decision support in considering other international activities and can lead to the identification of general or company specific strategies, development trends and action patterns for airports.

2 Airports and their Role in the Air Transport Chain

To better understand the issue of airport internationalization and the potential areas for internationalization projects airport operators might choose, the core activities performed by airports first need to be identified. Second, airport operators have to make deliberate choices to select which of these activities become part of an internationalizing project and how this should be done.

These generic activities are expected to be preferred in case the airport chooses to become engaged internationally. In general, airports are the interface between the surface transport and the air transport in any air transportation chain.

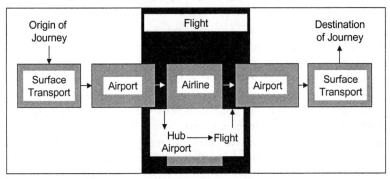

Figure 16-1: The Air Transport Chain (Passenger Transport)

The role as linking point between the different modes of transport puts the airport in a specific position, making it both an infrastructure and a service provider for a highly specialized demand. Fulfilling this function, an airport has to satisfy the specific needs of such diverse customers as airlines, ground handling agents, catering companies and of course passengers and freight forwarders, to name a few. Thus, the range of services offered and the commercial activities performed by an airport covers an extensive variety. Based on Porter's value chain concept, the following figure gives a brief overview of an airport's business complexity.[4]

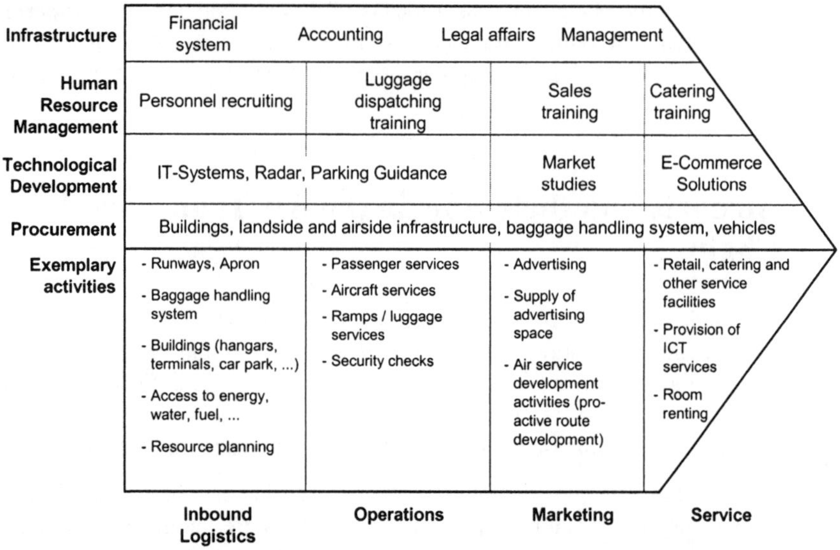

Infrastructure	Financial system	Accounting	Legal affairs	Management
Human Resource Management	Personnel recruiting	Luggage dispatching training	Sales training	Catering training
Technological Development	IT-Systems, Radar, Parking Guidance		Market studies	E-Commerce Solutions
Procurement	Buildings, landside and airside infrastructure, baggage handling system, vehicles			
Exemplary activities	- Runways, Apron - Baggage handling system - Buildings (hangars, terminals, car park, ...) - Access to energy, water, fuel, ... - Resource planning	- Passenger services - Aircraft services - Ramps / luggage services - Security checks	- Advertising - Supply of advertising space - Air service development activities (pro-active route development)	- Retail, catering and other service facilities - Provision of ICT services - Room renting
	Inbound Logistics	Operations	Marketing	Service

Figure 16-2: The Value Chain of the Airport[5]

The term "internationalization" covers a broad variety of activities in both its theoretical as well as practical interpretations. Our understanding of "internationalization of airport companies" comprises all activities in which an airport operator undertakes any international activity of any kind, which means any activity at any location outside his own airport and outside his home country. Within these activities, the type of projects realized, the functions carried out and the kind of participation involved may vary. The respective international companies have competencies in very different areas of expertise such as concession management, ground handling or airport management. These core competencies build the basis and the fundament of their projects. Thus, for each specific international project various companies usually join forces in a

---

[4] For a detailed discussion of the activities of an airport and of the differences and similarities between the value chains of airlines as well as of potential fields of cooperation between these companies refer to Albers et al. (Forthcoming).

[5] Adapted from Albers et al. (2005).

consortium, combining their respective capabilities to be able to deal with the actual tasks and requirements in the project's environment and framework in the best possible way.[6]

In general all activities performed by an airport along the value chain can be subject to an internationalization project. Depending on the activity being performed internationally, the project design can of course differ significantly, ranging from selling services to external clients (e.g. offering ground handling services at a foreign airport) to acquiring an entire airport abroad.

3 Internationalization of Airport Companies

The internationalization of firms has been widely discussed both in general terms as well as for specific industry sectors.[7] Nevertheless, especially in the field of airports, research is limited. The following analysis of this specific area of the aviation business approaches the issue based on an empirical study of the phenomenon of airport internationalization. Doing so we do not attempt to develop a comprehensive, specific theory of the internationalization of airport operators. Moreover our intention is to give insights into the decision mechanisms behind such investment projects.

3.1 Opportunities and Risks of Airport Internationalization Projects

The analysis of the airport companies' motives for entering into international projects discloses a considerable set of factors which enforce the companies' decision to become active in this business. These drivers can be differentiated

---

[6] To guarantee to serve all project requirements, a typical consortium would most commonly consist of the airport operator, investors, a construction company, concessionaires and other project-specific partners. These can include e.g. local lawyers, investors, construction enterprises or other parties with broad experience and knowledge of the local business and political environment. In some cases it might also be necessary to have a partner holding a "golden" share for juridical or political reasons.

[7] For a general overview of the most discussed theories of the international company refer to Perlitz (1995) or Kutschker & Schmid (2002).

into internal and external aspects. The most important targets within the internal dimension are to create additional value through international projects, the diversification of business risk, the utilization of revenues from other markets for financing and developing investments in the home market and thus the overall improvement of the own competitive position in the market. Moreover, the target to use the knowledge and experience gained in the home market to generate new revenues in international markets in order to increase the company's total business volume plays a vital role in the internally driven internationalization strategy.

In addition to these internal factors, the external drivers for the internationalization of airports include in particular the general globalization trend in the economy as well as privatization developments which have a major impact on the airport industry worldwide. This trend has quickly led to the emergence of several globally active airport operating groups who are expected to soon become powerful and dominant players in the world's airport markets. Thus, many airport operators are afraid of falling back in competition if they remain passive in the global playing field, including the risk of being taken over at some point by one of the new major airport groups.[8]

Airports use different forms and methods to internationalize their activities. A general categorization of different internationalization models can be based on the dimensions "duration" and "depth of involvement". Accordingly, the partnership between the single parties involved[9] can take the form of a joint venture with an equal distribution of influences and responsibilities (deep involvement of all partners, including major financial investments), the form of a consortium with different shares of the single partners (different depth of involvement, depending on the partner's importance within the consortium) or the form of an investor group with one leading partner who carries the overall responsibility for the project (with significantly less involvement by the other partners, sometimes being only sub-contractors). The duration of the internationalization models can vary significantly as well. While most projects

---

[8] A comprehensive analysis of these and other motives for the internationalization of airport companies can be found in Graham (2001), pp. 46 ff. See also Aaronson, chapter 14 in this volume.

[9] The group of partners involved very often does not only include private investment companies, but also government institutions who keep a certain share in the airport. In this constellation a very frequent cooperation model is the Public Private Partnership, in which the private and the public sector develop the project jointly. For examples of such models refer to Juan (1995).

are based on concession contracts under a pre-defined timeframe, the acquisition of airports has no predetermined end.

Within these general partnership models, the activities provided by the airport company, as its input into the cooperation, range from providing consulting services for management contracts, to the acquisition of operating concessions and taking-over shares of the airport.[10]

International projects bear a significant number of problems and risks, which by far exceed the respective potential threats and risks of the investor's traditional core operation. Again, these factors can be grouped according to their origin into internal and external risks.

The external, environment-related risks and influences mainly comprise the lack of experience and the insufficient evaluation of the situation in the target market, the choice of a local partner, neglected problems in the investor's adapting to the different business environment, communication problems and shortcomings due to the geographical and cultural distance between the home airport and the international project as well as economical, juridical and political restrictions. Take for example a possible investment of a European airport operator in a mainland Chinese airport – a challenge no European airport operator has taken so far. Thus, there is neither experience with the local, Chinese market situation nor with the respective business environment. The quest for a local partner is known to be extremely time consuming and would not be successful without regarding the political aspects in developing joint venture projects. Furthermore, the regulative framework is currently in an ongoing state of flux, making it almost impossible for a far distant investor to quickly realize important changes and progresses in the target country.

Internal risks and threats for international projects include lack of communication and homogeneity in the targets, expectations and duties related to the project between the responsible parties i.e. between consortium partners as well as within a single partner's organization and an insufficient or inadequate distribution of inputs and shares in the project.[11] For example, a company

---

[10] Nowadays a common approach is for the consortium to establish a joint venture which takes over or operates the target airport. Another increasingly important trend is the growing trade sale market, the sale of airport shares to strategic partners in which both the seller and the buyer are usually private investors.

[11] The case of the failed cooperation for the Berlin airports can be seen as an example in which the lack of an agreement on the distribution of risks between public and private partners led to the end of a major development project. See Frankfurter Allgemeine Zeitung (2003).

investing in such different regions as Latin America and Asia has to realize that the problems within the acquired airports – both economical as well as cultural – will usually be completely different. Thus, for each investment project specific and unique problem solving mechanisms and management approaches have to be developed, transferring the knowledge from one project to the other often fails. Also, since typically several investors jointly operate the acquired airport the management has to face the different ways of doing business and the different interests of each partner – both on a project as well as on a headquarters level. Furthermore, the project teams usually are not interconnected, limiting their chances to exchange their respective experiences to a minimum.

As one of the dominant fundamentals of all business partnerships, the financial structure of the cooperation and the distribution of shares between the partners have an important impact on the overall project success, reflecting the aspects mentioned above. In many cases, the well-balanced distribution of benefits and duties can resolve severe threats in the implementation phase well in advance.

3.2 Analyzing the Airport Internationalization Phenomenon – Methodology and Data Framework

A comprehensive picture of the total market for the internationalization of airports would be a very challenging, major task, by far exceeding the framework of this chapter.[12] Thus, this analysis focuses on four case studies of international airport operators of different sizes, which differ in strategy, international experience, and business environment.

The general basis for this study was extensive research of secondary sources, ranging from scholarly literature on the fields of the international firm to the aviation business in general and to airport management in particular. Additionally, recent publications in industry magazines, newspapers, newsletters and comparable media have been analyzed.

Moreover, we collected primary data based on interviews with various industry experts. These do not only include numerous consultants with vast experience in the field of airport privatization and internationalization, but also project managers and top management of the investigated airports.

---

[12] Bentley (2002) gives a detailed overview of the privatization and internationalization of airport companies worldwide. In this, he mainly presents the projects which have materialized in the different countries, rather than discussing the underlying investors' motives, intentions and experience.

We selected airports for our case studies based on a broad set of different criteria. To allow for representative statements on the phenomenon of airport internationalization operators with different backgrounds were selected. Thus, companies with a different culture and different geographical origins had to be chosen as well as operators of different sizes and with different ownership structures, since all these factors might have influences on their respective international behaviors. Additionally, all airports analyzed should still be active in the development of international projects and should be known to be currently searching for new investment opportunities. At the same time, a certain experience with internationalization projects was a prerequisite for this study. Finally, the availability of project managers for recent or ongoing projects and / or members of the responsible top management for personal or telephone interviews were selection criteria. Additionally, the number of airport operators analyzed was limited to a total of four to keep the study on a reasonable level.

Resulting from this selection process, the airport companies analyzed were Fraport AG of Germany, Flughafen Wien AG of Austria, Unique Zurich Airport of Switzerland and YVR Airport Services Ltd. of Canada.

On these four airport companies we have collected the available data and reports on their international activities as well as on their overall business performance – especially in the light of the influence of their international engagement. This basic data has been analyzed and led us to specific profiles for each operator. Based on these findings, questionnaires have been designed and mailed to our interview partners to allow them to prepare themselves for personal interviews. These were performed personally or by telephone to both confirm the findings from the secondary research and to clarify additional and open issues by widening the fields of discussion.

3.3 Case Findings

3.3.1 Fraport AG, Germany

Fraport AG is the operator of Frankfurt Rhein/Main International Airport, the largest airport in Germany and one of the largest in Europe. Since the initial public offering in June 2001, 29% of Fraport AG is in private ownership, the remaining shares are held by public entities, including the city of Frankfurt, the State of Hesse and the Federal Republic of Germany.[13] For the company, the

---

[13] A public sale of further shares is currently under discussion.

business units aviation, non-aviation and ground handling are all of approximately the same importance, all generating comparable shares of the total turnover. This structure shows the relatively high importance of the usually smaller ground handling segment, reflecting the core competency Fraport has developed in this area. Other core competencies are in the fields of terminal management and airport retailing.

Fraport's development is positively influenced by its advantageous intermodal connections to the German motorway and high speed railway systems as well as by the continuously growing traffic volumes. Nevertheless, competition from other airports for both transfer as well as low cost carrier passengers and competition within the own airport in the field of ground handling are building up a certain pressure on the airport's management. Furthermore, especially the increasing capacity constraints and the long planning and approval processes for the future extension of the airport's airside infrastructure create severe limits to the company's growth. Thus, internationalization is being seen as a potential source for additional business growth and as a measure of risk reduction by earning revenues in other, possibly less competitive environments.

Taking advantage of the general privatization trend, Fraport has acquired shares of several German airports,[14] before a step towards international markets has been made. In the field of ground handling, Fraport is successfully selling its services at airports in Austria, Portugal, Spain and other European countries. Customers for security services are mainly located in Germany, France and Great Britain. Under the roof of the strategic alliance with Amsterdam Schiphol Airport, Pantares, Fraport is also present with a ground handling joint venture at Hong Kong's Chek Lap Kok Airport.

For Fraport, internationalization projects became an interesting opportunity in the mid-1990s. Driven by the upcoming privatization trend, the expectations of significant revenue potentials in this business and the international engagement of other major airport operators, the management of Frankfurt Airport made the decision to become an active player in the market for airports as well. While the underlying intention was mainly to improve the own competitive and strategic position and to leverage the restricted growth opportunities at the home airport, the large amounts of free financial resources supported this decision.

This financial background was the basis for the strategy of very capital intensive international engagements. Mainly targeting an attractive return on investment Fraport entered into several large projects which required significant financial

---

[14] Within Germany, Fraport currently holds 73% of Frankfurt Hahn Airport, 51% of Saarbrücken Airport and 20% of Hanover Airport.

resources and high up-front payments while implying long payback periods.[15] In this context, the company has acquired stakes in the international terminal of Antalya, Turkey, the airport of Lima, Peru, and was part of the Philippine International Air Terminal Co. for the construction and operation of a new terminal at Manila Airport in the Philippines.[16] Moreover, Fraport has participated in numerous tenders for airport sales and contracts over the time. Whereas Antalya and Lima turned out to positively contribute to the company's development, the Manila engagement, especially, led to significant losses and the depreciation of the total invested capital.

While political disturbances led to the loss in Manila,[17] different problems occurred in the other projects. Of these, the bankruptcy of the local partner in the Lima consortium was the most severe disturbance, which led to lengthy negotiations on the funding and financing between the remaining partners. This caused significant delays in the overall project's development. Keeping in mind the long payback period of this kind of international engagement, the importance of a strong financial base becomes very obvious.

Originally, the financial background and the intention to take part in the competition with other airport operators in the field of international projects were the most important driver for any internationalization decision. In the early days of Fraport's international engagement, the company's organization was structured by business segments. Thus, every international activity was planned and realized in the respective operational division, supported by the common strategy to go global, but not systematically coordinated. However, all the projects required specific expert know-how in various areas and in particular financial resources, all being scarce resources in every organization.

Having learnt from the difficulties over the past years, Fraport started to rethink and reorganize its international activities. While especially cultural differences between the German operator's staff and the local partners at the international airports can be reduced by hiring international key personnel for each project and by taking the time to build up stable, trustful relations between the partners, political and economical risks in the target countries are now especially subject

---

[15] BOT projects usually have a running time of 15-40 years, with an amortization period of the invested capital of at least 10 years.

[16] Fraport holds 50% of the shares of the operating company at the international terminal of Antalya airport, plus additional 30% of the dividend rights. In Lima, Fraport holds 42.8%. The engagement in Manila came to almost 500 million Euro.

[17] An unclear and constantly changing contractual situation between the consortium and the government of the Philippines has finally led to the total loss of the investment.

to a far more intense evaluation before any activity towards a tender or project is started. Also, to improve the coordination between the different projects, a central coordination department for all international participations was established. This unit is not only responsible for future acquisitions, but also for ongoing issues, including the processing of experiences and lessons learnt and the transfer of the gained knowledge throughout the respective key staff.

In line with these developments and following the shareholders' interests, Fraport also changed its international strategy. While the financial resources are now mainly to be invested into the local developments at Frankfurt Airport, the focus for international projects is clearly set onto service and management contracts. Financial participations are considered only on a minor scale to support these other activities.

3.3.2 Flughafen Wien AG – Vienna Airport, Austria

Flughafen Wien AG, the operator of the Airport of Vienna, was one of the first privatized airport operators. In 1992, a first part of the company's shares was listed at the Vienna Stock Exchange. By 2001, market capitalization had increased to 50%.[18] Having established a stable position as a hub for East-West-traffic within and passing Europe, Vienna Airport demonstrated a successful and profitable business development. This provided the airport with a good financial basis enabling the company to fund a permanent participation in international projects, regardless of various costly improvement and development projects at the own airport. The airport sees its core competencies mainly in the areas of operational and handling know-how as well as in marketing.

Even during its time as a state enterprise, Vienna airport was not managed as a government authority, but with a strong focus on profitability and economic growth. Over a long period of time the position as the country's capital airport was sufficient to generate the expected development, but with a nonetheless limited local market potential, internationalization soon became a vital part of the company's strategy. Funded by the existing financial resources and supported by the starting privatization developments in the airport business, Vienna Airport entered into internationalizing its businesses.

The first cross-border activity started as early as 1992 with a long-term consulting contract with the Slovak airport of Bratislava. This project benefited especially from the cultural proximity between the partners, encouraging the

---

[18] The remaining shares are held by the City of Vienna, the State of Lower Austria and an Employee Foundation.

airport to further develop its international presence. In the following years, Vienna Airport participated in numerous tenders for airport projects worldwide, but lost most of them to other, financially stronger and larger competitors. Having experienced the significant costs involved in each tender,[19] the company changed its strategy from a frequent participation in open tender processes towards a very selective, targeted engagement in more promising projects. This resulted in successful bids for the development of a terminal at the Airport of Istanbul, Turkey, and for the Airport of Malta.[20] A third financial investment was the Airport of Ciudad Real, Spain – a greenfield airport project which has not yet been realized – but this investment was reduced to a consulting and training contract in the meantime with the option to receive a management contract as soon as the airport is operational. Another intended investment, 7% in the new Berlin Brandenburg International Airport in Germany, was postponed due to political decision difficulties in Germany and the resulting halt of the entire development. The most recent development has been the successful bid for the development of the Airport of Riga, Latvia.

Overall, these projects have shown a successful development so far. Following the establishment of a new CEO at the Airport of Malta, who is a Vienna Airport expert, additional technical service contracts could be closed with the airport exceeding the existing management contract. Furthermore, influence on the investment's development is being guaranteed by means of the permanent presence in the airport's organization. Based on advantageous contractual agreements the engagement in Istanbul exceeded the original expectations as well. The consortium won the contract due to the shortest offered total running time, with a local construction company carrying the largest part of the financial burden. Since this partner was able to complete the terminal construction in a far shorter than scheduled timeframe, Vienna Airport was able to generate higher than expected revenues due to the longer utilization period of the new infrastructure. Additionally, this engagement led to new projects for the Austrian airport.

---

[19] The financial investment in every tendering process for any bidder can reach up to 1 million Euros, resulting from the significant efforts in the fields of market research, engineering, planning and financing structuring necessary in order to prepare the expected tender documents.

[20] Vienna Airport holds 57.1% of the Malta Mediterranean Link consortium, which again holds 40% of the airport of Malta. The Istanbul project was a Build – Operate – Transfer (BOT) project in which the partners had to construct a new terminal, had the license to operate for a certain period of time before transferring it to the state. This engagement has been terminated already.

In general, the principal decision criterion for any international project for Vienna Airport was the profitability of the engagement. As a measure of risk reduction, no other criterion than financial results was used in the decision for any internationalization project the company participated in. Motives like establishing a certain image or completing the own business portfolio have not been decision factors for the internationalization strategy. Moreover, a cap of 200 million Euros was set for any financial involvement. This also limited the potential currency exchange and profit repatriation risks.

Besides this quantitative criterion, numerous qualitative decision factors turned out to be crucial for the project selection. A clear focus is set on the parameters used in the context of the partner selection process. Congruence of the partners' targets and complementary core competencies of the consortium members are regarded to be of utmost importance for the project success and for fulfilling the customer's expectations. Thus, the cooperation with construction companies as in the case of Istanbul are strict exceptions, since conflicts are difficult to avoid between the construction company's short-term interests in providing new infrastructure and the airport operator's long-term interest to profitably operate the facilities.

The benefits of a thoroughly completed partner selection are supported by a clear separation and delegation of tasks between the partners and an appropriate, reliable contractual framework. For this reason, for Vienna airport it is important to limit investments in regions and cultural surroundings in which the same understanding of the contracts can be expected and which provide the necessary political and economic stability. Thus, a long-term fruitful and successful cooperation is currently only seen in European environments, leading to a limitation of all international activities to the home continent.

From an organizational point of view, the relatively long history of international activities within Vienna Airport is reflected in a well settled organizational structure. From a strategic perspective, including mainly the acquisition of new projects, a central unit for participations is in charge of all international activities. After the respective contracts have been closed and the project reaches an operational phase, the responsibility is transferred to a dedicated subsidiary which takes care of all international participations on an operational level.

3.3.3 YVR Airport Service Ltd., Vancouver, Canada

Contrary to the two companies analyzed before, YVR Airport Services Ltd. (YVRAS) is not an airport operator, but the subsidiary of the state-owned operator of Vancouver Airport in Canada, Vancouver International Airport

Authority. In 1994, this authority established a holding company, Vancouver Airport Enterprises Ltd. (VAEL). At the same time, VAEL set up YVRAS as a subsidiary dedicated to market the organization's core competencies in the fields of operations and consulting. Today, YVRAS is a joint venture of VAEL and the international investment company CDC Capital Partners. Under this constellation, CDC Capital Partners provides the capital needed for the international expansion, while the airport contributes the necessary airport expertise. The common goal is to achieve a leading position in the field of international airport management.

Currently YVRAS is the only internationally active airport company in North America, being engaged in operations of over a dozen airports in various countries throughout the Americas. These engagements include both financial participations as well as management contracts, including all kinds of services in the fields of airport management and operation.

Since 1997, YVRAS has held 10% of a consortium including a Spanish construction company and two local partners running a BOT project to develop the airport of Santiago de Chile. Another participation, in which YVRAS holds 7.75% of a consortium with a European construction company and several local partners, won the concession to operate and expand six airports in the Dominican Republic. Additional projects include management contracts at Sangster Airport, Jamaica, and at the airport of the Turks and Caicos Islands. Within Canada, YVRAS is engaged in the development of the airport of Hamilton and Moncton and of several smaller airports in British Columbia.

Vancouver Airport started its internationalization strategy in the mid-1990s as a response to both the upcoming privatization trend in the airport industry and the move of the Canadian Government to sell off operations of their airports to private operating companies while keeping the ownership of the airports themselves. By this time, YVRAS was first founded to sell the specific expertise of Vancouver Airport to external clients. Since YVRAS did not have sufficient financial strength without a strategic partner, the partnership with CDC Capital Partners opened up the way towards developing a large scale international strategy by guaranteeing the mandatory funding for selling the airport's expertise in larger scale activities. After entering into this partnership, obtaining a critical mass of projects to efficiently and effectively utilize these resources was one of the core drivers of YVRAS' development strategy. International projects are regarded as part of a diversified product portfolio which allows the airport to both reduce its overall business risk and to increase its own value. Thus, the selection of any project is strictly based on its profit potentials.

The large variety of countries in which YVRAS is conducting projects does not only lead to a well diversified, risk reduced portfolio. It also creates a multitude of difficulties resulting from the various cultural and business environments the company has to deal with. Thus, planning, managing and staffing the projects undertaken has turned out to be a remarkable challenge, since the key staff used in any project do not only need their project relevant fields of airport expertise, but also language skills and cultural experience. Even these capabilities cannot eliminate what has been identified as the most critical source of risks in YVRAS' projects, which are uncertainties in the contract negotiations and closure. In particular a long period between being awarded a project and the final contracting can cause severe difficulties for any investor who cannot rely on an extensive financial reserve. YVRAS has experienced these effects in several cases in which over years the project financing remained unclear due to consortium internal difficulties – as was the case in their project in Jamaica and the Dominican Republic – and in their bid for Sharm El Sheikh Airport in Egypt. In the latter case a consortium including YVRAS was awarded the concession to develop the airport, but the Egyptian Government withdrew the project later on due to the upcoming crisis in the region – a move leading to significant financial losses for the successful bidders.

Besides these contractual and political risks, cultural problems are regarded as the most crucial aspect of any international project. On the one hand these comprise the internal structure in setting up the project consortium. Experience has shown that certain tasks need to be left with certain consortium members, regardless of their share in the investment. The case of Santiago de Chile has impressively proved this: in this project, the construction company was in charge of operating the airport, even though YVRAS had the mandatory expertise for successfully filling this position. The result was for example a threefold replacement of the CEO within a 5-year period. On the other hand, the different ways of doing business, especially in Latin America, cannot be underestimated and require special experience and know how.

In general the current international experience of YVRAS led to a certain risk aversion in the selection of new projects. From an organizational point of view, a matrix structure linking top management members who are in charge of their respective projects with specialists within the organization was intended to guarantee thoroughly selected and controlled engagements. Within these, more and more a contractual framework assigning all management and steering tasks to YVRAS and clearly defining the project's timeline and duration was sought to reduce the risk of failures caused by uncertainties in the project design. This might also be achieved by accepting majority participations in the future and by focusing on secondary sales, rather than BOT projects.

3.3.4 Unique Zurich Airport, Switzerland

Unique Zurich Airport is the operator of Zurich Airport, by far the largest airport in Switzerland and thus one of the larger central European hubs. Being in a financially very advantageous and strong shape, in 1999 45% of the company was sold in an initial public offering. This reduced the shares of the Local Government of Zurich to 49% and handed over large shares to Swiss banks and financial institutions. Nevertheless, the airport's success has always been highly dependent on the fate of the Swiss flag carrier and with the bankruptcy of Swissair in late 2001 and the ongoing difficulties of its successor Swiss Air Lines, the airport's situation has dramatically changed as well.

During the 1990s, backed by the vast financial and personnel resources, Zurich Airport started to target an international expansion in fierce competition with the other major airport operators, following the intention to globally market its expertise in the fields of airport management, marketing and environmentally friendly management. During that time, any kind of international activity was considered, including very capital intensive engagements. Due to the crises over the past years, this focus has completely changed towards a preference for consulting and management contracts which do not imply a financial involvement – or at least keep it at a very low level.

The first successful internationalization project was realized in a joint venture with a Chilean company in which Unique Zurich Airport is the majority shareholder. This consortium runs three regional airports in Chile. Due to the different structures of these airports and the low investment, this project has shown a positive development for the Swiss investor over the years, despite the economic turbulences South America saw during that time. The second major internationalization project for Unique Zurich Airport is the concession for the development and operation of a new airport for Bangalore, India. This was awarded to a consortium in which the airport cooperates with Siemens, Indian construction group Larsen & Tubro and Indian local authorities. Recently, with the airport of Isla de Margarita, Venezuela, another 20 year management contract was given to Unique Zurich Airport.

The experience made with the realized projects is as diverse as the economic and cultural environments they imply. The Chilean project, currently being the only international engagement of Unique Zurich Airport which is both operating (Bangalore has not yet been opened) and economically successful, benefited significantly from a good partner selection. The regional company being part of the consortium has proven to provide sound knowledge of the local market and its mechanisms and contributed the network which was needed to successfully close the crucial contracts for the project. Unique Zurich Airport thus concentrated on providing its reputation as airport operator, expertise and the project funding. This combination, based on focusing each partner's engagement on its respective core competencies, created the basis for a successful project realization.

Contrary to this success story, the case of Bangalore clearly demonstrated the difficulties and uncertainties of internationalization projects. Having learnt from the experience in Chile, the partner selection for the consortium has again resulted in significant advantages for the project, leading to an improved political and legislative basis for the project. Nonetheless, the project realization has severely suffered from too ambitious expectations on the air traffic development at the airport and from political problems. The latter mainly results from permanently growing demands and requirements by the Indian Central Government, which brought the project close to its termination. This again has shown the problems incurred in the contractual environment and the political uncertainties borne in projects in different cultural contexts.

The lessons learnt from these projects have led to a change in the project selection policy. In the 1990s, when Unique Zurich Airport started its international engagements, risk diversification and the exploration of growth opportunities exceeding the home airport were only two of the motives to go global. Using the international engagement as a marketing tool in the upcoming privatization, along with the wish to keep up with competing airports in the field of international activities and personal interests of management members were all driving forces in the general decision to participate in international projects and in the selection of specific tenders.

Following the difficulties in the realized projects and forced by Unique's own economic situation, the strategy shifted towards a very selective, proactive international expansion. Due to the limited availability of investment opportunities, a dedicated portfolio strategy was neglected. At the same time, the available financial and personnel resources were needed to a larger extent for the development of the home airport, limiting the budget for international projects. Thus, Unique Zurich Airport has turned towards an opportunity driven strategy, focusing on profitable projects which preferably do not imply a tendering process but result from informal contacts. This generally leads to a more efficient and more successful project realization.

The clear profit orientation of Unique Zurich Airport is also reflected in the organizational positioning of the international activities. These are completely planned, controlled and steered by a dedicated unit which directly reports to the company's CFO.

The key findings of our case studies are summarized in the following table.

	Fraport AG	**Flughafen Wien AG**
Strategy	- strategic change - shareholders' influence - risk aversion - focus on home airport	- private economy management approach - strict customer orientation - risk aversion
Competencies and Capabilities	- operational and handling know-how - ground handling know-how - financial resources	- operational and handling know-how - marketing - financial resources
Motives	- creating additional value - competition - privatization trends - imitation - capacity shortages	- creating additional value - growth strategy - privatization trends - financial resources
Concepts / Project type	- value creation and security orientation - management contracts - BOT projects - ground handling projects	- concept of minority shares - management contracts - project selection based on RoI - strategic fit of partners - optimal contract design

	YVR Airport Services Ltd.	Unique Zurich Airport
Organizational structure	- restructuring from segment to strategy orientation - establishing central body for internationalization	- phase 1: strategic planning - phase 2: operational implementation
Potential problems	- lack of strategic know-how - political, cultural problems - international failures	- classical internationalization problems - many unsuccessful project tenders
	YVR Airport Services Ltd.	**Unique Zurich Airport**
Strategy	- no stock quotation - risk aversion - critical mass through portfolio diversification	- strategic change - no further projects - planned (besides running projects) - risk aversion and internal focus
Competencies and Capabilities	- operational and handling know-how - operational management - financial resources	- operational and handling know-how - quality management - protection of environment
Motives	- value creation - national factors - privatization trends	- marketing aspects - value creation - personal preferences - risk diversification - privatization trends

Concepts / Project type	- management contracts - portfolio diversification - share holding - secondary sales	- management and consulting contracts - opportunity driven - shareholding very restricted - cooperation with local partners - utilization of informal contacts
Organizational structure	- matrix: strategy and operations experts in Vancouver and at project location	- strategic internationalization department, reporting to CFO
Potential problems	- duration and resource utilization - financing / funding - culture - personnel utilization	- severe internal problems (development of home base carrier) - insufficient private economy management style - limited experience in international airport development

Table 16-1: Comparison of Empirical Case Study Results [21]

3.4 Discussion

The common expectations on the future development of the airport industry in the late 1990s were summarized by Doganis, who predicted a far-reaching privatization of airports worldwide in line with the establishment of only a limited number of "global mega airport companies".[22] This consolidation trend has not yet materialized, mainly due to the incompatible interests of sellers and

---

[21] Source: Lufthansa Consulting (2003).
[22] See Doganis (1999), p.112.

investors, international restrictions and the limited influence and interest in airports by the private sector.[23, 24]

Nevertheless, several other developments and trends have influenced the international positioning strategies of airport operators in the past and will gain more importance in the future, as the case studies have shown. Within these, especially the increasing focus on value creation as a supreme business philosophy, the growing influence of shareholders, the growing importance of customer orientation in the management and increasingly intensive competition between investors play vital roles in the changing market environment.

Due to individual and unique project characteristics and environments, no single solution for the design of international airport projects can be identified as the best or most frequently used cooperation or shareholding model. However, we could identify some common features across our case studies. The utilization of the market for trade or secondary sales, the private market for airport stakes is increasing, reflecting the growing preference for risk-limiting, safer forms of internationalization as can be seen in YVRAS' strategy.[25] Furthermore, airport operators increasingly prefer shorter-term projects, often in line with the attempt to keep equity involvement at a minimum.[26] These trends clearly mirror a changing perspective on considering risk, a project's influence on the overall business portfolio and the project dynamics as discussed above. Thus, expectations for major international investment projects are currently rather low.

Several of these preferences are well reflected in the analyzed case studies. The external motives like the general privatization trends in the airport industry, specific national and political influences and the competitive pressure from other

---

[23] This limited interest by private investors is partly due to the extremely high acquisition prices for airports in the late 1990s, which have proven to be very disadvantageous especially following the economic and industry downturn at the beginning of the new century.

[24] So far, a consolidation within the industry in the form of closer cooperation between airports, e.g. by establishing alliances, has not been observed to a significant extent. Bennett discusses the advantages and disadvantages of the consolidation of airport companies in more detail. See Bennett (1999), but also Pal & Weil, chapter 17 in this volume.

[25] Numerous examples for transactions in the trade sale market, e.g. the sale of several airports in the United Kingdom in 2000, including Bristol, East Midlands and Bournemouth, can be found in Graham (2001) or Bentley (2002).

[26] This target is often realized through consulting or management contracts, as can be seen in the example of Vienna Airport's engagement at Ciudad Real Airport and their change from a financial shareholding strategy towards a consulting and management contract to leverage their financial involvement.

players in the market have become drivers for international projects of airport operators as well as the internal factors like growth and capacity restrictions at the home airport, portfolio diversification strategies and the target of creating additional value or increasing rates of return. The ownership of Hahn Airport by German airport operator Fraport is a good example for achieving these goals. This engagement allows Fraport to utilize additional capacities especially for time-consuming air cargo flights by handling selected cargo airlines at Hahn Airport, to take part in the fast growing low cost market by serving the major European low cost carrier Ryanair and thus to successfully serve markets which would be impossible in the highly congested main airport Frankfurt.

Other goals could be observed only to a minimal extent. One of these targets which shows only negligible empirical evidence is the achievement of synergies through international activities.[27] This mainly results from the uniqueness of the single projects and participations, leading to the inability to create economies of scale or scope to a mentionable degree.

The intention to create additional value through international projects has not always been realized, due to the high number of external influential factors impacting the project development and success. Thus, the very optimistic judgment of airport internationalization projects as commonly viewed in the late 1990s should be revised. This re-evaluation is also reflected in the increasingly critical perspective of competition forces impacting single airports.[28]

While some researchers postulate that a single company, namely the airport operator, should be capable of providing all the services along the value chain of an external airport development project,[29] this is almost impossible to realize. Thus, the cooperation of carefully selected companies is advantageous for a consortium or a joint venture, within which the partners best complement each other in fulfilling the functional requirements. In forming this cooperation the main focus is to optimally combine the complementary core competencies of the partners to guarantee a higher chance of success in both the preparation and implementation of the project than would be the case with the realization through

---

[27] Nevertheless, this target is one of the main drivers for YVRAS, as shown in the case study.

[28] Meeder has analyzed the current competitive situation in the airport industry using Michael Porter's Five Forces scheme. This analysis points to the increased competition between the major airports, the growing competition by other modes of transportation, the negotiation power of the airlines and the increased importance of regional airports for the overall market. See: Meeder (1999).

[29] For example, this postulation can be found in the Mercer study on Global Airport Management. See: Schneiderbauer/Feldman (1998).

a single company.³⁰ This approach allows the airport operator to focus and restrict their own services to specific capabilities³¹ while certain aspects of the overall management of the external airport are taken over by a partner company.

In the case studies analyzed, we found evidence for the importance of value based management concepts as well as for the influence of the shareholders on airport managements' decisions on internationalization projects. This is exemplified by the ongoing move towards a strict profit orientation in any international engagement, which has replaced the previous drivers of sole "me too" strategies lacking sufficient economic reasoning. The specific influence of the stock quotation and of the shareholders' preferences has been observed in various cases as well.³² Additional factors with significant influence on the business performance can be seen in the adaptation of a real private economy behavior in the aftermath of the operator's stock exchange listing. After some time of operating as "privatized authorities", true private enterprises started to emerge. Also the know-how gained in the own privatization process has proven to be useful in the internationalization projects. However, there are only very few cases in which a significant correlation or interdependence between the stock prices of an airport operator and its internationalization projects can be observed. The economic situation at the home base still has the highest impact on the stock prices and thus on the shareholders' decisions.³³

An empirical analysis of typical problems, following the approach by Meeder provides a theoretical basis for the general strategies for airport internationalization projects.³⁴ He distinguishes the dimensions of political, economic and technical risks. While the case studies indicate that technical problems only played a minor role in the project implementation – mainly due to adequate preparation and sophisticated, risk-limiting planning processes – economic and political risks cannot be neglected. Especially the latter have

---

[30] Unique Zurich Airport's successful project in Chile proves this.

[31] The core competencies airport operators are likely to provide in such a constellation include in particular operations, general management, handling issues as well as the transfer of airport specific know-how and consulting. Again, Vienna Airport's involvement in the Airport of Ciudad Real in Spain is a good example of this phenomenon.

[32] A very clear example is the shareholders' influence on the strategic development of BAA plc., who were forced by their shareholders to restructure their international activities due to losses.

[33] The most important determinants in this context include investment projects at the home airport as well as development strategies of the main carriers or political decisions impacting the airport.

[34] See: Meeder (1999), p. 190.

shown to be significantly important for the project's success. This can include potential problems such as local laws, governmental decisions or changes in the general political environment at the project location.[35]

While most previous projects have been located in very different geographical regions, all airport operators analyzed within the study showed unanimous expectations regarding the future potentials for successful airport internationalization projects. From a cultural and political perspective projects in Europe have been rated as especially preferable, while the most interesting growth potentials are expected from the Asian and especially the Chinese market. Projects in South America are generally expected to result in successful participations only in a mid-term perspective.

In the case studies, especially imitation strategies and strategies to create risk-reducing portfolios or to generate additional value for the own company have turned out to be the core driving factors for airport internationalization projects. Following these goals, the strategies chosen differ between the various companies analyzed. Regarding other important aspects on successful internationalization projects, all enterprises in the case study showed similar or even equal judgments – as can be seen best in the expectations regarding the future core market for airport internationalization projects or in the equal importance of setting up powerful, unique consortia for every single project.

3.5 Success Factors in Airport Internationalization Projects

Every international activity has to be understood as a unique project, embedded in an individual environment under specific conditions and with specific participating partners involved. Thus, it is difficult to define general success factors. Nevertheless, from the broad variety of factors a series of general success factors can be derived for the realization of international projects. These aspects are not only crucial for a successful preparation and implementation of the intended activities but also show common characteristics which allow for their general definition, abstract from a specific company context, and thus may be transferred to other situations.

The factors identified can be divided into two separate groups: company specific success factors related to the internationally active company itself and project specific success factors dealing with the project's prerequisites and the

---

[35] See the example of Unique Zurich Airport's involvement at the new International Bangalore Airport.

requirements for the companies involved. Both groups contain aspects which are mandatory for the successful implementation of an internationalization project and in this are often highly interdependent and may influence and necessitate each other.

Company specific success factors	Project specific success factors
- capital resources	- potential for value creation

- personnel resources & know-how
  ▪ operational expertise
  ▪ international, strategic know-how
  ▪ efficient organizational structure
	- contractual framework
	  ▪ long-term security
	  ▪ monitoring and influencing ability
	  ▪ customer orientation

- management approach with focus on value creation	- choice of partners
	  ▪ local cooperation partners
	  ▪ consortium management

- international experience (risk aversion)
  ▪ direct (own) experience
  ▪ indirect (others') experience
	- capability for adaptation and dynamic adjustments

Table 16-2: Success Factors for Airport Internationalization Projects

As can be seen from both the characteristics of the investing companies – all being major airport operators with significant business volumes in their home operations – as well as the characteristics of the projects realized, the fundamental prerequisite for every internationally active airport company is a sufficient capital base. This is not only mandatory for financing the project investments and shares, but also for funding the tendering process. In most projects, the partners opening their airport for international participation do not only expect consulting services or the transfer of know-how, but are in particular looking for the investors' financial support. Due to the long duration of the projects,[36] a long-term amortization period has to be considered as well as the

---

[36] Typically the duration of airport internationalization projects can be considered as ranging from at least 5–10 years. Most projects tend to show far longer durations, reaching up to 20 or 30 years especially if significant investment is involved.

underlying long-term resource utilization.[37] In addition to the seller's expectations, in many cases a capital participation in the company is even a formal prerequisite for entering into a project and at the same time secures a certain degree of influence and power in the partnership. Thus, depending on the investing company's strategy, a certain critical mass of projects is needed to guarantee an effective and efficient utilization of the available personnel resources.

These personnel resources and the related know-how should be embedded in an efficient organizational structure characterized by good internal communication, a coherent set-up and an effective performance. In this context, operational as well as strategic expertise is absolutely mandatory.[38] First, this includes airport-specific know-how which needs to be available in the form of experience with the core competencies of airport operations.[39] Second, at a higher hierarchical level, the tasks of acquisition, coordination and evaluation of the economic feasibility of all activities are needed to allow for a focused bundling of the operational core competencies. Thus, highly specialized strategic expertise is compulsory in the fields of international contract negotiations, legislative problems, terms of concession agreements and political matters.

Another basic prerequisite for an airport company engaged internationally is a private, market and profit oriented management, without which an economically efficient resource allocation and value creating realization of the project is difficult to achieve. A further factor representing an elementary advantage is past experience in international activities, since the information and knowledge gained from this and the resulting progress on the learning curve can be of significant value for new projects and allows for a dynamic adaptation of the

---

[37] The importance of the strong capital base for an investor is also reflected in the trend towards targeting more short-term projects. These do not only allow for a faster return on investment, but also reduce the overall capital base to be employed for the internationalization strategy, since in total less financial resources have to be employed at the same time.

[38] In both of these knowledge sectors the airport operator's reputation plays a major role, since this is a crucial decision factor in the partner or supplier selection especially in the field of internationalization of airport services. Consequently, reputation has to be regarded as one of the most important success factors in the area of airport internationalization.

[39] Within the category of operational experience and know-how there is a broad variety of single activities. For example, the airports analyzed in this study consider their specific strengths in such different fields as ground handling (Fraport AG), marketing (Vienna), management (YVRAS) or quality and environmental management (Unique Zurich), which they all use in their international projects.

project strategy.[40] In this context direct experience, gained by the respective airport company itself, can be distinguished from indirect experience, representing the experience made by other companies. Both kinds of experience show their influence on the airport's further international activities.[41]

The second group of success factors summarizes a series of characteristics and prerequisites of a specific project and environment. These are of elementary importance for the airport company, but can actively be influenced by the project management. A first mandatory prerequisite of any international project can be seen in its potential for value creation – a prerequisite which has not always been satisfied in the time when internationalization was more about image competition than about financially reasonable project acquisition. Every international activity should result in a higher total return on investment for the investing airport operator, thus justifying the typically risky and uncertain international activity. This success factor – the long-term security of the capital invested – should be guaranteed to the best achievable degree through a solid and detailed contractual framework. The experience from recent projects in particular has shown that besides the potential for an attractive value creation, a guaranteed, sufficient degree of capital security and overall transparency are mandatory prerequisites for international activities. In this contractual framework country-specific influences like the political situation and the developments in the capital markets play a significant role and allow an airport operator to achieve long-term, sustainable results. In this context, the power gained and opportunities to control and influence the project development with the focus on client needs have an increasing importance for developing and implementing a targeted organization and for successfully realizing the project. It is the client who decides on the choice of the investor and thus on the awarding of project contracts. Consequently, the client should be in the focus of any project realization.

Additional project related success factors are closely linked with these aspects. Choosing the consortium or joint venture partners should be driven by the focus on similar interests between the partners, compatible targets and underlying intentions as well as complementary responsibilities. In this it is of special importance to reduce or even eliminate the cultural distance and potential

---

[40] The majority of the analyzed projects have been characterized by negative experience, thus leading to a more restrictive, risk-averse project selection for subsequent activities.

[41] The importance of experience for the airport operators in the case study can also be seen in the category "potential problems" in table 16–2 above. For all companies analyzed, a development along the learning curve has been highly influential for their further activities.

regional uncertainties in the project's business environment by cooperating with a local partner. Thus, the process of choosing the right partner is a crucial success factor influenced by the airport operator.

The underlying success factors have to be regarded due to their crucial importance not only in the preparatory and start-up phases of an international activity, but during the subsequent implementation and the ongoing project work they also have to remain in the investor's focus. Empirical evidence has shown that a multitude of changes can occur throughout an international engagement which can have a lasting influence on the situation beginning as early as the start-up phase, leading to the engaged airport company's need to be prepared. In this context, the adaptation capabilities and the dynamic adjustments of an airport operator's operational and strategic plans under changing conditions can be regarded as success factors as well.

If the discussed factors are being considered in an internationalization project, faults in the preparation, the structuring and the coordination of the project – which have led to negative results in the past – can be avoided.[42]

4 Conclusion and Closing Remarks

The international activities of airport operators are a complex issue, influenced by a multitude of factors to be regarded in the preparation and implementation of an international project. Based on four explorative case studies, this paper provides an analysis of the internationalization of airports.

Internationalization strategies of airports have changed significantly over the past years. Following the "gold rush" of the 1990s in which image and competition between global airport operators had more relevance in the project selection process than the economic viability of the engagement, the aviation industry's worsening economic situation has led to a more conservative attitude towards international activities. Return on investment, value creation, risk aversion and avoiding contractual uncertainties have returned as most important decision criteria in the field of international airport projects.

Despite these general developments, the case studies have also shown the general success factors for internationalization projects in the airport business. The importance of know how, of efficient and effective partner and project selection

---

[42] See also Götsch & Albers' chapter 11 in this volume.

processes, of experience in different cultural and political environments and the need for extensive capital resources cannot be underestimated. All internationally active airport operators had to experience these aspects in long and expensive project development attempts.

Nevertheless, airport internationalization will remain a vital and important issue for the industry. However, the development of the "global mega airport companies" seems to be taking as much time as the long-awaited consolidation of the European airline industry.

Acknowledgements

This chapter is based on the findings gathered in a study on the internationalization of airport companies performed by Lufthansa Consulting GmbH in cooperation with the University of Cologne in 2003.

5 References

ALBERS, S.; KOCH, B.; RUFF, C. (2005): Strategic alliances between airlines and airports – theoretical assessment and practical evidence. In: *Journal of Air Transport Management*, 11(2), pp. 49-58.

BENNETT, P. (1999): Taking on the World. In: *Airline Business*, 15(12), December 1999, pp. 45–48.

BENTLEY, D.J. (2002): *World Airport Privatization Study: The Financing and Development of Airports at the Turn of the 21st Century*. London: Euromoney Institutional Investor PLC.

DOGANIS, R. (2001): *The Airline Business in the Twenty-first Century*. London, New York: Routledge.

DOGANIS, R. (1999): The Future Shape of the Airline and Airport Industry. In: Pfähler, W., Niemeier, H-M., Mayer, O.G. (Eds.): *Airports and Air Traffic – Regulation, Privatisation and Competition*. Frankfurt a.M.: Lang.

DOGANIS, R. (1992): *The Airport Business*. London, New York: Routledge.

FRANKFURTER ALLGEMEINE ZEITUNG (2003): *Privatisierung der Berliner Flughäfen entgültig gescheitert*. 23 Mai 2003. Frankfurt.

FREATHY, P.; O'CONNELL, F. (1998): *European Airport Retailing: Growth Strategies for the New Millennium*. Houndmills, London: Macmillan.

GRAHAM, A. (2001): *Managing Airports – An International Perspective*. Oxford: Butterworth-Heinemann.

JUAN, E.J. (1995): *Airport Infrastructure: The Emerging Role of the Private Sector*. CFS Discussion Paper Series Number 115, Washington: World Bank.

MEEDER, S. (1999): *Public Private Partnerships zur Finanzierung von Flughafenprojekten in Europa*. Frankfurt a.M. et al.: Lang.

PERLITZ, M. (1995): *Internationales Management*. 3rd edition, Stuttgart: Lucius & Lucius.

PORTER, M.E. (1990): *The Competitive Advantage of Nations*. London, Basingstoke: The Macmillan Press Ltd.

SCHNEIDERBAUER, D.; FELDMANN, D. (1998): *Global Airport Management: Strategic Challenges in an Emerging Industry*. Munich, Paris: Mercer Management Consulting.

17

EVALUATING AIRPORT COOPERATION AND ACQUISITION STRATEGIES

ANDREA PAL AND WOLFGANG WEIL

1	History of Cooperation in the Aviation Industry	410
2	Different Approaches to Cooperation Between Airport Operators	414
3	Investors Expectations Related to Airport Privatization Processes	421
4	Future Developments in Aviation Alliances	423
5	References	425

Summary:
As the airport business is becoming more and more multifaceted, business strategies are being put to the test. The current trend in airport management is to rely on commercial operations to contribute an increasing share to airport revenues. But such an approach has to be balanced with the basic needs of the airport's customers, the airlines and passengers. With airline alliances spreading across the world, what is the airport industries' answer to that challenge? Airport cooperation has quite a short history. But will airport cooperations and alliances prevail and be the right answer?

1 History of Cooperation in the Aviation Industry

1.1 Airline Mergers and Cooperation

Looking at the aviation market, cooperation between airlines has quite a long history. The US Airline Deregulation Act of 1978 resulted in a major restructuring of the airline industry in America in the early 1980s. A nationwide process of concentration through mergers and cooperation followed. The major players tried to establish a critical mass allowing for improved operational efficiencies and economies of scale.

The European carriers had difficulties following this process of concentration because they were limited to their relatively small domestic markets and because cross-border mergers and cooperation faced regulatory hurdles (e.g., the need for a domestic majority shareholder).

Nevertheless, the largest domestic markets in Europe, namely the UK, Germany and France, experienced some kind of concentration process during the past decade. The large home-based carriers were able to defend their territory by acquiring their smaller rivals, which gave them strength to act beyond their domestic markets.

1.1.1 Code-share Services

Besides so-called interline agreements, where airlines agree jointly to price a passenger's itinerary in case he is using more than one carrier during his journey, code-share services were the most visible form of cooperation between airlines. Code-share services are mainly marketing-driven. They enable a carrier either to offer more destinations to his customers and/or to increase the frequency between certain airports.

1.1.2 Airline Alliances

In the second half of the 1990s, the level of cooperation between international airlines changed dramatically. A new era began when the Star Alliance was established. Alliance partners agreed on extensive code-sharing between all of the members. Flight schedules were optimized to increase connectivity at the alliance's major hub airports. Alliances started to introduce the landlord concept, giving the major alliance partner at a specific airport increased management and operational opportunities: e.g., sole provider of certain facilities and services such as lounges and check-in areas. What seems to be a logical step to increase efficiency by realizing economies of scale means a higher degree of concentration for suppliers of the airline industry. Alliances negotiate tenders increasingly on an alliance basis rather than on an individual carrier basis. The latest development in this evolutionary process is the joint tender for a regional airplane by several members of the Star Alliance.

Today, three major airline alliances have become established around the world. They seem to work along similar business models, but a closer look shows that, even within one alliance, the level of cooperation among partners varies. They might be classified as a 'commercial alliance' or even as a 'strategic alliance' (see figure 17-2). However, alliances and other forms of airline cooperation only developed over the last two decades. There was no role model and, thus, alliances do not look alike.

1.2 Cooperation Between Airports

Airport cooperation has a far shorter history than alliances in the airline industry. They can be seen as a result of the intensified privatization efforts of governmental airport owners since the mid 1990s. Figure 17-1 gives an overview of these airport privatization projects. Governments at all levels have increasingly focused on the benefits privatizing public facilities and services can provide. While privatization is not always the most attractive option, few would argue that having the option is of great benefit, particularly in times of tight government budgets and taxpayer revolts.

In terms of alliances, the classical life cycle model applies (see Section 3.3). Airport cooperation and alliances are mainly in the start-up phase and/or alliance creation phase. Airline alliances are more mature and can be classified as being in the creation phase and/or alliance maintenance phase.

Due to their focus on serving a geographically determined market, airports emerge from a completely different historical background. Their main focus was to serve the demand rather than proactively developing markets and creating traffic.

Each airport operates within a unique framework, determined by factors like the regulatory system, ownership structure, local political situation, strength of the home-base carrier, captive market, catchment area, and attractiveness as a business and/or leisure destination. Furthermore, all of the world's airports differ in size and their airside and landside infrastructure. There are no identical twins among them. Recognizing these facts, it is obvious that looking for synergies between airports is a much more complex task than in other industries, for example, streamlining the production process of two automotive plants.

1995	1996	1997	1998	1999	2000	2001	2002
							Sydney
							Meilan
							2003
							Florence II
							2004ff.
			Argentina				Budapest
		Dusseldorf	Hanover			Turin	Cairo
		Bolivia	Wellington			Florence I	HongKong
		Bristol	Skavsta			Jamaica	Delhi
		Liverpool	Prestwick			East Midl.	Mumbai
London-		Sanford	NT Airports	Beijing	Lima	Oman	Chennai
City	Copen-	Perth	Hobart	Stewart	Hamburg	Newcastle	Calcutta
Vienna	hagen	Brisbane	Canberra	Costa Rica	Bristol	Bangalore	Shanghai
Bourne-	Belfast	Melbourne	Adelaide	Auckland	Plymouth	Hyderabad	Cyprus
mouth	Athens	Napels	Luton	Malaysia	Zurich	Malta	Milano
Cardiff		Birmingham	South Africa	Mexico I-III	Rome	Fraport	

Figure 17-1: Airport Privatization Projects 1995-2004

The baseline for successful cooperation between airports is the detailed understanding of each other's business processes and infrastructure requirements in a very special and unique environment. Despite all of these different conditions, there are many areas of potential cooperation for airports, such as planning and management of infrastructure and commercial operations.

1.3 Classification of Alliances

Airline Alliances can be classified in many ways. Depending on their extension they can be domestic, regional or global. Doganis classifies airline alliances into commercial alliances and strategic alliances (see figure 17-2), based on the type of agreement between alliance partners.[1]

---

[1] See Dognais (2001).

TYPE OF AGREEMENT	TYPE OF ALLIANCE	
Interline / Pro-rate	Commercial Alliance	Strategic Alliance
Mutual Ground Handling		
Frequent Flyer Programs		
Code-share		
Block Space		
Common Sales / Ticketing Outlets		
Schedule / Capacity Coordination		
Joint Engineering		
Joint Flights		
Franchising		
Common Branding		
Joint Cargo and Passenger Services Ventures		
Full Merger		

Figure 17-2: Alliance Categories[2]

The classification above is certainly not suited to distinguish alliances or cooperation modes between airports. Due to the fact that airports are seen as regional monopolies, they compete on a different level from airlines.

Airports, which join forces to achieve economies of scale on their supply side, can be classified as *shared-supply alliance*.

In *quasi-concentrated regional alliances*, airports within a specific region or country enter into an agreement to market their services jointly.

Complementary alliances describe airports joining forces to contribute their complementary strengths: e.g., aviation/non-aviation business skills or passenger/cargo management skills.

---

[2] Adapted from Doganis (2001), p. 66.

2 Different Approaches to Cooperation Between Airport Operators

The airport business is becoming more and more multifaceted. It is extending into real estate development and management, commercial and other ventures. These activities can be split into two main areas: the provision and management of airside/aeronautical facilities and services, e.g. runways, terminals, and the provision and management of so-called landside facilities and services e.g. retail management, parking facilities. Airside facilities by their nature are quite often considered monopolistic and are subject to a regulatory framework.

The current trend in airport management is to rely on commercial operations to contribute an increasing share to airport revenues. But such an approach has to be balanced with the basic needs of the airport's customers, the airlines and passengers. Shopping malls with aircraft parking stands attached are not the ultimate solution. Airports are a traffic center with complex procedures. These procedures should not be compromised by commercial activities. Best practice is to simultaneously develop aviation and non-aviation requirements in a hand-in-hand approach. Airports are becoming increasingly multi-modal intersections of aviation, train and car. From a user's perspective such intersections develop into Airport Cities with round-the-clock services and infrastructure.

2.1 Potential Areas for Cooperation

As alliances and cooperation between airports emerged mainly in the last decade, it was up to those spearheading this trend to analyze potential benefits. Some cooperation projects initially focused on the supply chain aspect. But airport purchasing is a very unique undertaking. Most of the capital expenditure of airport companies is related to construction of airside and landside facilities. Standardization is limited because most facilities are uniquely designed. Since a significant share of airports is still majority owned by governments, joint purchasing tenders are a complex issue, potentially outweighing price advantages.

With e-commerce developing, the idea of B2B e-commerce was adopted by some airports. The basic idea was to establish an online exchange, enabling member airports to buy goods more cheaply and easily. But little has happened since the launch of the idea about three years ago. The idea seems to be limited to operating cost rather than capital projects as mentioned above.

There are a number of potential benefits for airport networks and cooperation as depicted by Howe:[3]

1. Economies of scale in providing products and services to meet the needs of the users.
2. Achievement of a better operational environment for the airlines.
3. Enhancement of flights safety and provision of alternate airports during bad weather or emergencies.
4. Increased efficiency and cost-effectiveness.
5. Improvements in resource allocation due to overall investment planning.
6. Coordinated development of airport capacity.
7. Reduction of airport and airspace congestion through allocation of capacity and differentiated charges.
8. Best practices at any airport can quickly be adopted at other network airports.
9. Borrowing in capital markets at more favorable terms because of risk spreading.
10. Standardization of airport facilities at all network airports.

However, any two airport operators intending to cooperate have to analyze in-depth which of the listed benefits might apply for their unique situation and therefore can be realized.

So did two major European airport operators in 2000. The first formal cooperation between two airports that has been systematically developed besides international organizations such as the Airports Council International (ACI) or national organizations like the German Airports Association (ADV) is Pantares. Pantares was founded by Fraport AG (owner and operator of Frankfurt Airport) and Schiphol Group (owner and operator of Amsterdam Airport).

As the approach by Pantares was new, many parties became curious about how it works. At the practical level, the airports work on programs on how to maximize the delivery of services to the 100 million passengers they had in their alliance.

---

[3] Adapted from Howe (2000).

A number of task forces are dedicated to key projects. The main focus is on realizing advantages through joint procurement, joint development and implementation of IT solutions and a common system and database for e-business. Other aspects of this type of cooperation can be seen in the development of joint units for offering services like planning, engineering and facility management, or even ground handling.

Another promising area of cooperation between airports is an intensified focusing on customers, especially the passengers. In the e-commerce field, B2B services and offering of customized products for airport customers (airlines, passengers, forwarders etc.) can be developed. For airport operators it is very valuable to understand which expectations, customs and preferences passengers from other countries or cultural backgrounds have. Exchanging such experiences with other partner airports will result in improved customer services. Signage systems, announcements, target-group optimized retail outlets, and food and beverage areas are among the items addressed in such customer orientation.

The Pantares alliance also developed into joint airfreight activities. Schiphol Group and Fraport AG – through their joint subsidiary Pantares Tradeport Asia Ltd. – have combined their know-how and jointly own 37.5 percent of Tradeport Hong Kong. Other shareholders are two renowned Chinese companies. Tradeport Hong Kong, a new airfreight logistics center, opened in 2003 at Hong Kong International Airport (HKIA). This facility featuring advanced technology will serve primarily as a high-performance logistics hub for high-tech products being transported between China and Europe or China and America. Tradeport Hong Kong is a joint venture of European and Chinese companies owning and operating it.

In terms of size and function, the €50 million Tradeport Hong Kong complex is unique in the region. The individual levels feature special air conditioned and high-security (anti-theft) areas. Thus, Tradeport Hong Kong takes into account the high-tech industry's special requirements for logistics and security technology. Tradeport Hong Kong offers a wide range of value-added services such as packing, commissioning, quality control testing, assembly, and repair. Located at the cargo handling area only meters from the runway system at HKIA, Tradeport Hong Kong is the only on-airport logistics complex in Hong Kong. The facility is designed for rapid and effective processing of products, whereby time-to-market and volume-to-market factors are critical for the profitability of the product's life cycle.

One of the latest projects of the Pantares partners is the introduction of a biometric automatic border control system. First developed and introduced at Schiphol Airport the system is now being tested at Frankfurt Airport.

In the past operators of larger airports have developed unique skills. While some are more focused on planning and technical skills, others have concentrated on commercial development. However, all of them have quite unique knowledge about their regional markets. If these sets of skills are complementary a basis is given to join forces and offer this know-how in form of consultancy services to other airports. Fraport AG and the Shanghai Airport Group have responded to such a market opportunity by establishing the Shanghai Frankfurt Airport Consulting Services Co., Ltd. Using the unique strengths of both partners, plans call for marketing and providing aviation consulting services throughout the Chinese aviation market.

Summarizing, one can conclude that in the airport industry the focus is only on knowledge and information transfer. But this would not reflect the full picture. Unlike classical knowledge transfer, future cooperation between airports will be aiming at unleashing hidden growth potential. But how can the partners be sure that such concepts will be successful? The answer to this question might be risk sharing through equity participation.

2.2 Airport System Operators

There are a number of airport operators who manage several airports within a metropolitan area or nationwide. They were mainly established as authorities with the specific task of managing the airports of such regions.

Examples include BAA (originally founded as the British Airports Authority) which manages the three major airports in the Greater London area (Heathrow, Gatwick, Stansted), the three largest airports in Scotland, as well as Southampton on the south coast of England. The Port Authority of New York & New Jersey is responsible for managing the airports in the New York City area (JFK, La Guardia and Newark) as well as the seaports. The Airports of Thailand PCL (formerly the Airports Authority of Thailand) manages Thailand's international gateway, Bangkok International Airport, and four regional airports with international status.

Managing such a system of airports in close proximity offers a number of distinct advantages. From an organizational point of view, overhead costs can be reduced and economies of scale can be created. Some of these operators have started to standardize their construction projects, thus reducing costs for tendering, installation and construction projects.

A disadvantage of some of these airport systems – from an airline point of view – is the possibility of the system operator allocating traffic within the system.

But airlines have the tendency to regard traffic allocation as a limitation of free market entry.[4]

Establishing new airport systems in certain countries is difficult. Sometimes it is the trend toward deregulation that prevents one entity from having additional management influence at other airports within that country. Also the regional influence in the shareholder structure of many airports creates some sort of competition between potential partners.

The differing sizes of airports within one country might be seen as an argument that management methods and skills of a large hub airport are not useful for running smaller airfields. Fraport's experience does not support this argument at all. Through equity participation and/or management contracts, Fraport manages several German airports along with its 50 million passenger per year intercontinental hub at Frankfurt. These include the international regional airport of Hanover, as well as the smaller regional airport in Saarbruecken. Special concepts were applied by Fraport to develop the former US military airfield at Hahn into Germany's most successful conversion project and one of the leading bases in Europe for low cost carriers.

In-depth analyses of such airports' (hidden) potential leading to custom-tailored business development strategies are the key to a successful development. Fraport's German airport system demonstrates that airports of any size or with different market positions can be managed successfully under one roof.

2.3   Airport Operators Going Global – Fraport's Strategy

As mentioned above it is difficult or simply impossible for many airport operators to develop additional business within their domestic market. As the airport business is an international business, going global is not that much different from developing on your home turf. The basic rules are the same, but specific international features have to be anticipated.[5]

Airports going global generally act as private investors and should follow the same principles. A private investor is basically looking to get a return on the invested equity. An airport operator successfully investing in another airport is doing exactly the same thing.

---

[4] Especially the long lasting disputes between foreign airlines and the operators in London and Paris highlight the different interpretation of traffic allocation versus free market entry.

[5] See Koch & Buddes' chapter 16 in this volume.

But why do airport operators not invest in other industries and diversify their portfolio? The answer might be quite simple. Airport operators are not financial investors. If they engage themselves in equity participation, they stick to the business they can do best: i.e., airport management. This is the business they understand best and where they can add value to potential targets. History has shown that investments in other industries are problematic. It ties up resources, which the operator may very well need for his own expanding airport business. And, it is a difficult task to create the management skills required for a different industry.

Taking this into account, Fraport's strategy for potential investments is to evaluate the profit development potential of a target. Only if sufficient potential is identified, Fraport will proceed with the transaction. The main aspects Fraport considers are: additional revenue potentials from the aviation and non-aviation sectors, efficiency improvement, and asset investment management. The analyses of the target's market position and development potential gives a first indication for future traffic volumes and, therewith, for the expected revenues.

Although future growth in traffic will generate a significant increase in aeronautical or aviation revenues (e.g., landing fees), experience shows that the highest development potential for many targets lies in the area of non-aviation revenues. Quite often, this area is underdeveloped and shows the highest potential for further development.

Fraport applies a twofold strategy to get things moving for developing this key segment: using target-specific innovative terminal concepts, Fraport intends to create the right environment for passengers and visitors to spend more money in the airport facilities. Fraport also intensively works on increasing the airport's passenger numbers by analyzing the airport's current and potential flight schedule (timetable). Such potential depends on the airport's geographic location as well as on the market segment (hub airport, low cost carrier base, niche markets etc.).

Real estate development is another important area, but due to its less attractive image this is often an underestimated area for improving the target's non-aviation revenues. As many airports have underutilized land within their boundaries, optimizing land utilization and attracting new tenants can be a very profitable approach. Such approaches can even be realized at the operator's home base. Fraport has been successful with its strategy to convert and develop an area within its perimeter fence, which was returned by the US Air Force several years ago. Since the mid-1990s, the entire area has been developed as 'CargoCity South'. Meanwhile, an area of about 100ha has been developed completely, far ahead of anticipated forecasts.

Understanding and evaluating the operational processes and the implemented tools is the key to success and prerequisite for the development of concepts for efficiency enhancement. The results are higher quality and lower operational costs. This can be achieved as many concepts, routines and tools have already been developed and implemented in other airports operated by the Fraport group.

As an infrastructure provider, an airport has to invest significant amounts of cash in buildings, apron areas, taxiways and runways. Years of operation and steady traffic growth are needed until attractive ROE rates can be achieved. This fact requires an excellent management of the infrastructure investments. It means to "sweat" the assets already operated and increase capacity as much as possible by improving traffic flows and investing in new assets at the right time and at the right size.

An investor who is not familiar with the airport business cannot deliver this performance or is running into a high-risk investment by underestimating the complexity of the business model.

This leads back to the answer of the initial question: airports that are confident in their business have the necessary expertise to improve and develop other airport companies. If sufficient equity is available, there are no better investors for airport companies than other successful operating airport companies.

2.4 New Market Entries

During the last decade the airport market attracted a number of players who are new to the industry. Attracted by steady worldwide growth (until 2001) the aviation market seemed to guarantee good investment opportunities. The new market entries can be split into two main groups:

1. Financial investors, looking for interesting projects, started joining consortia and bidding for airport privatization projects around the world.

2. Construction companies or their affiliates joined the market, attracted by the large potential for construction contracts. Airport privatization projects in the form of BOT projects or concession agreements incorporating airport expansion programs are attractive for construction companies because they can deliver their basic construction services and products. They compensated for their lack of airport know-how, as described above, by having an airport operator join the group or consortium.

3 Investors Expectations Related to Airport Privatization Processes

Airport management was historically a public business because it was seen as transportation infrastructure. However, this view changed dramatically in the last decade.

The managing companies of Europe's two largest hub airports, BAA at London-Heathrow and Fraport at Frankfurt, went public. Many airport authorities around the world were "corporatized" in a first step. Airport companies were partially or wholly privatized.

At the beginning of this privatization process, the main players in the buyers team were the airports and the construction companies. Later new players stepped into the market, such as companies investing generally in transportation infrastructure or financial investors. In the meantime, local investors are playing a significant role because airports are strongly linked to the image and the national economy of the countries where they are located.

After 11 September 2001, the number of airport transactions slowed down noticeably. The main reason for this was that neither sellers nor the buyers had a feeling about how the industry would perform in the coming years. The crucial traffic forecasts could not deliver reliable parameters, thus making it impossible to calculate a fair price on either side. Buyers were afraid to pay too much and sellers were afraid of getting an inappropriate price for their shares.

After two years of uncertainty, it seems that the aviation industry is catching up again with a positive trend and even the future of several airlines is much more predictable. In this environment, an intensification of activities in the airport market can be expected.

	Option 1	Option 2	Option 3
Allocation of responsibilities			
- Ownership	State	State	Private sector
- Investment	State	Private sector	Private sector
- Management and Operation	Private sector	Private sector	Private sector
Common strategies for private participation	- Service concessions - Contracting-out - Management contracts - Multiple concessions	- Build-operate-transfer schemes - Long-term leases - Master concessions	- Wraparound additions - Trade sales - Build-own-operate schemes - Strategic buyouts (management-employee buyouts) - Capital markets

Table 17-1: Options for Private Sector Participation in Airports[6]

The process that the airport industry went through in the last decade can be defined as a learning curve. Many airport operators went into a frenzy as the new market opportunity of airport privatization projects started. One record price tag after another was placed on airport projects. Quite often it seemed difficult to justify the prices. But possibilities for comparison were limited. Then, after 11 September 2001, transactions became fewer and price building seemed to become more rational. This evolutionary process will certainly continue with

---

[6] Adapted from Juan (1996), p. 2.

both sides, sellers and buyers, learning from the past and applying more maturity to their strategies. Public sellers are thinking more and more of awarding concessions rather than selling shares. Buyers are becoming more cautious and selective in the target identification and evaluation process.

4 Future Developments in Aviation Alliances

It is certainly much easier to predict the future of airline alliances than that of airport alliances. With the merger of Air France and KLM, the 'Wings' grouping of airlines will likely be assimilated by the Sky Team alliance. The three major airline alliances Sky Team, Star Alliance and oneworld are now well established. Nearly all major international players are part of one alliance or another. The only major white spot left is China with its three airline groups. It seems logical that, sooner or later, each Chinese group will join one of the global airline alliances. Looking at existing code-share agreements, which have been a logical first step in airline cooperation, it is our best estimate that the Air China group will join Star Alliance, while China Eastern will likely cooperate with oneworld and China Southern with Sky Team.

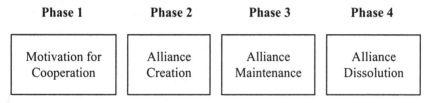

Figure 17-3: Strategic Alliance Life Cycle[7]

Certainly, airline alliances are now in the alliance maintenance phase (see Figure 17-3 above). Along with this "integration" of the Chinese groups, one can expect to see an ongoing process of regional carriers (in attractive markets) joining one of the global alliances.

In our opinion, the assumptions above seem logical and very likely. But history has shown that alliances and membership therein are not sanctioned forever. The former cooperation between Singapore Airlines, Swissair and Delta Air Lines

---

[7] See Sparkling & Cook (2000), pp. 69-94 and Scheuering (2002), pp. 26-28.

fell apart, with the result that today these carriers (or their successors) are spread over the three global alliances. Also the consolidation process in Europe – which has just started with the Air France/KLM merger – and the ongoing process of US carriers operating under bankruptcy protection law might stir up the alliance groups in the future.

But what is the essence of airport alliances? There are about a dozen global airport groupings, with some of them limiting their activities to their domestic market. It can be expected that those of the international groupings which have reached a certain critical mass will further develop and strengthen their market position. Other groups were quite active at the beginning of the global privatization process, but have not shown any interest in potential projects since then. In these cases, a secondary market might open up in the future.

With more airports being privatized during the next decade – with a focus on Europe and Asia – it is expected that five to ten large global airport groups will be established. They will (co-)manage the majority of strategically relevant airports around the world. The basic business strategies in the aviation and non-aviation sector of each group will be based on the major partner's visionary strength and implementation skills.

However, one development will be interesting to follow. Will airport groups develop parallel to airline alliances or will they diversify their activities across the airline alliances? Airports are becoming increasingly dependent – and hence vulnerable – on their home-base carriers and respective alliances. Zurich and Brussels airports are two victims of the airline concentration process. Growing along an airline alliance increases an airport's potential to negotiate with its main customers at eye level. The airline can consider that a higher degree of standardization in equipment and procedures throughout its network is a positive effect or benefit.

It is too early in the process of airport cooperation/alliance projects to predict the future of airport alliances precisely. But, just as airlines overcame their traditional competitor roles and formed alliances, it is unlikely that the major airports of the world will stand alone. The airport industry has a long way to go and it will take a lot of powers of persuasion, but it is the authors' belief that airport alliances and groupings will prevail.

5 References

DOGANIS, R. (2001): *The Airline Business in the 21st Century*. London, New York: Routledge.

HOWE, J. (2000): The Benefits of Airport Networks. *Airport World*, 5, p. 3.

JUAN, E.J. (1996): *Privatizing Airports – Options and Case Studies*. The World Bank, Note No. 82.

SCHEUERING, J.D. (2002): *Airport Alliances – Some remarks on a new phenomenon*. Erasmus University Rotterdam, Rotterdam School of Economics (RSE).

SPARKLING, D.; COOK, R. (2000): Strategic Alliances and Joint Ventures under NAFTA: Concepts and Evidence. *Proceedings of the Fifth Agricultural and Food Policy System Information Workshop*, University of Guelph.

18

THE ROLE OF INTERMODAL TRANSPORTATION IN AIRPORT MANAGEMENT: THE PERSPECTIVE OF FRANKFURT AIRPORT

HANS G. FAKINER

1 The Challenges 428
2 The Role of Intermodal Transportation for Frankfurt Airport 429
3 Fraport's Intermodal Product "AIRail".. 433
4 Future Perspectives of Intermodal Services for Fraport.............................. 441
5 Conclusions.. 446
6 References.. 447

Summary:
High-speed rail services are leading to intensifying competition among Europe's airports. This chapter discusses Fraport's intermodal strategy, exemplified by its role in the "AIRail Partners" alliance with Lufthansa and Deutsche Bahn German Rail. Products and services developed since 1999 (operational start of Frankfurt Airport's (FRA's) long-distance railway station) and their major impact on the airport's relevant market are described. Departing from the assumption that intermodality involving integration with high-speed rail links is economically viable only for hub airports, the opportunities and risks of future business strategies and means of overcoming currently discernible obstacles to extending Fraport's portfolio are examined.

1 The Challenges

Rod Eddington, the chief executive of British Airways, stressed the current strategic importance of intermodality when, just three years ago, he admitted to the Aviation Club in London, that "London Heathrow has fallen behind Frankfurt, Paris and Amsterdam in terms of the available network, if one includes the building of high-speed rail links."

In some ways this is a surprising assertion. On the one hand, we have the airport's original, traditional core function: Frankfurt Airport has had long-distance trains since the seventies, check-in and check-out at the departure station with through-checked baggage have been available in connection with Lufthansa's Airport Express for years, and comparable services existed between 1998 and 2000 for Saarbrücken (the Pendolino regional train service) and Cologne/Düsseldorf ("Moonlight Check-in"). On the other hand, we have the emerging strategic relevance of intermodality for airlines and airports, we are witnessing – here and there – the emergence of services such as airport expresses, there are code-sharing agreements between airlines and train operators, and there are high-speed rail links to various other European airports. And all of this is backed by international studies[1,2] and national and international policies. Two examples are the EU White Paper[3] and the 'Rail Air Intermodality Facilitation Forum (RAIFF, 2004)', launched by DG TREN for the purpose of developing recommendations and suggestions for political action at the EU level.

So, what is new in connection with intermodality?

Of decisive importance is the quantum leap in technology achieved by Europe's rail operators by introducing high-speed trains, in France beginning in the early 80s and in Germany starting in 1991 with the ICE. An overview of rail travel times in 2015, agreed on with German Rail (DB) and based on realistic plans for route extensions and operating procedures, makes clear how dramatically the regions of Europe will grow together: it will only take about 2¾ hours to travel from Frankfurt Airport (FRA) to Amsterdam or Munich, some 2½ hours to Brussels or Basle, and little more than four hours to Dresden or Paris!

---

[1] See IATA (2003).
[2] See ILS et al. (2004).
[3] See European Commission (2001).

European high-speed transport has two important consequences for airports and airlines. First, there will be increasing overlap of catchment areas served. More and more travelers will suddenly find themselves faced with a choice between more than one airport of departure. The European intermodal hubs will increasingly be in competition among themselves for departing passengers. As far back as 1996, a survey by the German Airports Association (ADV) stated that large numbers of affluent German travelers living in decentralized major conurbations would be lost to foreign airports – including originating traffic. Action was obviously required, and has been taken through our corporate strategy.

Second, high-speed transport meets a prerequisite for developing intermodal products and services. As trains – for distances between 100 and 300 km – become competitive with feeder flights, integrated air/rail products can effect a complete shift from short-haul flights to rail, thus freeing up slots at airports that have run up against the limits of their capacity.

Another challenge is posed to airports by the emergence of airline alliances wishing to redefine their regional spheres of influence within Europe. This means that airports must appropriately reposition themselves within the regions they serve. Intermodality provides the tools for this. Much is also changing within the alliances as hub systems develop, for example with Frankfurt and Munich as the main bases of the Star Alliance. No end to this trend is yet in sight. Airports will also have to accommodate these hub systems and compete successfully under the new conditions they create.

2 The Role of Intermodal Transportation for Frankfurt Airport

2.1 Fraport's Intermodal Strategy

The geostrategic location of FRA is extremely favorable: it is situated in the center of Germany and Europe, there is a decentralized settlement structure with agglomerations within a 200-km radius around the airport, and the distances from the airport to large cities are ideal for taking advantage of the high-speed rail networks. Germany's settlement structures may be a drawback where feeder flights are concerned, because in the competition between airline alliances they constitute an open flank that is also exploited by European hub carriers. But from the standpoint of intermodality, they have great potential for generating passenger volumes and are thus the basis for developing cost-effective air/rail services.

Fraport was quick to recognize the enormous potential increase in traffic volumes that FRA could realize with ICE trains stopping at the new long-distance railway station. Within an assumed travel time to the airport of one hour, the higher speed of the ICE alone is enough to double the radius of the airport's catchment area from 100 to 200 km. The number of people that can then be reached is impressively extended: for FRA it is 35 million, or 50 percent more than London, Paris or Amsterdam. This reveals a major aspect of intermodality: its effect on the position of FRA relative to its main competitors.

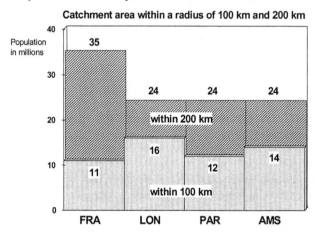

Figure 18-1: Market Potential of the Major European Airports

FRA is the central hub of the Star Alliance. The hub structure of Lufthansa, and especially the intercontinental nodes that are unique to Germany, are an important factor driving the expansion of Frankfurt Airport's catchment area, which now covers all of Germany and even extends into neighboring countries. Within this framework, high-speed rail traffic will substantially extend the catchment area further – at the expense of competing hub airports. But high-speed trains make economic sense only for hub airports, although operators and policymakers are being slow to recognize this fact. Only hub airports serve sufficiently large areas and thus command the market volume to provide intermodal services cost-effectively. And only hub carriers are in a position to shift feeder flights to rail – thereby freeing up slots – by implementing attractive air/rail services.

In 2001 intermodality became a vital element of Fraport's business mission, which states that "we make constant efforts to link means of transport". This reflects the company's intention of achieving a number of its strategic goals by connecting air and rail. They are:

- to increase FRA's capacity and thus the value of the company;
- to drive growth in support of its customers, the airlines;
- to strengthen customer loyalty in the various regions that make up FRA's catchment area; and
- to develop innovative "seamless travel" products and logistics services.

Specifically, this – in keeping with the company's "business mission" – yields the following strategic goals in connection with intermodality:

- to enhance Frankfurt Airport's hub function by establishing German Rail as a "zero-altitude feeder";
- to strengthen FRA's position within the hub system of the Star Alliance by connecting with its catchment area via a high-quality rail system;
- to compete with other intermodal hub airports by enlarging FRA's catchment area by linking the airport to European high-speed rail services;
- to generate slot capacity by shifting feeder flights to rail; and
- to strengthen multimodality in respect of access to and egress from the airport, to give people alternatives and thus maximize the reliability of the access modes, keeping in mind that car traffic also generates direct income (parking at the airport) for the airport operator.

2.2 Economic Feasibility of Intermodality from the Perspective of the Frankfurt Hub

A look at the market impact of intermodality reveals, first of all, that airports face varying consequences, and in some cases starkly contrasting commercial results. The shift from private cars to rail access results in a new modal split that affects different capacity aspects at the airport. The reduction in income from parking revenues is not too great, because cars and high-speed trains define catchment areas of quite different sizes and only directly compete with one another within a distance of 50 to 200 km from the airport. And a considerable share of the passengers who arrive by car are dropped off or picked up there in any case, so there is – in this respect – absolutely no effect on parking revenues.

Nor would that be a concern in the long term anyway, because the airport suffers from a shortage of parking spaces in the immediate vicinity of the terminals, and the airport train station is also generating a new market for the parking facilities by attracting non-flying passengers who depart from there by train.

The shift from air to rail will affect point-to-point traffic within Germany. There will be losses in this sector whether or not the airport has a long-distance train station. Rail and air transportation are to this extent competitors, but this fact is of no consequence for Fraport's long-range strategic plans, since the associated market effects are minor.

To the extent that the railway serves as a feeder for international air travel, it creates two opposing effects. First, there is a shift from the air (feeder flights) to rail, while travelers continue to be the airport's customers for their onward travel. And second, from the point of view of competition between airports, there is an extension of the catchment area. This latter effect has been estimated by several Intraplan studies at over four million pax per annum for 2015, depending on the intermodal product in question. This gain is greater than the loss in feeder traffic and certainly has great value for securing future potentials, since competing hub airports are now also developing intermodal systems and regaining some of the ground they had lost.

The relevance to income of losses in feeder traffic must be viewed in two ways: because passengers continue to depart on international flights from FRA, the loss of passenger and landing charges is restricted to the feeder flights. So this shift causes no change whatsoever in the number of "terminal passengers", which is relevant to the airport's retail activities. The number of people visiting and spending time in our terminals, restaurants and shops is the same whether they arrive at the airport by rail or by air.

The long-term profitability of intermodal products is secured by the fact that the extension of the catchment area brings the airport new passengers and thus income (passenger charges, retailing revenues).

Of relevance for the profitability of air/rail products is also the slot substitution effect: short-haul flights involving smaller aircraft are shifted to the rail and replaced by long-haul flights with larger aircraft. The associated impact on revenues from ground handling and airport charges only needs to be included in the airport's business plan if there are long-term capacity constraints.

For the airport, the direct costs incurred by intermodal services – i.e., the cost of integrating the station with the terminals – are not matched by any direct income, although there are secondary revenues produced by the above-mentioned effects on the market, not all of which are positive.

As far as spreading the financial burden is concerned, so far little progress has been made with any of the operators involved. Lufthansa and German Rail have agreed to bear some of the very considerable investment costs of linking the train station to FRA's fully automated baggage conveyor system in connection with developing the AIRail Service. At some point in time, especially if other airlines or train operators are to be included in the system, we will consider the possibility of a "zero-altitude passenger charge" and revise the present system of airport charges to compensate for the substantial costs of intermodality.

3 Fraport's Intermodal Product "AIRail"

3.1 Product Definition and Development

According to the definition[4] of the term "intermodal product", such a service involves linking a variety of partial services provided by different modes of transportation over the travel chain, with the aim of creating a transport service covering the entire journey while ensuring maximum smoothness of transfer between the different modes. These "partial services" could include, for example, passenger and baggage transport to and from the airport, reservations, booking, ticketing and sales, check-in, check-out and customs, information for intermodal travelers, and coordination of flight and train schedules. A basic distinction should be drawn between two levels of quality: "zero-altitude flights" with check-in at the train station of departure and through-checked baggage with integrated ticketing and sales via the airlines' reservation system on the one hand, and "normal access/egress by rail" without baggage service or station check-in on the other. The airport must provide premises, facilities and logistics for both submarkets.

The "zero-altitude flight" service has particular strategic significance. The "AIRail Partners" are Lufthansa (LH), German Rail (DB) and Fraport, a cooperative endeavor that introduced the AIRail Service to Stuttgart (in 2001) and Cologne (in 2003). This air/rail service is sold to consumers throughout the world in competition with "feeder flight" services. It is sold via computer reservation systems. The first screen displayed shows what can be sold. The ranking order of the connections between the points of departure and destination is based on the criterion of total elapsed travel time, which is decisively

---

[4] See Pousttchi (2001).

influenced by the minimum connecting time (MCT) between train and flight at the hub airport. LH, thinking of the potential network losses to foreign hubs if all flights to FRA were shifted to rail, regards a "zero-altitude flight" service as competitive only if a rail/air MCT at FRA of 45 minutes can be achieved. This is for both passengers and baggage, and both incoming and outgoing! Consequently, Fraport set itself the strategic goal of an MCT of 45 minutes, and has in fact achieved it with the AIRail Service. This is the main difference between this service and the LH Airport Express of the 1980s and early 1990s. However, this accomplishment has only been possible with a major technological and logistical effort.

Owing to its location, Frankfurt Airport has always been well-integrated in all long-distance transportation networks. The development of products and services for integrating FRA into Europe's core high-speed rail network joining Paris, Brussels, Amsterdam and the Rhine-Main area has proceeded as follows since 1998:

May 1999:	AIRail Terminal with the long-distance train station
June 2000:	Check-In T at the FRA train station
March 2001:	Extension of baggage conveyor system
March 2001:	AIRail Service for Stuttgart
August 2002:	New high-speed route between Cologne and the Rhine-Main area
December 2002:	New schedule of German Rail
April 2003:	Codesharing between German Rail and TAP Portugal
May 2003:	AIRail Service for Cologne
June 2003:	Codesharing between German Rail and American Airlines
September 2004:	Codesharing between German Rail and ANA
Since 1993:	DB Rail & Fly ticketing for airlines

Table: 18-1: Criteria for the Standardisation of Intermodal Products/Services

So, the intermodal provision at Frankfurt Airport consists of the AIRail Terminal, which is currently made up of the long distance train station, the Check-in T – to serve the "normal access/egress by rail" market – and the AIRail service, and of codesharing agreements between airlines and German Rail.

3.2 Market Success

When the long-distance train station was completed in 1999, all of southern Germany and the northern route to Hannover/Hamburg were linked to the airport directly, i.e., without requiring travelers to change at Frankfurt's main railway station. The market's response has been enormous. The new timetable of German Rail, introduced on 15 December 2002, and the new high-speed link between Cologne and Frankfurt/Main, caused the number of high-speed trains serving FRA to jump by a staggering 280 percent to 174 trains per day in 2003 (Paris CDG is served by 60 high-speed trains a day).

A second reason for this market reaction at FRA is that the airport is served mainly by luxury ICE trains, most of which are German Rail's ultra-modern, prestigious ICE 3 trains. Third, travel times for all connections from Brussels, Amsterdam, Hamburg/Münster and Dortmund will be reduced by around one hour by the new route from Cologne.

With its Check-In T (T for train) service, Fraport is making every effort to integrate flight check-in facilities into the train station wherever possible. The market response to this service, which is extremely convenient for consumers, has unfortunately been low. There is room for 28 counters; 10 were initially built, four of which are now operating. Lufthansa and Fraport are handling check-in for about 44 airlines. LH also uses its counters for boarding procedures for the AIRail Service.

Comparable facilities also existed at the long-distance train station at Düsseldorf, which is two kilometers from the terminals there. However, the 20 counters had to be closed in early 2004; the check-in counters for Heathrow Express at Paddington station in London were shut down too, except those of LH and the Star Alliance. Paris' (CDG) airport operator has been attempting to develop a similar service at its TGV train station with 10 check-in desks, but its introduction has had to be postponed because the airlines reacted cautiously due to the additional costs involved.

With its AIRail Service on the routes to Stuttgart and Cologne, Fraport, LH and German Rail have launched a "seamless travel" service including:

- check-in/check-out and customs facilities at the train station;
- integration into German Rail's ICE service;
- through-checked baggage; and
- integrated ticketing and international distribution by Lufthansa.

This service has been slow to establish itself. Although acceptance by users is extremely high, at over 90 percent, it has so far only taken 20 percent of the market away from air feeder traffic for Stuttgart. The situation is far better in the case of Cologne, where the service accounts for half of the passenger volume after just six months.

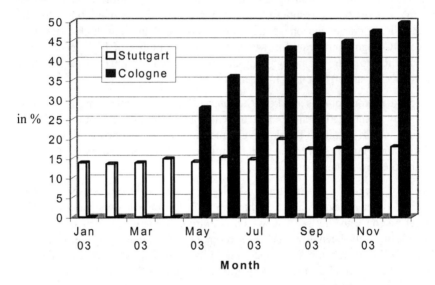

Figure 18-2: AIRail Market Shares

The reasons for this difference in popularity of the AIRail Service between Stuttgart and Cologne have to do with how well the train and flight schedules dovetail on the one hand, and with differences in the accessibility of the train stations and airports in these two cities. The challenge of harmonizing the AIRail train schedules with the departure times of Lufthansa's international flights is particularly thorny; it should be possible to at least partially overcome this obstacle with pragmatic solutions (by increasing the frequency of train services; currently there are trains to and from Stuttgart only every two hours).

The other AIRail routes (Bonn and Düsseldorf) will be tackled gradually over the next few years. Nuremberg will not be included in the service until the current train travel time of about two hours has been reduced sufficiently to compete with air travel. German Rail has not yet announced any plans to substantially decrease the travel times on this route.

One part of the AIRail Service is a codesharing agreement between Lufthansa and German Rail. Similar agreements now also exist with American Airlines

ANA and TAP Portugal, and others are in preparation. At Paris CDG, nine airlines have already concluded agreements of this kind with SNCF. This is strictly a sales instrument for marketing the railway connection worldwide via the airlines' computer reservation systems (GDS). German Rail has complied with its side of the deal by making available IATA codes for train stations. The airlines' reservation systems have not yet been networked with German Rail's, however. Each airline selects a number of trains and between 10 and 15 important train stations in the extended catchment area served by Frankfurt Airport (Lufthansa has so far only picked Stuttgart and Cologne). At FRA, a total of 120 trains each day now also have a flight number. These "zero-altitude flights" thus account for about 10 percent of the 1,200-plus flights each day. This significantly strengthens FRA's hub function.

The market has seen dramatic changes since the opening of the new train station on 30 May 1999: this station is now used by some 10,000 passengers a day, for a total of about 3.7 million passengers in 2003. The share of passengers choosing the high-speed train has thus doubled since 1998. And according to forecasts by Intraplan, the modal split of users of high-speed trains will double again by 2015.

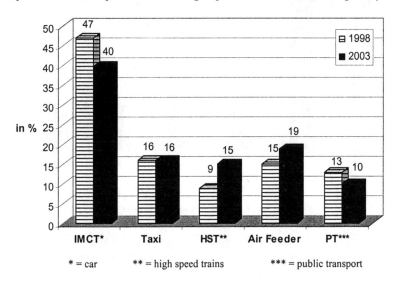

* = car   ** = high speed trains   *** = public transport

**Figure 18-3: Shifts among Modes of Transportation
(main modes of access and egress)**

The high-speed trains are winning passengers away from feeder flights (on the Cologne route alone, air traffic volumes dropped by half in 2003, although this decline is more than offset by the growth of the Star Alliance hub) and to a limited extent and for other reasons, from cars and local urban rail services ("S-Bahn"), and are also benefiting from the expansion of the airport's catchment area. Over a period of just five years, this has produced a globally unique situation in which more long-distance than local travelers choose the railway for going to and from Frankfurt Airport. A comparison with competing intermodal hubs elsewhere in Europe shows that nearly four to five times as many people use long-distance rail services to and from FRA than is the case with Zurich, Amsterdam or Paris.

An examination of the catchment area is revealing. It was already growing during the first year after the opening of the new station, an interesting fact from the point of view of inter-airport competition. The number of people traveling over 100 km to and from FRA has risen significantly:

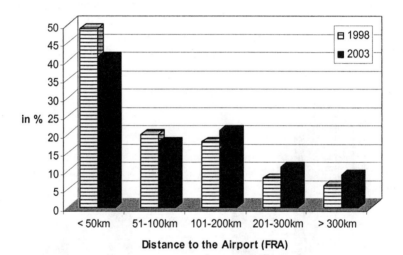

Figure 18-4: Growth of the Catchment Area

Intermodality is having a slot effect by shifting short-haul flights to rail and thus increasing slot capacity at FRA. This is relevant to the expansion plans of the airport.

The memorandum of understanding and associated contracts between the three AIRail partners – LH, German Rail and Fraport – focus on connections with Stuttgart, Cologne, Bonn, Nuremberg and Düsseldorf. The aim of the AIRail Service is to apply the standard of quality of air travel to train journeys as well, thus creating conditions in which short-haul flights to and from Frankfurt can be gradually reduced. LH took the first step in this direction in 2003 by canceling 35 percent of its feeder flights to FRA. Although the routes mentioned can potentially free up only five percent of the slots – not enough to render unnecessary the additional runway planned for 2009 – the effect on capacity is nevertheless important and will allow FRA to react flexibly to the market even before the runway system is extended.

Summing up, it may be stated that the HST intermodality has significantly impacted FRA's market situation within the space of only five years. This provides the first evidence in support of the thesis that integration into the European high-speed rail network can pay off for hub airports.

3.3 Managerial Challenges of Fraport's AIRail System

The managerial challenges for Fraport depend on its role within "AIRail Partners". Seamless travel requires investment into infrastructure and technics, especially for baggage handling systems, for bridges and routing systems to move passengers between trains and aircraft, for customer information systems, for IT-logistics.

The costs are considerable, since they include connecting the train station to FRA's fully automatic baggage conveyor system in the so-called "fly baggage room" above the railway platforms (for the "zero-altitude flight" market) and the check-in desks in the connecting building near the station (for the "normal access/egress by train" market). The long-distance train station is part of the AIRail Terminal, which also includes a planned superstructure, and in particular the facilities in the connecting building to Terminal 1. Fraport intends to move its air travel handling facilities as close as possible to the train platform.

The "zero-altitude flight" service calls for IT logistics and management of problems and disruptions. IT logistics are important because movements of containerized baggage must be registered and communicated to participating transportation providers for various purposes. It has to be possible to adjust the baggage handling logistics required at the start and end of each train trip to actual baggage volumes, which is why the "AIRail Partners" decided to install the Frankfurt Airport baggage reconciliation system at Stuttgart and Cologne railway stations as well. Another important capability is "tracking and tracing", in other words locating lost luggage. To accomplish this, there must be computer-supported registration of pieces of luggage at various points of the intermodal transportation flow so that movements can be subsequently reconstructed. A related issue here is determining which of the participating operators has caused a loss of baggage and can be held liable for it.

Various kinds of problems can occur at the interface between rail and airport. Trains can be late, they can leave without passengers' bags, or they can stop at a platform other than the one planned. This can have consequences for baggage handling at the airport train station. And it can have consequences for travelers if they miss a flight as a result. In such a case they will be taken care of by the airline, just like when problems of this kind crop up in connection with a flight-to-flight connection, which of course happens. But passengers must know what counter to go to in order to get assistance. And the airline's staff must be informed as quickly as possible so they can initiate suitable action (e.g., rebooking, overnight accommodation, etc.). Frankfurt Airport operates appropriate data systems on flight events (Info Plus), but these need to be extended by adding intermodal data and they need to be made accessible to all operators. At this time the intermodal data is still entered manually into the system. It is planned to network Info Plus with the corresponding system of German Rail (RIS), which is currently also being restructured and developed further.

The airport must also make sure it provides orientation and information systems for its final customers. These include intermodal information for information kiosks and internet portals. Fraport is collaborating with the Rhein-Main Verkehrsbund (RMV), the regional public transport authority for Frankfurt and the Rhine-Main area, to develop a web-based intermodal (integrated) and multimodal (covering all modes of transportation) flight information system. A basic module with planning data and routing information has been completed and integrated into the websites of RMV and Fraport AG. Additional modules are now being prepared, for example for incorporating data on late arrivals and departures (public transportation) and traffic jams on the roads. There will also be a flexibly configurable support and consulting function that customers can register for and take advantage of, possibly for a fee. The idea is to provide information on the entire travel chain. It no longer seems far-fetched to enable internet access on board aircraft.

4 Future Perspectives of Intermodal Services for Fraport

4.1 Extending the Market

Intermodality causes traffic flows to focus even more heavily on the hub. This happens at the expense of airports that lack a nearby railway station served by long-distance trains, and also at the expense of those that enjoy such connections but have little or nothing in the way of additional efficient air/rail services. It even, on balance, detracts from the volumes handled by competing hub airports whose market position is weaker for geographical reasons or because they have been slower in developing and implementing intermodal services.

Settlement patterns and relative distances critically influence the extent to which traffic flows center on a hub. And the geographical distribution of hub airports defines the competitive situation and which parts of their catchment areas are vied for by other airports as well. There are opportunities, and there are also limits. High-speed trains can currently compete with air travel by offering "zero-altitude flights" only within a radius of 90 minutes. And owing to competition with other airports, there are geographical limits to possibilities for extending catchment areas by offering "light" versions of intermodal services (such as evening-before check-in as at Magdeburg/Leipzig or Zurich or remote check-in without baggage transport such as in Brussels for Paris CDG) or via normal trains. And it is the railway that determines how strongly regions grow together over time: by building high-speed routes, by developing superfast trains, by planning networks, and by managing train traffic by means including advanced signal technology.

The development of intermodal services always requires various rail and air operators to collaborate. The strategic positions of airlines, rail operators, and airports depend not only on their geostrategic location and settlement patterns, but also on the contexts within which they compete. And these vary immensely, so it is only possible to attain a win-win situation with certain selected services and on certain routes.

Lufthansa, for example, competes with other carriers at FRA and also with other airline alliances and their hub systems within Europe. It will thus naturally refuse to team up with other airlines to jointly operate a remote common check-in service or an airport express. But from the airport's perspective, it would definitely be desirable to involve different airlines in developing services, not least in order to exploit the potential of secondary rail routes and gain the ability to operate cost-effectively. Conversely, the airport will naturally resist the wish of the rail operator to extend the same service to several airports in order to more fully utilize the capacity of its airport trains. This makes it plain that every air/rail service and every train connection involved must be subjected to a separate, trilateral strategic examination. Where FRA is concerned, there is a consensus in favor of the premium AIRail Service for the routes stipulated in the memorandum of understanding of 1998.

The competitive situations of the partners involved are in flux. The airline alliances are still evolving, and with them the hub systems of each alliance. Airports are now also entering into alliances, cooperative endeavors, and systems. The zones of influence within Europe are being redefined. And it appears to be only a matter of time before genuine competition makes its advent on the railways as well, especially for long-distance services.

There is development potential in the possibility of launching higher- or lower-quality services on the market. The premium version is the airport express, a high-quality shuttle exclusively for airline passengers. The trains can be specially equipped to meet the needs of air transportation. This involves not only catering to travelers' needs, but also logistics and specifically baggage handling, which in turn impacts both efficiency and costs. The schedule – assuming that there are sufficient available rail slots – can be optimally tailored to flight departure and arrival times. An appropriately marketed airport express has quite different effects from those of an AIRail Service integrated into the normal rail service provision.

The alternative targets regional markets with secondary potentials that can be tapped with "light" intermodal products. Here the focus is absolutely on cost-effectiveness: in addition to leveraging economies of scale and expanding the market by including all airlines in "common use self-service" (CUSS) check-in systems, there is considerable potential for cutting costs by implementing innovative ideas such as on-train check-in with baggage drop at the airport train station. For this to work, the following requirements must be met.

- The airlines must, at the very least, be willing to accept CUSS systems in connection with secondary connections and cease regarding on-ground check-in as a means of competition.

- Check-in capacities at the airport have traditionally been rented to airlines as counter space. A charging system on a "per pax" basis ought to be introduced instead. Shifting check-in capacities to remote sites and individual modes of transportation for reaching the airport – something the airport operator definitely considers desirable in order to free up more space for retail outlets in the terminals – has so far been prevented by cost considerations: the airlines would have to operate a check-in counter in the terminal and another one for the same flight in, for example, the long-distance train station. But things have started moving. Driven by new technologies, we now have Internet, phone and mobile-phone check-in as well as various pilot projects involving baggage CUSS machines (Lufthansa) and automated CUSS check-in machines with baggage drop counters (Amsterdam).

- Technical and operative issues still need to be resolved: miniaturization of servers, integration of barcode/RFID technology, check-in on trains via online links, even in tunnels, and Internet links between CUSS and airline systems, for example. These are recommendations for official policies to promote new technologies, as the RAIFF expert commission of the EU recently ascertained in its final report.[5]

There are thus definitely limits to how far the market can be expanded. Some of the obstacles can only be overcome in the medium term.

4.2 Exploiting Market Potentials

Train and flight schedules can only be harmonized within limits. The networking plans of rail companies and airlines are based on fundamentally different conditions and strategic considerations, and consequently it appears to be very difficult to specifically gear train timetables to connecting international flights. The relevance to cost considerations is apparent from the fact that the capacity utilization of individual AIRail trains running on the routes to Cologne and Stuttgart varies from under 20 percent to over 60 percent. The potential market can therefore be tapped more effectively by appropriately planning the frequency of rail services.

The travel time does not begin at a train station or airport. The accessibility of the train station in comparison with the airport is of major importance. By studying local potentials and systematically implementing measures to extend an intermodal service (by providing parking at the train station, integrating local

---

[5] See Rail Air Intermodality Facilitation Forum (RAIFF) (2004).

public transportation, etc.), it is possible to extend the air/rail travel chain so it comes closer to being a more attractive "home-to-hub" connection.

The worlds of rail and air transportation differ greatly, and so does the behavior of train and air passengers. One does not expect a check-in counter at a train station, and therefore one does not look for it either. One does not expect checked-through baggage transport in the train, and consequently one is not inclined to seriously consider a "zero-altitude flight" as a viable alternative for getting to the airport. Lufthansa reports that it has already spent several times as much on marketing the AIRail Service as it would on a new international route. It takes time and a considerable marketing effort to change people's habits.

The International Air Rail Organisation (IARO) estimates that there are about 180 variants being practiced at 70+ airports worldwide with a rail connection: there is a great variety of air/rail products, and there is duplication. Copenhagen, for example, has high-speed, regional and local services (and will soon get a metro). Frankfurt has high-speed, regional and suburban services, some with full integrated ticketing and full hold baggage with in-town check-in/out.

IARO – in a quick survey – identified the following types of passenger airport rail connections:

- those that have some intermodal integration and use high-speed, long-distance rolling-stock (AIRail, Thalys Air, TGV'Air, SBB + Finnair/SWISS);
- those that use high-speed, long-distance rolling-stock but have no integration (Düsseldorf, Copenhagen);
- dedicated airport expresses (there are 14 of these around the world, including the Heathrow Express, Gatwick Express, and Malpensa Express);
- underground urban and metropolitan railways (like in Stuttgart, Madrid, and Copenhagen);
- regional railways (like in Friedrichshafen, Southampton, Leipzig-Halle, and Copenhagen);
- those that have some degree of integration and use a mixture of rolling-stock (the Swiss rail/fly, and the Newark system in the USA).

The typology gets even more complicated if baggage and/or ticket integration is included.

4.3 Costs

Intermodal services – from the perspective of a company as a whole – struggle with the issue of cost-effectiveness. In the initial stages of offering a service, problems are posed by the additional costs typically caused by still-inadequate efficiency and a lack of capacity utilization. But air/rail services always incur additional costs that flight connections are free of: investments at the interfaces between rail and air transportation (networking of reservation systems, connection of the airport train station to the baggage conveyor system, the security costs incurred in conjunction with baggage transport, check-in/check-out systems at the train station where passengers initiate their journey, baggage handling at all train stations, etc.), the cost of decentralized customs processing at train stations, and additional marketing activities to encourage people to change their normal travel habits.

Baggage logistics on trains are in conflict with opportunity costs for special container compartments and efficiency, which drive the handling costs up in connection with quick change systems. Within the AIRail Service, baggage logistics is currently rather different on the Cologne and Stuttgart routes: for Stuttgart there is containerized baggage transport in dedicated baggage compartments on ICE trains, while for Cologne, the AIRail Partners are using a quick change system similar to the one that proved successful for the night mail network. The seats are provided with protective covers, and the baggage is loaded and unloaded by hand. From the business point of view, this avoids the opportunity costs that are incurred if baggage compartments have to be converted for container transport and are consequently no longer available for use by passengers, even when the train is no longer bound for Frankfurt Airport.

The costs of security logistics throughout the travel chain are covered by the service. The operative costs for decentralized customs processing are borne by the government in Germany. Concerning security logistics, which is also important for customs, a compromise has to be found between the requirements of the supervisory authorities (maximum security standards) and those of the providers of an intermodal service (minimum security costs). Since "9/11", it has been particularly difficult to settle these differences with the United States; this matter should be re-addressed at the EU level. For instance, United Airlines still has not been able to obtain permission to participate in the AIRail Service.

A service's cost-effectiveness can be ensured by charging travelers a price (like in Switzerland and at Leipzig), or else it can be offered free of charge (like the AIRail Service) because, in connection with the hub system, shifting flights to rail permits more comprehensive costing. The costs can also be reduced on a long-term basis: with support from the state[6] it is possible to develop efficiency-boosting technologies, for instance for CUSS check-in systems for long-distance trains, containerized baggage transport, or quick-change systems for baggage logistics in high-speed trains.

5 Conclusions

Politicians now place a high value on intermodality – at both national and EU level – because of its potential benefits for capacities and the environment. Studies commissioned by Fraport[7] have shown that Frankfurt Airport can expect

- the AIRail Service to free up about five percent of the slots now utilized at FRA, and
- networking of different modes of transportation (long-distance train station, new rail routes, the AIRail Service) shift about 11 percent of the traffic from road to rail, expressed as a percentage of the projected car traffic volume in 2015.

These studies take into account investments in infrastructure by all transportation providers, to the extent that they are planned and implemented by 2015. It is not possible to apply the mentioned results as such to other airports.

The extension of intermodal services at FRA is – as has been shown – subject to constraints, although these can be partly overcome.

The contribution of policymakers could be, as the RAIFF Commission concluded, to establish uniform conditions for participating operators throughout Europe (legal aspects, security standards, treatment of VAT, integration in the CRS of the airlines, typologization and standardization), and to make available funding to promote the development of technologies within the scope of existing programs or to reduce the costs of the pilot phase.

The industry, for its part, is urged to acknowledge the intermodal market and develop technologies that will help expand the market or help reduce the associated costs.

The transportation operators, finally, must face up to competition, improve their strategies, and coordinate their activities. FRA's unique advantages, especially in

---

[6] See RAIFF Commission (2004).
[7] See Schubert & Schwaibold (1999).

comparison with other airports with intermodal hub ambitions, are not outstanding enough for Fraport to ever relax its efforts to continually develop its intermodal services. Operators must strive to influence end-users by launching new or enhanced services that will motivate travelers to make different decisions.

These few comments may help to clarify the gap that still yawns between the strategic and political visions on the one hand and reality as it stands today. Rapid progress is not to be expected – but a start has been made!

6 References

EUROPEAN COMMISSION (2001): *White Paper. European transport policy for 2010: Time to decide*, Luxembourg.

IATA (2003): *Air/Rail Intermodality Study*, for the EU, ACI, ATAG, CER, UIC.

POUSTTCHI, P.F. (2001): *Kompetenzorientiertes strategisches Management intermodaler Verkehrsdienstleistungen*, Gabler, Wiesbaden.

RAIL AIR INTERMODALITY FACILITATION FORUM (RAIFF): *Final Report* (July 2004), initiated by DG TREN (unit F2).

SCHUBERT, M.; SCHWAIBOLD, C. (1999): *Fluggastprognose für den Frankfurter Flughafen unter besonderer Berücksichtigung der Transeuropäischen Netze* (Intraplan, Munich).

SILS ET AL. (2004): Germany, Babtie Spol., Czech Republic, Langzaam Verkeer, Belgium, ETT, Spain (2004): *Towards Passenger Intermodality in the EU*. For DG TREN, Unit G4. Dortmund.

Part IV

Air Cargo

19
REFLECTING THE PROSPECTS OF AN AIR CARGO CARRIER

ANDREAS OTTO

1	Introduction	452
2	The World Air Cargo Market	453
3	Competing in the Air Cargo Market	462
4	Successful Positioning – the Case of LH Cargo	464
5	Conclusion and Outlook	468
6	References	470

Summary:
This chapter reflects the opportunities and hurdles of the capricious air cargo market and describes the strategic requirements and pre-requisites a professional air cargo carrier must deliver in order to occupy a profitable position. The latter is exemplified with the premium strategy of Lufthansa Cargo AG.

1 Introduction

Looking back in history to the beginning of civil aviation one quickly learns that mail and parcels were air-transported on a scheduled basis long before passengers.

The first regular airmail services on the routings Vienna-Kiev, Berlin-Weimar, Paris-Brussels and Paris-Cherbourg operated as early as 1919. Three years later, the same could be observed in the United States.

Now, almost a century later, the relevance of air cargo has ceased in favor of the passenger business. However, with a contribution of 13% in 2002[1] to the total air traffic revenue, the importance of air cargo remains – be it as core business for dedicated air cargo carriers or as a second standing leg for traditional passenger airlines.[2]

Listening to conversations of modern businessmen, one quickly gets the impression that the 21st century has brought out an army of aviation experts. Opinions about strategic decisions of airline executives are exchanged, fierce discussions circle around the quality of the in-flight services of various airlines or the benefits of frequent-flyer programs.

However, the public awareness of the related air cargo business remains low. Is there a first-class service available for containers and pallets? Is there a frequent-flyer program for air cargo customers? I argue that there is.

As it is impossible to deliver a critical market reflection from all perspectives, this chapter shall serve to describe the market from the position of an air cargo carrier with its core competence on delivering airport-to-airport transport solutions.

---

[1] The revenue contribution of air cargo varies significantly between regions and airlines. With an overall revenue share of 25% Asian-based carriers achieve the highest, with an overall revenue share of 5% US-based carriers achieve the lowest portion.

[2] See Boeing (2003), p. 8.

2 The World Air Cargo Market

The world air cargo market can be described as a function of industrial production and global trade.

Governments increasingly seek to strengthen their economies by liberalizing their foreign trade. The foundation of GATT[3] in 1948 was one first cautious step towards economic freedom – though with limited success. A major achievement of GATT was the lowering of import tariffs among several of the member states by 40–50% to an average of 4–5%.[4]

A major milestone in the development towards an economic world order without any barriers to trade was definitely set by the formation of the WTO[5] in 1994. Formed out of GATT and the fellow organizations GATS[6] and TRIPS,[7] 146 member states (including China) joined until April 2003. The organization enjoys extensive authorities of legislative, judiciary and executive nature, WTO decisions and sanctions thus overrule those of individual nations.[8]

Thanks to the achievements of the WTO, raw materials and parts can now be purchased on an – almost unlimited – global scale, the production of goods can be relocated to those countries with comparative advantages; finished goods are available world-wide.

Complemented with the growing relevance of just-in-time inventory systems as a matter of cost saving and increasing customer demands, time-to-market became a critical success factor.[9] The relevance of geographical borders is ceasing; air transport became the vehicle of modern trade.

Today air cargo is a $46 billion business with 40% of the world trade value travelling on board commercial aircrafts.[10]

---

[3] Abbreviation for General Agreement on Trade and Tariffs
[4] See Grottendieck (2003).
[5] Abbreviation for World Trade Organinzation.
[6] Abbreviation for General Agreement on Trade in Services.
[7] Abbreviation for Trade Related Intellectual Property Rights.
[8] See World Trade Organzation (2003).
[9] See Pompl (1998), p. 89.
[10] See Saling (2004), p. 3.

Even though the demand for air cargo is mainly driven externally, the following factors on behalf of the air cargo capacity suppliers contributed positively to the growth in air cargo demand:

- An increasing deployment of wide-body-aircrafts for passenger traffic resulted in additional air cargo capacities.
- Technical innovations in the field of handling services and loading devices supported the development of new air transport solutions.
- A steady yield decline[11] increased the competitiveness of air cargo vs. the slower and cheaper freight transport by sea.[12]

However, not all shipments qualify for the speedy air transport. Strict safety regulations prohibit the transportation of certain goods, e.g. explosives. Others are too much of a commodity to be subject to the comparatively more expensive air cargo rates.

Air cargo shipments are usually characterized by a high-value and/or extreme time-sensitivity. Therefore the main demand derives from key industries, namely the pharmaceutical, high-tech, telecom, fashion and the automotive industries.

Due to the increase of high-value goods in the foreign trade within today's business environment, the demand for air cargo has experienced a steady growth. A continuation of growth with an average rate of 5.3% effective 2005 is estimated (see figure 19-1). The main driver for growth derives from Asia.

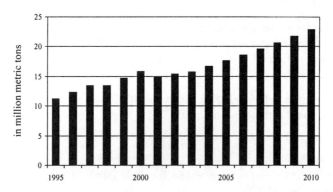

Figure 19-1: Estimated Development of the Air Cargo Market[13]

---

[11] See section 2.3 of this treatise.
[12] See Grandjot (1997), p. 26.
[13] See Lufthansa & Merge Global (2004), p. 22.

It is noteworthy to mention that the year 2001 has to be highlighted as an absolute exception. Nevertheless the slump in growth (almost - 6% vs. 2000) was rather a result of the slowdown of the US economy and the collapse of the so-called New Economy than of the aftermaths of 11 September.

The gross domestic product of countries correlates positively with their demand for air cargo. Not surprisingly, the market development depends heavily on the triangle Asia-Pacific, North America and Europe. The exports of these regions account for 88% of the total market volume in tons (see Figure 19-2).

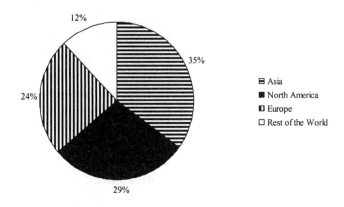

Figure 19-2: World-wide Demand for Air Cargo ex Origin[14]

2.1 Air Cargo Business is unlike Passenger Business

In 2001, little more than half of the world's airfreight moved in freighter aircrafts. Consequently, the network planning for almost half of the available capacity is dictated by the demands of the passenger market.[15]

There are a number of characteristics that apply to passenger and cargo airlines likewise. The investment into freighter and passenger fleets are comparable, both business models have to provide the same safety and security standards for crew and equipment and both suffer from the still incomplete open sky agreements among nations.

---

[14] Data Source: Merge Global (2004).
[15] See Kadar & Larew (2003), p. 4.

Trying to understand the air cargo business, however, requires forgetting about passenger business for the following reasons:[16]

- Airfreight does not fly on return tickets. The directionality of airfreight is in line with the economic trade flows and these are unequal. The satisfaction of the demand from Asia to North America for example would translate into a gross oversupply of the market in the reverse direction.[17] *The imparity of airfreight questions does not comply with the traditional return-flight philosophy of airlines.*

- Airfreight is a heterogeneous good and comes in numerous shapes, weights and values. Accordingly the various transport demands are manifold in comparison to the transport demands of passengers (one seat = one passenger). *The heterogeneity of airfreight bears a challenge for efficient load planning.*

- Three in-flight-products suffice to satisfy the demands of most airline passengers. Due to the heterogeneous nature of air cargo shipments, numerous different transport solutions must be available for air cargo. *The heterogeneity of airfreight translates into a significantly higher complexity of processes.*

- Cargo customers are concentrated, with a limited number of forwarding agencies accounting for the major share of air cargo demand,[18] whereas countless individuals and companies purchase passenger tickets. *The bargaining power of the customers in the air cargo business is significantly higher than in the passenger business.*

- Numerous companies can be involved in realizing the air cargo transport chain in order to fulfil the required transport, handling, warehousing and customs tasks. *Other than passenger airlines, air cargo carriers do not own the end customer relationship.*[19]

---

[16] See Maurer (2001), p. 86.
[17] In 2003, the demand for air cargo from the Asian-Pacific region to North America outnumbered the return demand by a factor of 1,9 (Source: Merge Global 2004).
[18] See Kadar & Larew (2003), p. 5.
[19] See Kadar & Larew (2003), p. 4.

2.2 The Air Cargo Transport Chain Consists of Several Logistical Tasks

The air cargo business is part of the integrated value chain in the global logistics market. Depending on the level of integration, up to four different types of logistics suppliers can be involved (see figure 19-3).

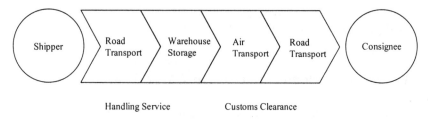

Figure 19-3: Transport Chain for Airfreight

- *Asset-based transportation companies*, such as trucking firms, airlines or integrators are taking care of the core transportation tasks – on the road and in the air.

- *Dedicated logistics firms*, such as warehousing companies are providing appropriate facilities and/or handling services.

- *Non-asset-based transportation companies*, such as freight forwarders are offering door-to-door transportation to their customers by coordinating the logistical tasks of the asset-based transportation and dedicated logistics companies.

- *Third party logistics providers/integrators* are taking over the complete supply chain management for their customers.

The different logistics providers depend on each other – more or less.

The global logistics market is a dynamic one, with a steadily increasing demand for integrated processes on behalf of the customers. Despite several horizontal and vertical mergers, however, there is yet no company able to supply all-logistics on a global scale.

Due to two reasons, it still remains open, whether an all inclusive offer would be accepted by the market:

- Are there enough synergies between the various elements of an all-inclusive offer or does the concept only increase the costs of complexity?
- Do enough customers really support this idea or do they rather fear to lose control over a key steering part of their business?

One company that seems to be convinced of the benefits of this approach on a global scale is Deutsche Post World Net.

2.3 Yield Development in the Air Cargo Market

From a carrier's perspective, certain factors had a negative impact on the yield development of the last couple of years, e.g.:

- exchangeability of capacity (commodity);
- government subsidies (flag carrier thinking);
- lack of professionalism among primarily state-owned carriers (no commercial targets).

In addition, one can observe an increasing trend to concentration within the core customer segment, the forwarding agencies and their strategic approach to further bundle bargaining power by the establishment of so-called consolidation hubs. [20]

Whereas the market for passengers experienced an average yield reduction of approx. 25% from 1985 - 2000, air cargo had to endure a price drop of almost 45% (see figure 19-4). Considering that the cost structure for employing passenger and freighter fleets and infrastructures are quite comparable, the competitive disadvantage for airlines with air cargo as core business becomes obvious.

---

[20] In 2000, 15 air freight forwarding agencies account for 61% of the worldwide marketshare, see Air Cargo World (2002).

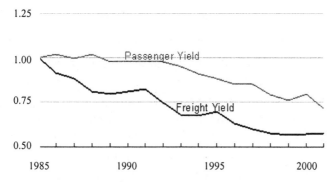

Figure 19-4: Development of the Air Cargo Yield, 1985-2001[21]

The relative stabilization of yields in the late 90s reflects the efforts of the leading carriers to differentiate their capacity with the development of premium services and thereby trying to improve their profitability.

So far, a small number of belly carriers have sourced out the entire freight management to dedicated cargo carriers. This business model actually describes a win-win-situation: The belly carrier receives fixed cargo revenues without any additional cost and the assigned cargo carrier gets access to additional routings and increased capacity without deploying additional aircrafts.

2.4 Not all Air Cargo is Alike

Despite the manifold perception of air cargo as a commodity business, there are numerous facets that identify this business as rather heterogeneous. This does not only apply to the infinite variations in size and weight of the shipments, but also to their specific transport demands. Whereas there is an obvious distinction in the required time of availability at the final destination for the different kind of shipments, further special features need to be taken into accounts for shipments that:

- are especially theft-endangered (e.g. microprocessors, computer chips);
- require a constant temperature management throughout the journey (e.g. pharmaceutical and biotech products);
- are especially shock-sensitive (e.g. semiconductors);

---

[21] Data Source: Boeing (2004), p. 8.

- need to be handled with caution (e.g. dangerous goods);
- require a live-sustaining of supporting environment (e.g. living animals).

In general, the market can be divided into three segments: *standard*, *express* and *specials*.

Standard cargo includes all shipments without any specific handling requirements on the ground or in the air. These shipments can be of any weight or dimension that is loadable without the deployment of specific devices.

The attribute express applies to those shipments that are highly time-sensitive, e.g. the time frame from the acceptance at the original airport to the availability at the final destination is less than 36 hours for an intercontinental transport.

Shipments with specific requirements regarding the preparation, loading or the actual air transport are considered special cargo.

The contribution of each segment to the growth of the air cargo market varies greatly. Whereas standard cargo still accounted for 82.1% of the total demand in 1997, a share of only 54.0% is anticipated for 2009. Growth takes place in the high value segments with estimated shares of 29.6% for special and 16.4% for express freight in 2009 (see figure 19-5).[22]

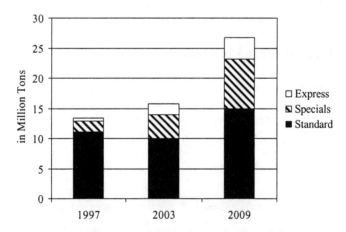

Figure 19-5: Segment Growth[23]

---

[22] See Lufthansa Cargo and Merger Global (2004), p. 54.
[23] Data Source: Lufthansa Cargo and Merge Global (2004), p. 30.

2.5 The Demand for Airport-to-Airport Products is Driven by Forwarders and Shippers

As air cargo carriers usually do not directly relate with the end customer, shippers usually purchase airport-to-airport capacity via forwarding agencies. About 90% of all contracts in the air cargo business are concluded between carriers and forwarders, only 10% with the actual shippers. This ratio is explained by the fact that forwarders are able to accomplish manifold logistical tasks, e.g. consolidation of shipments, customs clearance and to-door-delivery.

Nevertheless it is noteworthy to mention that an increasing number of shippers prefer to either purchase air cargo capacity directly from the airlines or at least to manage the three-fold relationship between forwarder, carrier and shipper. They also directly influence the product development of the carriers – sometimes even more than the forwarders. In addition, integrators and postal services complete the customer basis of an air cargo carrier.

Forwarding agencies purchase airport-to-airport capacity as part of the door-to-door-product they offer to their customers.

Integrators such as UPS, FedEx, DHL and TNT purchase airport-to-airport capacity as uplift for their intercontinental traffic in addition to their own freighter capacity.

With few exceptions (e.g. Deutsche Post World Net) postal services lack their own equipment and purchase airport-to-airport capacity for the air-transport of airmail and parcels.

An interesting side note is that for historic reasons airmail is actually not considered as airfreight. Accordingly cargo airlines always refer to *cargo and mail* when stating their freight performance.

2.6 Competitors in the Air Cargo Market

In general, one distinguishes between dedicated cargo carriers, so-called belly carriers, and integrators when looking at the different suppliers of air cargo capacity.

Within the segment of dedicated cargo carriers, a further distinction is necessary. Some airlines, such as Cargolux and Polar, rely solely on their freighter fleet; others, such as Lufthansa Cargo or Korean Airlines Cargo, are able to offer air cargo capacity in freighters as well as in the belly-hold lower decks of passenger aircrafts and are thus understood as combination carriers.

Those carriers with access to freighter capacity only seldom offer comprehensive transport solutions required for special cargo. Instead they concentrate on the segment of standard cargo and/or act as flexible production platforms for other carriers and forwarding agencies as they also offer charter and wet lease options.

Due to their ability to access freighter and belly capacity, combination carriers are able to offer a comparably impressive network and frequencies, which fits well with the demand of time-sensitive customers. Most of the combination carriers have also developed special handling processes and are therefore able to offer special transport solutions for handling-critical goods and to take advantage of the premium these customers are willing to pay.

Belly carriers sell air cargo capacity in addition to passengers and luggage transportation. With the transport of passengers being their core business, airfreight plays an inferior role and is therefore often regarded as an incremental contribution to the airline's bottom line only. They are thus usually dealing with standard cargo only. This undifferentiated approach to business has contributed significantly to the downward spiral in prices (see figure 19-4).[24]

The segment of the integrators occupies a dualistic role in the airport-to-airport market. On the one hand integrators purchase airport-to-airport capacity as uplift in the intercontinental traffic, on the other hand they appear as competitors. Whereas these market players are able to offer comprehensive, integrated transport solutions at an unbeatable speed for smaller shipments (<30kg) with access to own truck and freighter fleets, there are no market barriers to hinder them from accepting larger shipments on the airport-to-airport basis whenever capacity allows.

3 Competing in the Air Cargo Market

3.1 Success Factors for Marketing Air Cargo

Due to the different business models, the individual requirements of the above-mentioned customer segments differ largely. Nevertheless, the success of air cargo carriers relies significantly on their ability to deliver:

---

[24] See Kadar & Larew (2003), p. 4.

- *Speed:* the ability to deliver within an agreed time frame.
- *Global accessibility:* a global network combining the commercial capitals of today's business world.
- *Flexibility:* the ability to quickly respond to short-term booking requests e.g. providing sufficient capacity within an acceptable time frame.
- *Quality*: the ability to provide a global network with high frequencies, integrated information systems, customized transport solutions and a high level of service standards.
- *Reliability:* the adherence to deadlines and low failure rate.
- *Price/Performance-Ratio:* the ability to offer a differentiated, demand-oriented pricing policy.[25]

3.2 Different Strategies of Air Cargo Participants

The ongoing struggle for a profitable positioning in the air cargo market leaves few options for dedicated cargo carriers. In fact, only few strategic elements are available for a profitable positioning – cost leadership or focusing on differentiation and seizing synergies of scale.[26]

Air cargo carriers applying the strategy of cost leadership are aiming at maximizing profits by becoming the most cost-efficient capacity supplier of the industry and at the same time achieving average market yields. This strategy does not only require a stringent cost management, but also a favorable economic location as base. With respect to the complex processes, and in face of shipments with specific handling requirements, this strategy will most likely result in concentrating on standard cargo.

Focusing on the time- and handling-critical shipments of the express and special cargo segment and the deliverance of premium service describes the differentiation strategy. This strategy requires the stringent orientation of the organization along the high demands of premium customers. Nevertheless, cost management remains an important issue.

---

[25] See Beder (1998), p. 125 and Reifenberg & Remmert, chapter 23 in this volume.
[26] See Porter (1999), p. 38f.

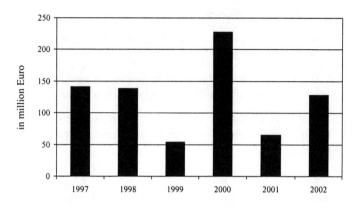

Figure 19-6: Operating Results of Lufthansa Cargo 1997-2002[27]

4 Successful Positioning – the Case of LH Cargo

One of the air cargo carriers successfully applying the differentiation strategy is Lufthansa Cargo (see figure 19-6). Formerly acting as cargo business unit of Lufthansa German Airlines, the company spun off in 1995, but is still wholly owned by the parent company.

In addition to a freighter fleet, which is presently made up of 14 MD-11SF and 8 Boeing 747-200 F/SF, the company has access to the belly capacity of the 332 passenger aircrafts of Lufthansa German Airlines (status 1/2004).

In response to the development of the logistics market, Lufthansa Cargo follows the strategy of differentiation through offering innovative, value-adding transport solutions along the shipper's and forwarder's demand.

4.1 Creating Product Value

The actual network of an air cargo carrier is the most important attribute of its product offering. As founding member of the first alliance in the cargo world (WOW), Lufthansa Cargo has gained access to the networks of the partner airlines Singapore Airlines Cargo, SAS Cargo and JAL Cargo. Complemented

---

[27] Data Source: Annual Reports 1997 – 2002, Lufthansa Cargo.

by several other bilateral agreements (e.g. with Cathay Pacific, LanChile, and Air China), the carrier is able to offer the largest world-wide cargo network to its customers. The majority of the process standards among the WOW-partners has been harmonized, enabling the alliance partners to offer air transport solutions on a truly global scale.[28]

Global networks alone, however, do not satisfy the high demands of premium customers. A premium air cargo carrier must also provide state-of-the-art facilities, loading devices and particularly a worldwide availability of specific handling processes.

With the launch of time definite services in 1998, Lufthansa Cargo pioneered in marketing three air cargo products; offering a guaranteed time of availability at the final destination within different time frames. Capacity access and performance guarantee[29] represent the major product attributes. The requirements of standard cargo are covered with the product td.Pro.[30] The products td.X and td.Flash provide the attributes time-critical shipments require.

The product portfolio of Lufthansa Cargo was quickly complemented with the offer of so-called service packages designed for specific, handling-critical shipments in 2000 and 2001. These comprehensive air transport solutions were developed in close cooperation with identified key industries.

The service-package safe/td2 for example was designed especially for logistic demands of highly theft-endangered and vulnerable shipments from the high-tech and telecom industry. The availability of special security areas at the airports, specific handling processes with defined hand-over processes and a completely documented transport chain guarantee the highest degree of protection for these shipments.

Another service package that can be clearly assigned to one specific industry is cool/td. The air transport solution for shipments that require a constant temperature management has been designed especially for the pharmaceutical and biotech industry. By implementing quality assured procedures in conjunction with state-of-the-art containers with an active cooling system, Cool/td is designed as a physical door-to-door process and guarantees an unbroken cool-chain from the shipper's to the consignee's premises.

---

[28] The WOW-network includes 523 destinations in 103 countries with 8 major cargo hubs (status 01/2004). See also Grönlund and Skoog, chapter 20 in this volume.

[29] If the agreed time of availability at the final destination cannot be met by Lufthansa Cargo, the customer receives a 100% reimbursement of the cargo fare.

[30] Standard cargo accounts for approx. 60% of all LCAG shipments.

With the recent launch of the customer-definite product cd.Solutions, Lufthansa Cargo has launched the third wave of its product differentiation and thereby complemented its core business, the airport-to-airport transport. The customer is now able to compose a tailor-made transport-solution from the different modules of the existing products and direct delivery-modes into the bonded warehouse.

With the launch of the time-definite services in 1998, Lufthansa Cargo pioneered in branding air transport solutions as products. This strategic decision changed the market perception significantly. Whereas the performance of air cargo carriers used to be perceived as an exchangeable commodity service, forwarders and shippers are now able to distinguish between standard and premium offerings.

The leading competitors adopted the philosophy of time-frames and product differentiation with a time-lag of two years. KLM and Singapore Airlines Cargo were the first to follow with the introduction of time-definite services in 2000 and a product launch in 2001.

4.2 Creating Brand Value

Knowing that the brand value of the parent company carries over to Lufthansa Cargo and using this effect, Lufthansa Cargo has developed a unique approach to branding and communication in the classical B2B-market.

Focusing on the identified key brand drivers "customer-orientation", "innovation", "quality" and "security", the company has designed an integrated communication strategy with specific concepts to address all relevant target groups, namely forwarders, shippers, employees, press and society.

Instead of using planes, loading devices or other production-oriented symbols as protagonists, customers are the focal point of the product advertisement campaign, represented by the specific requirements of their individual shipments. The product advertisement for live/td, Lufthansa Cargo's transport solution for living animals shall serve as an example (see figure 19-7).

Figure 19-7: Product Advertisement "Live/td"

The careful orientation of the communication and marketing strategy along the above-mentioned key brand drivers resulted in a positive emotional fuel of the brand Lufthansa Cargo.

4.3 Value Orientation in Sales

The worldwide sales organization of Lufthansa Cargo follows a value-oriented structure along specific customer segmentation criteria. A special business partner program – quite comparable to the frequent-flyer programs that passenger airlines offer to their premium customers – was established for the most valuable customers from the forwarding industry on global and regional scale.

Lufthansa Cargo has established a worldwide key account organization for the management of these partnerships. Their task is to create homogeneous and stringently coordinated relationships on all levels of LH Cargo and partner organizations and to build a fruitful foundation for a steadily growing and increasingly profitable joint business with a degree of non-exchangeability of the individual partners.

In response to the relevance of identified key industries for the air cargo business, Lufthansa Cargo has also established a key account organization for these stakeholders. Working in close cooperation with the product development department and the worldwide sales force, the expanded global industry team acts as a two-way transmission between shippers and Lufthansa Cargo. The bundled industry know-how of these employees has been incorporated into the development and further advancement of the described *Service Packages* and all other industry-specific transport-solutions.

A profitable positioning within this market, however, does not only require professional customer-relationship-management activities but the control of the sales unit costs. As a significant part of air cargo bookings are of a routine or standard nature, these transactions do not need to be assisted personally by cargo specialists. They require to be processed quickly and easily, with as little process costs as possible. Lufthansa Cargo reacted to these demands and established regional customer service and call centers for standard bookings.

While flanked by the options of electronic commerce, a comprehensive portfolio of electronic booking channels has been developed. This includes a web-based solution, electronic data change via so-called Cargo Community Systems (e.g. Traxon) and the neutral Internet marketplace GF-X that in addition offers a complete overview of the available capacities and rates of the participating airlines.

Time-consuming bookings by phone or fax or inconvenient inquiries about cargo capacities and rates became a thing of the past. The benefits of the electronic booking channels resulted in a growing acceptance among the customers. Endurance and performance proved to be critical success factors: whereas only 1% of all bookings were conducted via electronic channels in December 2001, the company was looking at a worldwide average of more than 35% by the end of December 2003.

5 Conclusion and Outlook

Terror prevention following the aftermaths of the tragic events of 11 September, has led to the implementation of several security measures in the airline industry. But not everything that is feasible for passenger airlines can also be applied to the cargo business. A 100 % x-raying of all cargo transported in passenger aircrafts, for example, would result in an enormous prolongation of time for ground handling and an abrupt jump in process costs which so far nobody is willing to pay. Higher rates and a longer transport time would translate into a significant loss of attractiveness versus sea freight.

One other scenario could be the total abolition of cargo from passenger planes. This decision would immediately reduce the available capacity to little more than 50% which would stand for an immediate relaxation of the competitiveness of the market and a significant disturbance of all production processes that rely on the air transport chain.

The responsibility for securing air cargo shipments, however, does not have to be taken solely by the airline within the small time frame for ground handling after accepting the shipments. For a maximum security, every participant in the transport chain must undertake all security precautions necessary in his area of responsibility. Lufthansa Cargo is therefore pursuing the approach of security partnerships together with forwarders, shippers and all other parties concerned (including governments and aviation authorities).

In addition to security a number of other factors will have high priority on the agenda of an air cargo executive, namely electronic commerce, quality, network and integration.

The availability of neutral Internet marketplaces and their exponential acceptance rates by the customers will significantly increase the availability of information within the air cargo market. Platforms such as GF-X will bring transparency into the jungle of available capacities and rates of the participating airlines. Finally it will be easy for the cargo customer to compare the actual offers.

Along with the increased transparency of offers, the degree of exchangeability among the bidding cargo carriers ascends thus disembouguing into an even higher relevance for true quality performance as the main differentiation criteria for the carriers.

Increased transparency and comparability will soon be applicable to the actual performance of logistics providers with the IATA interest group Cargo 2000 heralding a new era of quality management in the industry.

In the future, the delivered service performance of the participants will be audited against agreed and joint quality criteria. For the first time ever, cargo customers will be in the position to objectively compare quality and price-performance ratio. Needless to say that only carriers with a professional approach to business will be able to withstand the rigid quality standards and to win the customer's favor by continuously perfecting their operation.

Being already a unique selling-proposition for an air cargo carrier, the relevance of a global network will enjoy an even greater popularity along with the next steps of globalization. As there is no end for traffic regulations to be foreseen yet, air cargo carriers claiming to be true global players will have to find other ways to extend their own network, maybe by mergers and acquisitions wherever possible or by a powerful alliance with a joint understanding of how to do cargo business.

The air cargo market of the future will not be characterized by horizontal alliances alone. Despite few felicitous and several failed approaches to vertical integration of air cargo carriers in the past, there are several potentials to be tapped. Satisfying the ever-increasing specific logistic demands of certain industries for faster, more reliable and cost-effective services suggests the vertical integration of certain elements of the transport chain or a much closer cooperation between cargo carriers and forwarders or integrators in order to reduce the number of interfaces. Joint ventures for industry-specific air transport solutions could emerge as another business model of the future.

What about those carriers that have not found a professional approach to marketing air cargo capacity without standardized processes, a sophisticated quality management system and without any kind of specialization? Uncoupled from the threat that air cargo will be abolished from passenger planes for security reasons, cargo experts reckon that belly carriers, which regard air cargo as side product only, will vanish into thin air. As described earlier, these carriers will probably entrust their entire freight management (e.g. all sales and handling activities) to professional cargo carriers – like Lufthansa Cargo.

6 References

AIR CARGO WORLD (2002): Air Cargo Bottom Line. *Air Cargo World*, 92(11), pp. 62-63.

BEDER, H. (1998): Der Luftverkehr. In: Isermann, Heinz (Ed.): *Logistik: Gestaltung von Logistiksystemen*, 2nd Ed., Landsberg/Lech: Verlag Moderne Industrie, p. 125.

BOEING (2003): *World Air Cargo Forecast*. Seattle.

GRANDJOT, H.-H. (1997): Luftfracht. In: Bloech, J., Ihde, G. (Eds.): *Vahlen's großes Logistik Lexikon*, München: Franz Vahlen, p. 26.

GROTTENDIECK, M. (2003): *Akteure der Globalisierung*. Stuttgart, per Internet access on 14 January 2003. http://www.attac.de/stuttgart/dokumente/wto/asw-1wt2.doc.

KADAR, M.; LAREW, J. (2003): Securing the Future of Air Cargo. *Mercer on Travel and Transport, Speciality Journal*, Fall 2003/Winter 2004 (1) pp. 3-9.

LUFTHANSA CARGO AND MERGE GLOBAL (2004): *Planet Trends & Forecasts*, Frankfurt, pp. 22-54.

MAURER, P. (2001): *Luftverkehrsmanagement*. Munich: Oldenbourg, p. 86.

POMPL, W. (1998): *Luftverkehr*. Heidelberg: Springer, p. 89.

PORTER, M. (1999): *Wettbewerbsvorteile*. 5th Ed., Frankfurt, New York: Campus-Verlag, pp. 38-41.

SALING, B. (2004): The Building of an Industry. *STAT Trade Times*, 1.1.2004, p. 3.

WORLD TRADE ORGANIZATION (2003): per Internet access on 14 January 2004. http://www.wto.org/english/thewto_e/whatis_e/whatis_e.htm.

20

DRIVERS OF ALLIANCE FORMATION IN THE AIR CARGO BUSINESS

PETER GRÖNLUND AND ROBERT SKOOG

1	Introduction	474
2	History	474
3	Regulatory Environment	475
4	Specificities of the Air Cargo Business	475
5	Key Drivers for Alliances	478
6	Rationale for Cargo Alliances	481
7	Obstacles to be Aware of	483
8	Conclusion	484
9	References	486

Summary:
Air Cargo is a rather young industry that is still trying to find it's foothold. The players in the industry can be divided into two major groups: Carriers that have incorporated their cargo divisions employ a mix of market driven pricing, whereas the other group still considers cargo from a pure revenue stream and "cream on top" point of view. A situation that long term will not benefit either the carriers or the customers. In order to create some kind of economy of scale and a bigger network coverage than is possible for any single carrier, alliances are being forged.

1 Introduction

It is an undeniable trend in most lines of business today to partner up with companies that only yesterday were your competitors and a threat to your survival.[1]

This phenomenon is getting more and more prevalent in air cargo, too, a part of the transportation business that maybe is the one least known to the general public.[2]

Air Cargo, constituted by freight and mail together, represents less than 1% of the world's transportation work when putting air, rail, shipping and trucking together. The value, however, of this 1% cargo carried is around 10% of all goods commercially transported in the world.

Contrary to popular belief, most of the air cargo carried is loaded in the belly holds of passenger aircraft. Dedicated freighter aircraft at the moment carry less than 40% of the total cargo carried, a situation that very well may change in favor of freighters.

The Air Cargo business, when including both so-called integrators (i.e. FedEx, UPS, TNT and DHL) and scheduled passenger and cargo airlines, is nowadays a multibillion dollar business and has become an important part of most airline's revenue stream.

2 History

In order to understand today's developments and restructuring of the air cargo business, it helps to know about its roots and history.

In the early days of airlines, going back to the 1920s, most of the concessions granted by the authorities to operate a certain route or number of routes were

---

[1] See Hamel, Doz & Prahalad (1989).
[2] See Graumann & Niedermeyer, chapter 10 in this volume, but also Albers (2000), or Hastings (1999).

based on an absolute requirement that mail was carried on the aircraft in question.

However, the idea to carry goods as well developed very quickly. An aircraft, due to its oval aerodynamic design, generally holds an open space under the floor of the passenger's seats. Due to the small size of aircraft in the beginning of aviation, only smaller parcels and packages could be carried, but it was soon realized what a fast way of transportation it could be for spare parts and other urgent shipments.

WWII pushed the limits of aircraft size; the huge bombers proved that transatlantic flights with heavy loads were quite doable.

After the war, aviation literally took off.[3] Each country with self-respect wanted a national carrier to show the flag to the world. A group of American war pilots started what is often seen as the first cargo carrier and called it "Flying Tigers".

The real growth, however, was in passenger business and capacity grew quickly and in all directions across the globe.

This meant that cargo capacity in the belly holds was growing as well and soon most international airlines were owners of substantial volumes of aircraft space that sought its commercial use.

3 Regulatory Environment

Up until the mid-eighties, the price-setting mechanisms for air cargo were designed and controlled by IATA and the governments that were signatories to the Warsaw convention, the Haag protocol and other such international treaties intending to regulate air traffic from various aspects.

Prices were fixed in two yearly conferences organized by IATA and published, after government approval in the various countries affected, in the Air Cargo Tariff, a catalogue of prices everyone had to abide by.

Providing a very stable and predictable situation, this system was destined to collapse. When supply rapidly outgrows demand in a market economy, prices will come down. Discounts started to appear and when a number of states in the mid-eighties decided not any to require filings and approvals for cargo price adjustments any longer, price setting was turned loose!

---

[3] See Smith (1974), pp. 31ff.

4 Specificities of the Air Cargo Business

4.1 Distinguishing between Air Cargo and Passenger Transport

What makes cooperation important for air cargo in particular?

One of the more distinguishing factors between passenger business and air cargo is the importance and handling of the information flow pertaining to the service in question. A simple illustration is the comparison between Mr. Jones, traveling between Stockholm and New York via Copenhagen, and his cargo moving the same way.

> Mr. Jones is his own information processor when it comes to keeping track of the lead times necessary (when do I have to be at the airport?), the check points required (the check-in counter, the gate, etc.), the information on his ticket and on the monitors in the departure, transfer and arrival halls he passes on his journey. A seasoned traveler or not, Mr. Jones does most, if not all, of the computing required himself and also acts independently from others on the results in order to move through the transportation system.

> His cargo, on the other hand, is both deaf and dumb and will not move an inch unless prodded. Elaborate, expensive and complex, much more so than on the passenger side, IT-systems are therefore pivotal to the industry.

In SAS Cargo the value of the information flow needed to manage the flow of the physical goods is regarded to be at least 50% of the total customer value of the transportation service provided. Catering to a market and customers that, as mentioned in the opening paragraphs, is growing more and more global both with regard to movement of production facilities and to distribution of goods, requires from the air cargo companies the provision of matching global services.

Since no scheduled carrier has the ability to cover all the needs of all the customers of its own accord, co-operation and partnership becomes critical and the most critical area is timely and correct information about any customers' shipment status – its whereabouts so to say.

This is what makes air cargo alliances different from passenger alliances, the necessity not only to create the *appearance* of harmonized services but to actually make it happen through a long series of hands-on, practical joining of information *and* physical interfaces.

4.2 Emancipation of the Air Cargo Business

In 1995, Lufthansa's cargo division was incorporated, an event that was a clear signal to the industry and the market that air cargo had begun to grow out of its baby shoes. Lufthansa Cargo was from the very beginning set up as a fully fledged airline with own operating permit, own pilots and own aircraft.

As a consequence, a full profit and loss statement and a full balance sheet followed.

What has been slower in following though, are other airlines doing the same structural change in their business. Even today (2004), the number of cargo divisions that actually have been turned into legal entities of their own can easily be counted.

Even so, the trend clearly is towards incorporation of cargo units simply because it makes good business sense. Cargo follows its own business logic and requires a production system and pattern that is different from the passenger business.

Furthermore, with high demand from the market on "one-stop-shopping" and continuous adaptability to the global movement of factories, it is obvious that air cargo must be organized independently from the passenger traffic flows.

However, the cargo units that have so far been incorporated are fighting an uneven battle since the majority of carriers are still selling their cargo capacity on the margin of their overall capacity output.[4] Prices are set without reference to the cost of air cargo production. Handling, IT, accounting, financial costs and much more is simply absorbed by the airline itself and cargo revenues are only considered as the cream on top.

---

[4] See Shaw (1993), pp. 76ff.

5 Key Drivers for Alliances

Under the headline "Building a Strategic Alliance", section 6.2 below, you will find seven factors that constitute the organization of an alliance. These are somewhat specific for the individual case, once the decision has been taken, but it is of course equally important to identify and understand the fundamental driving forces for considering participation in an alliance in the first place.

5.1 Customer Value

Being able to create customer value that is greater than what can be accomplished on a stand-alone basis is the first and most important driver.

This includes primarily the following:

1. The traffic network must be organized in a way that allows for an increased number of destinations with as few, if any, route overlaps by partners as possible. Instead of flying wing-to-wing on city combinations, capacity must be redeployed to increase efficiency and reach of the total network.

2. The customer must experience at least the same quality and service as that of the most proficient of the partners in the alliance. Remember that air cargo customers have wide experience from a number of carriers and will immediately compare and become aware of any difference between the services of the alliance and the services of the individual carriers.

3. An alliance cannot afford to present the customer with logistical solutions that are more complicated than what one of the individual carriers already can offer. Simplicity is a key word and must be evident in all parts of the service. Booking export handling, tracking, transit handling, import services and finally billing must be as easy for the customer with the alliance as with only one carrier.

4. Finally the sum of the three items above has to lead to greater efficiency for the customer by using the alliance as a distribution system than any other system. Shorter lead times, improved information flow (quicker, more accurate, and consistently reliable), better value for money etc. are things that the customer will be looking for. These are obviously also areas that are

important to the alliance in order to continuously increase the competitive edge.

5.2 Synergies

For the partners and from an airline and capacity production point of view, certain synergies have to be realized, at least over time, in order to make alliance membership a viable proposition to a partner.[5]

1. In a market with ever falling yields and increased volumes, revenue growth is in itself an important goal. An alliance will provide added opportunities for utilization of unused capacity, thereby increasing the revenues for the partners.
2. Costs can be shared in a number of ways; almost everything from offices to marketing activities can be divided between the alliance partners creating more efficient use of funds.
3. The air cargo business has moved from being fairly "asset-light" in the early days to being more "asset-heavy". IT-systems, handling facilities, mechanization, aircraft and much more now weigh rather heavily on air cargo carriers balance sheets. Obviously, by sharing some of these investments, the financial costs for each partner can be lessened.
4. Procurement of e.g. handling and flown equipment (pallets, containers, etc.) can be made more efficient in more than one way. By combining purchasing volumes of the partners, more attractive prices can be obtained, the combined knowledge and experience of the partners should lead to smarter (and thereby more price-worthy) technical solutions and the "built-in" cross-utilization of equipment that follows from joint designs, will have an advantageous effect as well.

This taken together will form a sound economic basis for any alliance.

5.3 Improved Market Position and Customer Loyalty

As was stated earlier, the air cargo market is one of decreasing yields but increasing volumes and output of capacity. In such a business environment, the fight will be for market shares and market position, based in turn on customer

---

[5] See Albers (2000).

loyalty. An alliance must therefore provide the partners with a positioning opportunity that is unattainable from a stand-alone perspective. A strong market position with the ability to deliver according to market and customer demand puts any partner in what can become a very positive loop indeed.

5.4 Employee Opportunities

Even though it is a difficult promise to give, being employed within a greater corporate structure such as an alliance will in general provide better job security and career openings for the employees of the partners. The personal and individual contact network grows, learning/training possibilities increase, broadening of own experiences and knowledge can manifold etc. Being very much a "people's business", the air cargo community as such can only benefit from letting alliances provide its most important asset with these opportunities.

5.5 Shareholder Opportunities

Last but in no ways least, there is of course the driver for alliances called "shareholder value". We are back again to the making of choices being discussed earlier in this chapter. Owning shares in a company that, from the way the market is defined, has limited growth possibilities when trying to do it alone, probably does not hold too much perspective. Letting the value of those shares, on the other hand, be influenced by a partnership situation may of course contain an element of risk but also of growth opportunity.

Therefore, there has to be a clear business case for entering an alliance, as seen from the shareholders point of view. What is the market forecast on a macroeconomic scale, how will world air cargo capacity develop, what are the moves on the forwarder and customer arena, what further regulatory restrictions may be foreseen etc.

Such a positive business case is then the final but maybe most important driver for a cargo alliance to be formed.

6   Rationale for Cargo Alliances

Marginal pricing of substantial volumes of air cargo capacity is therefore one reason for any cargo carrier to try to solidify its market position; another important reason is the power of the integrators. Companies such as DHL, TNT and FedEx are expanding from their traditional business base of transporting small parcels and moving into the arena of scheduled carriers, cargo and passenger alike.[6] In other words, the cargo units now trying to survive as corporate entities are threatened from at least two sides.

What do you do when you find yourself in a difficult situation and see that others are sharing the same strategic challenges? You obviously have a choice to make: can I manage and survive on my own or is there strength in numbers? In other words, will it help to partner up with someone?

Again we have to return to where air cargo, as a still young business, currently is. The yields, earnings per unit produced, are consistently declining which puts emphasis on cost cuttings and a strong need for revenue growth.[7] Also, the industry faces large investments in IT-systems development, safety & security measures, cargo aircraft technology, airport and handling facilities improvements etc. All of this, of course, involves a great deal of risk, financial as well as strategic.

6.1   Staying Independent

To many companies, the spinal reaction to any outside threat probably is to try and maintain independence and integrity because of the assumed uniqueness of the own achievements, the name and brand that so much was invested in, the product portfolio and, of course, the possible trauma of having to deal with loyal staff and unions in an alternate scenario.

What else has to be considered is the market position of the company, the catchment area in terms of customers, volumes and revenues, the geography

---

[6]   See Doganis (1991) and Reynolds-Feighan (2001).
[7]   See Boeing (2002), pp. 2 and 7f. and Smith (1974), pp. 41ff.

(where is the company located), the financial situation, and last but not least, ownership.

If the combination of these factors is strong enough to face the competitive situation of today and tomorrow, assumed future volumes and revenues development etc., then it is close at hand to choose continued independence.

The question is, of course, how often this actually happens.

The entire airline industry, like the rest of the transportation business, distinguishes itself by functioning only through a mix of competition and co-operation between the companies involved. No single entity is big enough to service its customers with all the services and coverage demanded, especially not in an increasingly globalized world where manufacturing plants and markets move at a hitherto unforeseen pace.

Cost of capacity production has to be considered, too. There is no possibility of profitability without close to maximization of capacity utilization which in turn requires large volumes being into and out of the transportation network.

Therefore, standing alone is more often than not a very short-term option for any cargo carrier.

6.2 Building a Strategic Alliance

The other option should be rather obvious: find some partners with whom development costs, production costs, network and coverage etc. can be shared.

This, however, is probably easier said than done in most cases, there are a number of criteria to be met before a partnership can even be considered.[8]

First of all, can the corporate cultures be matched without any disturbing noises? It is not a question of being the same or acting the same in all respects, but it is a question of being kindred spirits when it comes to business focus and priorities, business conduct, market reputation and management style.

Second there should preferably be a natural fit of the respective networks, both geographically and technically. A complete imbalance between e.g. slimbody and widebody types of production is not a sustainable combination because of, among other things, the complications and costs in handling that it incurs.

---

[8] See e.g. Child & Faulkner (1998), Doz & Hamel (1998), Das & Teng (1997).

Third is finding ways of accounting for revenues and costs that are to be shared and find the keys to distribute this money.[9]

Fourth is to agree on how the co-operation development work is to be organized and driven.[10] Is it to be a traditional project assignment to a group of people behind the scenes or should it be organized in a way that immediately involves and commits the resources of the daily operations of the companies involved? A traditional project has the advantage of being able to move fairly quickly since it has the necessary resources available to 100% and little to distract it from its target. On the other hand, its conclusions and recommendations may eventually turn out to be too theoretical to fit the real needs and circumstances.

Basing the development on resources that remain in daily operations may slow things down a bit and will thus be a challenge to the tenacity and long term goal orientation of the companies engaged. It has a strong advantage though, all inputs to the development work are fresh off the shelf and shared knowledge grows parallel to changes in the market place, customer requirements etc.

Fifth is to be able to overcome maybe longstanding competition between the companies in question. Thus, it seems vital to realize and to understand that the long-term goal of close co-operation is more important and must not be disrupted by fights in local markets over prices or customers.

Sixth is trust and openness among the partners. No trust, no thrust.

Seventh is to be able to balance the six factors above while still maintaining a strong bottom line result for each partner, because nobody wants be partner with a weak company.

7 Obstacles to be Aware of

It is well known that this is not an ideal world. Not in a general sense and not for creating cargo alliances.

There are, of course, some stumbling stones that need to be dealt with and preferably before being stepped on.

---

[9] See chapter 11 by Götsch & Albers in this volume, or Albers (2000).
[10] See chapter 10 by Graumann & Niedermeyer in this volume, or Albers (forthcoming).

One is the regulatory environment that is not yet adjusted to the emergence of alliances and partnerships in a business little known and understood by competition law makers and enforcers. A number of things are not defined, the question of market dominance and when it becomes chargeable is one. Another is the customer value of "one-stop-shopping" when requested by the customer vs. the assumed negative consequences of cartels. Where, in fact, is the border line between being an alliance and being a cartel? No one seems able to provide an advance answer to that question, but it is highly probable that quite a number of people will know precisely once there is someone to accuse!

Another possible obstacle is, of course, what can come from within an alliance, the emerging inability to keep the co-operation going. One of the driving factors behind this could be lack of proven customer value, i.e. no extended network service, seams apparent in the seamless products, same lead-times as before etc.

The biggest obstacle, however, is time, or rather lack of it. With competitors and customers continuously growing bigger, any cargo alliance that does not get its act together soon will not survive in the shape and with the constellation that was intended.

8 Conclusion

Air Cargo is a rather young industry that is still trying to find its foothold. It is distinguished by a mix of market driven pricing for the carriers that have incorporated their cargo divisions and marginal pricing for the ones that still consider cargo from a pure revenue stream and "cream on top" point of view. A situation that long term will benefit neither the carriers nor the customers.

In order to create some kind of economy of scale and bigger network coverage than is possible for any single carrier, alliances are being forged.

In order to create, develop and maintain an alliance, the following "Tool Box" is required: [11]

- *Top/senior management commitment*

    Air Cargo, as most service organizations, is managed to a large extent by the values and policies that are expressed by its management. Moving a

---

[11] See also Devlin & Bleackley (1988), Dyer, Kale & Singh (2001), Spekman, Isabella & MacAvoy (2000), pp. 133ff., and Yoshino & Rangan (1995), pp. 109ff.

company into an alliance has both direct consequences and implicit effects on its employees and this can only be dealt with through very clear signals from management on why that particular choice has been made. The power of the line functions to, wittingly or not and unless convinced, impede the success of an alliance should not be underestimated.

- *Quick and effective decisions*

  Other air cargo alliances, actually among carriers and forwarders alike, will be watching what you are doing so do it fast in order not to lose the competitive edge.

- *Continuous and efficient communication to staff and customers about goals and progress*

  Seeing is believing. Air Cargo is a fairly small business community with a high degree of information leakage, it is therefore important to take the lead in informing about your activities instead of having rumors running around.

- *Personal trust, understanding and relations between top managers of the partners*

  Air Cargo is also very much a people-based business. A good personal understanding between the decision makers is extremely important to make an alliance work. Remember that we are talking about a long ranging co-operation but not a merger.

- *Understanding of and respect for cultural differences and similarities, both ethnic and corporate, also a willingness to learn about the same*

  The world of air cargo is in its nature international, but only when you actually have to be part of joint endeavors you realize what differences in culture you have to be able to overcome. In WOW at least four corporate cultures (actually six since SAS Cargo has three within itself) and a much higher number of ethnic cultures are coming together to accomplish a number of joint offerings.

- *Clear targets, measuring and follow up of performance, individual as well as joint*

  Measuring is the mainstay of every business and with an alliance the importance grows. The challenge is to find agreement on what and how to measure.

- *Cost control and the realization that alliance participation will generate costs you did not have before*

A strong alliance development organization, e.g. a project team, is important and one of the chief tasks is to engage in the financial side of the co-operation. Up front agreements on what costs are to be pooled and what not relieve much unnecessary strain on the alliance work.

- *Pragmatism, patience and the ability to maintain focus on the long-term goal*

Air Cargo as a service industry is not locked in by large local investments and the individual corporations are therefore fairly free to move between partnerships. It has to be remembered, though, that "remarrying" carries new entry costs and that if you do it too often the potential spouses will become more and more apprehensive over time.

The survival of these corporate co-operations will eventually be decided by the market forces, the management capability of the companies concerned and the regulatory authorities. The alliances that manage to use the "tool box" above may stand a better chance than others, but remember, it certainly is hard work!

9 References

ALBERS, S. (forthcoming): *The Design of Alliance Governance Systems*. Cologne: Kölner Wissenschaftsverlag.

ALBERS, S. (2000): *Nutzenallokation in Strategischen Allianzen von Linienluftfrachtgesellschaften*. Working Paper No. 101 of the Dept. of General Management, Business Policy and Logistics of the University of Cologne, Cologne.

BOEING (2002): *World Air Cargo Forecast 2002/2003*. Seattle: Boeing Commercial Airplane Group.

CHILD, J.; FAULKNER, D. (1998): *Strategies of Co-operation. Managing Alliances, Networks and Joint Ventures*. Oxford: Oxford University Press.

DAS, T.K.; TENG, B.-S. (1997): Sustaining Strategic Alliances – Options and Guidelines. *Journal of General Management*, 22(4), pp. 49-64.

DEVLIN, G,; BLEACKLEY, M. (1988): Strategic Alliances – Guidelines for Success. *Long Range Planning*, 21(5), pp. 18-23.

DOGANIS, R. (1991): *Flying Off Course – The Economics of International Airlines.* 2nd Ed., London, New York: Routledge.

DOZ, Y.L.; HAMEL, G. (1998): *Alliance Advantage: The Art of Creating Value through Partnering.* Boston: Harvard Business School Press.

DYER, J.H.; KALE, P.; SINGH, H. (2001): How to Make Strategic Alliances Work. *Sloan Management Review,* 42(4), pp. 37-43.

HAMEL, G.; DOZ, Y.L.; PRAHALAD, C.K. (1989): Collaborate with Your Competitor – and Win. *Harvard Business Review,* 67(1), pp. 133-139.

HASTINGS, P. (1999): In Partnership. *Cargovision,* 14(3), pp. 18-21.

REYNOLDS-FEIGHAN, A.J. (2001): Air-freight Logistics. In: Brewer, A.M.; Button, K.,J.; Hersher, D.A. (Eds.): *Handbook of Logistics and Supply Chain Management,* Amsterdam et al.: Pergamon, pp. 431-439.

SHAW, S. (1993): *Effective Air Freight Marketing.* London: Pitman.

SMITH, P.S. (1974): *Air Freight. Operations, Economics and Marketing.* London: Faber & Faber.

SPEKMAN, R.E.; ISABELLA, L.A.; MACAVOY, T.C. (2000): *Alliance Competence. Maximizing the Value of Your Partnerships.* New York: Wiley.

YOSHINO, M.Y.; RANGAN, U.S. (1995): *Strategic Alliances. An Entrepreneurial Approach to Globalization.* Boston: Harvard Business School Press.

21
AIRFREIGHT DEVELOPMENT SUPPORTING THE STRATEGY OF GLOBAL LOGISTICS COMPANIES

RENATO CHIAVI

1 Introduction 490
2 From Niche Products to Global Logistics 491
3 Learnings 511
4 References 514

Summary:
Airfreight is embedded in a framework of global economics, trade development and regulations. It is also the product of innovations in infrastructure and communications technology. Airfreight forwarding is and will continue to be a key driver in furthering globalization. On the way to realizing the one-stop shopping vision, global logistics companies will integrate airfreight into the overall supply chain strategy to meet the needs of multi-national customers.

1 Introduction

The aim of this chapter is to examine the airfreight business and its learnings for the strategies of global logistics companies. Besides theoretical aspects, historical and empirical evidence will support the argumentation.

What is the strategic framework of global logistics companies? The airfreight strategy of these companies has to be analyzed with a focus on the macroeconomic environment and changes in the business cycle (figure 21-1). Global and regional perspectives as well as regulations influence the strategic decisions of global logistics companies. The reason is that logistics has been transformed from basic transportation processes to a complex amalgam of network, process and information driven services, from A-to-B forwarding to an end-to-end supply chain approach and towards one-stop shopping. The product range has been developed towards integrated logistics products, services and value adds.

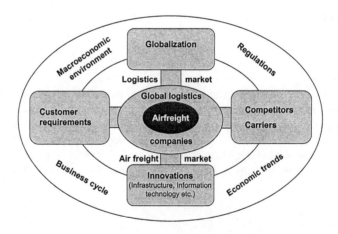

Figure 21-1: Strategic Framework of Global Logistics Companies

Airfreight, created as a niche product for the urgent delivery of expensive goods, has become irreplaceable. Today, it plays a vital role in supporting the globalization of production and distribution processes driven from international business opportunities.

2 From Niche Products to Global Logistics

An overview of the development of the economic, regulatory and institutional environment and the reactions of airfreight demand and supply provides an understanding of the progress of airfreight and its integration into the strategies of global logistics players.

2.1 The 40s and 50s: Airfreight as Niche Service

Economic growth and world trade accelerated in real terms during the 1950s under the regime of fixed exchange rates. One milestone of the process of liberalization starting to reduce international trade barriers were the General Agreement on Tariffs and Trade (1947) as a forum for negotiating lower customs duty rates and removing non-tariff trade barriers. A further one was the creation of the European Economic Community (EEC) chartered with forming a common market in 1958.

Airfreight was a sideline operation to mail and passenger traffic until 1941, when the "Big Four" airlines (United, American, TWA, and Eastern Airlines) formed Air Cargo Inc. to begin regular flights. By the end of the war, many of the airlines, including United and TWA, began their own independent airfreight services.[1]

After World War II, veterans of the US air force were given special preference to purchase or lease government surplus aircraft. This spawned airfreight pioneers like Robert Prescott, who founded the National Skyway Freight Corporation (NSFC with DC-3, C-46) with the famous Flying Tigers.[2] Two US Air Force pilots formed Seaboard World Airlines with an ex-USAAF C-54 transport freighter converted to civil standards. In 1947, the first charter flights to Europe and the Middle East using a DC-4.

In the post-war era, more than 100 veterans founded various small air cargo carriers, competing with the Big Four. They also fought against the Civil Aeronautics Board (CAB), which was responsible for pricing regulations and

---

[1] See US Centennial of Flight Commission: www.centennialofflight.gov.
[2] Historical details on: www.flyingtigerline.org/history.htm.

favored the large carriers. During this time, prices fell from 26 cents per ton-mile to 11 cents. To stop this destructive competition, the CAB installed the "Minimum Rate Order" in 1948, in force until 1953.[3]

The commercial airfreight business differed from the express business in several respects.[4] Compared to express, airfreight was product specific to each company and managed by the company itself or by the freight forwarder as broker.

From the start, airfreight rates were significantly lower than express: 61 cents per ton-mile for express compared with 26 cents for airfreight (1946). The main reason for the price difference was the operational involvement of normal airfreight versus faster express delivery. Another important difference was the necessary equipment on the ground that required significant capital investments.[5] Airfreight was therefore a niche service for expensive products.

Much of the early growth of airfreight stemmed from the involvement of the US military in the post-war period and in the Cold War. The needs to reduce inventory stocks and increase transportation speed generated investment in research and development. The standardization of pallet and container took shape and network infrastructure was installed (especially in the US with the Logistic Airlift "Logair").[6] The civil airfreight industry would ultimately benefit from the military advancements.

The International Air Transport Association (IATA), reformed in 1944, established parameters for a systematic organization, technical standardization and larger infrastructure. IATA's efforts promoted safe, regular and economical air transportation, as well as standardizing documentation and processes.[7] The Bermuda Agreement between the United States and Great Britain was the first bilateral air services agreement, establishing the concept that carriers, not the respective governments, should set fares and rates. This compromise between the open skies policy of the United States and the regulatory policy of UK induced the start of the regulatory phase.

With the Standard Agency Agreement (1952) the pattern for airline-agent relations was set. The International Civil Aviation Organization (ICAO), founded in 1944, worked on standardizing the international air transport network.

---

[3] 16 cents per ton-mile for the first 1,000 miles and 13 cents for the quantities above; for details about the airfreight history see Allaz (1998).
[4] See Allaz (1998), p. 321.
[5] See OTA (1982).
[6] See Allaz (1998), pp. 301ff.
[7] See IATA (www.iata.org).

The volumes of international air forwarding grew at a double digit pace. Technological innovation led to the development of turbo-propeller aircrafts in the early 1950s and jet propulsion later in the decade.

Airfreight forwarders (AFF) started as intra-US transport organizations in the early 1940s. For economic reasons, airfreight forwarders acted as consolidators. IATA price policy differentiated between General and Special Commodity Rates.

Demand for air cargo was facilitated by the fact that passenger capacity was stimulated during the 1950s. In the middle of the decade, the production of the first jet aircraft started. Pan American World Airways ordered 44 aircraft in 1955 (B707 and DC-8). Ocean freight and project companies began to offer airfreight services with first freighter schedules (pallets). The freight was transported in passenger aircrafts, loaded into the baggage holds by hand.

2.2 The 60s: Airfreight as Ad Hoc Services

The 1960s turned out to be the period with the highest average growth of world real GDP. But the expansive fiscal and monetary policy produced increasing inflation and generated an increasing imbalance between global participants. The growth of international trade was reflecting changes in world growth. And international air transport expanded along this favorable environment with double-digit growth.

IATA's activities concentrated on technical work, regulations and automation in airline operations and safety. The agency agreement was reformed and the regimentation of air passenger and cargo was split. With the assistance of the International Federation of Freight Forwarders' Association (FIATA),[8] the cooperation between the airfreight industry and the agents was professionalized.

Airfreight capacity was expanded by the development of new aircraft. For example, Lufthansa entered the jet era in 1960 with the B707. The international network was further developed. State owned carriers invested in modern network and equipment. Airfreight forwarders offered ad hoc airfreight services. Rapid industrial development drove demand for faster transport modes. Passenger

---

[8] The non-governmental FIATA was founded in Vienna/Austria in 1926, and represents today an industry covering approximately 40,000 forwarding and logistics firms.

carriers converted into freight carriers. This development was supported by the fact that infrastructure capacity was restricted in ocean freight networks.

2.3 The 70s: Airfreight as Intercontinental Product

High inflation rates and low economic growth characterized the 1970s. The Arab oil embargo in 1973 caused an increase of the oil price from 3$ to 12$ per barrel, triggered massive and widespread inflation (figure 21-2) and depressed the business cycle. In the same year, flexible exchange rates were introduced. But this did not, as expected, reduce the dependency on the oscillation of the international business cycle. The volatility of the exchange rates, nevertheless, was identified as a new source of instability in the following years. The EEC introduced the European Monetary System (EMS) in 1979 with the objective to reduce this volatility by a system of semi-fixed exchange rates.

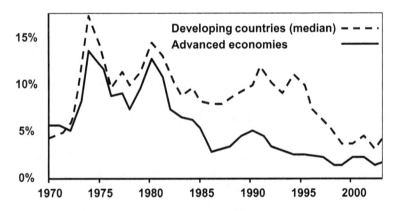

Figure 21-2: Consumer Prices (annual change in percent)[9]

Airbus was established as a consortium of French, German and later Spanish and UK companies in 1970 to compete against the US aircraft manufacturers. The first Airbus was launched in 1972 (A300) as passenger aircraft. The first aircraft, convertible from passenger to cargo (A300 C), was delivered in 1980 and the pure cargo version (A300 F) in 1991.[10]

---

[9] See IMF (September 2003): figures estimated for 2003.
[10] See Allaz (1998), p. 485.

IATA established the ULD (Unit Load Device) Control Agreement in 1971. By the Show Cause Order (SCO) two categories were created: the Trade Association as obligatory and Tariff Coordination as optional. The first cargo airline concepts were introduced by passenger carriers (e.g. Alitalia, Air France, and Lufthansa) and independent operators (e.g. Cargolux, IAS, German Air Cargo – now a subsidiary of Lufthansa – and British Air Fary). The airfreight market was characterized by an imbalance from the demand side and ocean port congestions in some developing countries. Real rates decreased slowly but remained at historically high levels.

The international fleet was a combination of propeller (CL 44) and jet aircraft. With the technical innovation of widebody aircraft, capacity was further expanded. Passenger aircraft had containerized freight sections (DC-10 CF) and new cargo-only aircrafts were introduced. Lufthansa started in 1972 with the first nose lifting cargo-only aircraft, the B747 F. The reduction of on/off-loading time was a further element in reducing transportation costs. Nevertheless, the handling equipment was not very sophisticated and forklifts were primarily used.

During the 1970s, significant regulatory changes took place in the US.[11] Until the 1970s, fares, rates, routes and services were regulated by two laws: the Civil Aeronautics Act of 1938 and the Federal Aviation Act of 1958. Following the oil embargo, prices per passenger-mile increased by one third from 1973 to 1978.[12] Overcapacity occurred during the 1974/75 recession as previously ordered widebody aircrafts were delivered and entered service.

These factors caused a political turnaround. In 1977, cargo carriers were allowed to set market prices and fly any domestic route. In 1978, the Airline Deregulation Act in the US gave the same freedoms to passenger airlines.

This regulatory change boosted the express business in the US. United Parcel Service (UPS), founded in 1907, became the world's largest parcel delivery company. During the 1950s, UPS started air services in the US. Europe was added to the route network in the mid 1980s; DC-8F, B747-100F and B757 (freighter version) were used.

Federal Express (FedEx), founded in 1971, two years later already had a fleet of 60 Dassalt Falcon 20 jets. FedEx expanded into larger planes, which stimulated daily package growth by 43% from 1975 until the end of the decade.[13]

---

[11] See Goodman (2000).
[12] The changes of US airfreight rates didn't equal inflation, so cargo as well as passenger rates decreased in real terms.
[13] Total package volume: 11,000 in 1975 and 68,022 in 1980.

In the US airfreight market, UPS and FedEx expanded from documents to small parcels. They also expanded intercontinentally with integrated customs clearance also for parcels with low commercial value as well. Customs documents had to be produced as part of the systemized procedure at origin.

DHL, founded in 1969, was a pioneer in the air express industry with its innovative idea of sending out documentation in advance of arriving cargo, thereby speeding up the process of importing goods. The DHL network grew rapidly from Hawaii to the Far East and Pacific Rim, then to the Middle East, Africa and Europe. Within the US, DHL used Swearingen Metro II and Cessna 402 twins with other scheduled airlines carrying the bulk of the packages. In just four years, the company had expanded to provide services to 3,000 customers. By 1977, it had extended its range of services and started to deliver small packages as well as documents.

Cargolux Airlines International began its operations in 1970 as an all-cargo airline. It became Europe's largest cargo airline using CL-44J turboprops. Another new freighter company was Tradewinds that flew cargo flights from Gatwick in the early 1970s with CL44-D4, later replaced with B707s in the 70s.

On an intercontinental basis, airfreight grew during the decade. The volume forwarded on the North Atlantic route grew from 1970 to 1980 at a rate of 7.7% per year (5.4% on cargo-only flights).[14]

2.4 The 80s: Structured Airfreight

OPEC increased the oil price from 13$ to 32$ per barrel in the early 1980s, dealing another blow to the purchasing power of the industrialized countries, and re-igniting inflation pressures. Ongoing concerns over "stagflation" (stagnation with inflation) changed the political climate. Ronald Reagan, elected US President in 1980, changed fiscal policy in 1981 by reducing government spending and cutting taxes. After a slowdown in 1982, the economy began to grow again in real terms. Deregulation became an issue in other countries like England ("Thatcherism") as well, starting to change the political minds in favor of reinforcing the markets.

---

[14] Freight forwarded by IATA companies cumulated in both directions: 411,000 tons (247,000 or 60% on cargo airplanes) in 1970 and 860,000 (386,000 or 45%) tons in 1980; see Allaz (1998), p. 475.

Taiichi Ohno and Shigeo Shingo from the Japanese automotive industry developed the "Toyota Production System", known as just-in-time concept (JIT). During the 1980s, JIT became a catalyst to optimize time, assets and productivity. At that time, US industry came under pressure not just by the Japanese competition but also by high interest rates and respective inventory costs. Since then, and partly by the adoption of JIT concepts, lower cost structures have been achieved by enhanced logistics management.

During the 1980s, steps were taken towards strengthening intra-regional liberalization. The Cooperation Council of the Arab States of the Gulf (GCC: Bahrain, Kuwait, Oman, Qatar, Saudi Arabia, and the United Arab Emirates) integrated their economies by eliminating barriers to the free movement of goods. In 1989, the US and Canada entered into a Free Trade Agreement. An inter-regional agreement was signed between the US and the European Civil Aviation Conference (ECAC) in 1982, authorizing the resumption of tariff discussions for North Atlantic routes.

New types of aircraft such as the B747 have expanded airfreight capacity. The use of dedicated loading and off-loading devices and system-oriented handling of airfreight, further reduced loading times and costs. Specialized designs such as nose-loading were introduced to handle oversized pieces. Airfreight was forwarded by freighters and by belly aircraft, which carried passengers and cargo. New growth patterns for passenger traffic increased airfreight capacity and the airfreight network.

The express business grew fast during this decade. Federal Express expanded its fleet by adding Boeing 727s, DC-10 and Boeing 747 widebody freighters. In 1989 FedEx acquired Flying Tiger Line[15] with its huge fleet of DC8-63Fs and B747-200s to become the world's largest full-service, all-cargo airline. Included in the acquisition were routes to 21 countries, a fleet of B747 and B727 aircraft, facilities throughout the world and Tigers' airfreight expertise. UPS started with its own fleet in 1982 (with 24 B727 QC) and rapidly expanded its engagement by additional aircraft capacity, 1985 with the B747 F and 1987 with the cargo-only version of the B757 ("Package Freighter").

Until then, integrators operated by converting passenger planes into cargo aircraft. UPS was the first company to order dedicated freighters when it ordered 20 all-cargo planes with an option for a further 15 planes.[16] Competition was fierce among the major express operators. DHL operated worldwide with the

---

[15] Including Seabord World that had been acquired by Flying Tiger in 1980.
[16] See Allaz (1998), p. 538.

B727 and on long-haul routes with DC8-73 freighters. By 1999, Airbus A300F aircraft had been delivered for the US routes.

These carriers were called integrators because of their systematized operating procedures and door-to-door pick-up and delivery service.

Airfreight was a time-definite product mainly between airports and offered little value-added to justify the fast transit times and premium prices. Delivery and clearance procedures were very cumbersome. In response, integrators began to standardize procedures for smaller packages.

2.5 The 90s: Customers Going Global

Despite the slowdown in 1990-91, the decade became known as the "Roaring Ninetics".[17] The level of growth did not reach that of the 1960s but continuously accelerated during the whole decade. Inflation, which had peaked in the early 1980s, began to decelerate rapidly in the 1990s (figure 21-2). While the world economy was still going strong Asia suffered a setback in 1998 after the Thai currency collapse the year before. The effect may be seen in the development of the global economy, trade and airfreight (figure 21-3).

During the decade, the major economies got impulses from further deregulation and privatization of state-owned companies. Fiscal and monetary policy seemed to have learned their lessons and supported sustainable, non-inflationary growth. It also contributed to a reduction of the amplitude of the business cycle as concluded by an OECD study.[18]

The reduction in inflation was also due to the increased service sector and the decrease of stocks, which were partly caused by improved methods of inventory control and increased outsourcing of logistics. The use of expedited, or time-definite, transportation services led to the reduction of storage and capital in transit. This trend may be observed by comparing inventory data with GDP: the reduction in the value of inventory in the USA declined since 1992 from 7.2% to 4.8% in 2003.[19]

Trade liberalization continued globally. One interesting example for liberalization in an emerging country is India, which was in a serious foreign

---

[17] See Stiglitz (2004).
[18] See Dalsgaard et al. (2002).
[19] Statistical data about inventory of total manufacturing are available on: http://www.census.gov/indicator/www/m3/hist/m3bendoc.htm.

exchange crisis in 1991. Its trade policy at the time was among the most restrictive in the non-socialist world. The newly elected government reduced the fiscal and current account imbalances and introduced structural reforms in the country's trade and fiscal regimes. While analyzing the effects of liberalization, empirical studies show the positive effects of this policy on the country's economy.[20]

On a regional level, liberalization of European countries was supported by the Treaty of Maastricht 1992 creating the European Union (EU). USA, Canada and Mexico signed the North American Free Trade Agreement in 1994 (NAFTA). On a global basis, GATT became the World Trade Organization (WTO) in 1995, based on a new umbrella agreement for trade in goods. Liberalization had a positive growth impact on globalization, world trade and worldwide transportation.

The trends of liberalization, the attractiveness of low cost countries and enforced JIT strategies empowered a new process of global relocation of production and distribution, as well as an enhancement of logistics concepts. In a German survey, companies expected to decrease the vertical range of manufacture by concentrating on their core competencies and outsourcing more of their logistics processes.[21] About 23% of the trading companies said they were actually adopting Efficient Consumer Response concepts. Other studies of the 1990s indicated that certain combinations of scheduling and routing rules were successful in lowering manufacturing and logistics costs while maintaining high levels of customer service.[22]

During this decade, Internet technology initiated new perspectives in generating and handling information and performing communication. E-business, based on EDI and Internet technology, started its growth path to become a part of B2B and B2C commerce.[23] Buying and selling with this new technology immediately created fulfillment opportunities for all modes of transportation.

Efforts in the airfreight business concentrated on refining and globalizing the network. The hub-and-spoke system became a preferred method for keeping line-haul costs low during periods of slack demand, while optimizing growth trends during periods of high demand.[24]

---

[20] See Go & Mitra (1998).
[21] See Baumgarten, Darkow & Walter (2000).
[22] Several researches have studied this subject; see e.g. Ruiz-Torres & Tyworth (1997).
[23] See Colin (2001).
[24] See Barla & Constantinos (2000).

In the 1990s, the airfreight industry, stimulated by strong customer demand, heavily increased capacity through new types of aircraft with a wider fuselage. FedEx, for instance, introduced Airbus A300 and A310 freighters replacing the older Boeing 747s. Another example is Cargolux enlarging its fleet with 10 B747-400Fs and becoming Europe's largest cargo airline. During the 1990s, customer demand stimulated the emergence of a new concept, the ACMI providers (Aircraft, crew, maintenance and insurance) "wet" leasing airplanes. ACMI services increased air capacity, especially in long haul, intercontinental markets serviced by widebody freighters by 24% per year during the decade.[25]

As with the decades before, cargo yields fell continuously as a result of increased productivity and pressure from the demand side (figure 21-3). During the 1990s, this situation did not hurt the airfreight business because volume demand generated sufficient growth.

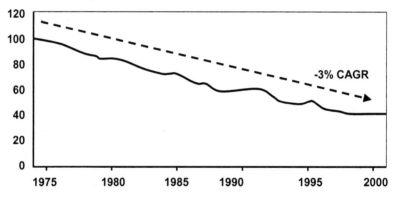

Figure 21-3: Indexed World Airfreight Cargo Yields (1974 = 100)[26]

Macroeconomic forces, market threats and opportunities as well as business concepts like JIT strategies drove customers to go for a higher level of global presence and involvement. The complexity of the logistics processes along the supply chain grew exponentially in the meantime.

After periods of diversification, companies went back on the road of driving their own core competencies. Outsourcing of logistics started to become a trend. Logistics companies, aware of this trend, started to define "one-stop shopping" as their vision towards fully integrated logistics services along the whole supply chain of the customers.

---

[25] RTK (billions): 1.522 (1990) to 13.335 (2000); see Boeing (2002).
[26] Figure based on Kadar & Larew (2003), p. 3.

Logistics companies started to extend their product portfolio in order to meet the customers' needs. But there were just a few big companies like Deutsche Post, FedEx and UPS that started to bring this vision to reality on a global basis by systematic acquisition of companies covering the missing products, services and network.

In 1998, Deutsche Post acquired Danzas, the Swiss logistics provider with a long forwarding tradition and operations in 52 countries. Later that year, Deutsche Post acquired a 25% share of DHL, and in 1999 purchased Air Express International (AEI), the leading US airfreight provider. With the integration of AEI, "Danzas AEI Intercontinental" became the leader in the international airfreight business with a market share of 6.4%.[27]

2.6 Starting the 21st Century: Overcapacity and Grounding

The economy had a weak start to the new century. The meltdown of many Internet and telecommunications companies in 2000 triggered a devastating stock market decline and a global recession. Corporate governance scandals in the US depressed business and investor confidence. The worldwide downturn, which began by many estimates in early 2001 did not truly end until the middle of 2003.

Despite the extremely negative economic environment and unsettled expectations, structural changes continued to stabilize and strengthen regional markets. In January 2002, the Euro became the single currency within twelve participating member states of the EU. The trend towards further global liberalization nevertheless suffered setbacks, cumulating in the failure of the trade Ministerial Meeting of the WTO in Cancun in 2003 near the bottom of the business cycle. The GCC members, introducing common external tariffs, launched a customs union in 2003. The GCC common market should come into effect in 2007. The GCC also initiated steps toward monetary union coming into effect by 2010 and formally pegged their currencies to the USD.[28]

Liberalization of aviation services got a new impulse from the first multilateral Open-Skies agreement in 2000 between the US, Brunei, Chile, New Zealand and Singapore. It mirrors the US bilateral agreements, which permit unrestricted or very liberalized international air service. This model could become a standard for the global marketplace.

---

[27] Measured as IATA turnover 1999.
[28] See IMF (September 2003).

The agreement is also liberalizing the traditional ownership requirement, thus enhancing foreign carriers' access to investment. As aviation is governed by thousands of bilateral agreements between more than 180 countries, this multilateral agreement could heavily simplify the procedure of negotiation.[29]

The IATA rate framework continues to be the benchmark for prices on many routes provided by scheduled passenger carriers in conjunction with airfreight forwarders. But as the IATA license is no longer a requirement to act as an airfreight forwarder, fixed tariffs are no longer being observed.

The economy and especially air transportation have been hit by several unforeseen events: terrorism, the Iraq war and SARS to name the most important. Terrorism became a new dimension on 11 September, having an impact on the economy and security, especially in the airline industry.

Terrorism and the measures taken to combat it have caused a shift in handling security issues. First of all there was a major effect on the demand side. The risk of the airway as mode of transportation was actually increased but also overestimated by an overshooting of negative expectations. On the other side there was a lack of infrastructure and procedures to upgrade security needs in the short term. The major consequences were the introduction of new rules in airports and for importing goods into the US. Airfreight rules (known and unknown shipper concept, x-ray of cargo, etc) had been introduced in early years after the Lockerbie incident.

In 2003, European airplane manufacturer Airbus delivered more jets (305) than its competitor Boeing (281). Airbus had expected delivery for 2004 of "close to 300" and seems to have received more orders and delivered more aircraft than its competitor.[30] Boeing was, according to its own statements, harder hit than Airbus by the slump in the civil aircraft market that followed 11 September.

Capital-intensive markets like the airline industry have been hit tremendously by the recession. During the 1990s, a glut of passenger airplanes with higher cargo capacity were ordered and delivered. The effect was a further acceleration of overcapacity resulting in an even higher pressure on rates.

An additional negative effect was the further attack of the new low cost airlines. Carriers started to reduce their lanes, networks and schedule density. Many

---

[29] See US Department of Transportation.
[30] Statement of Philippe Camus, co-chief executive of European Aeronautic, Defense and Space Co., reported by y KOMO Staff & News Services on 15 January 2004 (www.komotv.com/boeing/story.asp?ID=29282).

freighters disappeared from the market and their planes were stored in the US desert.

Nevertheless, freight-only companies like Polar Air Cargo[31] emerged. Lean organization and small overhead costs allowed them to work with profit. Other freight airlines operated with a different concept. They concentrated on a few customers and vertical alliances with direct shippers and forwarders and still operate on a relatively healthy basis (e.g. Cargolux with its 11 B747-400F).

An IATA Interest Group of major airlines, freight forwarders and ground handling agents has initiated Cargo 2000. The goal is to implement a quality management system for the worldwide air cargo industry by implementing processes that are measurable and improve the efficiency of the supply chain. The 30-member group has re-engineered the transportation process from shipper to consignee.[32]

FedEx and UPS acquired several companies during this period to broaden their service portfolio.[33] In 2001, UPS acquired Fritz Companies, a global freight forwarding, customs brokerage and logistics company. FedEx integrated the Tower Group, a leading provider of international trade services, specialized in customs brokerage and international freight forwarding, and formed "FedEx Trade Networks Transport & Brokerage". Besides the standard product portfolio it offers value-added services like information tools (Global Trade Data) that allow customers to track and manage imports.

Deutsche Post went public in 2000 changing its brand to Deutsche Post World Net (DPWN). Two years later, DHL became a wholly owned subsidiary of DPWN. In a short period of a few years, DPWN changed to a multinational group that provides integrated mail, express, logistics and financial services worldwide with about 380,000 employees. With the motto "One brand – one face to the customer" DPWN has further pushed its "one-stop shopping" approach in 2003 by rebranding DHL as the umbrella brand for Express, Freight, Logistics and Solutions. With the acquisition of Airborne Express, the fourth-largest parcel carrier and third-largest air express carrier in the US, the market presence of DHL in the US has been significantly upgraded.

---

[31] Polar, founded 1993 has been acquired by Atlas 2001 and is using exclusively B747-100F and B747-200F.
[32] See IATA (www.iata.org).
[33] See Polito (2003).

2.7 Quantitative Analysis of the Past

The accepted quantitative measures of the international airfreight business are Freight-Ton Kilometers (FTK) and Revenue-Ton Kilometers (RTK).[34] In figure 21-4 growth rates of FTK are compared to world GDP and world manufacturing export volumes, representing world trade.

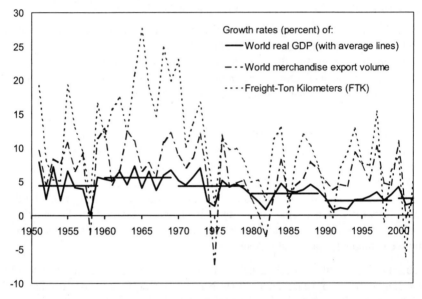

Figure 21-4: World Growth, Trade and Airfreight Development[35]

In an empirical UK study, a simple linear regression model indicated that growth of airfreight is likely to be 3.18 times GDP growth.[36] Applied to the FTK data plotted in figure 21-4, the regression model[37] reveals a coefficient or elasticity of 2.40 for the whole period of 1951 to 2002 (Table 21-1). The average growth

---

[34] RTK are commonly used exchangeable with FTK but can include passenger weight for total revenue.
[35] World real GDP and world trade figures are based on data of World Trade Organization WTO (www.wto.org) and FTK are based on data of Air Transport Association of America ATA (http://www.airlines.org/econ).
[36] See MDS Transmodal et al. (2001), p. 16.
[37] The model: $(\ln(FTK) - \ln(FTK_{-1})) = f(\ln(GDP) - \ln(GDP_{-1}))$.

rate[38] for FTK is 9.76% with a standard deviation of 6.22 compared to the lower GDP growth rate of 3.75% with a 1.85 standard deviation. RTK shows similar results. RTK growth has an average rate of 9.08% with a standard deviation of 5.57% and is likely to be 2.16 times GDP. The volatility of the airfreight business, induced by the business cycle, does not seem to be much stronger on airfreight volumes than on revenues.

Period	FTK = f (GDP)				FTK
	Coefficient	s	p	R^2	ø%
1951-02	2.40	0.33	0.000	0.51	9.76
1951-59	1.62	0.65	0.031	0.51	10.21
1960-69	0.70	1.33	0.614	0.03	16.75
1970-79	2.43	0.74	0.011	0.58	10.53
1980-89	2.18	1.01	0.063	0.37	7.13
1990-99	2.06	1.80	0.287	0.14	6.43
1951-70	1.84	0.61	0.008	0.33	14.01
1961-80	2.64	0.68	0.001	0.46	13.34
1971-90	2.23	0.41	0.000	0.62	7.94
1981-00	1.76	0.78	0.037	0.22	6.95

Table 21-1: World Real GDP Impact on Airfreight (FTK)[39]

Has this effect changed over the years? Table 21-1 lists the results of the analysis of the decades and overlapping 20 years' periods.[40] The GDP coefficient or "multiplier" started on a lower level through the airfreight "take-off phase". During the "growing 60s", other influences seem to take the lead and prevent GDP from reaching significance.[41] An explanation of this phenomenon may be

---

[38] The average growth rate ø% = ln(x)-ln(x_{-1}) as used in the regression model corresponds to the Compound Annual Growth Rate (CAGR) but avoids errors caused by random deviations of the start and end points of an observation period.
[39] Statistical significance: * = p < 0.05; average annual growth rate ø% (see footnote 5).
[40] The larger database increases the probability to get significant results.
[41] Neither do results reach significance if the period is enlarged to the whole growth phase 1959 to 1973.

the increased influence of other factors like transportation costs and relative prices. During the period of accelerating inflation and growing instability, GDP changes regained influence on a significantly higher level. The strength of GDP impact on the airfreight oscillation decreased considerably in the succeeding periods, observable also at the decrease of the percentage of the variance explained (R^2). A reason for this may be that structural changes like deregulation and stronger competition brought the airfreight business closer to changes of demand. A further effect could be the increased globalization of production and distribution processes, linking the volatility of airfreight tighter to the business cycle.

The results of the analysis show four characteristics. First, the average growth rate of freight forwarding over the entire period is significantly higher compared to world trade and GDP. Second, airfreight forwarding is strongly related to the business cycle and trade development. Third, airfreight development shows a considerably higher volatility compared to real GDP. Fourth, the relative volatility seems to have decreased over the last three decades.

Sophisticated econometric models define airfreight demand as a function of economic activity as well as transportation costs, exchange rates and relative prices.[42] The empirical results reveal evidence of a significant influence of all of these variables on the airfreight business.

2.8 2004 and Beyond: Towards Global Supply Chain Logistics

A worldwide economic recovery began to take shape in March 2003 soon after the Iraq war began. The real economy began a major upswing in the second half of the year, with corporate profits and sales rebounding strongly. Consumer spending and industrial productivity, which stayed strong in the US during the recession, remained buoyant.

However, much uncertainty remains. There are concerns about companies regaining pricing power for their products. There are worries about a resurgence of inflation due to the recent upward price spikes in oil and other commodities. Fears of terrorism, reinforced by the 11 March railroad attacks in Madrid, will continue to dog the global economy.

---

[42] See e.g. MDS Transmodal et al. (2001), Boeing (2000) and Airbus (2002).

In order to gauge the health of world trade, and by extension of air trade, we have to rely on forecasts of experts and on what we have learned from recent developments. Assuming no extraordinary negative impacts will occur, the world economy is expected to grow on the long-term by about 3% per year. Nevertheless, the possibility of a high volatility cannot be dismissed. Estimates for world trade will show a significantly larger growth rate than the GDP development.[43]

GDP Growth	% change 2001-2021
China	6.2%
Asia	3.3%
Latin America	3.8%
Africa	3.7%
Middle East	3.9%
Europe	2.3%
North America	2.7%
Japan	1.7%

Table 21-2: Forecast for Regional Real GDP Growth[44]

The growth of the airfreight business will vary by region. The predominant growth impact will come from Asia, especially China, after it enters the WTO in 2007 (table 21-2). The industrialization of the developing countries and regions will create new local purchasing power, thus increasing domestic demand for new products.

This development calls for increased capital investments, which will be directed mainly into the regions with lower labor costs. The preferred regions will be Asia, Latin America, Central and Eastern Europe and probably a few African Nations. The 15 "old" European countries seem to be the losers, with many experts forecasting relatively low-growth for those economies due to higher labor rates and less flexible cost structures.

---

[43] See Airbus (2002).
[44] See Boeing (2002).

The additional ten nations joining the single market of EU in May 2004 might trigger domestic opportunities during the initial period. Some of these countries are low cost labor producers and might attract investments.

Russia is outperforming the EU countries. In 2002, Russia's GDP grew by 4.3%, surpassing average growth rates of all other G-8 countries, and marking the country's fourth consecutive year of economic expansion.[45] The growth was fueled primarily by energy exports, particularly given the boom in Russian oil production and relatively high world oil prices. The Russian economy holds much promise, but complex regulatory and an uncertain political will for liberalization may temper growth prospects there.

Liberalization of trade and globalization will continue because free trade is the best chance to generate wealth in emerging regions. One example is the Free Trade Area of the Americas (FTAA) that will expand NAFTA to Central and South America. Negotiations will be completed by 2005.

We will see stronger co-operation between the states and international organizations. The ICAO has currently 188 Contracting States that are obliged to comply with the minimum safety, security and environmental standards established by the organization or to inform other countries of variations.

Liberalization will continue in the airfreight and passenger industries. One example is the Pacific region with its 14 Forum Island Countries, having 25 bilateral Air Service Agreements and 26 with the rest of the world.[46] The process of establishing a single aviation market started with the Forum Aviation Policy Meeting in 1998 to build the "Pacific Islands Air Service Agreement" (PIASA): ministerial approval (September 2001); agreement on final text of PIASA (October 2002); PIASA opened for signature (August 2003); entry into force after 6 ratifications (early 2004); Phase 1: (six months after entry into force); Phase 2 (one year after entry into force); Phase 3 (two and a half years after entry into force. The full implementation of PIASA will take place in 2007, meaning a nine to ten years' process of building a single market.

The airfreight business will develop further with growth rates significantly superior to what is assumed to be adequate for the GDP development of the next years (table 21-3).

Airfreight capacity, counted as numbers of air freighters, is forecasted by Boeing to grow by an annual compound growth rate of 2.8% from 2001 to 2021. But as

---

[45] US Energy Information Administration (www.eia.doe.gov/emeu/cabs/russia.html).
[46] See Ferguson (2003).

the average volume per aircraft will increase at a higher rate, total available capacity, measured in ton-kilometers (ATK), will increase in the same time by 4.7%.[47]

Forecast	Time-line	GDP	Airfreight					
			FTK				RTK	Yield
		World real GDP CAGR	Total CAGR	Express CAGR	General Cargo CAGR	Special CAGR	CAGR	CAGR
Boeing	2002-2021	2.9%	6.4%	na	na	na	6.5%	na
Airbus	2000-2020	3.3%	5.5%	na	na	na	4.7%	na
Merge Global	2003-2009	na	5.5%	12.6%	2.7%	8.3%	na	0.3%

Table 21-3: Forecasts[48]

Airbus is forecasting a total net increase of about 90 freighters per year until 2020 with a shift in the capacity per freighter.[49] The A300-600F is already in service, the A380F will enter service in 2008 and plans already exist to launch the A330-200F. With an average growth of 4.7% over the forecast period, Airbus controls a substantial part (2000: 48% to 2020; 43%) of freight being transported by belly carriers.[50]

The regional growth opportunities for airfreight will also vary significantly. Figure 21-5 lists the top ten air lanes, including intra US and intra Asia, with a total share of 64% of the world market. It is clear that the GDP growth potential of the regions will determine the future of the regional airfreight business.

---

[47] Airfreighters categories 2001 (2021): small (less than 30 tons: B727, B737, DC-9/MD-80, BAe 146): 39% (29%); medium standard-body (30 to 50 tons: B757, DC-8, B707): 22% (11%); medium widebody (40 to 65 tons: B767, A300/A310, DC-10-10; L-1011): 17% (34%); large (more than 65 tons: B747, MD-11, DC-10-30, A380): 22% (26%); see Boeing (2002)
[48] See Boeing (2002), Airbus (2002) and Clancy and Hoppin (2004).
[49] See Airbus (2002).
[50] See Airbus (2002); Boeing (2000) is expecting a higher share of over 50%.

The long-term potential will definitely be in China, where massive free-market changes are taking place. The US domestic market, already highly developed, will grow more slowly (3.5% instead of 7%). Nevertheless even with a twofold growth rate Asia would need 55 years to equal the size of the US market.

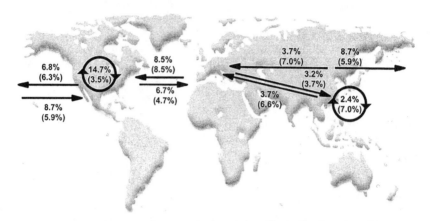

Figure 21-5: Top Ten Airfreight Lanes in Market Shares and their Forecasted Annual Growth Rates (CAGR 2000-2020)[51]

Structural market changes may occur through increasing use of e-Business platforms like Global Freight Exchange (GF-X) and Cargo Portal Services (CPS). They will not alter the business fundamentally, but may increase productivity and perhaps cause some pressure on rates.

Another market impact will come from technical innovations like Radio Frequency Identification (RFID) and Global Positioning Systems (GPS). These innovations will support cargo movement and enhance security.

One-stop shopping and all-embracing supply chain management are grand strategic visions, important first steps towards this goal have been made.

Multinational companies (MNC) are forcing the logistics industry to meet their supply chain needs on a worldwide basis. It is reasonable to assume that only a few large logistics providers will have the tools to meet these needs. And airfreight will continue to play a pivotal role as more leading-edge logistics concepts integrate airfreight into the general supply chain decision process.

---

[51] See FTK figures for market share (2000), see Boeing (2002).

The overall strategy of global logistics providers is to focus on customer needs and reorganize their internal structures to meet those needs. Starting in January 2004, Deutsche Post World Net has inaugurated a Global Customer Service Center to ensure a uniform customer approach. DHL is developing a sales strategy for its largest clients with its international Global Customer Solutions (GCS) organization. A special function within GCS is assigned to the Global Customer Manager, who serves as the customer's sole contact person, combining resources with product and industry expertise.

3 Learnings

3.1 Changing Patterns

The analysis of the past shows the core factors influencing the behavior of companies engaged in the airfreight business and the changing patterns of what we defined as their strategic framework.

Macroeconomic variables like GDP and trade directly influence airfreight demand. Simple empirical modeling could not explain changes in demand of every period observed. Other influences like transportation costs and relative prices as well as structural changes have gained influence. Globalization and deregulation have increased the possibilities of the whole industry, but not stability of business cycles and therefore volatility of the airfreight business. This asks for advanced modeling of airfreight demand and a more sophisticated company strategy.

In the take-off phase airfreight started as a niche business in a mainly national (US) environment, driven by pioneers entering and forming new markets for express and commercial airfreight. Aviation authorities were originally formed to regulate passenger aviation and to control quantities (routes) and prices (rates) of the airfreight market. But the practice of market regulation has changed over time towards more liberalization of the national, trans-national, trans-continental and intra-regional markets. Deregulation policy has, in many cases, freed global rates from bureaucratic control.

As the commercial airfreight forwarding industry is a broker-driven business, many other restrictions and boundaries influence this market and its participants' behavior. The liaison to the passenger aviation industry is a story of its own, characterized by the need of capacity sharing and the wish of independence. Airfreight companies had to cooperate to get capacity from carriers at best conditions, fixing their engagement for an adequate timeframe to reduce risk. Today's global logistics players have developed models of collaboration to handle this complex interaction of capacity planning and pricing, with the inclusion of further variables like reliability and security.

Capacity development in the medium and long run depends on technical innovations. These innovations primarily concern aircraft and ground infrastructure, but increasingly involve the connection to other modes of transportation, as airfreight is becoming a more integrated part of the supply chain processes. The difference in long-term perspectives of capacity expansion and short-run changes in demand is another source of instability. In the strict sense, it is the difference of the expectations of the market participants that creates the volatility. We can observe this in the recent ordering behavior of no-frills airlines. It is possible that the capacity will be delivered in a new phase of lower growth as observed some years before and provoke another phase of overcapacity with shrinking yields. Airfreight and logistics companies as brokers do not own aircraft. Nevertheless, they cannot inhibit disturbances to spill over to their own business. So they have to consider this phenomenon in their strategy of contracting with carriers.

The steady decline of airfreight yields during the whole observation period is the moderate sign of an intensive competition to increase productivity and the pressure of the customers to hand-over the benefits.

The demand for airfreight has dramatically changed, as customers go global not just for selling their products but also for producing them. Managing the supply chain, worked out in the regions and between continents, is also going global on an accelerating speed. Logistics companies are required to mirror or trump the customer's presence and offer a network, local knowledge and relations covering the globe.

3.2 Takeaways for Strategic Decisions

Relative to other transport modes, airfreight will continue to be priced at a premium. It will always be the mode of choice for moving urgent shipments over long distances. A second consideration for the selection of airfreight is the trade-off between transit time and capital costs. If an extra dollar spent on airfreight results in more than a dollar savings in inventory carrying costs, the airfreight investment is money well spent.

Key takeaways for the airfreight strategy of global logistics companies may be summarized as follows:

The macroeconomic environment has a predominant impact on the performance of the airfreight business through multiplying fluctuations of the business cycle, world trade as well as costs, currencies and relative prices.

The regulatory framework must inject adequate measures to stabilize the industry in an increasingly unstable world.

Globalization has created tremendous opportunities for market participants, but has also created a tighter dependency between customer and carrier and their performance.

Global production is highly labor cost-oriented and companies increasingly relocate production to drive down labor costs. In this scenario, logistics providers are not creating a market, but are instead following a trend governed by global customers.

The airfreight business still operates in a fragmented market that is dependent on the availability of airfreight capacity. Passenger aircraft routes that enable global cargo transportation mainly define this capacity.

Innovations such as new aircraft, airport infrastructure or e-market places need time to be developed. The sum of these factors creates a positive long-term force for airfreight activity.

Wherever goods are produced and have to reach consumers, logistics providers will be involved. But global logistics today is a process and information driven business, with transportation just one part of the picture. Logistics providers must continuously reengineer their products and processes to increase productivity and lower costs in order to be successful in the market.

A few big global logistics providers are well on their way to realize the one-stop shopping vision. They incorporate their airfreight business into an offering of global supply chain logistics with numerous value-added solutions. The goal: serving the dynamic needs of multinational customers.

Acknowledgements

The author would like to thank Christian Jäggi for his assistance.

4 References

AIR CARGO MANAGEMENT GROUP (ACMG; November 2003): *International Airfreight and Express Industry Performance Analysis 2003*. Seattle, Washington.

AIR TRANSPORT ASSOCIATION (2004): *The Airline Handbook – Online Version*. www.airlines.org/publications.

AIRBUS (2002): *Global Market Forecast 2001-2020*. Ref. CB 390.0008/02, Blagnac Cedex, France.

ALLAZ, C. (1998): *La grande aventure de la poste et du fret aériens du 18e sciècle à nos jours*. Paris: Institut du Transport Aérien.

BARLA, P.; CONSTANTINOS, C. (2000): The Role of Demand Uncertainty in Airline Network Structure. *Transportation Research, Part E: Logistics and Transportation Review*, 36(3), pp. 173-180.

BAUMGARTEN, H.; DARKOW, I.-L.; STEFAN, W. (2000): Die Zukunft der Logistik – Kundenintegration, globale Netzwerke und e-Business. In: Hossner, R. (Ed.): *Jahrbuch der Logistik 2000*; Düsseldorf: Verlagsgruppe Handelsblatt; pp. 12-23.

CLANCY, B.; HOPPIN, D. (2004): *Cargo Doesn't Care – The 2003 MergeGlobal World Airfreight Forecast*. MergeGlobal, Inc., www.aircargoworld.com.

COLIN, J. (2001): *The impact of e-commerce on logistics*. Paper presented at the Joint OECD/ECMT Seminar, Paris, June 5-6, www1.oecd.org/cem/online/ecom01/-Colin.pdf.

DALSGAARD, T.; ELMESKOV, J.; PARK, C-Y. (2002): *Ongoing changes in the business cycle – evidence and causes*. OECD Economics Department Working Papers No. 315.

DEL POLITO, G.A. (2003): *Competition within the United States Parcel Delivery Market*. Association for Postal Commerce, Arlington.

FEDERAL AVIATION ADMINISTRATION (FAA, 2004): *Aerospace Forecast Fiscal Years 2003 – 2014*, www.api.hq.faa.gov.

FERGUSON, E. (2003): *The Implementation of the PIASA Agreement, presentation at the Conference on Air Transport in the Pacific*. The Pacific Economic Cooperation Coucil PECC, November 12-13, Noumea, New-Caledonia, www.pecc.-org/community/second-airtransport-program.htm.

Go, D.S.; Mitra, P. (1998): *Trade Liberalization, Fiscal Adjustment and Exchange Rate Policy in India*. Paper presented at the conference on "Trade, Growth, and Development" in honor of Professor T.N. Srinivasan, March 27-28, at Yale University.

Goodman, W.C. (2000): Transportation by air: Job growth moderates from stellar rates. *Monthly Labor Review*, 123(3), pp. 34-48.

International Air Transport Association (IATA): www.iata.org.

International Civil Aviation Organization (ICAO): www.icao.int.

International Monetary Fund (IMF) (2003): *World Economic Outlook*.

Kadar, M.; Larew, J. (2003): Securing the Future of Air Cargo. *Mercer on Travel and Transport*, Fall 2003/Winter 2004 (1), pp. 3-9.

Levine M. (1987): Airline Competition in Deregulated Markets: Theory, Firm Strategy, and Public Policy. *Yale Journal of Regulation*, 4(2), pp. 393-494.

MDS Transmodal, Roger Tym & Partners, ACMS (2001): *UK Airfreight Industry Study*. Undertaken for the Department of the Environment, Transport and the Regions (DETR) Aviation Policy Unit. Ref: 26199r_August2001.

Morlok, E.K.; Nitzberg, B.F.; Balasubramaniam Karthik (2000): *The Parcel Service Industry in the US: Its Size and Role in Commerce*. University of Pennsylvania, Philadelphia, PA 19104-6315.

Office of Technology Assesment OTA (1982): *Impact of advanced air transport technology, Part 2: The Air Cargo System*. Background Paper, Government Printing Office Washington, D.C. 20402 NTIS.

Organisation for Economic Co-operation and Development (OECD, 2003 and 2004): *Main Economic Indicators*.

Ruiz-Torres, A.J.; Tyworth, J.E. (1997): *Simulation based approach to study the interaction of scheduling and routing on a logistic network*. In: Andradóttir, S.; Healy, K.J.; Withers, D.H. and Nelson, B.L. (Eds.): Proceedings of the 1997 Winter Simulation Conference.

Stiglitz, J. (2004): *The Roaring Nineties – Seeds of Destruction*. London: Penguin Books.

US Department of Transportation: *International Aviation Developments (Second Report): Transatlantic Deregulation The Alliance Network Effect*, ostpxweb.dot.gov/aviation.

Yergin, D.; Vietor, R.H.K.; Evans, P.C. (2000): *Fettered Flight: Globalization and the Airline Industry*. Cambridge: Cambridge Energy Research Associates.

22

INTEGRATOR NETWORK STRATEGIES AND PARAMETERS OF AIRPORT CHOICE IN THE EUROPEAN AIR CARGO MARKET

BENJAMIN KOCH AND ANDREAS KRAUS

1 Introduction 518
2 Liberalization as the Initializing Factor 520
3 Basic Integrator Air Network Patterns 521
4 Operational Parameters 523
5 Airports as Service Providers 529
6 European Integrator Network Strategies 531
7 Changing Parameters of Airport Choice? 535
8 References 537

Summary:
Within a relatively short time period, integrator transportation has become an essential part of the Intra-European transport system. As air transportation holds a key function in the Integrators' time-guaranteed freight delivery services, their network design has to reflect a specific set of operational and infrastructure-related parameters. This chapter elaborates on the structure of the networks and the major providers' different network strategies. Consequences for integrators' choice of airports are drawn and an analysis of the main challenges of the changing European market environment is conducted.

1 Introduction

The supply side of the air cargo transportation market is divided into four different kinds of airlines: (1) combination carriers, carrying both passengers and cargo, (2) scheduled all-cargo airlines, (3) all-cargo charter airlines and (4) integrators.[1] While the first three categories of air carriers are traditional airlines, operating flights predominantly for the purpose of transporting goods (or passengers as well) from airport to airport, the latter group expanded its function and services along the whole transportation chain. Thus, integrators do not leave the air cargo transportation chain as a sequence of independent, single activities, but integrate it into a holistic, comprehensive concept, covering the transportation from door to door – from the shipper up to the consignee.

While the concept of the integrated transportation chain allows control of all activities from origin to destination and not only parts thereof, it reduces the clients' costs by diminishing the profit seeking party to only one provider. On a parallel basis, integrators strive for an optimization and standardization of the physical goods flows by establishing transparent and efficient logistics systems. Typical disadvantages of this business model can be identified in the large transport volumes necessary to activate economies of scale and scope, and to take advantage of the benefits of highly standardized processes. Furthermore, customer-specific solutions cannot be offered to a large extent, since this counteracts the target of standardization.

In order to guarantee the efficient provision of services, the integrators permanently need to adapt their product specifications to the changing market environment. This is mainly driven by altering customer requirements, including not only the expectation that the shipments are delivered to every desired place at a reasonable price and with continuously shorter total transportation times. Exceeding this, additional value adding services are required, often covering the outsourcing of logistics functions such as warehousing, logistics process design, and management or even services such as repair shops for computer equipment.[2]

Besides these customer driven influences, changes in the regulatory framework pose the need for adaptation of the integrators' business strategies as well.

---

[1] The term "express carrier" is used as a synonym for "integrator" in this article, referring only to the four service providers mentioned.
[2] See Doganis (1999), p. 14.

Deregulation in the air transportation and mail markets, environmental and noise emission issues or changing regulations on the transport of specific shipments such as dangerous goods determine the market development additionally.

Even though the integrated transportation concept covers different modes of transportation, the fast carriage by aircraft always enjoyed a key importance in the integrators' transportation activities. In a well-functioning integrator, air transport system additional capacities can be provided relatively easily as long as supplementary or larger aircraft are required, which can be leased-in at short notice.

Contrary to this, the decision on the choice of airports to be served as part of the integrators air services demands for a longer planning time frame, as airports form an important node between the integrators' closely coordinated land and air transportation systems. Conflicts can occur in this key function, when the integrator cannot fully control the physical handling at an airport because of legal restrictions or non-existent facilities. This is why integrators tend to invest in their own air cargo facilities, once the cargo volume justifies their establishment. On a parallel basis, such an engagement raises the inertia of the airport choice for the integrator company significantly.

In Europe, the integrator market is dominated by four major players: DHL, Federal Express (FedEx), TNT, and United Parcel Service (UPS). Only these companies maintain their independent air transportation networks in Europe, most of them even being connected to their own global air networks.

An enlarging European market puts new pressure on the existing set-up of integrator air services in Europe. As customers demand the same transportation quality to and from the new European Union member countries, geographical coverage and services provided have to be re-allocated. Thus, the European integrators have to re-evaluate their current system not only regarding the general performance, but also with regard to how they might be able to serve all markets within an expanded air service system.

In this chapter we discuss the impact of the changing business environment on the European integrators and their air transportation networks, with a focus on airport choice. We start with a brief description of the liberalization of the European postal services sector to introduce the integrators' market and its development before a basic framework for the design of air networks is sketched. Then the operational parameters for the integrators' air transportation systems are elaborated and the different strategies of the European market players are discussed. Finally, the results are transferred to provide an outlook into the future development of the integrator networks in Europe.

2 Liberalization as the Initializing Factor

A stepwise liberalization of the European postal services sector[3] paved the way for the supply expansion of express carriers' services in the European marketplace. The transportation of packages on international services meant the market entrance into the European postal scene in the late 1970s and early 1980s. Since that time the new suppliers have expanded their service portfolio continuously. This trend was and still is only limited by national regulations to secure the former monopolist national postal service providers. This is why the express carriers' political power groups have been and still are lobbying for a complete liberalization of the European letter carriage market.

As a consequence, the liberalization pressure has forced the traditional European postal service providers to react. Initiatives to privatize the national postal service providers unleashed the influence of national governments on the market positioning of these companies. The national foci – unknown to the international nature of the transportation business – were waived in order to act as international logistics companies. The strong competition from the integrators led to the co-operation of the former Dutch postal services with TNT and the former German postal services with DHL. The express carriers were eventually acquired by these postal services – today, only UPS and FedEx remain independent from former national postal service providers.

The integrators' product had a significant impact on customer behavior: from a niche market position in the early 1980s the continuous expansion of the geographical market coverage in Europe and the addition of new customized services (e.g. less weight restrictions and special purpose transportation) led to a significant redistribution of market shares in the European air cargo market.[4] Especially in the 1990s a dramatic expansion of the cargo volumes transported on the air services of integrators in Europe took place (see figure 22-1).

Figure 22-1 furthermore indicates that the post 9/11 aviation crisis initialized a fundamental restructuring of the air cargo supply in Europe: traditional air cargo network carriers reduced capacity, by (1) reducing passenger flights [less belly

---

[3] See e.g. Schwarz (2004).
[4] A clear distinction between general cargo and express shipments does not exist. It is estimated that integrators transport as much regular cargo as express consignments. See Grin (1998), p. 84.

cargo capacity onboard passenger aircraft was the consequence] and (2) the withdrawal of freighter services from Intra-European services and the reshuffle of cargo volumes to road feeder services. As a consequence, air cargo volumes transported on other scheduled freight air services decreased as the supply side restructured on a parallel basis. Intra-European transportation demand for express air cargo was less negatively affected from 9/11 effects and remained relatively stable compared to the heavy effects on the passenger market.[5]

Furthermore, all relevant industry forecasts expect a continuous market growth of the integrator business in Europe.

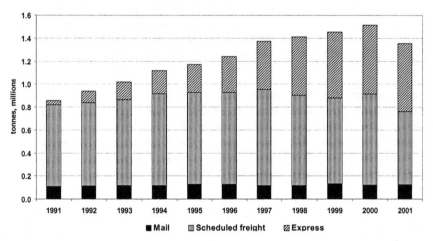

Figure 22-1: Transported Air Cargo Volumes in Europe (in million tons)[6]

3 Basic Integrator Air Network Patterns

Basically three general air route network types can be distinguished, the first being a network consisting only of direct links between all service points. The second network type is the so-called "hub-and-spoke-network", in which the direct links between the single points in the system are replaced by indirect services, which are all routed via the hub as the central transit point. Finally, a third variant combines a number of hub-and-spoke-networks with each other.

---

[5] See Hätty & Hollmeier (2003), pp. 51ff.
[6] Adapted from Boeing (2002), p. 37.

Besides these basic network types, combined networks can be found as well, showing structures based on both a hub-and-spoke-network and direct services circumventing the hub.[7]

The first network provides the best service coverage of all network types, connecting every point within the network directly without a transit point in between. On the other hand, it also implies a significant need for transport capacity and prohibits taking advantage of economies of scale and the chances of combining traffic flows. Hub-and-spoke networks always require a transit at the hub, but allow a bundling of traffic flows and a reduction in the overall capacity needed to satisfy the demand, since the average capacity utilization can be optimized. For this reason, most networks of scheduled airlines are nowadays designed in a hub-and-spoke-structure, providing the capability of serving a high number of destinations and services while maintaining high average capacity utilization.[8]

Several airline operators have enlarged their networks towards a multi-hub-system.[9] Multiple reasons led to such a decision. The first results from the target markets served. As soon as a certain number of destinations is not only located in a relative geographical proximity to each other, but also generates sufficient transportation demand, a bundling of these traffic flows via an additional hub bears significant efficiency potential.

Additionally, operational and financial aspects can make the set-up of a multi-hub-system a preferential alternative. Hub operations require large facilities, since a hub airport can only lead to an efficient transit of passengers or cargo if fast and reliable onwards connections are guaranteed. This requires a large number of aircraft and their loads to be handled at the same time at the hub airport, leading to the need to establish appropriate facilities. In order to avoid congestion at existing facilities or to keep the traffic volumes at the hub airport at a reasonable scale, a split between several hub locations can be a useful option.

The diverse geographical distribution of population and economic activity in the European Union leads to a heterogeneous and dispersed demand for passenger and cargo air transportation, resulting in considerable complexity of origin/destination air transportation patterns in Europe.

---

[7] See Delfmann (2004), Pfohl (2004), pp. 106ff.
[8] See Delfmann (2004).
[9] This effect cannot only be seen with single airlines, but especially in airline alliances. In these cases, the main hubs of the various alliance partners are interlinked while maintaining their specific roles as regional or local hubs in their respective home markets.

Thus, following US aviation industry references, European network carriers in the passenger business implemented the hub-and-spoke network patterns to cover this specific demand and provide a wide portfolio of city-pairs.[10] In contrast, real intra-European air cargo networks rarely existed: some European countries (e.g. Germany) maintained a national network for airmail flights, and the large scheduled all-cargo carriers ran their European feeder networks to predominantly bundle air traffic demand for their long-haul services. But dedicated networks for general air cargo within Europe lacked economic viability.

Following the liberalization of European national postal markets, the integrator companies widened their geographical coverage. New to the European market, the market expansion was mainly driven by extended air services to ensure the integrators' guaranteed transportation times. A new quality of intra-European air cargo networks was created as the number of cities connected to the base airports continued to grow. The international set-up of the integrators' activities led to a development of intra-European express hub-and-spoke systems.

Due to the fact that the geographical coverage focused on Western Europe, the distance pattern and the derived flight times allowed for a single-hub structure. A central European hub location can facilitate a geographic coverage of almost all economic centers within a flight time of two hours. Within a multimodal integrator network, these hub locations also serve land transportation, mainly trucks, that integrate short distance locations to the airport.

4 Operational Parameters

The integrators' fundamental decision to specify the central hub location is mainly driven by identified market demand.[11] For the final choice and design, a large variety of operational parameters have to be taken into account that a selected airport should satisfy. Subsumed are all issues regarding the actual flight operations, the handling of shipments and aircraft.

---

[10] See Hanlon (1997), pp. 70ff.

[11] The choice of a hub airport location shows similarities to a fundamental model of location theory, the Weber (1909) or "minisum" problem. Its objective is to locate a plant in the plane by minimizing the weighted sum of Euclidean distances from that plant to a finite number of sites representing the markets where the plant purchases its inputs and sells its outputs. See also Delfmann (1987).

The operational parameters determine the general framework for the air network, they include a multitude of factors, which can be changed in a relatively short term. They can be internally driven, i.e. by the integrators' strategies, or externally influenced, e.g. by the legislative framework. Thus, the decision making process to develop an air transportation network has to be regarded in a wider context, ranging from legislative, environmental and general political issues to the definition of the aircraft fleet utilized and the daily flight plan operated.

The choice of the optimal location forms the basis for the establishment of a hub-and-spoke network. The hub is not only the point where the central distribution of the majority of shipments will take place, but in a much earlier and more fundamental stage it is at least the nucleus for all flight operations. The overall air network including the selection of the aircraft fleet employed is based on and designed around this focal point. Therefore, being a highly important parameter in the network design, the hub selection itself cannot only be regarded as operational parameter in our sense, but has significant consequences on the overall business and network strategy. With this decision, a general, static framework is set for the establishment of the air network. This leaves the fleet selection, the fleet assignment and the fleet utilization as the only real operational parameters allowing to optimize the network or to adapt it in the short term to a changing market environment.

Aircraft selection is the longest-term decision in this group of variable influences. This is not only due to the relatively long lead times when aircraft are ordered, but also to the high investment involved in the acquisition of a fleet. In order to avoid the financial risks involved in developing an own aircraft fleet and to maintain a certain flexibility regarding aircraft types and numbers utilized, some express carriers have decided to have parts of their network operated by partner or full-charter airlines. In either case, the decisions necessary for this basic set-up imply a long-term, strategic planning horizon and thus are not discussed at this point as operational planning parameters.

Due to their different underlying strategies, the different express carriers also operate very different fleets. While some companies try to stick to a homogenous fleet and thus to keep their aircraft interchangeable, others operate large fleets of aircraft of very different sizes and characteristics which are assigned to the single routes following specific parameters.[12]

---

[12] In Europe, TNT and UPS operate only a very limited range of aircraft types (especially Boeing 757, Boeing 767, Airbus A300 and British Aerospace 146), while

4.1 The Role of Fleet Assignment for Network Operations

Based on an existing aircraft fleet – regardless whether consisting of own or chartered aircraft – the fleet assignment pre-defines the operational performance on an everyday level for the whole network. The main decisions in fleet assignment are which aircraft to assign to operate which route and whether this assignment should allow for flexible changes.

In general, fleet assignment allows an airline to optimize its operations by using the most appropriate aircraft available for a specific route. The main basis for this decision is the demand for capacity on this route and technical parameters, particularly the aircraft's range and its take-off and landing performance. While for passenger airlines aspects such as the service comfort offered[13] or the intended aircraft rotation[14] have a major impact on their fleet assignment, the factors regarded by the express carriers predominantly result from operational needs. In general, all flights operated are return flights between the hub and either another hub or a spoke airport. One-stop flights between the hub and more than one spoke airport are common to consolidate a critical mass of cargo volume. The main driver is the capacity needed on the respective route.

On the other hand, a central continental hub results in long flight distances to the more remote airports (see figure 22-2). This makes aircraft range an important issue especially for routes with limited demand, not allowing the economical utilization of large aircraft. Closely interrelated is the decision between jet and propeller aircraft, with the latter often being the right choice from a capacity point of view. While the utilization of propeller aircraft might thus satisfy the needs regarding the optimal capacity, the problem of flight length can make this choice impossible. Due to the tight timing requirements at the hub airport, implying a timely arrival and departure of all aircraft in a rather short time frame, the slower propeller aircraft might not be capable or suitable for longer routes.

DHL uses a large variety of aircraft ranging from the Swearingen Metroliner and Convair aircraft to Boeing 757 and Airbus A300 as well.

[13] Especially the decision between turboprop and jet aircraft has gained an increasing importance over the past years, prohibiting the airline from changing from a jet to a turboprop aircraft on a specific route due to the customers' expectations.

[14] Due to most airlines' network structure many flights are not return flights but imply that the aircraft operates beyond its destination airport to another airport. This might result in the need to operate a larger aircraft on a flight even though for this single flight a different aircraft type would be the operationally recommended choice.

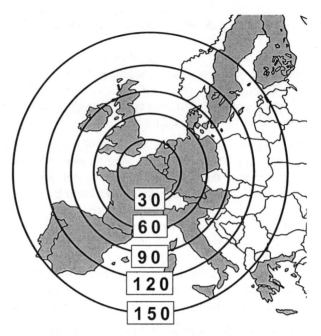

Figure 22-2: Flight Times from the Central European Hubs (minutes)

Finally, the route's function in the network context plays a major role for the aircraft assignment. If the route is the link between two hubs – either on the same or on different continents – the choice of the high capacity aircraft is the standard.[15] For routes between the hub and a spoke airport there is a broad variety of options, allowing the utilization of about every aircraft depending on the parameters described above.

Based on these factors, the assignment of the available aircraft is conducted in a flexible way to take into account the scheduling time frames and the differences in the routes' importance reflected in the average capacity demand. Thus, a demand driven, flexible operational fleet assignment in the day-to-day operations is an industry standard. While this implies a flexible response to changing transportation demand patterns, it also allows the gradual development of spoke airports by initially operating small aircraft and replacing them with larger aircraft once a certain demand level has been reached.

---

[15] See Lin, Lin & Lin (2003), p. 257.

This development brings a change from the transport of uncontained, loose parcels to the preferred transport of containerized shipments, resulting again in different requirements from the airports involved.

The utilization of different aircraft types within the network is supported by the integration of other aircraft operator partners as well. Additionally, these partners, originating from different countries, allow the tackling of traffic rights restraints. For routes to/from destinations outside the liberal European Union aviation market, the express carrier might not be entitled to operate their own flights because of bilateral air transport agreement regulations. By engaging a partner for these routes it is not only possible to assign the adequate capacity, but sometimes also to explore the market at all.

4.2 Fleet Utilization as a Crucial Factor for the Profitability of Airline Operations

Aircraft only earn money for any company whilst in the air. This applies to the express carriers as well as to all other airlines. Thus, besides the optimal operation of the hub and the efficient allocation and assignment of aircraft to different routes the utilization of the aircraft fleet exceeding the regular night operations is an important issue for all companies.

In general, each aircraft fleet's utilization is determined by the structure of the air network served and the timing therein. In the case of the express carriers, this is exclusively driven by the tight timing of the processes at the main hub. Every flight is coordinated in line with the needs of the hub airport operations and is thus limited to a tight time window as well. This usually results in a very low number of flights operated per aircraft – usually only one return flight between the hub and the spoke airport or a second hub per night – with rather short flight times of usually less than two hours.

Another factor limiting the overall fleet utilization is the typical flight frequency of five flights per week. The full network is operated only during weekdays, at the weekend only very few major routes are served, due to the low demand. Thus, the majority of the aircraft fleet is parked over the weekend, resulting in a low utilization of the aircraft. During the week, the effect of low utilization is worsened further by the strategy to only increase the capacity provided on a route by assigning larger aircraft rather than adding additional frequencies to serve increased demand.

Typically during daytime the integrators' aircraft are parked at the spoke airports. This as well is due to the hub network mechanisms, since the parcels from the spoke airports have to be transported into the hub to start typical night operations. As the spokes tend to be smaller airports and thus often do not provide the necessary facilities, this daily layover time cannot be used efficiently for aircraft maintenance activities. For this reason, some companies slightly change their flight network for the weekend to bring their aircraft to a larger, central airport providing the facilities and staff for the mandatory checks and repairs.[16]

According to the operational model chosen, the problem of the low fleet utilization differs significantly. As soon as the express carriers co-operate with partner airlines, the risk of low utilization is on the partners' side. The integrators only generate the demand for night flights, the challenge to secure the aircrafts' utilization during the day and thus to increase the revenues earned fully remains with the partner airlines. Since these partners are often charter cargo carriers, they try to sell their aircraft capacity during the off-times to the market, thus keeping their utilization higher. On the integrators' side, several other solutions enable them to increase their fleet's utilization during the times when the hub is operating only at a low level. Besides the attempt to offer regular commercial charter flights as well, a part of the fleet is also employed during daytime to leverage problems or shortcomings that occurred during the night time peak operation. For example, flights between the hubs and sub-hubs are operated to transport parcels, which have not arrived in time for the night operations. Additionally, some of the flights between the intercontinental hubs are operated during the daytime, relieving the flights which are integrated into the night peak operations.

Nevertheless, the low fleet utilization in combination with the high number of aircraft needed to provide reliable and efficient peak operations and a working hub system at night is one of the most fundamental problems for the express carriers' profitability.

---

[16] For example, DHL reroutes the network in Spain during the weekend to Madrid to allow all maintenance to be done during the time at the airport.

5 Airports as Service Providers

The role of the airport infrastructure within the express carriers' air transportation networks is defined through the capacities required for aircraft movements and the daily cargo throughput volumes. Two principal airport functions can be differentiated within these route networks: hub airports and spoke airports. The integrators' requirements differ significantly between these two functions (see table 22-1).

For all airports an intermodal accessibility is of major importance, since the air operations have to go in line with the ground transportation. The parcels are delivered and picked up by trucks, which require the adequate facilities at the airports allowing a loading and unloading of the trucks in relative proximity to the express carriers' sorting and commissioning area. Exceeding this common requirement, the operational and design parameters for the two kinds of airports differ significantly.

An allowance for night flight operations is an essential prerequisite for the hub. The peak operation in every express carrier's hub takes place during the night to allow for a late pick-up of the parcels in the evening at the customers' premises and a timely delivery to the recipient in the next morning. With regard to the flight times between the hub and the spoke airports, the authorization for night flight operations is not always mandatory for the spoke airports since the evening departure has to be early enough to allow for the transfer to the hub, while the return flight arrives in the early morning. For the spoke airports the distance to the city centers is of far greater importance since tight acceptance times at the airport have to be guaranteed. Additionally, the market at the spoke airports has to have sufficient local demand to economically operate the flights.

In terms of the infrastructure to be provided, the role of an airport has a significant impact as well. A large runway is only needed for the hub, allowing for intercontinental flights to other hubs. The spoke airports are usually being served only by smaller aircraft or below maximum payload, making a shorter runway sufficient for the integrators' operations. Comparable requirements determine the need for apron space. Since spoke airports are usually served only by one aircraft at a time, the need for apron space is very limited. Only at hub airports, serving a large number of aircraft during the peak operations which all perform unloading and loading procedures on a parallel basis, vast apron areas and sufficient loading equipment are needed to allow for an efficient and timely processing of all shipments.

Since many of the smaller spoke airports are served with small aircraft not capable of transporting containers, at these stations only a limited set of handling equipment is needed. Also the installation of major sorting and commissioning areas or terminals is not needed unless the airport is being developed into a major station for the carrier or even into a sub-hub. On the other hand, the most expensive investment at a hub airport is the automated sorting and commissioning terminals which build the essential backbone for the peak night operations of any integrator hub. This also implies sufficient runway capacity to allow for a timely operation of all flights within the set timeframe.

Thus, the requirements an airport has to fulfill to satisfy the needs of an express carrier highly depend on the target position in the network. While for hub airports several complex and costly investments are needed, from an infrastructure point of view almost every airport could be used as spoke airport.

Requirement	Hub Airport	Spoke Airport
24 hours operation	Yes (high share of night operations)	Not necessarily needed (no / few night operations)
3000+ m runway	Yes (for intercontinental flights)	No (mainly short distance flights)
Extensive apron space	Yes (for cargo transfer during peaks)	No (only limited number of aircraft)
Highly automated terminal/warehouse	Yes (for smooth peak operations)	No (no cargo transfer)
Equipment for containers	Yes (intercontinental aircraft)	Not necessarily needed (only for large aircraft)
Intermodal traffic access	Not necessarily needed	Yes (proximity to origin and destination of shipments)

Table 22-1: Integrators' Infrastructure Requirements to Airports (Examples)

6 European Integrator Network Strategies

Comparable to most passenger airlines, the standard network structure of the European integrators is the hub-and-spoke-network with a single-hub. This does not only allow for an efficient central sorting and distribution process in the main hub within a single major facility, it also allows the bundling of major traffic flows transported by larger aircraft at a very efficient cost structure. This strategy is followed by three of the major four players in the market, with their hubs located in Paris/Charles-de-Gaulle in France (FedEx), Liège in Belgium (TNT), and Cologne in Germany (UPS). These hubs are also the transfer points to their intercontinental air networks. A different strategy, based on a multi-hub concept, is followed by competitor DHL.

In order to further illustrate the differences and benefits between the single networks types the network structures of TNT and DHL are discussed in more detail below.

TNT operates a central air cargo hub at the airport of Liege in Belgium. This hub is not only the central transit and commissioning point for all air operations within Europe, it is also the interface to the central trucking hub, which is located at Arnhem in the Netherlands. While the general system of TNT's network follows the typical hub-and-spoke structure (see figure 22-3), especially for the longer routes and some smaller markets, a specific operational pattern has been introduced. These stations are operated with a one-stop-service pattern, meaning that a single flight often services more than one airport before returning to the hub.

TNT's network is fully designed to serve the operational requirements of the hub. The timing of all flights between the hub and the spoke airports is coordinated with the peak time which requires all parcels to be delivered to the hub around midnight. During the following two to three hours all aircraft are unloaded, the shipments are sorted and assigned to their destination flights before the aircraft are reloaded and return to their spoke airports where they arrive in the early morning hours. For the one-stop-flights, the arrival at the final destination is usually later than at the other spoke airports, due to the operational pattern chosen. This also leads to a different customer service quality, since deliveries cannot be provided as quickly as if the locations were served with a direct connection from the hub.

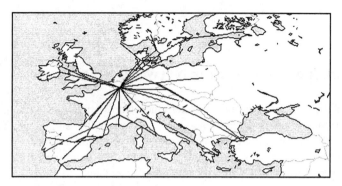

Figure 22-3: TNT's European Single-Hub Air Network[17]

Differing from this approach, DHL has established a multi-hub network for the European market (see figure 22-4). DHL's main hub is located at Brussels-Zaventem in Belgium, additional sub-hubs are in Bergamo (Italy), Cologne/Bonn (Germany), Copenhagen (Denmark), East Midlands (UK) and Vitoria (Spain). Within this network, all traffic flows are consolidated in the large trunk routes between the sub-hubs and between the sub-hubs and the main hub. Only the regional distribution is completed immediately within the respective sub-hub's network.

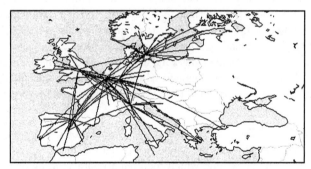

Figure 22-4: DHL's European Multi-Hub Air Network[18]

As in the standard network configuration, all flights within the network have to be coordinated in line with the operations at the main hub. Still this is the point where the main traffic volumes are redistributed and where the majority of the transfers to the intercontinental flights take place. While the function of the

---

[17] Based on TNT data (2002).
[18] Based on DHL data (2002).

operations at the sub-hubs is the same as in the main hub, the timing has to follow the main hub's needs. Figures 22-5 and 22-6 illustrate the effect on the sub-hubs using bank charts. These show the number of aircraft arriving at the airport on the right, the number of aircraft departing on the left side.

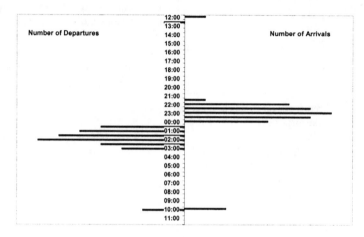

Figure 22-5: Operational Pattern at DHL's Main Hub[19]

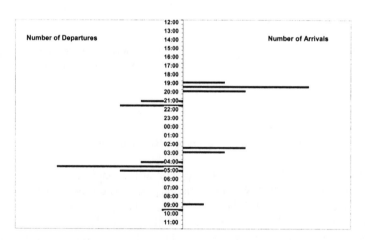

Figure 22-6: Operational Pattern at DHL's Sub-Hubs[20]

---

[19] Schematic illustration based on DHL flight schedules and data (2002).
[20] Schematic illustration based on DHL flight schedules and data (2002).

Time	Main Hub	Sub-Hub
07:00 p.m.		Aircraft arrival from spoke airports
07:30 p.m.		Unloading of aircraft; sorting, commissioning of shipments; loading of aircraft
08:30 p.m.		Aircraft departure to the main hub and other sub-hubs
10:00 p.m.	Aircraft arrival from sub-hubs and spoke airports	
11:30 p.m.	Unloading of aircraft; sorting, commissioning of shipments	
00:30 a.m.	Loading of aircraft	
01:00 a.m.	Aircraft departure to sub-hubs and spoke airports	
02:30 a.m.		Arrival of aircraft from main hub and other sub-hubs
03:00 a.m.		Unloading of aircraft; sorting, commissioning of shipments; loading of aircraft
04:00 a.m.		Departure of aircraft to spoke airports

Table 22-2: Timing Scheme of DHL's Hub Operations[21]

While there is only one major peak operation at the main hub around midnight, the sub-hubs have to deal with two smaller peaks. These are scheduled before and after the main hub's peak in order to consolidate and redistribute the parcels arriving from or leaving to the spoke airports served. For both kinds of hubs there are additional flights during the day, which on the one side are the intercontinental services in the main hub, and on the other side are additional services between the sub-hub and the main hub or other sub-hubs, covering

---

[21] Schematic illustration based on DHL flight schedules and data (2002).

shipments, which could not be handled within the previous night's peak operation. The processes in the two hubs are sketched in table 22-2.

While gaining a dense and more efficient service pattern for the sub-hub regions on the one side, a significant degree of operational planning, co-ordination and controlling complexity has to be dealt with compared to a single-hub network on the other side.

7 Changing Parameters of Airport Choice?

Airport night curfews are incompatible with an express carrier hub operation. This conflicts with the fact that the airports in the densely populated Central European region are facing increased pressure on unrestricted flight operations during the night. The dilemma between optimal location and uncertain future for the hub investments has proved to be a major concern to the stability of the express carriers' hub and spoke systems in Europe. The most prominent location shift took place in 1998 when TNT chose the former military airbase of Bierset (today Liege Airport) for its hub operation to replace Cologne/Bonn Airport because of its uncertain night flight operations perspective. The competitor UPS remained in Cologne/Bonn and has increased its operations significantly since then.

The most prominent current discussions are taking place at Brussels Airport, where the expansion plans of DHL's main hub will challenge the imposed ceiling of 25,000 night flights a year. DHL's growth plans forecast 34,000 night flights by 2012, accommodating an increase in air cargo traffic from 312,000 tons yearly to an expected sum of 781,000 tons annually by 2012.[22]

As night flight constraints traditionally held the position of being the most important factor threatening the stability of an express carrier's hub location in Europe, the enlargement of the European Union puts new pressure on the existing network structures. So far the focus of the express carriers' networks has mainly been Western Europe (see Figures 22-2, 22-3, and 22-4). Following the integration of the new EU accession countries, an increased demand for express transportation to/from the other EU member states is expected. Even though the express carriers already inaugurated initial flights to important destinations such as Warsaw or Budapest before the countries joined EU membership, the integrators will have to implement structural changes to their networks to cover

---

[22] See Turney (2004), p. 12.

the new transportation demand. Because of the long distances to the hub, the two hour flight time threshold will be challenged by the integration of additional Eastern European airports.

With its multi-hub structure, DHL is in a good position to cover the East European market and has taken respective action already: a decision for an upgrading of Leipzig/Halle airport in Eastern Germany to a sub-hub can be expected after the European Commission approved a grant for the establishment of an air logistics centre operated by DHL at Leipzig/Halle airport.[23] With its spacious layout, additional runway capacity and unrestricted 24 hours/ 7 days flight operation, Leipzig/Halle airport is in good shape to facilitate DHL's expansion to Eastern Europe. Figure 22-7 shows that most of the East European agglomerations can be served from Leipzig/Halle within two hours flight time. Furthermore, an additional sub-hub eastwards from the European main hub can efficiently ease capacity restraints at Brussels, by providing direct more sub-hub–to–sub–hub services.

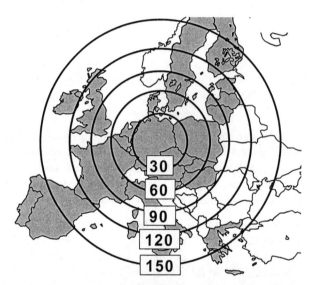

Figure 22-7: Flight Times from a Potential Sub-Hub in Leipzig/Halle, Germany (minutes)

The express carriers' network developments in North America and the Far East have shown that on a parallel basis to the air cargo traffic expansion a tendency

---

[23] See European Commission (2004).

towards dedicated airports for express carriers can be observed. Examples are Subic Bay (FedEx) and Diosdado Macapagal International Airport (UPS), both located in the Philippines, or Memphis, the US hub of Federal Express.

In Europe, only Liege has a comparable high degree of specialization in the express carrier business, all other main hubs serve the full variety of civil air transportation segments. This specialization is highly effective for the express operator, allowing this key airport client greater influence on airport infrastructure planning and management.

On the other hand, a one-sided specialization on the express carrier business limits the positive effects of a co-operation with other cargo airlines and service providers. A new policy of co-operation can be observed between traditional air cargo carriers and express carriers, e.g. sharing capacity on long-haul services or making use of the fast Intra-European express air carriers' connections for urgent deliveries.

The European integrator industry, due to its complexity, the high investments (especially in the air network) and the tough competition between the four major providers is currently under no risk of a new market entrant. Nevertheless, in addition to the new developments mentioned, the dynamic nature of the integrator business calls for a constant review of the network structure and the choice of airports to secure competitive advantage.

8 References

BOEING (2002): *World Air Cargo Forecast 2002/2003*. Seattle.

DELFMANN, W. (1987): Das Steiner-Weber-Problem. *Das Wirtschaftsstudium (WISU)*, 16(6), pp. 291-293.

DELFMANN, W. (2004): Hub-and-Spoke-System. In: Klaus, P.; Krieger, W. (Eds.): *Gabler Lexikon Logistik*, 3rd Ed. Wiesbaden: Gabler, pp. 193-194.

DOGANIS, R. (1999): *The Importance and Impact of the Express Industry in Europe*. A Report for the European Express Association, Brussels http://www.euroexpress.org.

EUROPEAN COMMISSION (2004): *Commission approves aid for air logistics centre operated by DHL Airways GmbH in Leipzig/Halle*. Press Release IP/04/502, Brussels, 20 April 2004.

GRIN, B. (1998): Developments in Air Cargo. In: Butler, G. F. and Keller, M. R. (Eds.): *Handbook of Airline Marketing*, New York: Aviation Week Group, pp. 75-93.

HANLON, P. (1997): *Global Airlines – Competition in a Transnational Industry*. Oxford: Butterworth-Heinemann.

HÄTTY, H.; HOLLMEIER, S. (2003): Airline Strategy in the 2001/2002 crisis – The Lufthansa Example. *Journal of Air Transport Management*, 9(1), pp. 51-55.

LIN, C.-C.; LIN, Y.-J.; LIN, D.-Y. (2003): The Economic Effects of Center-to-Center Directs on Hub-and-Spoke Networks for Air Express Common Carriers. *Journal of Air Transport Management*, 9(4), pp. 255-265.

PFOHL, H.-C. (2004): *Logistikmanagement*. Berlin et al.: Springer.

SCHWARZ, K. (2004): *Briefpoststrategien in Europa*. Cologne: Kölner Wissenschaftsverlag.

TURNEY, R. (2004): Whither Wallonia? *Air Cargo World*, 7(4), pp. 12-13.

WEBER, A. [translated by Carl J. Friedrich from Weber's 1909 book] (1929): *Theory of the Location of Industries*. Chicago: The University of Chicago Press.

23
ESCAPING THE AIR CARGO BAZAAR: HOW TO ENFORCE PRICE STRUCTURES

KORNELIA REIFENBERG AND JAN REMMERT

1	Introduction	540
2	The Air Cargo Price Bazaar	541
3	Why Change Pricing?	542
4	Requirements for Air Cargo Pricing	544
5	A Concept that Balances Flexibility and Standardization	547
6	Success Factors for the Rate Card Design	549
7	The Rate Card Development Process	552
8	Conclusion	554
9	References	555

Summary:
The air cargo pricing process is heavily dominated by individual negotiations and the objective to increase load factors by any means. New concepts are needed to standardize the current pricing processes and to focus more on profitability aspects. In defining clear pricing rules, the effort involved in price and revenue management can be reduced. This chapter first identifies the problems in the current pricing methodology and why the air cargo sector needs new pricing concepts. Subsequently, a possible approach that meets the requirements of the air cargo business is presented, describing the main challenges of its implementation.

1 Introduction

In the current market situation of the air cargo business, it is nearly impossible to make reliable forecasts. Instead of the expected growth figures, stagnating tonages and revenues are reported again and again.[1] The market is still waiting for a significant recovery from the shock of 2001. However, optimism for a return to better times does exist.[2] Airlines are starting once again to expand their capacity, either by ordering new aircraft or by reactivating those stored in the past months.

Unfortunately, a lack of growth coupled with increased capacities are making the pricing situation in the air cargo industry even worse, leading to a continuous downward price spiral. The risk is great that such a market situation will ignite a price war (see Figure 23-1).

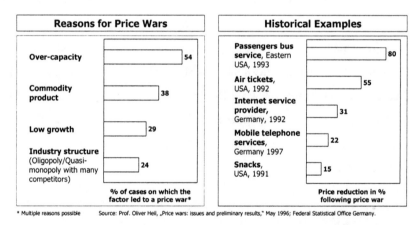

Figure 23-1: Reasons for Price Wars and Historical Examples

Price wars can occur in any industry and they all share one trait: hardly anyone ever wins.[3] Although managers in the air cargo carriers' headquarters may be

---

[1] See for examples Lufthansa Cargo AG/MGI (2004), S. 22. For the rate development, see also DVZ (2004a), p. 7.
[2] See DVZ (2004b), p. 7.
[3] See Wübker (2001), p. 74.

aware of the risk of price wars and the need to avoid them, their knowledge has in fact only little influence on the actual prices offered in the markets. The available pricing instruments in the air cargo market are multifarious, and it is predominantly the sales representatives who are free to choose one of them.

The TACT rules, which are the gross rates published by the IATA, give the first clue as to how much to charge for an air cargo shipment.[4] But, similar to the prices you find on the list in hotel rooms, only rarely will this be the actual price on the bill.

The official standard price lists of air cargo carriers already include a first discount on top of the TACT rates.[5] Because electronic distribution makes price comparisons between carriers easier, this published rate sheet is rendered obsolete to the majority of forwarders. They confront the airlines with even cheaper offers and demand further discounts. As a result, business with regular customers is usually concluded with individually negotiated special rate agreements. For the short-term ad hoc business, even lower spot rates are offered. From this point on, carriers are stuck in the air cargo bazaar.

2 The Air Cargo Price Bazaar

The customers – agents and air cargo forwarders – are in a comfortable position. They are aware of the air cargo carriers' dependency on load factors and the pressure on them to offer individual, reduced prices so not to lose business. Operating in a market with a limited number of key players makes it possible for customers to check each carrier's price offer before deciding on the best. The best is not always the cheapest offer, as in some cases a long-term business relationship outweighs price importance. But even these relationships are at stake if the competing carrier offers 30% lower prices. And as long as carriers still have free capacities, hard discounting or even dumping become a part of day to day business, especially for companies with a weak market position.

Furthermore, the relatively low variable costs in the air cargo business 'naturally' boost load factor optimization as a main target of business: even a price of a few cents is favored over the alternative of selling nothing. Consequently, lowest fill-up rates and special promotions are common business.[6]

---

4   See www.IATA.org. For a structural explanation see Shaw (1993), p. 135.
5   See Shaw (1993), p. 135.
6   See DVZ (2004a), p. 7.

A large amount of surcharges and fees might help to maintain the price level, as the price sensitivity for these charges is lower than the degree of sensitivity for the rates themselves. But charges and fees are also a source of great annoyance when the surcharge amount becomes clear.

In the business-to-business sector in general, offering discounts to hard negotiating customers is nothing special. But developing thousands of customized rates for every lane, product and weight segment, customer by customer, is not normal. Why doesn't this industry have a standardized pricing approach? The answers to this question are diverse. The first reason is that every forwarder is different. Depending on the individual tonnage and product mix per destination, the forwarder requires specific service and pricing solutions that perfectly fulfil his needs. The second reason is that every flight is different. The specific capacity situation of each flight determines the level of the short-term spot rates. Third, the cargo business is characterized by intense personal relationships between forwarders and the airline's sales representatives. Offering a specific, tailor-made price list only for one specific customer is exactly what the forwarders demand. But the main reason why the air cargo industry does not use standardized tariffs is the one which can be attacked the least: it is simply the way the market structures have developed and been established in the last decades.

3 Why Change Pricing?

"If I have 2000 customers on a given route and 400 different prices, I am obviously short 1600 prices." (Robert L. Crandall, former CEO of American Airlines.)

So why change pricing, when individual negotiations give you the opportunity to offer a price to a customer that exactly equals the willingness to pay? The following section will explain why there are several important reasons for changing air cargo pricing.

The *lack of cost consideration* is the first motive for changing pricing. For most air cargo companies, the clear assignment of single cost categories to a certain shipment is still a black box – only known to a limited number of specialists in the controlling department. The sales force usually has no information at all about the costs that different shipments can generate. As a result, they are quite in the dark regarding the required average minimum prices per flight. Designing optimal prices requires knowledge about variable costs. Without a clear pricing guideline, developed systematically and with the carrier's cost structure in mind, the sales force has no chance to find the optimal prices.

The second reason for changing pricing is *workload reduction*. When a customer asks for a quote on the spot market, the sales representative first checks and blocks open capacities (sometimes through the central revenue management department) and then calculates the potential price he will offer. After he has done that, the sales representative waits until the customer chooses a carrier. This process can be repeated over and over again if the customer does not accept the offer but later contacts the sales representative again for a better rate.

Similarly, more work is generated when customers request an individually designed special rate agreement. Both the sales and accounting representative invest great effort throughout the year into the negotiations and processing of the individual rates.

A third reason for changing pricing is the fact that in many cases today, *prices are not connected to a promised volume within a defined period*. As a result, forwarders often do not know – or are at least not interested in – the conditions that led to their individual price level. This makes the negotiations difficult for the air cargo carrier if the forwarder delivers decreasing volumes or starts to cherry-pick by giving attractive business away to competition.

Furthermore, the customers' volume development is frequently not monitored in a systematic way by the airlines. This lack of price controlling and general discount rules deteriorates the relationship between price levels and lowers the value that a customer represents for a carrier.

Such a deterioration already becomes a problem when the market is fragmented and the negotiation power of every customer is still limited. But carriers should be aware that fragmented price authority on the customer's side will soon be a thing of the past.[7] Customers will increasingly build up purchasing cooperations to enforce lower prices. From then on it will be crucial to apply, communicate and monitor clear discount rules. Otherwise the carriers have no means to resist the power of their customers.

The fourth reason for a change is to *improve the price setting competency*. The price setting process is usually local. In other words, the various sales areas have full price authority. For the major carriers, this applies to the hundreds of stations where good sales people with strong pricing competencies are necessary. But how many companies are so blessed to have so many sales representatives who

---

[7] See Klaus (2003), p. 72.

are specialists in price optimization? And who has the capacity to thoroughly optimize e.g. a million prices p.a. in a network of 150 stations with 5 products and 5 weight classes in 2 seasons for 100 customers?

All in all, these four reasons make it clear why new ways of pricing in the air cargo business are so crucial. A compromise must be found between individual negotiation and complete standardization, between load factor optimization and yield maximization and between profitability and rate attractiveness.

4 Requirements for Air Cargo Pricing

The air cargo pricing bazaar can be escaped by various means. But whatever the details of a carrier's final price system are – they all have to meet certain requirements which will decisively determine the success of such a system: encouraged customer loyalty, differentiated discounts, increased standardization but limited transparency between customers, and time-based price differentiation.

4.1 Encouraged Customer Loyalty

In a market like the air cargo industry, the possibility of generating new revenue through natural market growth was limited in the last years. As a result, it is crucial for the carriers to expand their market share by increasing their customers' share of wallet and attracting business from competition. Customer loyalty will consequently be a major success factor of the future.

The main reason for the weak customer loyalty is cherry-picking behavior: for each shipment, forwarders select the carrier who offers the best prices, the fastest connection or the best service. Intense business relationships that drive forwarders to focus on only one airline are rare. The conclusion airlines draw based on this behavior is understandable: it is not enough to be the most attractive supplier only on average. For every single shipment or market segment competition starts again.

Instead, stronger customer loyalty would create more opportunities for the air cargo carriers.[8] Once the company is selected as the preferred supplier, the for-

---

[8] See Butscher & Clark (2002); Butscher (1999).

warder no longer picks the best offers from the market, and the competition focuses on the carriers' overall offer instead of each single shipment.

4.2 Differentiated Discounts per Lane, Product and Weight Band

Air cargo shipments have many particularities and thus require more flexible discount guidelines than those applied in other industries. Simply assigning a customer to a certain discount level, say 20%, and then granting this discount on all shipment types the customer ever sends, will neither meet the needs of the customer nor the cargo airline.

The price structure will have to take into account that differing lanes, products or weight bands require differentiated discounting. A flight to a top destination where capacities are scarce, for example, gets by with low discounts even for large customers. Similarly, small and light weight shipments below 50 kg or premium products do not necessarily call for any discounts at all.

Correspondingly, lanes running below capacity, or products and weight bands with higher competitive pressure require surpassing discount levels. A pricing guideline should therefore offer the possibility of varying discount levels without losing control of a customer's overall average discount.

4.3 Standardization Combined with Customization

One of the major challenges for a company's pricing and discount concept is to strike a balance between customization and standardization. Customization is necessary to serve the customers' individual needs as much as possible; standardization is necessary to optimize the efficiency of the internal processes and to reduce the costs of price design (see Figure 23-2).

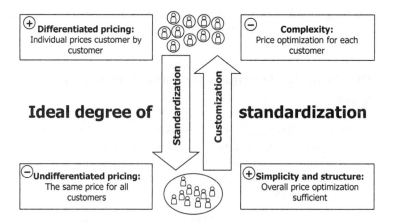

Figure 23-2: Customization vs. Standardization

A systematic discount guideline always entails a standardization of pricing rules and price levels. At least for small and medium-sized customers, finding one standardized pricing scheme which still matches to their willingness to pay is necessary and critical – the reasons have been explained in the previous section.

Nevertheless, such a structure also requires enough customization so that the varying needs of customers and market segments are addressed. If a customer, for example, delivers much of their business to Asian countries, but rarely sends anything to the US, he will demand an attractive discount on shipments to Asia, whereas the prices for the North Atlantic region are much less important. For a second forwarder with the same size, this might be reversed. A good price and discount system will have to take these particularities into account.

4.4 Non-transparency

The air cargo business is made up of a low number of carriers, each of them facing hard competition. This oligopolistic market structure requires that price levels and price structures are concealed as much as possible.

The option of simply designing four or five official price lists, including different price levels, and distributing these lists among the forwarders, would quickly make the price structure completely transparent. A possible competitor would only need to convince a few forwarders to handing over the cargo airline's offer to him to get a complete overview of the company's price and discount structure.

In light of this, the number and variety of a carrier's price lists have to be much larger.

4.5 Time-based Price Differentiation and Yield Management

Even the most customized price lists cannot necessarily take into account that market prices can vary over time. Even outside of the Christmas season can market prices be 50% or more above the prices during the rest of the year. Demand for cargo capacities is usually higher on the weekends, thus prices might be lower during the week as well. Finally, also randomized variations of capacity utilization which result in higher or lower fill-up rates in the spot market can be seen.

Ideally, a cargo airline's consistent price structure should also be able to consider these "seasonal" price variations. Otherwise, the price list offered to a forwarder will include realistic prices only for a certain share of shipments – but never for all of them.

5 A Concept that Balances Flexibility and Standardization

One concept can offer a means out of the pricing bazaar: the rate card concept.[9] This concept integrates all requirements of the discount system listed above. Once implemented, it stimulates higher sales, pricing and billing efficiency.

---

[9] For examples see www.dhl.de.

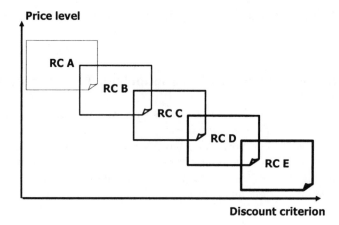

Figure 23-3: The Rate Card Concept

The concept's foundations lie in the rate card system offered by many Courier/Express/Parcel-service providers (CEP) today. It becomes innovative through the transfer into a different industry with a few similarities and a lot of differences to the CEP situation. Because the concepts of the CEP industry and the air cargo industry still are similar, the price system for the air cargo industry will be called the "rate card system".

A rate card system is defined by a set of complete price lists with differing discounted rates for destination, weight class and product combinations. Correctly designed, each rate card offers the same or lower rates than the previous one (see figure 23-3).

These rate cards are assigned to the customer, ideally following a strict, quantifiable logic and a predefined process for potential exceptions. If the customer is supposed to receive a rate increase or reduction, this will be achieved not by changing rates for selected destinations but by switching the customer from one rate card to another. Doing this, no customer individual rates will emerge and the price performance relationship of customers can still be monitored.

This price bundling approach is rather different from the current pricing situation in the industry. Although the rate cards are an obvious simplification, as each customer's price level can be determined only by his rate card information, they are still able to meet all requirements listed in the previous section.

6 Success Factors for the Rate Card Design

How can it be that such a simple concept like the rate card approach is able to take into account the complexity of the air cargo sector? Of course, it depends on the implementation. A price system is always as good or as bad as its design. The success of the rate card system depends on the design of the following elements.

6.1 Encouraged Customer Loyalty

One major element of the rate card system is the clear relationship between a customer's rate card assignment and his value according to the discount criterion: the more attractive a customer is for the airline (discount criterion), the more attractive the rate card is to be offered (rate card assignment). In light of this, the definition of the discount criterion determines how much the system can increase customer loyalty.

The easiest, but at the same time most inefficient way to define a discount criterion might be to apply the general customer segmentation approach (e.g. A-, B-, C-customers). Although such a criterion can ensure that the customers' prices decrease with increasing customer attractiveness, it definitely will not encourage customers to stick to only one supplier. Even consolidating 100% of the business to one single carrier will not lead to an upgrade for most of the small and medium customers. And as the largest customers are already in the level A, they are not motivated to be more loyal.

By choosing the customer's share of business as a criterion, loyal customers would directly be rewarded: the more they concentrate their shipments on one carrier, the higher the discount they receive.

But in this pure form, a discount system would only reward customer loyalty and neglect other important aspects, like the customers' size. Therefore, choosing a discount criterion that integrates numerous factors into one discount value can be a suitable solution that supports both customer size and loyalty.

What is crucial for strengthening loyalty is that the forwarders' performances, their development over time and the fulfillment of promised volumes are strictly monitored by the airline. Furthermore, the monitoring activities will have to be actively communicated to the forwarders, so that they know: sending more or less than originally intended or promised will directly lead to a positive or negative change in their discount level.

6.2 Differentiated Discounts per Lane, Product and Weight Band

The necessity to differentiate discount levels is the main reason why in industries like the logistics business, rate card concepts generally are superior to other discount guidelines. The large number of prices for all kinds of weights, origins, destinations and products, all with different margins, market situations and competitive pressures, makes the segment-wise optimized discount approach necessary. Thus, a simple and easy-to-use discount guideline, for example one that provides only maximum discount levels for different revenue categories, would be suboptimal. Such an approach would lead to the same discount levels for both premium and standard products; for low weights, the same level as for heavy weights, and so on.

Instead, the rate card concept includes detailed prices per market segment. If the competitive situation for only one specific segment changes – e.g. an airport's departure and arrival capacities were increased or decreased – the rate card prices can be adapted to the new situation for this segment only – without im-pacting other prices.

6.3 Standardization combined with Customization

The degree of customization or standardization in pricing depends directly on the number of rate cards that a company offers: offering the same rate card to all customers is the one extreme towards absolute standardization; offering each customer an individually designed rate card is the other extreme towards absolute customization.

In light of this, the decision regarding the degree of customization depends on the rate card design and number.

But is it advisable to strictly offer all prices as a fixed bundle in a rate card? Definitely not. Without losing the advantages of the standardized system, world wide rate cards can just as well be divided according to several dimensions. For instance, continents, destination rankings, products or weight classes could be possible criteria for dividing a rate card into different independent elements.

For example, the customer's shipment volume to continent A leads to a certain rate card level, whereas the volume to continent B can lead to another one. When rate cards are divided into separate elements for each continent, the final rate card for a customer is a combination of these elements.

The dimension for rate card fragmentation should be chosen according to the customers' heterogeneity. When customers, for instance, differ mainly by their shipments' average weight, it might be recommendable to vary the customer's rate card assignment for different weight break prices.

If the most important differences between forwarders are their shipments' destination areas, rate cards should rather be divided according to destination areas, like continents or country groups.

6.4 Non-transparency

Exactly the same approach of slicing rate cards into several elements and combining those elements variably by customer creates the required non-transparency in the price system. Having a set of, for instance, seven rate cards at hand, with each of them divided into five different elements, this would already lead to $7^5 = 16,807$ possible rate card combinations. In fact, nearly every customer will receive a completely individual rate sheet.

6.5 Time-based Price Differentiation and Yield Management

The described rate card system only builds the foundation for a new pricing scheme based exclusively on the customers' value for the airline company. This basic scheme can be further improved when the prices become variable relating to different flights, weekdays, or months. But how can large price lists like those of rate cards, which only are defined by the carrier once or twice a year, be varied on a day by day basis? The answer is very easy: by pre-defining specific add-ons and add-downs for different weekdays, seasons, or capacity situations during the rate card design process.

Similar to when a customer pays an add-on for a certain additional service, like transportation in cool containers, add-ons can also be charged for certain months or weekdays, when capacities are scarce. Such a set of pre-defined time-differentiated rates could look like that illustrated in figure 23-4.

This add-on approach makes it possible to even implement a genuine yield management structure in the air cargo business, similar to those currently used in the passenger airlines sector. By defining different reservation classes and the respective add-ons, every flight can be categorized by its degree of capacity utilization. The category assignment might even vary over time, depending on the flight's development of capacities.

Additional agreements	EUR/kg
1. Weekdays	
For flights on day 1-3	-0.15
For flights on day 5-7	0.20
2. Transit services	
Connection flights or road feeder services	0.10
To-door delivery (for shipments > 4 tons only)	0.30
Pick-up service (for shipments > 4 tons only)	0.30
3. Special services	
Dangerous Good	0.25
Perishables	0.05
Safety transport	0.50
(...)	

Figure 23-4: Example Weekday Pricing

Compared to today's purely capacity-driven spot market in the air cargo industry, the "rate card plus add-on" model offers not only differentiated prices for the flights' various free capacities, but it also delivers a fair tariff foundation based on the customers' value to the carrier. As a result, capacity-driven pricing basically also provides more attractive prices for more attractive customers, thus integrating the customer-specific differentiation approach into the short-term pricing strategy.

7 The Rate Card Development Process

Rate cards and promoting the general ideas behind them are one thing, but determining their detailed content is quite another. When implementing a rate card concept, certain questions arise: how many cards are needed? What is the difference of price levels between the different rate cards? What are the exact prices included in a rate card? Which criteria are used for assigning a customer to a rate card level? These questions will be addressed subsequently.

How many rate cards are needed? When finding the right number of rate cards, a trade-off must be made between complexity and flexibility. Too many rate cards require that even more effort be spent on rate card design, assignment to customers, adaptation to later price changes and data storage. Fewer rate cards lead to a situation in which rather different customers might get the same rates. Only an in-depth analysis of the degree of the customers' heterogeneity and of the current variance of prices can help resolve this dilemma. One important criterion which must be considered here is that each rate card should differ significantly from the next. A customer who (from his perspective) has been offered a better rate card must be able to recognize a difference between the rate cards at first glance. However, the rate cards' differences are mostly necessary for top destinations, whereas other destinations can do with a less obvious rate reduction from one card to the next.

What is the difference of price levels between the different rate cards? The first step is to divide the one-time customers from the regulars. As the former are usually not as price sensitive as the latter, they should be given a rather high rate level. The latter, even when comparably small, are currently accustomed to receiving discounts and will not be ready to do without. A noticeable difference between rates cannot be avoided here. For all other rate cards, the distances could be become shorter (see figure 23-5).

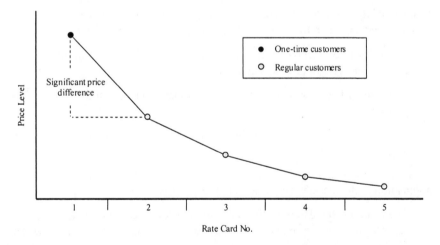

Figure 23-5: Price Differences between Rate Cards

What are the exact prices included in a rate card? Finding the right prices is a major multidimensional optimization problem. The solution can be found in the past data of a corresponding period. The prices paid by forwarders in the past for specific shipments and flights, combined with a reliable forecast of the future market development, build the basis for simulating the effects of different future price levels and price structure scenarios.

Earlier in this chapter, the rate cards were assigned to customers according to a strict logic. What does such logic entail and what are the requirements?

The logic behind assigning rate cards has to be in line with the companies' targets. A focus on revenue is best coupled with a logic that is revenue-based. A focus on customer retention inevitably means that "share of wallet" is a discount criterion. Other criteria might include the amount of premium products purchased, tonnage, or a combination of several factors in one calculation figure. The result is that the more valuable the customer is from the carriers' point of view, the better the customer's rates are. If used proactively in sales talks, this fact can help increase customer loyalty in rate negotiations: "The more you send with us, the better your rates will be in the future."

8 Conclusion

Compared to the current, bazaar-like price development processes, a more systematized approach like the rate card concept offers many advantages:

- The sales force receives clear information on the required minimum prices per customer/flight/product category.
- The standardized approach leads to a workload reduction in the price setting process.
- The customers' price level is directly linked to their volume/ performance/ attractiveness.

However, the actual design of such a system is crucial for a successful application in the market. The following prerequisites have to be met:

- The criterion for offering different discount levels should include a loyalty aspect, in order to make sure that loyal customers can expect further discounts.
- The discount levels should not be applied flat across the whole tariff, but should take into account different competitive situations per destination, weight segment or service level.

- The rate card concept should combine standardized and easy-to-develop tariffs with a customized approach that selects single elements out of these standardized tariffs and combines them individually.
- The tariff structure has to make sure that it won't become transparent to competition.
- A price differentiation and yield management approach has to be applicable additionally to the rate card approach.

Several companies have already made the shift from rather individual pricing behavior to a consistent price system. Sooner or later this will also be the case for the air cargo business. Some agents have already demanded this kind of pricing, and they will insist that the carriers respond.

Taking into account that pricing is one of a company's most important profit drivers,[10] it is hard to understand how a large and important industry like the air cargo sector is still unable to enforce their price structures. In this business, a price deterioration of only 1% could very well lead to a profit loss of about 10%-15%. It is no longer sufficient that pricing rules and decisions are made at the local level. The entire price setting process, currently operating like a bazaar, must be systematized. This is a task that requires not only the support of one or two carriers, but also the concerted action of all key players.

9 References

BUTSCHER, S.; CLARK, L. (2002): Loyalty is worth paying for. *Better Business*, 101, pp. 17-18.

BUTSCHER, S. (1999): Using Pricing to increase customer loyalty. *Journal of Professional Pricing*, 1(8), pp. 29-32.

DOLAN, R.J.; SIMON, H. (1996): *Power Pricing*, New York: The Free Press.

DVZ (2004a): Kampf um Marktanteile drückt die Preise. *Deutsche Verkehrs-Zeitung*, 41/42(58), p. 7.

DVZ (2004b): Cargolux fliegt Rekordgewinn ein. *Deutsche Verkehrs-Zeitung*. 49(58), p. 7.

---

[10] See Dolan & Simon (1996), p. 3; Simon (1992), p. 4.

KLAUS, P. (2003): *Die "Top 100 der Logistik": Marktgrößen, Marktsegmente und Marktführer in der Logistik-Dienstleistungswirtschaft*. Hamburg: Deutscher Verkehrsverlag.

LUFTHANSA CARGO AG/MGI (2004): *Planet, Global Airfreight Outlook*, Frankfurt, p. 22 and p. 56.

SHAW, S. (1993): *Effective Air Freight Marketing*. London: Pitman Publishing.

SIMON, H. (1992): *Preis Management*. Wiesbaden: Gabler.

WÜBKER, G. (2001): Preiskriege und Strategie. *Frankfurter Allgemeine Zeitung*, 238, p. 74.

24

ASSESSING AIR HUBS: THE HAYES-WHEELWRIGHT FRAMEWORK APPROACH

SIN-HOON HUM AND SWEE-KOON TAN

1 Introduction...558
2 Air Hubs in Asia 559
3 Hayes and Wheelwright Framework..560
4 Three Case Studies as Illustration..563
5 Conclusion 580
6 References..581

Summary:
We introduce to the aviation literature the use of the Hayes-Wheelwright (HW) Framework, which is often used in the operations literature, to evaluate the strategic effectiveness of air hubs. While providing information of three major air hubs on the Kangaroo and Euro-Sino routes using the case-based approach, we attempt to apply the HW framework and illustrate its use in the aviation industry. We demonstrate the usefulness of the framework as a strategic effectiveness audit instrument for air hubs, and highlight how it points to the direction for progress of such air hubs.

1 Introduction

Since the advent of modern day commercial aviation, air hubs have sprouted in Australasia at strategic locations for four main reasons: geography, economic activity, population, and tourism.

In terms of geography, many countries in the region are islands or archipelago nations. Furthermore, there are vast distances between countries within the region, which extends from Kabul in Afghanistan to Papeete in Tahiti (a distance of 16,000 km) and from Tokyo in Japan to Sydney in Australia (a distance of 7,800 km). About 60 percent of the city pairs receiving through-plane service are at least 2,000km apart. Second the growth in external trade has led to a need for more air transportation services. Third, even though the region's per capita volume of air traffic is only about half the world average, the size and growth of population point to an increasing demand for air transportation services. Fourth, tourism has become a vital economic enterprise for a number of countries in this region, and air transportation is a more integral part of tourism in this region than in other regions in the world.

Within the Australasia region, of particular economic significance is the route between Australia / New Zealand and Britain, often known as the "Kangaroo Route". This route, approximately 17,000 km in distance, is lucrative because it connects the populous Oceanic continent with the major gateway into Europe, viz Heathrow Airport in Britain. Within this route are major alternative stopover points, namely Singapore, Hong Kong, Kuala Lumpur, Bangkok, and Dubai. Similarly, these five major points, or a subset of them, may also be seen as increasingly significant as the economic linkages over the Euro-Sino route grow.

In this chapter, we introduce to the aviation literature the use of the Hayes-Wheelwright Framework (1984),[1] which is often used in the operations literature, to evaluate the strategic effectiveness of three of the air hubs used as stopovers along the Kangaroo and Euro-China routes. While also providing information on these major air hubs using a case-based approach, we attempt to apply the HW framework and illustrate its use in the aviation industry.

This chapter is organized as follows. We provide in the next section a brief description of the economic significance of air hubs in Asia. This is followed in

---

[1] See Hayes & Wheelwright (1984).

section 3 by an explanation of the HW (1984) framework in its original form which is used in an operational management context. We also explicitly adapt the HW framework and explain how it can be used in analyzing air hubs. Then in section 4, we provide three case analyses of major air hubs, using the HW framework as the conceptual tool to evaluate the strategic effectiveness of each of the hubs. Using internationally recognized airport codes, these air hubs are: Kuala Lumpur International Airport ("KUL") in Malaysia, Suvarnabhumi International Airport ("BKK") in Thailand, and Hong Kong International Airport ("HKG") in China. We then conclude the chapter with a brief summary and direction for further work on this topic.

2 Air Hubs in Asia

An air hub does not exist as an end in itself but rather as a means to an end. The primary purpose is for flights to begin, end, or transit. By doing so, the air hub generates business revenues through the provision of flight-related services such as, *inter alia*, in-flight catering services, fueling and re-fueling services, parking services, ground handling services, and aircraft cleaning and maintenance services. Additionally, the air hub collects fiscal charges, such as airport tax and landing fees, on behalf of the authorities.

On the retail business end, the air hub would provide rental space for businesses to retail goods and services to travelers, and with this, collect rentals for the use of such space. On the commercial business end, the air hub is broadly involved in operations relating to logistics management, warehousing, freight forwarding, and cargo hubbing.

To illustrate the economic significance of such retail businesses, it has been reported[2] that HKG achieved sales of US$210m^3 in 2001.

Aside from such direct activities, the presence of an air hub contributes to downstream effects on related industries such as air freight / logistics hubbing and aircraft parts manufacture and repair. Indirectly, suppliers of goods and services to the abovementioned direct activities stand to benefit too. Other activities outside the air hub, such as retail shops, hotels, travel agents, and sightseeing attractions, stand to gain from the influx of tourists and visitors.

---

[2] See Tung (2001).
[3] Throughout this chapter, the exchange rates used to convert local currencies to US$ are as follows: US$1.00 = M$3.80, Baht 39.70, HK$7.80.

In total, all these air hubbing and related activities / businesses generate significant revenues for the country and provide employment opportunities to the indigenous population. The indigenous population gains through increased consumer welfare.

To illustrate the economic significance of an air hub, in Singapore, the direct impact of the airline industry was 5.5% of the GDP in 1999 or US$4.7bn (S$8bn) of goods and services. The indirect impact was 3.7% of GDP or US$3.1bn (S$5.3bn) of goods and services. The total impact can best be described as US$0.6 (S$1) out of every US$6.5 (S$11) of GDP was contributed by SIN. Out of every seventeen jobs in Singapore, the local airline industry created one job.[4]

Estimates by various aviation authorities of growth in air travel are encouraging. IATA forecasts that passenger volume will grow by an average of 4.7% per annum to 685 million in 2007. The highest growth area is the Asia-Pacific, which will also see air cargo volume soar. Boeing[5] estimates air travel in Asia to grow 5% to 8% per annum over the next 20 years, higher than the global average of 5%.

Given the significance of an air hub to a country's economy and image, and the expected opportunities in the foreseeable future, its strategic effectiveness is of paramount importance. While an air hub cannot change its geographic location significantly, it can raise its attractiveness status by being more strategically effective. In this chapter, we illustrate the use of the HW framework to analyze the strategic effectiveness of three air hubs that are often used as stopover points along the Kangaroo Route, and which are becoming more significant as the Euro-Sino air traffic increases.

3 Hayes and Wheelwright Framework

Hayes and Wheelwright[6] proposed a four-stage framework to analyze the strategic effectiveness of a manufacturing operation. We propose the use of this framework to analyze the strategic management capabilities of the three air hubs considered in this chapter.

---

[4] See Chow, Ong & Lai (2001).
[5] See Boeing Corporation (2003).
[6] See Hayes & Wheelwright (1984).

The HW framework postulates a continuum of strategic effectiveness (of any operation) that ranges from the least strategic Stage One through to the most strategic Stage Four.

In Stage One, external experts are used in making decisions about strategic manufacturing issues. Internal management control systems are the primary means for monitoring manufacturing performance. Manufacturing is kept flexible and reactive. The perspective taken is to monitor performances and to detect negative variances when they arise. The position is one of minimizing operation's negative potential, that of being "internally neutral".

In Stage Two, the company strives to achieve parity with its competitors. In this aspect, the industry standard is followed. Capital investment is the primary means for maintaining this parity. Hence, the position is one of being "externally neutral" vis-à-vis competitors.

In Stage Three, manufacturing investments are screened for consistency with business strategy. Changes in business strategy are automatically translated into manufacturing implications. Longer-term manufacturing developments and trends are systematically addressed. In this stage, operations provide credible support to the business overall. Such an operation can become "world-class". This is the stage of being "internally supportive".

In Stage Four, efforts are made to anticipate the potential of new manufacturing practices and technologies. Manufacturing is centrally involved in major marketing and engineering decisions. Long-range programs are pursued in order to acquire capabilities in advance of needs. This proactive stance allows the operations function to go beyond "world-class", becoming "externally supportive" to the business as a whole.

3.1 HW Framework for Air Hubs

In the context of this chapter, the framework postulated by Hayes and Wheelwright[7] can be thought of as whether the air hub operations are making a contribution to the national business strategy or not. To assess this, there are numerous issues involved, and hence, it is much more complicated than a simple "yes-no" answer would suggest. These four stages of development in an air hub fall along a continuum. Hence, progress along this continuum is more of a step-by-step incremental process rather than a jump from one stage to another. To

---

[7] See Hayes & Wheelwright (1984).

apply this framework to air hubs, we suggest identifying the characteristics of each of these four stages in air hubs, as follows.

In Stage One, the air hub is internally neutral and relies on outside experts for advice. Due to the sophistication in technical know-how needed as well as strategic issues to consider, external consultants are employed to advise the senior management of the air hub on turnkey developments. In the development of the air hub, capacity and facilities are likely to be kept flexible so that it does not get "locked in" to the wrong set of facilities and processes. It is likely that the equipment and technology used are the norm, as the management seeks to minimize their own internal operations' negative potential.

In Stage Two, the air hub seeks parity with its major air hub competitors. Hence, it tries to follow widely-known or widely-recognized industry practices. One such industry practice could be that its workforce follows industry-wide unionized agreements. In planning its operations, the air hub uses a business cycle view. In terms of facilities, equipment, and capacity planning, the air hub employs sensible timing in these capital investments by looking ahead to what the competitors are doing and following suit.

In Stage Three, the business and operations of the air hub is used to provide credible and significant support to a higher-level strategy. This higher-level strategy is likely to be formed at the highest level of government, given the significant infrastructure investment involved as well as the significance of the air hub towards the overall national development. In this stage, decisions made at the air hub must be consistent with the overall higher-level strategy. Likewise, the higher-level strategy provides guidance towards longer-term developments in the air hub. While the business premise of the air hub may not be viewed as a significant source of potential competitive advantage, its very existence is primarily for the purpose of supporting the higher-level strategy.

In Stage Four, while still supporting the overall higher-level strategy, the air hub is able to anticipate the next phase of macro-level developments and plans to have the needed systems, processes, facilities, capabilities, and workforce, in place long before their implications are fully apparent. The air hub is likely to develop long-range business plans as this is a means for attaining a significant advantage in the positioning of the nation vis-à-vis its external environment. Hence, the air hub is a significant source of potential competitive advantage in itself. In this stage, the air hub is likely to be engaged in process innovation in-house while also producing in-house state-of-the-art equipment, systems, processes, and intellectual capital.

3.2 Usefulness

Quite clearly, with its description of the characteristics of the four stages of strategic effectiveness, the HW Framework can be used as an audit instrument to evaluate the strategic effectivessness of an air hub. Air hub management can use the framework to identify where its air hub lies on the HW Framework. Once it audits itself and identifies such a position (say, Stage 2), the HW Framework then clearly points out the natural direction for strategic progress for the hub (namely, to attempt to then move into Stage 3 and beyond). In this chapter, we demonstrate such an application of the HW Framework to three major air hubs.

4 Three Case Studies as Illustration

The main air hubs in Asia, together with data on key parameters, are as indicated in table 24-1. Due to space constraints, we will provide case studies of three of these air hubs, namely Kuala Lumpur International Airport, Suvarnabhumi International Airport, and Hong Kong International Airport. (A longer version of this chapter is available where similar case studies of the other two hubs, Singapore and Dubai, are included as well).

Air Hub / Country	Commencement Date	Passengers Handled (pa)	Passenger Handling Capacity	Cargo (tonnes pa)	Airlines Served
Chek Lap Kok / HK	July 1998	26.8m	45m	2.67m	74
Changi / SG	July 1981	24.6m	44m	1.61m	68
Dubai / UAE	April 2000	18.06m	22m	0.94m	105
KUL / MY	June 1998	17.5m	25m	0.59m	48
Suvarnabhumi / TH	September 2005 (expected)	NA	45m	NA	> 68

Table 24-1: Major Air Hubs in Asia

4.1 Kuala Lumpur International Airport

4.1.1 Brief History

The planning and development of Malaysia's new Kuala Lumpur International Airport ("KUL") at Sepang began in early 1990 when it became evident that the existing Sultan Abdul Aziz Shah International Airport (formerly known as Subang International Airport) had limited expansion capability to meet long-term increases in passenger and cargo demand. The Malaysian government, therefore, decided to build a new airport at an alternative site to accommodate both the increase in passenger and cargo volumes and the growing demand of the tourism and services sector.

KUL was built on 10,000 hectares (25,000 acres) or 100 sq. km. of agricultural land, once thick with rubber and palm oil plantations, which makes it one of the largest airport sites in the world. KUL was completed in four and a half years, with round-the-clock construction work (making it one of the fastest airports ever built) undertaken by an international workforce of 25,000 people (the largest number of workers for a Malaysian project) at a cost of about US$3.5bn. The large size of land designated for the airport would allow the airport to expand as needed to meet present and future air traffic demands. KUL commenced operations in June 1998.

4.1.2 Technology, Equipment, Standards and Facilities

With a rambling roof resembling white Bedouin tents, the five-level KUL has a 130 metre tall air-traffic control tower, the biggest column less hangar, the longest baggage conveyor belt system, biggest passenger lounge and the capacity for handling 25 mppa.

KUL is equipped with an advanced computerized system called Total Airport Management System ("TAMS"). This system interfaces and integrates the majority of electronic information within the airport, facilitates the maximum flow of information for operations and management, and provides a high level of service, safety and comfort to users of the airport.

The system integration supports the business goals of providing an efficient, cost-effective operation of KUL. As both a primary node on the Multimedia Super Corridor ("MSC") and as the precursor to the next generation of airports in Malaysia, KUL uses advanced airport technologies and is envisaged to be a model for future IT enterprise solutions for Malaysia and the world.

TAMS encompasses:

1   Centralized databases that expedite information sharing among various airport systems.

2   Centralized Management Information System for management of the airport's human, physical, financial and information assets.

3   Centralized Airport Operation Center that co-ordinates and expedites airport functions during normal and emergency situations.

4   These systems are all connected by a network infrastructure of fiber optics and copper-based telecommunications backbone used by all airport voice, data, and video users. However, each can operate independently in the event of a central control system failure.

5   The integration of these primary groups of sub-systems at KUL brings a new level of functionality and service to international airports and will be the catalyst that will ensure KUL as a major hub in this region.

There are seventeen sub-systems in TAMS, including the US$6.3m (M$24m) Olympex Flight Information Display System, Gate Allocation System, Runway/Taxiway Lighting Control System, Passenger Check-In System, Baggage Handling System, Automated Warehouse System, Facility Management System, Building Management System, Closed Circuit Television, Meteorological System, Air Traffic Management System, Apron Services Management System, Noise Monitoring System, PABX System, Point of Sale System, Card Access System, and Trunked Radio System.

TAMS was expected to provide a proving ground for electronic commerce for other similar enterprises. The Crisis Control Centre (CCC) at KUL could be expanded to manage a national crisis. It was envisaged that intelligent airport systems will come on line as technologies mature.

As a major component of the MSC, TAMS megabit fibre-optic network and information infrastructure was designed to provide a platform that will allow for easy integration and link to the other components of the MSC.

The KUL airport planners envisaged KUL to be a regional hub, with the flexibility to expand well into the next century. The development was envisaged to be carried out in three phrases:

Phase 1 (1998-2003)

Scheduled for completion in 1998 and operational in June 1998, Phase 1 called for the construction of facilities to handle 25 mppa (about 80 flights per hour) and 1.2 million tonnes of cargo per annum. The major facilities include

- Two full-service 2.5-mile parallel runways (4000m x 60m).

- A mega terminal building with a satellite – 83 aircraft stands (contact and remote).

- Sixty contact piers, 20 remote parking bays with 80 aircraft parking positions, one mega terminal, one satellite, two runways and other facilities are available to accommodate a throughput of 25 mppa.

The runways are on a staggered configuration 2,535m apart to allow for simultaneous operation. The runways are equipped with Category II navigational and lighting aids and a full system of twin parallel taxiways and rapid-exit taxiways. This will ensure that runway capacity is maximized. The airport has been designed to accommodate all the latest types of civil aircraft, with built-in safeguards for the future introduction of larger, new generation aircraft. Complemented by a taxiway system for the efficient and expeditious flow of aircraft on the ground, these two runways are capable of handling up to 120 aircraft movements per hour. Both runways are lit by edge, centreline, touchdown zone, threshold, wingbar, runway end, and PAPI lights, which allows for all-weather navigation.

The 241,000 sqm main terminal building was designed to allow for expansion into the next two phases. There are a total of 216 check-in counters along 6 check-in islands. All domestic, Singapore and mixed flights are boarded from a pier connected to the main terminal, while all other international flights will be served by a four-armed satellite building located in the main parking apron. An automated train shuttle system links the terminal and the satellite building. This system is designed for a maximum waiting time of five minutes.

The US$86.8m (M$330m) baggage handling system is fully automated. The operational concept is based on a 24-hour daily operation providing check-in service anytime and anywhere. It incorporates an automatic bar-code sorting control, four-level in-line baggage security screening and high speed conveyor belts. The baggage handling system is designed to operate efficiently and safely, transporting inbound and outbound bags between the Main Terminal Building and Satellite Building. It can handle 4,020 pieces of baggage per hour; the baggage waiting time is about 25 minutes.

Adjacent to the main terminal building is a five-star hotel, The Pan Pacific.

Phase 2 (2003-2008)

The airport is envisaged to handle 35 mppa by 2008. Designs for construction of a second passenger terminal have been completed in 2004, according to Bashir

Ahmad, Managing Director of Malaysia Airports Holdings Berhad. Construction is to start in 2005 with operations expected to commence by 2007. This new terminal is expected to double KUL's handling capacity to 50 million passengers a year. It was reported that, depending on demand, there is a possibility of building a no-frills terminal to cater for low cost carriers. Additionally, the airport plans to call for tenders to build a new RM 150m satellite building (where aircrafts park for boarding and disembarkation) which will take 36 months to complete. Talks are underway with the national carrier, Malaysian Airline System ("MAS"), to prepare the airport for the Airbus A380 when MAS takes delivery in 2007.

Phase 3 (2008 & beyond)

There are plans to expand the airport so that it can handle 45 million passengers per annum by 2012. There is sufficient land and capacity to develop facilities to handle up to 100 mppa, four runways by the year 2020 and two mega-terminals, each with two linked satellite buildings.

Once all three phases are developed, the airport's backyard will include hiking trails for jet-lagged travellers, golf courses, a theme park, a shopping center, hotels, and a wetlands nature preserve.

4.1.3 Strategy

KUL is part of a bold vision of ex-Prime Minister Datuk Seri Dr Mahathir Mohammad to transform Malaysia into a developed nation by 2020. This is indeed laudable because the vision tantamount to imposing from the top an economic order that relies on creative and disciplined bubbling from below. The responsibility of managing KUL lies with Malaysia Airport Berhad ("MAB" or "The Management"). It is useful to note that MAB is 70%-owned by the Malaysian Government.

KUL was planned as a primary node in the MSC. The MSC covers an area of 15km by 50km, encompassing Kuala Lumpur City Centre, Putrajaya, Cyberjaya and Sepang. It was envisaged that the MSC be a hothouse of technological activity – a sort of tropical Silicon Valley. KUL, being a node in the MSC, was to serve as a link to the world-at-large, with the hope that it would bring in the tourists, could ring in foreign direct investments, and be the conduit for inbound and outbound air cargo. As an anchor for MSC, it highlights the importance of international transport networks as a complement to the MSC's "virtual network" of very high capacity fiber optic cables.

In line with Datuk Seri Dr Mahathir's Vision 2020 (a vision for Malaysia to attain developed nation status by 2020), the MSC would be the means to achieve higher productivity and high value-added economic activities; it can be inferred that KUL would be integral to this strategy. The large area of the MSC would be connected by an integrated road and rail system that is detailed in the Draft Structure Plan Kuala Lumpur 2020, also known as KL20. The KL20 embodies the vision of turning Kuala Lumpur into a multi-cultural and thriving metropolis, in short, a world class city.

The MSC was developed to serve the regional and world markets for multimedia products and services by inviting world-class multimedia corporations to locate their business units and research and development facilities in this area.

The MSC was envisaged to have:

1   A world-class physical and information infrastructure.

2   Fast access to the country's transportation facilities.

3   Global megabit fibre-optic telecommunications links.

As documented above, much has been done to enact Dr Mahathir's plan, also known as Vision 2020. Valuable resources have been invested in the airport in the form of advanced equipment and software, and there are plans to invest even more in the later phases of the airport's development. KUL is a reflection of Malaysia's desire to leave the past behind and become a modern, developed nation. As mentioned by Bashir Ahmad, CEO and MD of state-owned MAB, "We need to build an airport that can cater to the next 100 years in terms of capacity and design."

4.1.4   Analysis of Kuala Lumpur International Airport using HW Framework

In illustrating the use of the HW framework on KUL, we note the following:

1   KUL has been experiencing service disruptions well past what can be described as teething problems and far too serious to dismiss as hiccups. Some of these problems include baggage handling breakdowns, jammed aerotrains, computer system failures, and reports of pilferage of passengers' baggage by luggage handlers. It was also reported that during peak periods, the baggage handling system often stops due to bags being wrongly placed and hence, cannot be read by scanners (i.e. staff-related reasons). To alleviate the problem of system breakdown, MAB planned to obtain 1,800 bag-carrying containers, costing US$33,158 (M$126,000).

Further on, in January 2004, the much touted US$86.8m (M$330m) baggage handling system again broke down, resulting in 22 flights being delayed and the luggage of 585 passengers failing to reach their destination. Seven carts had to be repaired while five other carts had to be replaced. The main cause of the breakdown was that the sorter link gave way under heavy loads. The first time that the baggage handling system broke down was during Hari Raya 2002 in which 15 carts were damaged. After the second incident of breakdown, it was also reported that the Malaysian Transport Ministry were looking into upgrading the baggage handling system.

Also, there were reported instances of power failure in April 1999, August 2003 and September 2003, resulting in hiccups in baggage transfer, ticketing, and passenger check-in. Flights had to be re-scheduled. The Management then announced that it would spend US$2.35m (M$4m) on upgrading its backup power supply systems.

Lastly, it was reported that touts and scalpers accost and fleece unsuspecting passengers while soliciting for business openly at KUL. Travellers at KUL have voiced their grouses regarding poor service quality, traffic congestion, indiscriminate parking, dirty toilets, and lack of language proficiency amongst airport staff. There have also been reported instances of a series of security breaches.

2   During the SARS outbreak in 2003, it was reported that KUL would install thermal imaging scanners in three stages. It is useful to note that SIN Changi was already using this technology in April 2003.

3   Travel Trade Gazette Asia (1-8-2003) reported that KUL would benefit from five additional key enhancements in the following aspects: accessibility, technology, connectivity, facilities for arriving and transiting passengers, and shopping.

The above three points would suggest Stage One characteristics. MAB adopts a reactive approach to solving operational problems: it is "internally neutral".

4   KUL's Engineering Division received the MS ISO 9001:2000 standard, awarded by the Standards Industrial Research Institute Malaysia. This was the sixth time that KUL has received this award. The award was based on achieving a quality system standard that covers the operations, maintenance, and management of KUL's infrastructure and facilities including the electrical power system, airfield ground lighting system, and communication system.

5   Due to the SARS outbreak in 2003, KUL followed the lead of other air hubs in the region by reducing its landing and parking charges by 50% for one year from 24 April 2003. Also, MAB implemented a five-year fee waiver on landing and parking fees on new flights into KUL.

As at April 2004, KUL has the lowest landing charge amongst all international airports in Asia. Charges for international airports in Asia range from US$588 for KUL to US$10,000 for Narita. SIN charges US$2,059 (S$3,500). It was able to entice five airlines to commence or resume its services via KUL in 2004.

It is interesting to note that even after reducing its fees, KUL still had quite a lot of un-used capacity. In 2002, it achieved a net profit of $47m, on a 4% increase in turnover.

6   In the 2003 IATA Global Airport Monitor,[8] KUL ranked fifth worldwide in terms of Overall Passenger Satisfaction.

The above points 4, 5, and 6 would suggest Stage Two characteristics. MAB attempted to seek parity with its major air hub competitors.

While it is difficult to say with absolute certainty where KUL lies in the HW (1984) continuum, it does seem that KUL exhibits characteristics of both Stage One and Stage Two. In light of the above mentioned anecdotal evidence, it would suggest that KUL is attempting to move into Stage Two of the HW framework (commitment of economic resources into infrastructure, systems, and processes) but given the teething operational problems, it is still not quite at Stage Two yet.

In their Country Briefing dated 18 March 2004, the Economist Intelligence Unit stated that the MSC has never realized its potential in spite of the laudable plans and huge investments made; the project has been slow to start and has not attracted a sufficiently large critical mass. Hence, KUL is unlikely to be in Stages Three or Four.

4.2   Suvarnabhumi International Airport

As at the date of the writing of this chapter, Thailand's Suvarnabhumi International Airport ("BKK") is still in various stages of completion.

---

[8] See International Air Transport Association (2003).

Construction, which started in November 2001, is expected to be ready for testing by June 2005 and operationally ready by September 2005.

BKK is often regarded as the "New" Bangkok International Airport. It is located at the Bang Phli District of Samut Prakan Province, east of Bangkok. The land area is around 32 sq km. Upon completion, it is expected to have a first phase passenger terminal capacity of 45 mppa and an initial air cargo capacity of 600,000 tonnes per year. The longer-term plan is to expand the airport to accommodate a passenger capacity of 100 mppa and an air cargo capacity of 64m tonnes per year.

Designed to be a world-class airport, it will have the single largest terminal ever built, with an advanced IT-based control system, state-of-the-art baggage handling system, dual runways (in a later report, it was said that BKK is likely to have four runways) for all types of aircrafts (including the wide body Airbus A380), all-weather navigational aids, and ICAO Category 10 Fire Safety Standards. Originally, it was expected to cost around US$3.0bn (120 bn Baht); this has since been reduced to US$2.27bn (90bn Baht), with the remainder slated for expansion of the runways and future capacity expansion plans. Eventually, BKK will have a capacity for 100 mppa and will have four runways with which to do this.

Additionally, to cater to visitors, facilities will be built around the airport. This is the "Suvarnabhumi Aerotropolis". This is a long-term plan which runs to 2035. It covers land use around the airport to accommodate more visitors, in response to the policy of developing Thailand as a regional aviation hub. It also involves the construction of more facilities and utilities and transportation links between Suvarnabhumi Airport and Bangkok metropolis, industrial zones on the Eastern Seaboard, and the upper and western areas in the central region.

BKK is the brainchild of Prime Minister Thaksin Shinawatra, who is also the Chairman of the airport project, aided by the Transport Minister Suriya Jungrungreangkit.

To emphasize the importance of this project, the development of BKK is known as the "National Agenda". BKK is placed as the first priority of the Thai Government's national agenda with the objective of establishing it as the premier aviation hub of the Southeast Asian Region from 2005. When completed and operational, BKK is expected to be a contributing factor in supporting and stimulating rapid development of Thailand's economy, especially tourism. Other sectors which are expected to benefit from this goal are aviation-related industries such as aircraft repairing and maintenance, viz the Thai Aviation Industries Company. Prime Minister Thaksin Shinawatra has said that he wants

Thailand to be the tourism hub of the region. He hopes to attract as many as 20 million visitors annually within a time span of five years. In line with the plan to expand the current landing capacity of BKK, Thailand has an "open skies" policy to facilitate movement of goods and travellers, opening the way to commerce and industry.

Much is anticipated of the airport. The progress in the construction of BKK augurs well for Thailand. Geographically, Bangkok is in a very good position to serve both the Europe-Asia and the Europe-Australia/New Zealand (Kangaroo) routes. The Bangkok – Hong Kong corridor is a very popular gateway as is the Bangkok – China gateway. Bangkok is also an attractive tourist destination.

We have applied the HW framework as much as we can based on anecdotal evidence that we were able to gather. Given that the new BKK airport is building capacity and adopting industry-wide technology and processes to compete with the major airports around the region, and that it is an upgrade from the current BKK airport, it is expected to be in Stage Two of the HW framework. However, the new BKK airport could experience the same problems as KUL since actual operational conditions have not yet been experienced, and these may alter its classification on the HW framework.

4.3   Hong Kong International Airport

4.3.1   Brief History

The construction of China's Hong Kong International Airport ("HKG") at Chek Lap Kok was one of the greatest engineering feats of the twentieth century. The core program cost more than US$20bn and involved four major sponsors, ten separate projects, 225 construction contracts and over 1,000 critical interfaces.

The motivation to build HKG was mainly due to the previous Kai Tak Airport having reached maximum capacity, and the dangers that it posed to residents and businesses, being located right smack in the middle of residential and commercial buildings.

The initial deliberations, feasibility studies, and planning sessions began in the 1980s. When, on 11 October 1989, the then Governor of Hong Kong, Sir David Wilson, announced the plan to build the airport at Chek Lap Kok, there were only about seven years left till the date (originally 30 June 1997; upon further high-level negotiations, the official opening was on 2 July 1998 by Chinese President Jiang Zemin) of the handover of Hong Kong to the Chinese government. Due to political reasons, it was decided that the airport had to be completed within the remaining time left (reclamation work on the two islands started in 1992). Hence, that which would take between ten and twenty years to build under normal conditions would now have to be completed within seven years at most. To further compound the time constrain, there were no suitable sites in Kowloon and Hong Kong Island. Hence, the selected site at Chek Lap Kok had to be built from scratch by merging two smaller islands, namely Chek Lap Kok and Lam Chau. At the height of the dredging operations to create the site on which HKG would be built, more than half of the world's dredging tonnage was at this site.

In itself, constructing the airport within the tight time frame is an impressive feat. However, this is only one of ten major projects in the Airport Core Program. The other projects include constructing the Tung Chung New Town, North Lantau Highway, the Airport Railway, the spectacular Tsing Ma Bridge, and the West Kowloon expressway. Altogether, brand new 35km of highways, two tunnels, a high-speed rail system, an extension of the Mass Rapid Transit line, and the world's longest double-decker suspension bridge – this has to be able to withstand Class 10 typhoon-type winds – had to be built from scratch.

Working around the clock on all parts of the airport infrastructure and IT systems, as well as the arterial transportation infrastructure, the construction crews managed to complete the project on time and within budget. For these achievements, HKG was voted as one of the top 10 construction achievements of the twentieth century by an international group of building specialists.

Upon completion of HKG, the most critical part of the entire operation – that of re-locating major equipment and personnel from Kai Tak Airport – was still outstanding. So essential is Hong Kong as the region's air hub in that there had to be no interruptions to the hundreds of flights that take off and land everyday. The new airport had to open up right on time. It had to be a seamless move. To ensure the utmost precision, the move was planned like a military operation.

In the crucial 7 hours from 11.30pm on Sunday 5 June 1998 to 6.30am on Monday 6 June 1998, the world's largest re-location operation swung into action. This involved more than 1,000 vehicles moving thousand of tonnes of heavy equipment, 70 barge sailings, and 1,000 policemen controlling the traffic. These crucial hours, dictated by the last flight out of Kai Tak Airport and the first in-bound flight into HKG, came to be known as 'The Night' in Hong Kong.

4.3.2 Technology, Equipment, Standards and Facilities

By starting on a clean sheet of paper, HKG had the clear advantage of determining which facilities, designs, layouts, and equipment would be needed for a world-class airport of the twenty-first century. A blank sheet of paper presented planners with the opportunity to create an airport that is functional, practical, and aesthetically beautiful. From building the foundation of the airport platform from the sea bed up, to determining the highest standards of service and convenience for passengers, the result has turned out to be world-class in most aspects.

HKG is one of the most spectacular airport complexes in the world. The airport is a planned development that is expected to continue evolving and growing. Currently, it has a capacity of 45 mppa, up from the original 35 mppa when it was first completed, and cargo capacity of 3m tonnes per year. Current usage is around 27 mppa. The intended capacity is 87 mppa and 9m tonnes of cargo annually, by 2020.[9]

The futuristic US$1.29 bn Y-shaped terminal building is 1.27 km long and has a total area of 550,000 sqm. The huge size of the terminal building can best be described as "jaw dropping". It is purported to be the world's largest enclosed structure. To facilitate negotiating through this big area, there is an internal shuttle train, also known as an automated people mover, which shuttles between East and West Halls, departing in frequencies of less than 3 minutes. This is augmented by 74 moving walkways, 98 lifts, 70 escalators, and 9,900 luggage trolleys. Passengers not on the move will be able to rest in one of the 15,000 lounge seats and check their flight details on one of the 2,000 display screens.

The nine levels of the terminal include 288 check-in counters capable of handling 13,680 pieces of luggage an hour. The US$71m Baggage Handling System which has 12 luggage reclaim carousels will be able to handle 13,680 pieces of luggage an hour. The average time of the first bag from the aircraft to the carousel is 20 minutes.

To cater to the needs of passengers, the retail and catering area, known as SkyMart, recently went through a US$76.9m (HK$600m) renovation and was re-opened in March 2004. The SkyMart has 160 shops, including 25 top-line brand names lining a 200-metre boulevard in the East Hall, plus 40 restaurants catering to a variety of tastes and budgets. Hence, it would not come as a complete surprise to note that HKG is one of the top 10 international airports in the world in terms of retail sales.

In the restricted area, the airport is also equipped with wireless internet facilities in its two internet lounges called Cyber Zone. The novel Cyber Zone combines entertainment and retailing with free internet access, while the multimedia

---

[9] See Airport Authority (2001).

lounges are installed with additional workstations that provide up-to-the minute worlds news reports and financial data. A wireless service for personal laptops equipped with LAN is available. Also, two children's play areas, sponsored by Cathay Pacific Airways Limited, keep tiny travellers occupied.

Adjacent to the Passenger Terminal is the Ground Transportation Centre. This is the central focus of transport connections to other points in Hong Kong and the Pearl River Delta area. The centre covers an area of 55,000sqm. The airport is accessible by a variety of transport modes such as the Mass Transit Railway Tung Chung Line, public buses, the high-speed Airport Express railway which takes only 23 minutes to reach downtown Hong Kong, and high-speed ferry services to Tuen Mun, as well as taxis.

The airport has two 3.8km runways which are equipped with CAT II (south runway) and CAT IIIa (north runway) Instrument Landing Systems, and capable of handling 49 flights per hour. The airport handles about 440 to 460 flights per day. The highest number of flights handled was 560, and the highest daily passenger throughput was 117,000 passengers.

4.3.3 Strategy

The strategy of HKG is to be the pre-eminent gateway to Hong Kong and China, through which passengers and air cargo flows. Additionally, it supports the local Hong Kong economy. To understand this strategy, a revisit on the history of Hong Kong is appropriate at this juncture.

Hong Kong is often considered as the bastion of capitalism. From 1898 to 1997, it was a British colony. Economic policy in British-ruled Hong Kong was based on the principle of positive non-interventionism. In particular to the focus of our chapter, Hong Kong was to remain a free port, in pursuit of a free-trade policy. This being so, the government imposes minimal restrictions to imports and exports.

In recent years the direction of Hong Kong's trade has veered markedly away from America, traditionally its main market, and towards China. Between 1986 and 1999 the America's share of Hong Kong's domestic exports fell from 41.7% to 30.1%, while China's share rose from 11.7% to 29.6%. China is now the source of about 61.1% of Hong Kong's re-exports and the buyer of about 33.9% of them. Because of the re-export trade, China is easily the most important source of imports.

More than one-sixth of Hong Kong's workforce is engaged in trading, and collectively this sector contributes nearly 20% to GDP. If on top, we add the related services, such as financing, insurance, transportation, advertising and marketing, the total impact of Hong Kong's trading and trade-related activities make up a substantial chunk of its economic activities. Services contribution to Hong Kong's GDP now stands at 85.6% in 2000, from 74.5% in 1990. Hong Kong is probably the most service-oriented economy in the world.[10]

Given the importance of trade and services to the Hong Kong economy, and hence the significance of ports and airports, the importance of having a world-class airport cannot be overemphasized. By 1997, the former Kai Tak Airport was already bursting at its seams in terms of passenger traffic and cargo volume. The situation at Kai Tak was further compounded by a midnight-to-dawn curfew on landings and takeoffs as it was located in the middle of a populous residential and commercial district. If nothing was done to alleviate this situation, the lifeblood of Hong Kong would be choked at this bottleneck.

In the past decade, Hong Kong's externally-oriented economy has been hit by a variety of uncontrollable events ranging from the Asian Financial Crisis of 1997, to more recent events relating to SARS, 11 September, and downturn in its major export markets. Concurrently, China's entry into the World Trade Organization in late 2001, Beijing's successful bid for the 2008 Olympics, and the expansion of export oriented manufacturing industries in China, together with development of ports and airports in the Guangdong province, has both provided opportunities as well as competitive challenges to Hong Kong.

In his 2001 Policy Address,[11] Chief Executive of Hong Kong SAR, Mr Chee-Hwa Tung recognized the need to restructure the economy. This would include, inter alia, enhancing co-operating with the Guangdong Provincial Government, especially the cities in the Pearl River Delta area, so as to achieve a 'win-win' situation, striving to take on more value-added activities, building infrastructure which would enhance the economy (this includes those relating to tourism, business, transportation, ports etc), promoting Hong Kong as a logistics hub by developing an inter-modal system that will speed up the flow of goods and information.

In the Hong Kong 2030 Planning Vision and Strategy public consultation paper,[12] it was suggested that "Hong Kong should position itself as 'Asia's World City'. It must establish itself as one of the world's great international cities and a leading city in Asia, making best use of its developing relationship

---

[10] See Credit Lyonnais Securities Asia (2002).
[11] See Tung, C. H. (2001).
[12] See Hong Kong SAR Government (2003).

with the Mainland (i.e. China). To be a successful world city, Hong Kong must strive to achieve a strong internationally oriented service economy and gain access to a workforce with specialized skills and knowledge that, in turn, attracts other skilled people and specialized resources to reinforce our competitive position. It must develop world-class hard infrastructure (such as transportation and telecommunications) and soft infrastructure (such as education and training), and must be able to provide a high quality of life that enables retention and attraction of the best of talen...".

In relation to the development of "hard" infrastructure,[13] we note that there are plans to develop ports on the western waters of Hong Kong and to develop logistics parks in Chek Lap Kok, Tsing Yi, and North Lantau. It was postulated that HKG should become an aviation hub of Southern China and a crucial node in the multi-modal transport link between Mainland China and the rest of the world. Taken as a whole, the infrastructure and related activities would enhance Hong Kong as a major cargo, logistics and supply chain management centre in Asia. In a nutshell, this was to be the strategy of HKG.

4.3.4 Analysis of Hong Kong International Airport using the HW Framework

To support the higher level strategy of Hong Kong, we note the following events/trends at HKG:

1   HKG carried out a comprehensive study on passenger and cargo patterns, and completed in 1994. This study indicated faster growth than originally anticipated. Hence, construction of the second (north) runway and completion of the Northwest Concourse of the Passenger Terminal Building were advanced and brought into operation in May 1999 and January 2000 respectively, increasing HKG's passenger handling capacity to 45 mppa. In short, a pro-active approach in long-term airport planning[14] was taken to ensure that facilities are provided in-time to meet the demands of passenger traffic and cargo volumes.

2   HKG has seen an increase in transfer passengers as airlines in general and home carriers in particular, take advantage of HKG's new capacity and scheduling flexibility to promote fast and convenient interconnections. As a consequence, there has been an increase in the relative proportion of transfer passengers (about 31% at HKG versus 23% at Kai Tak) versus

---

[13]   See Airport Authority (2001).
[14]   See Airport Authority (2001).

Origin / Destination passengers. On the whole, there has been an increase in the number of aircraft movements, from 167,377 in 1999 to 186,910 in 2003, reflecting the inter-connectivity of flights and HKG's air hub status. As a reflection of passenger satisfaction, HKG has won the "Airport of the Year" award from SkyTrax, a UK-based independent aviation research institution. Additionally, in the 2003 IATA Global Airport Monitor,[15] it ranked third amongst international airports in terms of Overall Passenger Satisfaction. It was also amongst the top three in terms of ground transportation to / from the airport.

3   There has been increased usage of its air cargo facilities. Cargo volumes increased about 8% each year from 1999 (1.97m tonnes) to 2003 (2.67m tonnes). HKG has been the world's busiest airport for international air cargo since 1996. For the second consecutive year, it was named the Cargo Airport of the Year 2003 by Air Cargo News, an independent air cargo trade publication.

4   Going forward, passenger growth is expected to be in the region of 5% pa, while that of cargo and express cargo are 6% pa and 12% pa respectively. The forecast horizon is over the next 20 years. To meet the expected demand on passenger and cargo facilities, HKG has plans to invest in infrastructure that will allow it to progressively expand the capacity to 87mppa and 9m tonnes of cargo per year. Additionally, logistics and warehousing facilities are planned on North Lantau.

HKG's investments in infrastructure well ahead of demand which is in-line with what competing air hubs are doing, is anecdotal evidence that it is at least in Stage Two of the HW continuum.

5   Recognizing that Hong Kong's future is increasingly tied to its hinterland, and spatially it is embedded in its regional metropolitan space in the Pearl River Delta area, in October 2003, the Airport Authority Hong Kong ("AA") signed a Letter of Intent with Shanghai Airport Authority to strengthen exchanges and closer co-operation between the two cities and airports.

Within the Pearl River Delta area are 5 airports within 100km of each other. The AA has spearheaded an alliance with the other 4 regional airports to explore greater efficiency. An early hopeful sign is that the 5 airports have put forward a united front internationally. The other four regional airports are: Shenzhen, Guangzhou, Zhuhai, and Macau. In July 2001, a joint

---

[15] See International Air Transport Association (2003).

statement was issued by delegates from these airports, collectively known as "A5", pledging to work together in the many areas of common interest, including cross-boundary issues, tourism, inter-modal transport links, future developments in China's domestic and international aviation industry, and China's accession to the WTO.

6 In September 2003, additional accessibility to HKG was created with SkyPier, which is a connection with ports in the Pearl River Delta area. High-speed ferries will shuttle passengers to SkyPier where they will board buses for the airport, bypassing immigration and customs formalities, and cutting by up to half the current four hours transit time. Initially, ferries will operate to four ports – Shekou, Shenzhen, Macau and Humen in Dongguan, extending later to Zhuhai, Guangzhou and Zhongshan.

From points 5 and 6, we see that decisions made at the airport are in-line with the economic strategy of Hong Kong SAR; hence, HKG has lent support to higher level strategy and this suggests it has reached Stage Three on the HW continuum.

The future of Hong Kong and indeed, that of HKG, is about "flow". Specific to HKG, it is about flow of people and trade, and the major trade-related services such as import / export, logistics, and connectivity through the air hub. While Hong Kong's aim is to be a leading city in Asia, to this end, HKG has an important part to play. It is to be the conduit through which economic and social opportunities flow through to residents and the global community.

In spite of its successes to date, HKG's future success as an air hub in support of Hong Kong's goal is dependent on its ability to manage its relationship with the General Administration for Civil Aviation of China ("CAAC") which controls routes starting and ending at Mainland China cities, hubbed through HKG. Specifically, this relates to air traffic rights for both passenger and cargo traffic going through HKG.

The Guangdong province provides competitively challenging opportunities for both Hong Kong and HKG as an air hub. Guangdong is purported to generate 40% of China's GDP and receives 30% of its foreign direct investments. It is clear that Hong Kong and Guangdong actually form one metropolitan region. But administratively, it remains messy with multiple authorities all looking out for themselves. How AA and the airport authorities in the Guangdong area can work together will determine whether they can define a clear vision for transport and trade flows in South China that will enable them to continue to stay ahead of other regions in China and beyond, and hence, move into Stage Four on the HW continuum.

5 Conclusion

In this chapter, we proposed the use of the Hayes-Wheelwright Framework (1984) to analyze the strategic management capabilities and effectiveness of three major air hubs in Asia. From anecdotal evidence collected, we may conclude that HKG is in Stage Three, while BKK shows promise to be in Stage Two and KUL is seeking to move out of Stage One.

In this way, we have demonstrated the use of the HW Framework in auditing the strategic effectiveness of air hubs by identifying the position of the hubs on its 4-stages continuum. Once the air hub management recognizes the stage of strategic effectiveness for its hub, it can then use the characteristics described for the next stage in the HW Framework to guide the next phase of strategic progress for its hub.

Going forward, one further area of work which may be built on this chapter is the more formal application of the HW framework for such assessments of air hubs. In this chapter, we have attempted merely to introduce the HW framework and some of the key air hubs in Asia, and illustrate the use of the framework for strategic assessments. Such assessments can be made more rigorous by first creating an air hub audit instrument based on the HW framework, and then using it empirically to collect common data from the different air hubs. This will then allow a more rigorous analysis of a common set of data and hence make the HW framework-based assessments more robust.

Acknowledgment

This chapter is partially funded by the National University of Singapore Research Grant, R-313-000-003-112.

6 References

AIRPORT AUTHORITY (2001): *Hong Kong International Airport Master Plan 2020.* Hong Kong SAR.

BOEING CORPORATION (2003): *Current Market Outlook.* Seattle, USA.

CHOW K.B.; ONG C.H.; LAI, J (2001): Economic Impact of Singapore Changi Airport. *Working Paper Series*, Centre for Business Research and Development, Faculty of Business Administration, National University of Singapore.

CREDIT LYONNAIS SECURITIES ASIA (2002): *Hong Kong SAR, Ports, Airports, and Bureaucrats: Restructuring Hong Kong and Guangdong.* Hong Kong SAR: Credit Lyonnais.

HAYES, R.H.; WHEELWRIGHT S.C. (1984): *Restoring our Competitive Edge.* New York: Wiley.

HONG KONG SAR GOVERNMENT (2003): *Hong Kong SAR, Hong Kong 2030 Planning Vision and Strategy: Stage 3 Public Consultation.* Hong Kong SAR: Hong Kong SAR Government.

INTERNATIONAL AIR TRANSPORT ASSOCIATION (2003): *Global Airport Monitor.* United Kingdom.

TUNG, C.H. (2001): *Policy Address 2001.* Hong Kong SAR Government.

Part V

Airport and Airline Strategy Impacts

25

THE IMPACT OF AIRPORTS ON ECONOMIC WELFARE

HERBERT BAUM

1	Airports as a Center of Economic Activity	586
2	Empirical Evidence for Impacts of Airports	587
3	Methodological Approaches	588
4	Quantifying the Economic Impacts of Airports	590
5	Results	601
6	References	602

Summary:
Airports make considerable economic contribution to the prosperity of a region. This prosperity impact becomes apparent in increases of created value added, income and employment. Airports produce positive economic effects in terms of investment factor through the operation of the airport and through location advantages for the users of air transport. The advantages for users become apparent in increases in productivity, cost reductions and market expansions. These growth effects impact positively on tax revenues for the Exchequer. Opposed to these positive effects on economic welfare are contractive impacts due to external costs of air transport. The economic development potential a region possesses is reduced in case of restricted capacities at the airport. These restraints negatively affect the creation of value added and productivity. Empirical studies show that considerable gains in growth and productivity can be obtained by eliminating existing restrictions of capacity..

1 Airports as a Center of Economic Activity

In highly-developed national economies air transportation becomes increasingly important due to proceeding globalization of the economy and ever more intensive division of labor. The internationalization of the economy causes a growth in air traffic, on the other hand effective air transport connections are an important pre-condition for regions to keep up with international competition.

Air traffic will experience further growth in the future. For the passenger sector annual growth rates of 4 % are forecasted. By the year 2015 passenger volume will have doubled compared to 1997. With 6 % per year the forecasted growth of air cargo is even higher.

Airports unfold their impacts on economic welfare as a gateway for air traffic. They are a center of economic growth generating increases in added value, income and employment. Airports have the following impact channels:

- Airports create economic activity as an investment factor. Investments are both made by airport operators and by other companies at the airport.

- In addition, airports function as an economic factor which is the result of the services of the airport. These services are performed by airport operators, airlines and other companies at the airport (e.g. gastronomy, shops, banks, travel agencies). The economic effects generated by input-providing industries for the activities of such companies have to be added.

- Airports eventually unfold economic effects as a location factor (so-called catalytic effects). They result from the high-quality air transport connection which is used by economy and population. They are reflected in productivity gains, market expansions, cost reductions, increases in competitiveness, fostering of structural change, settlings of companies and lead to gains in productivity, employment and income at companies in the region.

- These positive effects on economic welfare are counteracted by losses of growth and employment due to damages caused by air traffic (emissions of noise, air pollutants and carbon dioxide). These external damages reduce the productive capacity and thus added value and employment by diminishing resources. These contractive effects have to be considered in an assessment of the effects of airports on the economy.

2 Empirical Evidence for Impacts of Airports

Several empirical studies dealing with the impact of airports on economic welfare can be found in the international research literature.[1] In the majority of cases these studies only refer to the impacts of airports as an economic factor that means it is examined which impacts are generated by the operation of and the investment in airports including the inputs. The direct, indirect and induced (secondary effects resulting from the spending of additional income) effects are quantified. The production effects relating to inputs are measured with Input-Output-Analyses. These impacts are summarized as "multiplier effects". The multiplier specifies how much added value is created by inputs and secondary effects for a unit of added value created at the airport. In the same way multipliers for other economic factors (income, employment) are established. Such multipliers have an average value of 1.5.

Until now it has not been considered in the literature how airports impact on the economy as a location factor. This relates to the effects for economy and population based on the better accessibility of the region due to the use of the airport. Such improvements of accessibility are reflected in productivity increases, cost reductions, increases in sales quantities and settlings of companies that means in economic advantages which result in increases of added value, income and employment. The particular difficulty in measuring these indirect productivity effects in contrast to the direct ones is the reason for the fact that these impacts have been neglected until now. Previous studies only mention that such local economic effects occur and that they cannot be quantified. The existing studies emphasize the influence of location factors by demonstrating the importance of airports – also in comparison to other location factors (e.g. labor market, connection to the road network, tax burdens) – in rating scales and by verifying whether companies regard these criteria as fulfilled.

An essential methodological and empirical advancement in the quantification and monetary evaluation of airports as a location factor has been made by the Institute for Transport Economics at the University of Cologne for the last years. Studies carried out in the context of mediation and planning approval procedures

---

[1] Hübl, Hohls-Hübl & Wegener (1994), Kaspar (1970), Wilbur Smith Associates (1993).

for the extension of Frankfurt Airport have constituted the starting point.[2] These studies explicitly dealt with the impacts of the airport as a location factor for the regional economy. Based on different methodological approaches a quantification of the impacts as a location factor could be achieved. The results are taken up in the following sections.

3 Methodological Approaches

In order to quantify the economic welfare effects of airports different methodological approaches are employed sometimes alternatively, sometimes in a complementary way.

- Part of the impacts of airports can be determined by means of statistical analyses. This is especially the case for the direct added value and employment effects at the airports for both the airport operator and for other companies at airports. Thus, employment and income can for example be analysed with workplace counts.

- Questionnaires are used widely in order to analyse the current situation and to forecast future developments. They are taken into consideration for facts and developments for which statistical data is not available. The impacts induced by the extension of an airport can e.g. be assessed by means of questionnaires, e.g. cost reductions, increases in sales, settlings of companies. Interviews are carried out both in the population and in the economy, among others interviewing of airport users or of companies.

- Part of the impacts arising from the operation of an airport can be calculated empirically with the help of secondary data sources, e.g. the indirect impacts concerning the inputs needed for the production of goods and services of the airport. Analytical methods and statistical tables exist which enable the user to obtain results in this context. In some cases Input-Output-Analyses are used for the calculation of added value, employment and income effects of airports. These analyses are based on Input-Output-Tables which contain the value of input and output flows of goods and services between the sectors of a national economy. For each economic sector the production values of other economic sectors are shown which are used as an input for the production of the sector. On the other hand, the tables specify the outputs

---

[2] Mediation group Airport Frankfurt/Main (2000); Baum, Kurte, Esser & Probst (1999); Baum, Esser, Kurte & Schneider (2003).

and the intended use of the produced goods and services. Input-Output-Calculations enable the quantification of economic effects (e.g. changes in employment figures), which are caused by external changes (e.g. demand increases due to the production of certain goods and services). Investments and continuous costs of airports are understood as impetus causing primary direct effects in the companies, which are producing the goods and services, and primary indirect effects in the input-providing companies. Input-Output-Tables are published in irregular intervals by the Federal Statistical Office of Germany.[3]

- The economic impacts of airports are sometimes quantified by econometric calculations based on secondary statistical sources. This method helps to overcome the weaknesses and uncertainties of data collected in questionnaires. The aim is an objective approach with the help of statistics.

- Econometric approaches are applied e.g. for the quantification of the impacts of the air traffic connection on the prosperity of the region. For the econometric tests time series and cross section calculations are carried out. The economic development of regions with and without airport connection is compared. The differences in the economic prosperity indicate the airport-induced impacts. Employment levels and employment development, share of economic sectors in the national product, labor productivity and added value are some of the indicators used in these calculations. Likewise econometric tests can be used in order to forecast the regional economic development.

- The "structural coefficients method" is an approach, with which a breakdown and differentiation of the impacts of airports can be conducted. In this approach, the airport impacts are analyzed and explained with coefficients from official statistics. Secondary statistical structural coefficients are applied on the global production and employment effects with which a subtly differentiated statement of the impacts can be achieved.

The airport-related employment effects are differentiated regarding quality and stability of employment, share of temporary employment contracts, share of employed women, share of night-work, shift work or work at weekends, full-time or part-time employment, apprenticeship places etc. For the evaluation of the impacts such qualitative specifications are of vital importance. The sets of variables, which are applied to measure the global impacts, come from research on labor market or employment or from

---

[3] Statistisches Bundesamt (2002).

business panels. Such structural characteristics facilitate forecasts of future employment, expected changes in employment contracts and developments regarding qualitative aspects of labor.

With the above-mentioned different methodical approaches empirical statements about the economic impacts of airports are derived in the literature. Problems and restrictions have to be considered when interpreting the results. In the following part of the chapter, an overview of different empirical results regarding the economic impacts of airports is given.

4 Quantifying the Economic Impacts of Airports

In the following, empirical knowledge about the impacts on economic welfare is presented. The impacts of airports as an economic factor and as a location factor are to be considered. The empirical effects are analyzed for concrete airport projects.

4.1 Impacts Resulting From Airport Operation

Positive effects in terms of added value, income and employment result from the operation of an airport (including investments). These impacts concern the activities of companies operating at the airport. Frankfurt airport (Frankfurt / Main) is taken as an example for these impacts. The economic effects are shown for the actual development of the airport and for a development taking into account the planned extension schemes. The determination of the impacts is achieved with a workplace inquiry in the year 2001, the regional effects for Hesse are analyzed with an Input-Output-Analysis.[4]

The following economic aspects are quantified:

- Direct effects: Income and employment of workplaces located at the airport.

- Indirect effects: Income and employment, which is generated in companies providing the input for the workplaces at the airport.

- Induced effects: Income and employment which result from the spending of the directly and indirectly created income.

---

[4] Rürup, Mehlinger, Hujer & Kokot (2001).

In the year 1999, 440,000 flight movements, 45.6 million passengers and 1.6 million tons of air cargo were registered. The employment at the airport amounted to 64,531 persons. The income generated at the airport totalled 2,319 million Euro. Furthermore, regional effects have to be added. In the year 1999 they added up to 82,529 employees in Hesse (of which 44,591 were indirect ones, 37,938 induced ones) and an income of 2,768 million Euro (1,199 indirect, 1,569 induced).

Besides the status quo of the economic impacts it is of major interest which effects will arise in the future owing to an extension of the airport. Taking the extension plans for Frankfurt Airport as an example, several studies have been carried out on that behalf. These studies were carried out in the context of the mediation process (1999/2000), the regional planning procedure (2001/2002) and the official approval of the plan (2003/2004).[5]

For the extension of Frankfurt Airport investment costs of 3 billion Euro have been estimated. The impacts have been forecasted until the year 2015. Two different scenarios have been assumed:

- Scenario "Extension of the airport (Construction of a runway)".

    The scenario presumes that a new runway and half of a new terminal will be operating by the year 2006. The second half of the terminal will be put into operation by 2012. This scenario corresponds to an extension which meets the demand development. The function of the airport as a hub will remain unchanged. A night-flight ban between 11 pm and 5 am will be introduced.

    In the year 2015, 656,000 flight movements are expected for this scenario. The passenger volume will be 81.5 million passengers, the air cargo volume will amount to 2.8 million tons.

- Scenario "Non-extension of the airport (Utilization of existing capacities)"

    In this scenario the runway and the terminal will not be constructed.

    In the year 2015, 500,000 flight movements are expected. The passenger volume will equal 58.1 million passengers. The air cargo is forecasted with 2.79 million tons.

The air cargo is of comparable volume in both scenarios. This can be ascribed to the implied night-flight ban in the extension scenario. Notwithstanding this fact, there are significant differences in the number of flight movements and the

---

[5] Baum, Esser, Kurte & Probst (1999); Rürup, Mehlinger, Hujer & Kokot (2001); Baum, Esser, Kurte & Schneider (2003).

passenger volume. Thus, the comparison shows that the number of flight movements in 2015 exceeds the number of flight movements in the extension-scenario by roughly 30 % compared to the scenario without extension, with regard to the passenger volume by 40 %.

The following extension-related impacts result for income and employment by 2015 (table 25-1): The employment will experience an increase to 94,961 employees in the extension-scenario. In the case of non-extension, the number of employees will only rise to 76,587 persons. Concerning the income generated at the airport, an increase of 4,110 million Euro is expected for the extension-scenario, 3,309 million Euro in the case without extension.

	Employment in Hesse (persons)	**Income in Hesse (million Euro; price level: 1993)**
Extension	143,155	6,910
Non-extension	115,216	5,525
Difference in absolute figures	27,939	1,385
Difference in percent	24.2 %	25.1 %

Table 25-1: Direct, Indirect and Induced Employment and Income Effects for the Extension-scenario and the Non-extension-scenario[6]

It can be shown that the employment and income effects are roughly 25 % higher for the extension-scenario compared to the case without an extension.

4.2 Impacts Resulting from the Airport as a Location Factor

The impacts resulting from the airport as a location factor cover the effects which occur in the vicinity of the airport for companies using air traffic. These impacts are advantages which result from better accessibility and better connection to air transport. They are reflected in productivity gains, cost reductions, sales increases and new company settlings.

The location factor effects are identified in inquiries questioning companies in the surrounding area of the airport about the effects of the air traffic connection

---

[6] Bulwien, Hujer, Kokot, Mehlinger, Rürup & Vosskamp (1999).

and the possible changes due to an extension or non-extension of the airport. With these surveys a forecast of the impacts for the future (2015) is made as well.

Such an analysis has been performed for Frankfurt Airport in the context of the mediation process.[7] The survey differentiates between the sectors industry, trade, transport companies and other service providers. Different development scenarios have been presented to the companies. The companies were asked to assess the development of passenger and air cargo traffic via Frankfurt Airport, of air cargo and passenger transport costs as well as of turnover and employment for these scenarios. Table 25-2 gives an overview of the data collected for the identification of the impacts resulting from the airport as a location factor.

- Expected passenger and air cargo volumes.
- Future turnover and employment development.
- Possible relocations of traffic (other airports, rail or road transport).
- Expected increase of travel costs if traffic is relocated (e.g. tickets, costs of accommodation, expenses).
- Extension of travel time of employees.
- Shortfalls in sales volumes for the company, for example because employees cannot reach customers in time or the premises of the company are more difficult to reach.
- Resettlements of companies or parts of companies (e.g. administration, production, distribution, logistics).
- Internal adjustment procedures due to possible cost increases and losses of turnover (e.g. use of information and communication technologies, changes in the vertical range of manufacture, changes in the organization of transport processes, decrease of number of supplier companies).
- Potential of attenuation of cost increases and turnover losses.

Table 25-2: Data Collection Program for the Empirical Determination of the Impacts of Airports as a Location Factor[8]

The location impacts of the airport have been analysed for different scenarios. The following scenarios have been considered for Frankfurt Airport:

- Scenario A: Unconstrained development of the airport.

---

[7] Baum, Esser, Kurte & Probst (1999).
[8] Baum, Esser, Kurte & Probst (1999).

All capacity shortages (air cargo and passenger transport) are removed.

- Scenario B: Limitation of the hub function of the airport.

    The airport loses its hub function due to a reduction in the number of flights which are taken over by other airports.

- Scenario C: Loss of possibility to fly time-critical cargo transports.

    The number of freight flights are cut down. Thus, the storage time of the freight at the airport is prolonged. Time-critical goods (which have to be delivered over night respectively within 24 hours) cannot be transported anymore.

The economic impacts of these three scenarios refer to the surrounding area of the airport that means Hesse, parts of Rhineland-Palatinate, Bavaria and Baden-Württemberg. The impacts are displayed in table 25-3:

	Employment (1,000)	Gross Value Added (Billion Euro)
Scenario A	1,991.9	217.1
Scenario B: Deviation from scenario A in absolute figures	- 50.9	- 6.4
Scenario B: Deviation from scenario A in percent	- 2.6 %	- 3.0 %
Scenario C: Deviation from scenario A in absolute figures	- 30.6	- 2.6
Scenario C: Deviation from scenario A in percent	- 1.5 %	- 1.2 %

Table 25-3: Impacts of Frankfurt Airport as a Location Factor (2015)[9]

In Scenario A (no capacity shortages) the overall employment in the surrounding area of the airport equals 1,991.9 million and the Gross Value Added amounts to 217 billion Euro. The case of non-extensions and subsequent loss of a hub function (Scenario B) brings about an employment decrease of 50,900 persons (= -2.6 %) and a decline of the Gross Value Added of 6.4 billion Euro (= -3.0%).

---

[9] Baum, Kurte, Esser & Probst (1999).

The restriction with regard to (the possibility of) flying time-critial goods (Scenario C) results in a reduction of employment by 30,600 persons (= - 1.5%) and in a decrease of Gross Value Added of 2.6 billion Euro (= - 1,2 %).

The estimation of the airport-induced impacts is based on forecasts. The actual validity of the results can only be verified by ex-post examinations. Such an ex-post examination has been carried out by the planning organization of the "Äußerer Wirtschaftsraum München" (greater Munich economic area) and the ifo Institute for economic research for the surrounding area of the Munich Airport.[10] In this report, the results of previous studies on the economic impacts of airports are compared with the actual status of development. The forecast horizon in a 1989 published ex-ante analysis was the year 2000. The comparison with the actual development shows:

- The predicted increase in the number of inhabitants in the vicinity of the airport corresponds to the actual increase (rising from 340,000 in 1987 to 420,000 in 2000).

- The forecasted employment growth exceeds the actual developments. While the prognosis estimated an increase of 80,000 employees, the actual employment grew by 100,000 employees.

- The forecasted number of workplaces at the airport (approx. 20,000) corresponds to the real figures.

4.3 Quality Analysis of the Employment Effects

The quantitative impact analyses of airports show the total employment and added value effects. They do not contain statements about the changes in type or structure of the employment. However, such changes in the quality of employment are of vital relevance for the affected employees and employers as well as for the decision-makers in politics and administration. These changes might indicate for example which requirements will be demanded in the future of employees (qualification level, working-time regulations).

By means of splitting up the labor market effects by quality criteria the information contained in the quantitative analysis can be significantly increased. In this context, different aspects play a role, e.g.:[11]

---

[10] Planungsverband Äußerer Wirtschaftsraum München, ifo Institut für Wirtschaftsforschung (2002).
[11] Baum, Schneider, Esser & Kurte (2004).

- Special characteristics of the workplace (e.g. wages, working times, especially night and weekend shifts, qualification requirements, work contents).
- Aspects concerning the work environment and the conditions on the labor market (e.g. apprenticeships, on-the-job training, employment chances, workplace conditions, co-determination, safety at work, employment protection).
- Subjective satisfaction with the employment as a measure of the quality of the workplace.

There are various criteria and indicators used for characterizing and identifying the quality of work. The following table 25-4 gives an overview:

- Quality level
- Crowding-out effects (e.g. due to staffing with over-qualified employees)
- Women's employment
- Stability and safety of created employment
- Night shifts
- Full-time or part-time employment
- Use of temporary labor contracts
- Apprenticeship places
- Rate of organization of new employment
- Recruitment rate of long-term unemployed
- Recruitment rate of elderly employees
- Recruitment rate of low-qualified employees
- Stability of labor contracts
- Employment at minimum wages
- Fictitious self-employment

Table 25-4: Quality Attributes of Labor[12]

The structural characteristics of the workplaces at the Frankfurt Airport have not been ascertained in questionnaires. This study simply incorporated the overall effects so that an ex-post analysis will be necessary in order to classify retrospectively the employment effects regarding their quality characteristics.

---

[12] Baum, Schneider, Esser & Kurte (2004).

For this qualitative distinction secondary sources of official statistics as well as of labor market and employment research are used.[13]

The first step in order to further characterize the employment with regard to its quality aspects is to work out the current quality structure of employment in the studied region Hesse resorting to secondary statistical data material. When the sectoral employment effects resulting from the extension of the airport and the quality structure of the labor market in Hesse are assembled, assertions can be made regarding which quality aspects benefit in particular from the employment effects and which quality aspects show a disadvantage. For the deduction of the qualitative employment effects the share of the individual quality criteria in all workplaces is determined in the status-quo-analysis of Hesse. These shares are combined with the shares of each sector in the employment effects induced by the airport extension. The combination of the shares shows as a result the relation between the share of the respective quality criterion for the airport extension-related employment effects and the share of this criterion for all employees in Hesse. Table 25-5 exemplifies the relative deviation between the average of the labor market in Hesse and the average of the employment effects of the airport extension for selected quality criteria (qualification structure and full-time respectively part-time employment).

With regard to the qualification level it becomes evident that the number of higher-qualified employees in relation to the newly-created overall employment arising from the airport extension is slightly below the Hessian average whereas the share of low-qualified employees shows a positive ratio compared to the average of the federal state of Hesse. Thus, given the status quo conditions, the extension of the airport privileges those economic sectors which show a higher share of low-qualified employees.

The employment effects are offset to some extent as far as part-time employment is concerned. Considering part-time work with more than 15 hours per week the relation of the employment effects out of the airport extension is negative compared to the Hessian average. The share of part-time employment with less than 15 hours per week in the newly-created jobs is on the other hand above state average. On the whole, the share of part-time employment resulting from the airport extension is slightly below the Hessian average (status quo).

---

[13] Bundesanstalt für Arbeit, Landesarbeitsamt Hessen (2003); Institut für Arbeitsmarkt- und Berufsforschung der Bundesanstalt für Arbeit (2002).

	Qualification Structure			Full-time and part-time employment				
	High Qualified	Low Qualified	Trainees	Full-time	Part-time < 24 h	Part-time 15-24 h	Part-time < 15 h	Without agreement
Industry	8,270	3,140	390	10,500	240	470	470	120
Building Sector	1,530	300	170	1,820	60	80	40	0
Trade	6,470	2,410	420	6,510	740	1,120	840	90
Business Services	14,260	4,950	490	15,760	790	1,380	1,570	200
Transport, Information transmission, Finance, Insurance	13,900	4,970	930	14,450	1,390	2,180	1,390	390
Total	44,430	15,770	2,400	49,040	3,220	5,230	4,310	800
Share in employment effects (extension)	71.0 %	25.2 %	3.8 %	78.3 %	5.1 %	8.4 %	6.9 %	1.3 %
Share in Hesse	71.7 %	24.2 %	4.1 %	78.0 %	6.0 %	9.0 %	6.0 %	1.0 %
Deviation betw. empl. effects and total Hessian average	- 0.7 %	1.0 %	- 0.3 %	0.3 %	- 0.9 %	- 0.6 %	0.9 %	0.3 %
Ratio extension to Hessian average	- 1.0 %	4.1 %	- 6.5 %	0.4 %	- 14.3 %	- 7.2 %	14.7 %	27.8 %

Table 25-5: Employed Persons Resulting from Construction, Operation and Location Factor Effects (in 2015) Divided into Quality Criteria (Status Quo)[14]

By using this method, the impacts of the employment effects caused by the airport extension on the quality of the employment can be analysed: It can be expected that sectors with an above-average share of women's employment benefit from the airport extension. Those sectors are brought forward by the extension which possess a share of employment with a dismissal risk below average, i.e. the risk of being made redundant in these jobs is attenuated. The airport extension impacts positively and over-proportionally on jobs with temporary labor contracts i.e. new employment is mainly created in sectors in which more frequent use is made of temporary labor contracts.

---

[14] Bundesanstalt für Arbeit, Landesarbeitsamt Hessen, Rürup, Mehlinger, Hujer & Kokot (2001); Baum, Schneider, Esser & Kurte (2004).

4.4 Economic Losses due to External Costs

The positive impacts of airports in terms of value added, income and employment are counteracted by contractive effects due to external effects arising from air traffic. For the assessment of welfare effects induced by airports it is mandatory to deduce the external effects from the positive expansive effects. The balance of the two components gives the net result.

Contractive effects due to the external costs of air traffic occur in so far as external costs absorb resources which are no longer available for the maintenance of other production activities. Thus, external costs diminish the macro-economic production potential which results in a decrease of value added, income and employment.

The external costs of air traffic consist of noise, pollution and carbon dioxide emissions. For different carriers the level of external costs is assessed in different studies.[15] The calculations are made in general for traffic performances (km per person and in km per ton in freight traffic). Such quantifications do not attribute external costs to the respective airports.

There are, however, individual studies which undertake an estimation of the external costs of airports. A study on the Köln/Bonn Airport conducted in 1995 quantified external costs amounting to 265 million Euro due to traffic performances.[16]

The actual external costs caused by the operation of an airport are lower. Without air traffic transportation would partly be carried out by other means of transport. In this case these external costs would originate with other carriers. The calculation of the external costs requires the deduction of the costs. The calculation supposes that without air traffic the European passenger and freight travel would be replaced by continental transport carriers (rail, road), that intercontinental passenger flights would be replaced by (sea-going) ships and that 25% of the intercontinental freight traffic would be carried out by (sea-going) ships. In case of carrying out European continental traffic via rail external costs of the value of 67 million Euro would be caused. If the European continental transport was carried out via road external costs of 277 million Euro would be caused. In order not to underestimate the external costs of air traffic it is assumed that European air traffic would be replaced by the relatively

---

[15] INFRAS & IWW (1994).
[16] Kurte (1999).

environmentally- friendly carrier rail. Given this assumption external costs of 197 million Euro per annum remain for the Köln/Bonn Airport.

4.5 Fiscal Impacts of Airports

Airports generate fiscal impact through the gain of tax revenues owing to economic activities at the airport and to location effects for the region. The fiscal impact is an important indicator for the regional economic impetus generated by the airport. The fiscal impact cannot be characterized as allocative as it does not increase the creation of value added and therefore does not contribute to the growth of the airport. The fiscal effects rather represent transfers in so far as tax revenues generated within the state are withdrawn from private economy. Notwithstanding this fact, the fiscal effects are part of the regional economic balance of airports.

Previous studies on the impacts of airports neglected the fiscal effects to a large degree. The extension plannings for the airport Frankfurt containing an extensive economical analysis did not take up the fiscal impact either. A quantitative assessment only exists for the Köln/Bonn Airport.[17] The fiscal impacts are identified by determining the losses in tax revenues that would result if the Köln/Bonn Airport did not exist. The modelling of this without situation makes it possible to determine the losses in tax revenues due to the loss of location-related effects. The difference drawn from the scenario without air traffic helps to determine the decreases in economic activity and in economic decline for the region, both resulting in losses of tax revenues.

The following types of taxes are taken into account for the quantification of the fiscal impact: income tax, revenue tax (corporation and trade tax), value added tax, fuel tax. A relation is drawn between the economic activities in the region and the gains in taxes by applying average rates of taxation. This relation is based on the quantification of the decline in economic activities (income, profits and consumption) in a situation without air traffic.

The income and profit effects the airport generates for the region Köln/Bonn amount to between 0.30 and 0.33 billion Euro with regard to the income and between 1.49 and 1.94 billion Euro with regard to profits. The losses in tax revenues per year without airport-related performances amount to between 39 and 44 million Euro in income tax, between 111 and 147 million Euro in revenue tax, between 15.3 and 16.9 million Euro in value added tax and 2 million Euro in

---

[17] Kurte (1999).

fuel tax. The losses in tax revenues excluding airport-related performances total between 215 and 269 million Euro for the Exquecher. The losses in tax revenues do not represent losses in revenue for the city Cologne alone but overall losses in revenue for the Federal Government, the federal states and communities.

5 Results

The analysis has given evidence that airports make considerable economic contribution to the prosperity of a region. This prosperity impact becomes apparent in increases of created value added, income and employment. These effects result from the fact that airports perform several economic functions. Airports produce economic activity in terms of an investment factor which generates productive impact with the ordering firms and input-providing companies. Airports do operate furthermore as locally concentred economic factor generating performances in production through the operation of the airport. Finally airports produce economic impact in terms of location advantages. Owing to the accessibility to air traffic air travel users draw advantages in form of increases in productivity, cost reductions and market expansions. These growth effects impact positively on tax revenues for the Exchequer. Considering their economic importance the effects derived from investment can be characterized as temporary ("passing effect"). Once the investments are realized they stop having an impact whereas the effects resulting from the operation of the airport and its location are permanent for both population and economy. As to the strength of impact the location effects rank before the effects related to the operation of the airport.

The economic development potential a region possesses is reduced in case of restricted capacities at the airport which affects the creation of value added and productivity. Empirical studies show that considerable gains in growth and prosperity can be obtained by eliminating existing restrictions of capacity.

6 References

BAUM, H., ESSER, K., KURTE, J., PROBST, K.M. (1999): *Der Flughafen Frankfurt/Main als Standortfaktor für die regionale Wirtschaft. Wertschöpfungs- und Beschäftigungseffekte alternativer Beschäftigungsszenarien*, Untersuchung für die Mediationsgruppe Flughafen Frankfurt/Main, Köln.

BAUM, H., ESSER, K., KURTE, J., SCHNEIDER, J. (2003): *Standortfaktor Flughafen Frankfurt Main- Bedeutung für die Struktur, Entwicklung und Wettbewerbsfähigkeit der Wirtschaft der Region Rhein-Main*. Unterlagen zum Planfeststellungsverfahren, Köln.

BAUM, H., SCHNEIDER, J., ESSER, K., KURTE, J. (2003): *Folgewirkungen einer Großinvestition auf die regionale Entwicklung – am Beispiel des Ausbaus des Frankfurter Flughafens*. Untersuchung für die Hans Böckler Stiftung, Köln.

BULWIEN A.G. (2003): *Der volkswirtschaftliche Nutzen des Flughafens Frankfurt/Main*, Synopse, München.

BULWIEN, H., HUJER, R., KOKOT, S., MEHLINGER, C., RÜRUP, B., VOSSKAMP, T. (2003): *Einkommens- und Beschäftigungseffekte des Flughafens Frankfurt/Main- Status-Quo-Analysen und Prognosen*, München, Frankfurt/Main, Darmstadt.

BUNDESANSTALT FÜR ARBEIT, LANDESARBEITSAMT HESSEN (2003): *Arbeitsmarktatlas Hessen, Teil 5:* Beschäftigungsentwicklung nach Wirtschaftszweigen von Juni 2000 bis Juni 2002, Frankfurt/Main.

BUNDESANSTALT FÜR ARBEIT, LANDESARBEITSAMT HESSEN (2003): *Betriebspanel Report Hessen 2002*, Teil 1: Ältere Beschäftigte in hessischen Betrieben, Frankfurt/Main.

BUNDESANSTALT FÜR ARBEIT, LANDESARBEITSAMT HESSEN (2003): *IAB-Betriebspanel Hessen 2001*, Abschlussbericht, Frankfurt am Main.

HÜBL, L., HOHLS-HÜBL, U., WEGENER, B. (1994): *Der Flughafen Hannover-Langenhagen als Standort- und Wirtschaftsfaktor*, Hannover.

INFRAS & IWW, *Externe Kosten des Verkehrs*, Zürich-Karlsruhe.

INITIATIVE LUFTVERKEHR (2004): *Masterplan zur Entwicklung der Flughafeninfrastruktur zur Stärkung des Luftverkehrsstandortes Deutschland im internationalen Wettbewerb*, Berlin.

INSTITUT FÜR ARBEITSMARKT- UND BERUFSFORSCHUNG DER BUNDESANSTALT FÜR ARBEIT (IAB), Beruf im Spiegel der Statistik, Beschäftigung und Arbeitslosigkeit 1996-2002 (www.pallas.iab.de).

KASPAR, C. (1970): *Die Bedeutung des Luftverkehrs für die Schweiz unter besonderer Berücksichtigung des Flughafens Zürich*, St. Gallen.

KURTE, J. (1999): *Produktions- und Beschäftigungseffekte von Verkehrsflughäfen am Beispiel des Flughafens Köln/Bonn*, Diss. Köln.

MEDIATIONSGRUPPE FLUGHAFEN FRANKFURT/MAIN (2000): *Bericht Mediation Flughafen Frankfurt/Main*.

PLANUNGSVERBAND ÄUSSERERWIRTSCHAFTSRAUM MÜNCHEN, IFO INSTITUT FÜR WIRTSCHAFTSFORSCHUNG (2002): *Der Flughafen München und sein Umland, Grundlagenermittlung für einen Dialog*, Teil 1 Strukturgutachten, Juli.

RÜRUP, B., MEHLINGER, C., HUJER, R., KOKOT, S. (2001): *Einkommens- und Beschäftigungseffekte des Flughafens Frankfurt/Main- Status-Quo-Analysen für 1999 und Szenarien*, Darmstadt, Frankfurt/Main.

STATISTISCHES BUNDESAMT (2002): *Input-Output-Tabelle 1997*, Wiesbaden.

WILBUR SMITH ASSOCIATES (1993): *The Economic Impact of Civil Aviation on the US Economy*.

26

LEARNING FROM THE AIRLINES? A COMPARATIVE ASSESSMENT OF COMPETITIVE STRATEGIES FOR RAILWAY COMPANIES

CHRISTIAN KAUFHOLD AND SASCHA ALBERS

1 Introduction..606
2 Competitive Strategies in Passenger Air Travel 606
3 Competitive Strategies in Passenger Rail Travel 621
4 Summary..632
5 References...633

Summary:
European railway companies are confronted with first entries into their formerly protected monopoly markets of long distance passenger travel, requiring them to adapt their competitive strategies – a situation the European airline industry has been confronted with several years before. In order to arrive at viable strategy recommendations for railways we compare post-regulation airline and railway market and product characteristics. We argue that the airlines reacted primarily by two major measures: the reconfiguration of their route networks and an increased application of cooperative strategies. Although an analysis of the underlying industry conditions shows that these strategy components are not easily transferred to the railway sector, some selected elements of these concepts promise to contribute to a more competitive positioning of the former railway monopolists.

1 Introduction

Nearly all former monopoly industries are undergoing deregulation processes – at least in the European Union – with the railroad sector among the most traditional industries with a long historical record. Alas, especially in the long distance passenger railroad travel segment, incumbents are facing first market entrants. For the former state companies intramodal competition is practically unknown terrain and thus considerable revenue losses are predicted. However, while this situation is new to players in this industry, companies in other formerly highly regulated environments have encountered a comparable situation before.

In this chapter, the strategic challenges for former state railway companies are analyzed referring to such a reference market: the market of passenger air transport. The considerable similarity of both, market and legal conditions as well as the underlying production principles promise to yield valuable insights for railway companies in their current situation. Competition in the airline sector has greatly increased since its deregulation. The actors in this industry can thus be expected to have obtained a certain amount of experience regarding their strategic competitive behavior.

The analysis is organized in two parts. First, airlines' reactions following the deregulation of the European air transport market are examined. Subsequently, their behavior is assessed with regard to its applicability to the railroad sector and recommendations for railways are deducted. The chapter does not provide an economic assessment of the course and scope of deregulation activities in this industry.

2 Competitive Strategies in Passenger Air Travel

The deregulation of the European air transport market, starting in 1987, increasingly exerted competitive pressure on incumbent airlines, pushing them to find appropriate strategic market positions. Today (2004), basically two broad competitive orientations of airline companies can be observed:

- Product differentiation as *network player*. Passenger air transport is a fundamentally homogenous product, rendering the pursuit of a differentiation strategy based on objectively measurable product characteristics problematic.[1] Partial differentiation is achieved by the establishment of comprehensive route networks with a high number and frequency of offered relations. Typically, such a network is configured as a hub and spoke system.

- Cost leadership as *point-to-point player*. Considering the difficulties in differentiating air transport services through different product characteristics as well as the high capital intensity in this industry, cost leadership strategies promise the achievement of a favorable competitive position in the airline sector. Economies of scale, frequently seen as important when pursuing this strategy, can not be realized in air transportation once a relatively small company size is achieved.[2] However, economies of scope, but especially economies of density are attributed to hub and spoke network operations.[3] Alas, these come at significant costs of complexity cost leaders in the airline business typically avoid by offering point-to-point services.[4]

In order to allow for the evaluation of competitive strategies[5] as they have evolved in the air transport sector with regard to their transferability on the railway sector, an identification and detailed assessment of their characteristics needs to be performed. The following review will roughly be based on Porter's generic value chain model[6] in order to show both similarities as well as differences in the underlying value creating processes. The primary activities are defined here on an intermediate level of abstraction, allowing for an industry specific analysis but at the same time avoiding a focus on one single mode of transportation. We have divided the major strategic decision areas into (a) design of the route network, (b) selection of the fleet structure, (c) design of the distribution system and (d) production characteristics (see Figure 26-1).

---

[1] Safety and speed are basic requirements to a marketable product in this industry and therefore hardly differ between the players' services.
[2] See Bailey & Panzar (1981); Burton & Hanlon (1994), p. 218; Hanlon (1999), pp. 45f.; White (1979).
[3] See Caves, Christensen & Tretheway (1984); Hanlon (1999); Nero (1999).
[4] It has to be mentioned that low cost airlines do not operate hub and spoke networks in their original sense. Their network structures may sometimes appear to be designed according to this concept. However, the apparent "hub" airport is not being used as a connection point; connecting flights are not intended.
[5] These strategic alternatives correspond to Porter's competitive strategies. See Porter (1980); Porter (1985).
[6] See Porter (1985).

Potentially relevant supporting activities will be taken into account where appropriate.

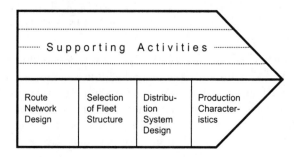

Figure 26-1: Strategic Decision Domains for the Comparative Analysis

2.1 The Incumbents: Product Differentiation in Passenger Air Travel

For new competitors, positioning as a network player has hitherto turned out to be nearly impossible. Since such a strategy requires a broad market presence, it is ideal for airlines that have already been active prior to market deregulation as state monopolists and therefore developed a sophisticated route network.

2.1.1 Route Network

In the course of market deregulation, the route networks of virtually all US and later also European based incumbent airlines have evolved into hub and spoke networks (see figure 26-2 for an illustration of Western airline's network reconfiguration).[7] The popularity of this concept on both sides of the Atlantic appears to indicate its economic suitability for the given competitive context.[8]

---

[7] See Aberle (2000), p. 182; Bailey, Graham & Kaplan (1985), pp. 73ff.; Butler & Huston (1999), p. 52; Button (2002), pp. 177ff.; Dennis (2000), p. 75; Kahn (1988), p. 318.

[8] Even though in European cross border transportation this development could also be explained by historical reasons (since the national flag carriers targeted international destinations mainly from their home hub airports), this is not possible for the US market. The North American airlines started from a completely different situation and even for the European context it seems unlikely that the "grown" hub and spoke

However, difficulties are associated with this network pattern as well. We will shortly discuss both, cost as well as revenue effects of a hub and spoke network configuration.

Cost Disadvantages of Hub and Spoke Layouts

The availability of a high number of attractive connections at the hub is crucial for the passenger in order to accept the hub and spoke layout. In order to create the required number of connections within a minimal connecting time, inbound and outbound traffic at the hub are concentrated in rather short time intervals (banks), resulting in additional costs:[9] within these banks, an appropriate turnaround capacity has to be available at the airport assuring a quick handling of travelers, baggage and aircraft.[10] Additional employees and sophisticated equipment are necessary. Such a complex system relies on smooth interaction of a number of processes, rendering it increasingly vulnerable. Delays due to technical problems or weather conditions are quickly passed on throughout the entire network.[11] Furthermore, it seems obvious that indirect service on formerly direct relations will significantly shorten the length of each individual flight and thus lead to a disproportionate rise in production costs which are a function of starting and landing procedures to a significant extent, far less on flight distance. Finally, airlines operating in a hub and spoke configuration have to arrange the overnight stabling of their airplanes. Efficient maintenance, fleet and crew planning will be easier with the planes stationed at the central hub, whereas higher load factors make a positioning in the spoke airports also worth considering.[12]

Cost Advantages of Hub and Spoke Layouts

The efficient use of existing fleet capacities is a crucial success factor for companies in the transport industry. The funneling of all passengers through one central connection point allows the airlines to consolidate all traffic from different origins but with the same destination.[13] Airlines thus have to service

---

structure would have been forced if economic reasoning had pointed into a different direction. See Bailey, Graham & Kaplan (1985), pp. 11f.
[9] See Dennis (2000), pp. 75 f.; Jäggi (2000), p. 127; McShan & Windle (1989), p. 212; Nero (1999), p. 227.
[10] See Dennis (2000), pp. 76, 80f.; Hanlon (1999), pp. 154ff.
[11] See Hanlon (1999), pp. 134f.; Jäggi (2000), p. 128.
[12] See Hanlon (1999), pp. 135ff.
[13] See Burton & Hanlon (1994), pp. 218f.; Delfmann (2000); Dennis (2000), p. 82; Hanlon (1999), pp. 154f.; Levine (1987), pp. 442ff.

less relations, have a higher load factor on the remaining spokes and can either use bigger, more efficient aircraft or offer higher frequencies. In any case, higher traffic densities lead to significant reductions in unit costs.[14]

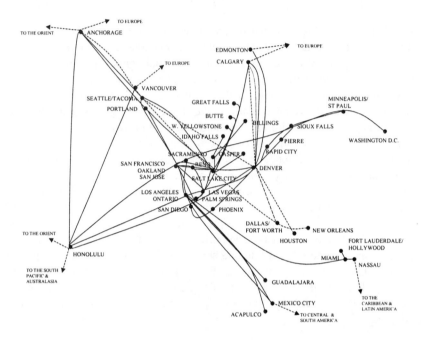

Figure 26-2: Western Airline's Route Network *prior to* US Deregulation[15]

---

[14] See Doganis (1991); Hanlon (1999). The amount of such gains from density in airlines' hub and spoke networks have been subject to several empirical studies. See e.g. Antoniou (1992); Caves, Christensen & Tretheway (1984); McShan & Windle (1989); Brueckner, Dyer & Spiller (1992). However, several obstacles make the calculation of exact cost advantages very difficult, some authors even consider it to be entirely impossible. See Hansen & Kanafani (1989).

[15] Source: Adapted from Williams 1994, p. 17

COMPETITIVE STRATEGIES FOR RAILWAY COMPANIES 611

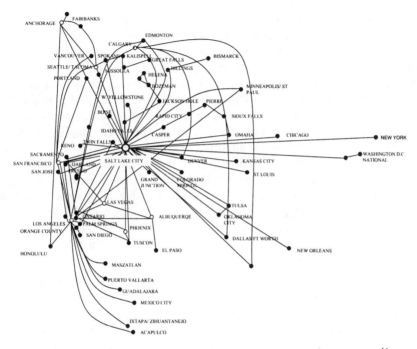

Figure 26-3: Western Airline's Route Network *after* US Deregulation[16]

Another potential cost advantage of a hub and spoke structure arises from the possibility to centralize support functions like maintenance and service of the airplanes, but also their coordination as well as that of the crews at the hub. Unfortunately, this option comes with certain disadvantages again, since the characteristics of the traffic flows make it more profitable to start and land the first and last flights of one day not in the hub, but the spoke location.[17]

Evolution of the Hub and Spoke Concept

The incumbent airlines' introduction of hub and spoke networks can be seen as a reaction to increased competition by market entrants servicing point-to-point relations, but also and perhaps predominantly by intensified price competition among the established network players. Travelers were apparently not willing to pay the mark-up for the incumbents' differentiated network products. Therefore, after already sophisticated refinements of the processes within their hub and

---

[16] Source: Adapted from Williams 1994, p. 17
[17] See Hanlon (1999), pp. 136f.

spoke systems, network players realized that their costs and therefore also fare levels still did not match the expectations of the majority of their customers. Furthermore, it appears that the remaining price pressure can no longer be met with the present level of service integration. Therefore, the first network carriers have already begun to lower their network costs, even at the expense of individual flight connections.[18] This emancipation of spoke flights spoils the connection quality within the hub where customers have to accept longer waiting times due to non-synchronized starts and landings. This in turn implies that such a strategy can only be pursued at airports with a sufficient amount of traffic, i.e. if the number of incoming and outgoing flights is sufficiently high that individual banks can hardly be distinguished.[19] Besides a more efficient production of the transportation service within the individual spokes, such a policy allows for a smoothened traffic load at the hub resulting in a lower capacity demand and therefore lower costs. It also helps stabilizing traffic flows at the hub airport, since problems on one spoke relation will no longer directly affect the entire network.

With their "de-peaking" concept, network carriers aim to exploit further cost potential that could allow for further price reductions in their competitive environment. However, these cost reductions are inseparably tied to a lower degree of network connectivity and thus slowly approach the production model of cost leaders in the market which also provide "uncoordinated" services from their base airports.

Revenue Effects of Hub and Spoke Layouts

Apart from the cost effects, a strong hub airport is seen as an appropriate means to dominate the regional market around the hub.[20] Facing the strong presence of an airline at its hub, competitors have practically no possibility to gain a significant share of the market in its vicinity.[21] This is caused by the insufficient

---

[18] The conventional hub and spoke structures sometimes require long ground times of the airplanes in spoke locations in order to hit the next wave at the hub airport. By emancipating these spoke flights from the hub processes, their efficiency can be significantly raised. See Dennis (2000), p. 82; Hanlon (1999), pp. 133f.; Jäggi (2000), pp. 123f.; Nero (1999), p. 226.

[19] Due to this observation, this policy is referred to as continuous hubbing or de-peaking. See Dennis (2000), p. 77.

[20] Such airports are pictorially described as fortress hubs. See Aberle (2000), p. 183; Argyris (1991), p. 33; Brueckner & Spiller (1994), pp. 381ff.; Dennis (2000), p. 80; Jäggi (2000), pp. 123f.; Nero (1999), p. 226.

[21] On a route level, air transportation markets are often referred to as the perfect example for the viability of the "contestable markets" theory (see Baumol, Panzar

availability of attractive slots for starts, landings and terminal access, but also by the great selection of destinations any carrier can offer at its hub, since it operates direct flights to practically all destinations available. This fact provides the hub airline with a regionally dominant market position.

2.1.2 Fleet Structure

Traditionally, incumbent airlines showed a rather heterogeneous fleet structure. However, faced with emerging competition after industry deregulation, their purchasing decisions increasingly focused on efficiency goals. This efficiency orientation was passed on to aircraft manufacturers who were eager to meet their clients' changing expectations, helping airlines to reduce the problems arising from a heterogeneous aircraft fleet. The modular construction of aircraft using the same cockpit, cabin elements or engines in several models lowers both the price for new planes as well as the maintenance costs of one airline for its fleet. However, the more aircraft types an airline employs, the more complex its planning procedures will become. A network carrier will always be more exposed to this effect than a point-to-point player due to its strategy of linking a great number of heterogeneous destinations on a national, continental and intercontinental level. For low cost airlines, the ability of their aircraft type to link a certain city pair can easily lead to the decision of whether or not to neglect this relation altogether.

2.1.3 Distribution Systems

With the liberalization of air transport markets, the computer reservation systems (CRS) suddenly were at the heart of competition between airlines since they offer market transparency, especially with regard to market shares and ticket prices.[22] As creation and evolution of CRS require significant financial expenditures, they were soon considered as effective competitive weapons for the incumbent carriers which, as owner of their systems, also set the rules for the

---

and Willig, 1982), i.e. the competitive behavior of market actors simply because of the threat of entry of potential other players into their contestable markets. However, on today's hub and spoke focused markets, the applicability of this theory is very limited. Since the network structure is the base for an improved service in spoke cities, the political judgment of the resulting situation is extremely difficult. See Burton & Hanlon (1994), p. 224; Butler & Huston (1989); Butler & Huston (1999), pp. 52f.; Hanlon (1999), pp. 41f.; Levine (1987), pp. 405ff., 444ff.

[22] See Burton & Hanlon (1994), p. 214; Duliba, Kauffman & Lucas (2001), pp. 705f.; Hanlon (1999), pp. 67ff.; Levine (1987), pp. 458ff., 464; Pompl (1998), pp. 256ff.

display and order of flights and carriers on travel agents' terminals. The American and European competition authorities reacted by issuing codes of conduct in order to prevent carriers from abusing their CRS in upcoming competition. However, the effectiveness of this measure is still under debate and provides another reason for small and low cost carriers to support common, so-called "no host" systems and, on the other hand, focus very much on direct internet and call center based ways of distribution.

However, meanwhile the incumbents have started to link their CRS to their internet sites, creating direct contact to business as well as private clients and thus establishing an additional distribution channel besides the traditional travel agencies. This allows them to reduce sales commissions, leading to reduced costs.

2.1.4 Specificities of Production: The Alliance Trend

The carriers' second fundamental reaction to market liberalization is the widespread establishment of multilateral alliance networks.[23] Cooperation among airlines spans virtually all possible activities with varying intensity.[24] Obviously, in the airline industry, alliances are seen as crucial means to achieve competitive advantage for the participating carriers in the current industry environment. We will analyze the rationale for airline alliance formation in more detail as we aim to assess the transferability of this strategy to the railway sector. As competitive advantage can be gained by either lowering costs or increasing revenue (or both), we will consider the benefit creation logic of airlines alliances from both, the cost and the revenue side.[25]

Cost Effects

In general, airlines expect their alliances to yield potential cost savings in four areas:[26]

---

[23] See Burton & Hanlon (1994); Button, Haynes & Stough (1998), pp. 99ff.; Button (2002), p. 179; Janic (1997), pp. 171f.; Oum, Park & Zhang (2000), pp. 16ff.; Schnell (2000), p. 21.

[24] They go from interlining and code sharing up to flight plan and capacity coordination, integrated brand appearance or the foundation of joint ventures. See e.g. Oum & Park (1997).

[25] See e.g. Albers (2000); Ebers (1997).

[26] See Button, Haynes & Stough (1998), pp. 117f.; Doganis (2001b), pp. 77ff.; Oum & Park (1997), p. 141; Oum, Park & Zhang (2000), p. 13.

1   Synergies should be created by staff savings and further avoidance of redundant processes.

2   Cost differentials among the partner airlines allow them to shift processes to the most efficient company in the respective area and equally profit from the lowered cost level.

3   Additional advantages can be achieved in purchasing by bundling of market power and common aircraft specifications.

4   Finally, revenue effects (to be discussed below) affect the cost structure as well. The transportation of more passengers through the partners' joint networks results in reduced production costs per passenger. These so-called economies of scope and density are influenced to a great extent by the system characteristics of incumbents' hub and spoke configured networks.

In most cases, cost savings potential of airline alliances, however, is expected to be rather limited. This is mainly due to the small amount of costs actually not depending on the level of production in this industry.[27] Especially the expensive aircraft fleets show hardly any additional cost advantages beyond a certain size.

Revenue Effects

In practice, cost advantages resulting from airline alliances have been second to revenue effects.[28] The combination of airline route networks allows to offer a much more comprehensive set of origins and destinations and to generate a higher amount of traffic within the airlines' systems by feeding passengers into each others' networks. Offering as many global destinations as possible is a crucial competitive factor for a transportation mode that has practically no intermodal competition on relations of more than 500 km due to its technical abilities. In addition to this more quantitative argument of carrying more traffic, another revenue-focused argument is the possibility of avoiding (mostly national) regulation barriers that still block access to many potential markets.[29]

---

[27] See Burton & Hanlon (1994), pp. 217f.

[28] See Burton & Hanlon (1994), pp. 213, 218; Doganis (2001b), p. 76; Morrish & Hamilton (2002), pp. 403f.; Oum & Park (1997), pp. 140f.; Oum, Park & Zhang (2000), pp. 12f.

[29] See Button, Haynes & Stough (1998), pp. 121f.; Schmidt (1993), pp. 43, 45; Schnell (2000), p. 18.

2.1.5 Support Functions

The incumbent airlines' relevant support functions basically cover two main areas. On the one hand, these are the instruments of yield management that are further developed here than in any other industry. On the other hand, loyalty programs play a significant role in passenger air transport.

Yield Management

In order to react to growing price pressure in their respective markets, airlines, besides cost reductions, sought additional ways to increase profitability of their existing operations. Business customers are willing to pay higher fares and need more flexibility than leisure travelers. Airlines began to use this situation in order to sell comparable seats at different prices by artificially differentiating their product.[30] Even without significant structural or process improvements profitability was already improved by this basically simple measure of price differentiation. Additionally, the precise adjustment of prices relative to demand and incoming bookings allows carriers to stabilize demand *over time*. Of course, not all customers are flexible enough to avoid higher ticket prices by changing their time of traveling, but within certain limitations demand can very well be influenced.

Loyalty Programs

Increasing presence of competitors at many airports and the principally greater choice in routing options through competitors' hubs involved a higher freedom for customers. The carriers reacted by introducing loyalty programs[31] which raised switching costs for their customers and allowed them to conserve traffic they would otherwise have lost to other airlines even on airports with a relatively weak own presence. The effect of loyalty programs was even stronger on the respective airline's hub airport where its already existing dominance could still be increased.[32] The development of the different loyalty programs is still pushed forward by their owners and they have sometimes virtually reached the status of independent currencies, offering the customer a great variety of products from affiliated companies in many different industries.

---

[30] See Bailey, Graham & Kaplan (1985), p. 47; Belobaba (1987); Hanlon (1999), pp. 189ff.; Jäggi (2000), pp. 144ff.

[31] See Gudmundsson, Boer & Lechner (2002); Hanlon (1999), pp. 53ff.; Levine (1987), p. 414.

[32] See Hanlon (1999), pp. 59ff.; Levine (1987), p. 453.

2.2 The Entrants: Cost Leadership in Passenger Air Travel

No incumbent airline shifted to a cost leadership strategy against new competitors entering their markets after deregulation. Of course prices were lowered and incumbents even entered into price competition with their new competitors, but this was on selected routes only. This made the cost leadership approach very promising for new players, as can be seen from the successful market entries of Southwest Airlines (1971) in the USA, later Ryanair (1991), Easyjet (1995), Debonair, Virgin Express (both 1996), Go (1998), Buzz (1999), Hapag-Lloyd Express and Germanwings (both 2002) in Europe.

2.2.1 Route Network

A first important cost driver the cost leaders try to avoid are relatively high airport fees at the big hub locations. The fees at these larger airports are several times higher than at smaller, secondary spots which are targeted alternatively.[33] Here, airlines do not only find free capacities in the vicinity of the large cities, but very often also support of local authorities who are interested in improving traffic connection of their respective regions. Linking these locations with direct point-to-point services, cost leaders avoid complex route systems (especially hub and spoke). Connecting flights, characterizing features of hub and spoke systems, are intentionally neglected in this concept. Luggage and passenger handling are therefore no factor at the destination airport, possible conflicts between connecting relations cannot occur. In addition to a lower complexity of planning and scheduling procedures, aircraft utilization is also increased significantly since ground times are shortened.[34]

2.2.2 Fleet Structure

The composition of its aircraft fleet is a further significant factor of an airline's cost structure. The efficiency of each single aircraft, especially its age and suitability for the respective destination, are decisive factors. Depending on route length and number of passengers, aircraft are specialized and thus cannot be employed for alternative routes or traffic areas. Cost leaders concentrate their routes on a limited geographical area allowing them to employ a simple fleet structure, mostly featuring only one aircraft type that has possibly already proven

---

[33] See Doganis (2001b), p. 141; Gilbert, Child & Bennett (2001), pp. 304ff.
[34] See Gilbert, Child & Bennett (2001), p. 303; Lawton (2002), pp. 51ff.

to be technically and economically reliable.[35] Besides the reduced planning complexity, this allows for the harmonization of processes such as aircraft maintenance or the education, training and staffing of technical, cockpit and cabin crew.

2.2.3 Distribution Systems

A third area offering new competitors a significant cost advantage over incumbent airlines is ticket sales. Newcomers typically offer return flights on one city pair not as combined, but as two separate tickets. This, together with the absence of connecting flights, simplifies the booking procedures that are mainly – if not exclusively – executed via internet and call centers. Customers combine their "package" on their own and book their flights online. The issuance of paper tickets and sometimes even boarding passes is often omitted. This process simplification already provides a measurable savings potential. Furthermore, a part of the distribution costs is shifted to the customers: ticket booking by phone is offered through rather expensive telephone connections, whereas internet bookers are frequently profiting from rebates. Direct ticket sales allow the airlines to cash in the revenues already before the actual flight date, by parallely avoiding the problem of "no show passengers" that booked but do not use a seat in their planes.[36]

2.2.4 Product Characteristics: "No Frills"

The product offered by cost leaders does not differ from the product incumbents offer in terms of safety and travel time – but the service on board is limited to a minimum. The tighter installation of more seats already separates the product from the two or three class cabin concept of established carriers. The traditional offer of complementary meals and drinks as well as commercial giveaways known from air transport is abandoned, only snacks and simple "meals" can be bought from the cabin crew.[37] This helps the new players to keep their logistics costs down and their ground times short.

---

[35] See Doganis (2001b), p. 141; Easyjet (2003); Gilbert, Child & Bennett (2001), p. 308.

[36] See Doganis (2001a), p. 64.

[37] See Bailey, Graham & Kaplan (1985), p. 104; Doganis (2001b), p. 140; Gilbert, Child & Bennett (2001), p. 305.

2.2.5 Support Functions

The airline business is not only very capital but also personnel intensive. In the field of personnel costs, the newcomers possess an advantage compared to the incumbents since they typically begin their operations with a comparably lean personnel structure. Furthermore, they enlarge the traditional tasks of their flying crew.[38] Next to the staff structure itself, unions' influence affects the cost level in the company. Airlines whose employees are not represented by unions have a significant cost advantage within their industry. Employees' identification and commitment to their employer is supported by employee ownership which is also advantageous for the company as it helps to reduce personnel costs' impact on liquidity, a weak spot of this volatile and capital intensive business. Finally, recently established companies hardly offer traditional social benefits that are another cost disadvantage for established incumbent carriers.[39]

Except for the multi-functional work organization, the advantage of new competitors regarding the entire block of staff costs is not due to their business model but their young age. As time goes by, this advantage is likely to deteriorate since social benefits might win over aggressive entrepreneurship in the eyes of most employees. However, the dynamics of a young company in the phase of its market entry still are an important factor in the competition with established organizations that are used to protected monopoly markets.

2.3 Summary

The results of this section are summarized in Table 26-1. The fundamental redesign of route networks, which is at the heart of the competitive positioning as a network carrier, has a crucial impact on both costs as well as the regional presence of an airline. It is a reaction to the intensifying competition among incumbents, but to the appearance of entirely new players as well. However, after the industry-wide establishment of hub and spoke systems, network adjustments are still in progress. Some of the incumbents are continuously reducing their differentiation level in favor of further cost reductions and thus move closer towards the competitive position of the point-to-point players.

---

[38] See Gilbert, Child & Bennett (2001), p. 303.
[39] See Bailey, Graham & Kaplan (1985), pp. 96ff.

	Cost Leadership	**Product Differentiation**
Network layout	Independent point-to-point connections	Integrated hub-and-spoke system
Fleet setup	Very simple structure, usually one aircraft type only	Heterogeneous structure, wide variety of different aircraft sizes and types
Distribution system	Internet or telephone booking, route by route (no return tickets)	Integrated computer reservation systems via travel agents handling complex flight patterns
Product specificities	Transportation service only, one class cabins, no frills	Integrated full services supported by means of alliances
Support functions	Task integration, lean personnel structure within young organizations	Powerful yield management systems, loyalty programs for frequent flyers

Table 26-1: Main Features of Incumbents' and Newcomers' Airline Strategies

The comprehensive cooperation activity has been advanced as a second important strategy element of incumbent airlines. Coming in various forms and intensities, airline alliances generate some cost savings, but hitherto have been mainly employed to generate additional revenues for the participating airlines. Due to the production conditions in this industry and especially its cost structure, it seems viable that even in a world without regulatory barriers alliances might maintain a prominent role.[40]

---

[40] See Oum & Park (1997), p. 134.

3 Competitive Strategies in Passenger Rail Travel

The analogies between air and rail transportation mentioned at the beginning suggest – at least to some extent – the transferability of airlines' competitive strategies to passenger rail travel. We will follow this path of argumentation and subsequently assess and discuss this transferability thesis in detail. Should there be no major differences in the relevant fields, our analysis of post-deregulation airline strategies promises to provide guidance and reference for the former state railway companies with regards to their future market positioning.

Here as in the airline case, the analysis of specific policy decisions will be following the industry-related forms of Porter's generic strategy alternatives.[41] Our first focus will be laid on product differentiation strategies which can be pursued with two emphases, since the variable traveling time that might be the basis for a respective differentiation strategy comes into account:

- *Product differentiation through high speed services*: for such a strategy, the majority of travelers would have to appreciate the potential time saving as a substantial point of differentiation and show a corresponding willingness to pay for this premium. Such a strategy would not only strengthen a player's intramodal position, but also provide the possibility to compete with individual traffic and air transportation on many relations. The former European state railways would be able to make use of the high speed equipment they have already established in the competition with new market entrants.

- *Product differentiation through network services*: according to the differentiation strategy option known from the airline industry, the former state railways could define their product through a comprehensive set of possible origins and destinations, focusing mainly on the network character instead of high speed traveling. This strategy also suits the former monopolists very well, since it enables them to bring into play their well developed route systems. However, with the absence of a clear high speed orientation, this approach would have to focus mainly on efficiency and cost arguments in the intermodal competition against air and individual transport. This, in turn, is the point where a product-oriented player would have to expect massive pressure from intramodal low cost competitors, rendering such a positioning much more difficult.

---

[41] See Porter (1980); Porter (1985).

Although the network idea is not in the center of the first approach described, a high speed player would have to maintain a certain degree of route system integration, without which it would probably not be able to achieve a satisfactory use of its trains. The experience of big railway companies shows that only few selected relations provide sufficient own traffic demand during certain time periods. Furthermore, taking into account today's infrastructural conditions in most European countries, adding more stops to one connection does not substantially lengthen the time of the journey, since it is not possible to bypass most knots anyway. In general, such an approach depends heavily on the intended improvements of the European high speed railway networks in the future and could hardly be pursued entirely already today. Therefore, for the former monopolists primarily the second strategic alternative, positioning through network services, requires further analysis.

3.1 Product Differentiation in Passenger Rail Travel

Due to the deregulation and already occurring market entries in long distance passenger rail travel, the former monopolists are increasingly set under pressure by new competitors. Facing this threat, they have to adjust their strategic focus quickly and adequately. Above, the underlying industry rules as well as the expected entry strategies of new competitors have shown great similarities between the airline and railway industry. Now, the single policies of airline companies – especially the adjustment of route networks and the comprehensive trend to cooperate – will be evaluated regarding their applicability for the railways.

3.1.1 Route Network

The most important change in the route networks of major airline companies following market liberalization was the establishment of hub and spoke systems. This innovation provided them with significant economies of density within their networks, thus lowering their production costs significantly. But on the other hand, as we have shown above, this network type also has considerable influence on the market power of those players in certain parts of their networks and therefore plays a key role in the protection of these markets against new competitors. Under these circumstances, such a concept must be extremely interesting for former state railway companies, who, besides all potential difficulties linked with it, expect comparable advantages from this route network design.

Hub and Spoke Layouts in Passenger Rail Travel

The price that has to be paid for the production advantages in hub and spoke networks is an increased length of the journey due to the longer distance between origin and destination when linked via the hub and also the time necessary for waiting and connecting at the hub. While the longer distance must not necessarily have a negative impact on the overall travel quality perceived by the passenger regarding the high travel speed of airplanes, having to change them during the journey certainly does. But usually there are no alternative direct connections, since the hub and spoke system allows airlines to service city pairs that would otherwise not have enough demand to justify their connection without linking them to the hub. Consequently, there will hardly be any superior intramodal service on this destination; whereas intermodal competition for airlines does not exist on routes above 500 kilometers.

The intermodal competitive situation of European passenger rail companies differs from the situation airlines were put in after their industry deregulation. For years and years, railway market share facing air and street transportation has been stagnating at about three percent. Every adjustment regarding price, route or service concepts will be subject to a direct comparison especially with the motorized individual transport as the major intermodal competitor. Facing these considerable intermodal competitors, the railway companies can only make use of policies in intramodal competition that will work related to other modes of transportation at the same time. It is this very condition that does not allow a direct application of the airlines' hub and spoke structures. As we have shown above, the airlines cannot reach the efficiency improvements in their networks until the travel times for their passengers are increased. These passengers have neither intramodal nor intermodal alternatives on most relations which is why they have to accept these solutions that are still the best ones they can get. If the railway companies tried to increase the usage of trains and networks with this policy significantly, they would have to impose the same extra travel time via certain hub locations on their passengers. Here, the acceptance problem would be exactly the other way round: changing trains at the hub station would certainly take less time than changing planes at an airport. But the fact that the actual way taken by the train via the hub is considerably longer than the direct connection would mean a great competitive disadvantage for the railway compared to individual transportation. The latter is a strong intermodal competitor making the realization of potential production advantages through a hub and spoke structure impossible for the railways.[42]

Evolution of the Hub and Spoke Concept

The recent activities of the big airlines regarding their route system adjustments result in a reduced degree of network integration even from the network-focused players, who are thus trying to achieve additional cost savings. Besides all stated difficulties in the application of airline policies to the railway industry, this "de-peaking" trend appears to be directly applicable here. It is based on the facts that passengers are not willing to pay for the high degree of integration within the incumbents' networks and that the daily amount of connecting flights at the major airports is big enough to achieve acceptable connection times even without directly matching the respective arrival and departure times of individual aircraft. A transformation of this situation to rail travel is easily possible; the connections in major train stations of one network – many of them traditionally previewed even without the explicit use of hub and spoke structures – could be planned separately without taking account of one another in an integrated approach allowing for more efficient train circulations. A necessary condition is the existence of a high minimum level of connections, especially regarding to their frequency, in order to lose as little connection quality as possible. This condition helps the former state companies, who can make effective use of their strong market presence at this point.

Revenue Effects

Domination of major system knots will be much more difficult for a railway company than for an airline, mainly due to the existing alternative stations in many major cities. Since the hub and spoke system as such turned out to be

---

[42] Nevertheless, the bundling effects responsible for the production advantages of such configurations still hold for the railways, although they must be exploited rather differently. The journey of an airline passenger through a hub and spoke network leads her to a location – the hub – she did not want to go to and where she does not want to stay longer than absolutely necessary. The same is true for every intermediate stop of a passenger train that lies between the origin and destination of a passenger. Obviously, the same bundling effects will occur as used by the airlines in their hub and spoke networks, namely the consolidated transportation of passengers with different origins and destinations in the same means of transportation. It is true, of course, that the consecutive analysis of the stops of only one train on the one hand, and especially the great amount of these knots to be considered on the other hand move the problem at hand quite far away from the actual hub and spoke conception known from air transportation.

difficult to apply for railways in general, offering a superior connection portfolio at one more knots and thus defending at least regional subnets appears to be difficult, too. At the same time, the possibility to achieve premium prices at these hub locations is diminishing as much as competitors gain access to these knots.

But knot access is not the only planning constraint for railway companies. In contrast to the airline industry, spoke access is also scarce here. If one company intends to offer a certain relation, it has to possess knot capacity at origin, destination as well as intermediate stations and, adding to this, access to all rail connections affected for the required type of traffic at the required point in time. Depending on how these capacities are being allocated in the respective national market, an incumbent company is very well able to dominate certain relations, regional or even entire markets and block them against new competitors. They could also compensate the revenue losses they suffer at major knots relative to the airlines on selected connections. But such a mixed calculation depends to a very high degree on the political capacity allocation procedure in that market.

3.1.2 Fleet Structure

The variety of locomotives and cars of most national companies in long distance passenger rail travel is rather limited. If they are using different devices at all, then those are mostly models of different generations of development. Compared to the airline industry, a much more differentiated range of relations and applications can efficiently be serviced with one and the same equipment. The railway companies are also interested in a high flexibility of new developments; but this refers more to the suitability of their train systems to both short and long distance travel. Additionally, the car manufacturers, which hardly had to meet economic demands in the monopoly history of the big railroad companies, now increasingly have to integrate efficiency criteria into their developments. In the airline industry, this led to a modular construction of most vehicles allowing for the use of one element in several products. The same development can be expected for the railway sector.

Taking an international perspective, the picture looks rather different. Here, the vehicle situation can hardly be called homogenous at all, since the various technical infrastructure conditions in European countries make very different vehicle abilities necessary. Overhead and break systems, signal and security management, even the gauges are far from being harmonized all over Europe. Therefore, the question of whether or not a set of vehicles can be used in several national railroad networks is of decisive importance. Both regarding the conquering of new markets in neighboring countries, as well as possible alliances with other players – this question has to be addressed later on – such technical aspects are the base for future purchasing decisions.[43] The only short and mid term answer to them appears to be the use of multi system vehicles, which, unfortunately, only extend compatibility to two or three different systems. The necessary harmonization of national standards is a political issue and can not be solved with the instruments of corporate competitive strategies.

3.1.3 Distribution Systems

While the airlines had developed their strong computer reservation systems on their own at that time, for the railroad companies, these are now already available on the market and can be used by them without major adaptations.[44] The according companies' ability to find solutions for neighboring markets quickly is a threat to possible development intentions of any railroad company in this direction. But the possibility to influence directly their own programs was and still is an important advantage for the companies that are maintaining their own reservation systems over those competitors whose flights they are also keeping in their databases. The same would certainly be true for the distribution of other companies' tickets by the incumbent railroads. However, taking into account the significant investments a strategic advantage comparable to that of the airlines seems only hardly achievable in the mid and long term. Computer reservation systems will certainly be necessary for establishing pricing systems comparable to those in air transportation; but they will hardly provide any additional strategic advantage.

3.1.4 Product Characteristics: A Trend to Cooperate?

We have above attributed a significant revenue potential to airline cooperations. Consequently, we will subsequently assess cost and revenue potentials of railway alliances in a liberalized European market.

Cost Effects

With staff synergies, consolidated common processes performed by the most efficient partner, the bundling of market power and, finally, the lower production

---

[43] See Lenke (1994), p. 83.
[44] European railroad companies did make use of simple electronic booking systems in the past, but those have never gained the power of comparable systems from the airline sector. They did not, for example, allow for a sophisticated management of different space contingents. But what is even more crucial is that they provided only very little information concerning passengers' preferences that is so valuable for the airlines.

costs per seat due to an increased passenger volume, all four areas of cooperation analyzed in the airline case should hold a certain savings potential on the rail as well. They are not airline specific, but can actually be met in a number of different industries and lead to an increasing number of cooperative relationships there.[45] But just like it is the case for the airline business, the cost cutting potential in these areas is limited. The railroad companies are facing the same cost structure, characterized by a comparably small fraction of variable costs. Most of their costs are directly linked to the actual transportation service like staff, vehicle and way costs. Beyond a certain critical size that all state companies should have reached long ago, no major savings can be expected from a combined production with another player. We therefore propose to neglect a detailed analysis of every single cost area in favor of the revenue effects.

Revenue Effects

The revenue effects have been identified to be a crucial factor for the intensive cooperation in air transportation. The set of origins and destinations an airline has to offer in order to position itself as a comprehensive network player, taking into account the aircraft's technical possibilities, must be comprehensive and truly global. The railroads should also be able to enlarge the number of city pairs they can link by cooperating, but will still not be able to compete against air transportation on relations that exceed about 500 kilometers. Consequently, such a strategy finds its limitations in the technical possibilities of the underlying means of transportation. Therefore, the revenue effects that arise from a mere addition of new nodes to an existing route network will, due to the limited travel distance of the railways, be smaller than for the airline sector.

Furthermore, within their smaller scope of activities, the railroads find national markets to which they do still not have unlimited access. Entering into such geographically neighboring markets and servicing their traffic can be done via cooperation with partners that are already active in those markets. Especially the close proximity of numerous cities render this possibility interesting for the European theatre today. Next to the still unsatisfactorily established cross border relations, especially a closer cooperation between suburban and long distance travel companies appears to be promising. The regional players could profit very much from a close integration of selected long distance connections into their own offers and establish strategic partnerships.

---

[45] See Lenke (1994), p. 80.

3.1.5 Support Functions

Yield Management

Due to the similarities between production conditions in air and rail travel, a conversion of highly sophisticated and successful yield management instruments promises considerable revenue potential on railroads as well.[46] However, the airlines' yield management systems are based on a couple of characteristics of their pricing and distribution logics that are not traditionally known in land-based passenger transportation. One of them is the freedom of any player to define separate price categories. This does not only allow to consider the actual booking situation and capacity utilization, but also to take its competitors' behavior into account and react appropriately to any single maneuver. Many railways do not have this freedom because they have based their pricing on a defined set of rules.[47]

A second fundament for effective yield management is the distribution of an exactly defined seat on an exactly defined flight, whereas the pricing systems of many railroads historically define only the transportation service on a particular city pair, leaving it up to the customer to finally choose both the day and the train on short notice. Train tickets can usually even be exchanged and used by any traveler. These essential differences do not allow to create contingents in given capacities and consequently also their synchronization with changing demand, one necessary condition for all yield management activities.[48] Trying to reduce the planning uncertainty concerning passengers' traveling behavior will result in a reduced flexibility for these customers to enter a long distance train at the next station on short notice. The railroad would lose a significant competitive advantage over its intermodal rival, individual transportation, whose biggest strength lies exactly here.[49] The possibility of smoothing the demand through appropriate price setting, something the airlines also reach via their yield management systems, will suffer from these conditions.

---

[46] See Daudel & Vialle (1989), pp. 132-133; Meffert, Perrey & Schneider (2000), pp. 25ff., 35.
[47] See Ehrhardt (2002); Krämer & Luhm (2002).
[48] See Meffert, Perrey & Schneider (2000), p. 35.
[49] One good example for the resistance this fact can evoke due to public expectations are the difficulties for the former German state railway company in introducing a new pricing system, that brought about tariffs linked to exactly defined trains and generally neglected a pricing logic based on kilometer distances for the first time.

Loyalty Programs

Facing market entries of new competitors, loyalty programs are another option to defend own market share. For a passenger who frequently uses the service of an incumbent company, changing to a smaller competitor will be linked with a loss of rebates and discounts for that particular travel; her changing costs will be higher.[50] This is a policy that is not linked to the production or distribution characteristics of the two means of transportation covered here; therefore, obstacles for its applicability on the railroad are obviously not present. Loyalty programs have in fact already been introduced by some railroad companies.[51] Taking into account the obvious difficulties for railroad players to establish a dominating service at selected network hubs, their meaning will likely increase a lot in the future.

3.2 Cost Leadership in Passenger Rail Travel

The market entries of several cost leaders in the airline sector find great public interest. Occasionally, their competitive strategy is already celebrated as a revolutionary change also applicable for the railway companies. If and how the most important elements of this strategy, the design of route network, fleet composition, distribution systems and other factors can be transferred onto the railway sector is subject to our subsequent assessment.

3.2.1 Route Network

Comparable to the airports' situation, practically no major European train station offers free capacities for additional arrivals or departures at interesting times of the day. New competitors will consequently have to prevail upon the incumbents with regard to this "slot issue", if they want to lead their services over the big system hubs. This possibility, however, is very often limited by existing legal conditions. It appears likely that the newcomers will follow the airlines' example and use alternative nodes, especially since an emerging competition between secondary stations for additional traffic seems possible. On the one hand, due to their limited travel distances, railway companies are less flexible regarding the geographical location of such alternative stop points. On the other hand, they

---

[50] See Coyne & Dye (1998), p. 106.
[51] See for example the bahn.comfort program in Germany or the programme Grand Voyageur in France.

find a much wider variety of potential alternatives than the airlines do, especially in the suburbs of the larger cities.

The cost leader airlines connect their system nodes via point-to-point relations. As has been indicated above, this is mainly due to the more complex processes associated to passengers' changing of aircraft. These conditions apply only partly in the railroad context. Here, travelers take care of their luggage themselves and, even in very large train stations, have much shorter switching times than at any well organized airport terminal. These differences represent a less significant process-simplification potential through point to point traffic than in air transportation. Furthermore, coordination requirements for integrated service are very high in both industries and conflicting connections are possible on the ground as well as in the air. However, the most important advantage of direct airline connections, the possibility of increasing operating hours of transport vehicles, promises a cost advantage to the railroad companies comparable to the one realized by point-to-point airlines already today.

Thus, point-to-point traffic is a valuable foundation for a cost leadership strategy for railroad companies, too. But an unchanged application of this concept is not conceivable since the airlines carry their passengers only to one single destination on every flight. In contrast, railroads have the possibility to integrate several additional stops into their routes. Alas, which and how many of these intermediate stops are beneficial to the railway company depends on the particular line and can not be answered on the general level we are pursuing here.

3.2.2 Fleet Structure

The airlines' requirements that their fleet offers them the highest possible flexibility can be directly applied to the conditions in rail travel. However, the vehicles' suitability for certain distances does not have a comparable weight; it hardly plays any role for modern railways. The companies profit far more from the precise adjustment of their trains according to the volatile demand that is so characteristic for this industry. The currently observable trend to use more and more fixed train units can be put into question since the use of classic locomotives allows at least for some flexibility by the variation of cars used in every specific train. Concerning the heterogeneity of their fleets, the railway companies are usually facing far less complexity in long distance travel today than the airlines. However, especially potential cost leaders are likely to look for vehicles from the suburban context that can very often be operated at significantly lower costs than the actual long distance trains. Here, the variety of models to choose from is much higher.

3.2.3 Distribution Systems

The distribution process structures of a transportation company depend very much on the services it plans to sell. Consequently, distribution systems can not be significantly simplified without an appropriate adjustment in the company's service range. The simple sales policy of cost leader airlines is based on the separation of return tickets into two different products as well as neglecting every kind of connecting flights and therefore allows for the simple booking routines. Every passenger is able to plan her journey independently without the need of additional assistance. First experiences with such distribution channels on the rail have turned out to be successful in adopting the airlines' policies. Thus, for clearly defined products, the distribution concept of cost leadership airlines promises analogous savings potential for railway companies.

3.2.4 Product Characteristics

Concerning the actual "transportation service", most innovations of low cost airlines refer mainly to processes aboard the airplanes and to the design of the aircrafts' cabins. These modifications are not specific to airlines and can be applied by railroad companies practically with no changes in an analogous way. The installation of additional seats at the cost of wardrobes or any other room elements and the substitution of compartment by open space cars can dramatically increase the number of paying passengers per car or train.[52] The abandonment of traditional dining-cars and comparable services can lead to another cost advantage over the incumbent players; free food and beverages have, except for the premium segments, never been common on the rail anyway.

3.2.5 Support Functions

The cost advantages of point.to-point competitors entering into the air transportation markets in the fields of staffing, administration and all supporting functions were basically due to their young age. In a personnel intensive business, their lean structures including multifunctional work organizations have a positive impact on their efficiency. This area, too, promises a comparable potential in the railroad business, since it can not be found to be depending on airline specific criteria. The enlargement of job profiles such as stewards, technicians or drivers by elements of passenger service or vehicle cleaning as well as during the turnaround of the trains at their destinations could lead to

---

[52] See Vieregg (1995), pp. 136ff.

substantial savings. Only the luggage handling and check-in processes will not play a comparable role in this industry. Finally, all outcomes of union influence and the degree of employee company ownership affect the cost structures of railroad companies in the same way as those of the airlines analyzed earlier.

4 Summary

Because of the strong product orientation of former state railways, the pursuit of a cost leadership strategy is as little an option for these companies as it was for incumbent airlines. They possess neither the relevant competitive cost structures nor would this strategy allow them to leverage the competencies they have built up during their past market activity. A more detailed analysis of necessary policy decisions for former state railways regarding a potential positioning as product differentiator or network player illustrated that the approaches known from air transportation are not easily transferable to the rail context without some adjustments.

With regard to the design of the route networks, the relatively comparable production conditions in both industries imply that the incumbent railway companies will need to commit to significant price reductions facing point to point competitors with a clear cost focus. The hub and spoke concept turns out to be not operational on the ground, mostly due to direct competition from motorized individual transport modes on the one hand, but also to the great importance of selected intermediate stops of long distance trains on the other. Hub domination following the airports' "fortress hub" example is not possible for railways resulting in the imperative for the railways to defend their spokes. A network player will have to ensure that her integrated traffic is perceived as superior to other connections by the legal institutions.

This requirement contradicts the "de-peaking" idea, that, as a younger element of the hub and spoke approach, is applicable also on the ground. With its realization, the former state railways could profit from their size and realize cost savings in their train circulations that are crucial in potentially upcoming price wars. However, they would have to lower their degree of network integration and consequently weaken an important factor for the dominance of their services on heavily used spokes.

In order to increase and improve exploitation of their network potential, the product-oriented airlines commit to revenue driven cooperations through which they are trying to establish a truly global reach of their services. This second option neither allows for a direct policy translation from the air to the ground, since the railroads' field of activity is technologically limited to about 500 kilometers of traveling distance. Within these limitations, however, traffic potentials can be observed in cross border relations as well as through the integration of suburban and long distance volumes. Of course, such a hybrid strategy is not only promising for former monopolists, but will probably be the preferred starting point for new competitors looking for first successes in the long distance business out of an established metropolitan service. Therefore, an early selection of appropriate potential partners in the short distance area appears to be crucial.

Parts of the changes observed in airlines' other process areas should be applicable in railway transportation with only minor adaptations. This is true for the optimization and flexibilization of vehicle fleets as much as for the design of yield management instruments, the underlying pricing systems and loyalty programs. The conditions for the use of computer reservation systems and internet distribution should also be the same for airlines as for the railway companies with one exception; the latter will hardly attain developers' status and therefore lack strategic influence.

5 References

ABERLE, G. (2000): *Transportwirtschaft. Einzelwirtschaftliche und gesamtwirtschaftliche Grundlagen.* 3rd Ed., Munich, Vienna: Oldenbourg.

ALBERS, S. (2000): *Nutzenallokation in Strategischen Allianzen von Linienluftfrachtgesellschaften.* Working Paper No. 101 of the Department of Business Policy and Logistics at the University of Cologne, Cologne.

ANTONIOU, A. (1992): The Factors Determining the Profitability of International Airlines: Some Econometric Results. *Managerial and Decision Economics*, 13(6), pp. 503-514.

ARGYRIS, N. (1991): Costs and Benefits of Airline Mergers and Strategic Cooperation – an Economic Analysis. In: Dagtoglou, P. D. (Ed.): *Airline Mergers and Cooperation in the European Community*, Athens: Sakkoulas, pp. 27-39.

BAILEY, E.E.; GRAHAM, D.R.; KAPLAN, D.P. (1985): *Deregulating the Airlines.* Cambridge, MA: MIT Press.

BAILEY, E.E.; PANZAR, J.C. (1981): The Contestability of Airline Markets During the Transition to Deregulation. *Law and Contemporary Problems*, 44(1), pp. 125-145.

BAUMOL, W.J.; PANZAR, J.C.; WILLIG, R.D. (1982): *Contestable Markets and the Theory of Industry Structure*. New York, San Diego, Chicago et al.: Harcourt Brace Jovanovich.

BELOBABA, P.P. (1987): Airline Yield Management. An Overview of Seat Inventory Control. *Transportation Science*, 21(2), pp. 63-73.

BRUECKNER, J.K.; DYER, N.J.; SPILLER, P.T. (1992): Fare Determination in Airline Hub-and-Spoke Networks. *Rand Journal of Economics*, 23(3), pp. 309-333.

BRUECKNER, J.K.; SPILLER, P.T. (1994): Economies of traffic density in the deregulated airline industry. *The Journal of Law and Economics*, 37(2), pp. 379-415.

BURTON, J.; HANLON, P. (1994): Airline alliances: cooperating to compete? *Journal of Air Transport Management*, 1(4), pp. 209-227.

BUTLER, R.V.; HUSTON, J.H. (1989): How Contestable are Airline Markets? *Atlantic Economic Journal*, 17(2), pp. 27-35.

BUTLER, R.V.; HUSTON, J.H. (1999): The Meaning of Size: Output? Scope? Capacity? The Case of Airline Hubs. *Review of International Organization*, 14(1), pp. 51-64.

BUTTON, K. (2002): Debunking some Common Myths about Airport Hubs. *Journal of Air Transport Management*, 8(3), pp. 177-188.

BUTTON, K.; HAYNES, K.; STOUGH, R. (1998): *Flying into the Future. Air Transport Policy in the European Union*. Cheltenham; Northampton: Edward Elgar Publishing.

CAVES, D.W.; CHRISTENSEN, L.R.; TRETHEWAY, M.W. (1984): Economies of Density versus Economies of Scale: Why Trunk and Local Service Airline Costs Differ. *RAND Journal of Economics*, 15(4), pp. 471-489.

COYNE, K.P.; DYE, R. (1998): The Competitive Dynamics of Network-Based Businesses. *Harvard Business Review*, 76(1), pp. 99-109.

DAUDEL, S.; VIALLE, G. (1989): *Le Yield Management. La Face Encore Cachée du Marketing des Services*. Paris: InterEditions.

DELFMANN, W. (2000): Hub-and-Spoke-Systeme. In: Klaus, P.; Krieger, W. (Eds.): *Gabler Lexikon Logistik. Management logistischer Netzwerke und Flüsse*, 2nd Ed., Wiesbaden: Gabler, pp. 189-190.

DENNIS, N. (2000): Scheduling Issues and Network Strategies for International Airline Alliances. *Journal of Air Transport Management*, 6(2), pp. 75-85.

DOGANIS, R. (1991): *Flying off Course – The Economics of International Airlines*. 2nd Ed., London, New York: Routledge.

DOGANIS, R. (2001a): Survival Lessons. *Airline Business*, 17(1), pp. 62-65.

DOGANIS, R. (2001b): *The Airline Business in the Twenty-first Century*. London, New York: Routledge.

DULIBA, K.A.; KAUFFMAN, R.J.; LUCAS, H.C. JR. (2001): Appropriating Value from Computerized Reservation System Ownership in the Airline Industry. *Organization Science*, 12(6), pp. 702-728.

EASYJET AIRLINE COMPANY LTD. (2003): *Unsere Flugzeugflotte*, internet site on 24 July 2003 (http://www.easyjet.com/DE/unsere/aircraft.html).

EBERS, M. (1997): Explaining Inter-organizational Network Formation. In: Ebers, M. (Ed.): *The Formation of Inter-Organizational Networks*, Oxford, New York: Oxford University Press, pp. 3-40.

EHRHARDT, M.R. (2002): Das neue Preissystem im Personenverkehr der DB AG. Hintergrund und Analyse. *Internationales Verkehrswesen*, 54(1), pp. 23-27.

GILBERT, D.; CHILD, D.; BENNETT, M. (2001): A Qualitative Study of the Current Practices of 'no frills' Airlines Operating in the UK. *Journal of Vacation Marketing*, 7(4), pp. 302-315.

GUDMUNDSSON, S.V.; DE BOER, E.R.; LECHNER, C. (2002): Integrating Frequent Flyer Programs in Multilateral Airline Alliances. *Journal of Air Transport Management*, 8(6), pp. 409-417.

HANLON, P. (1999): *Global Airlines. Competition in a Transnational Industry*, 2nd Ed., Oxford et al.: Butterworth-Heinemann.

HANSEN, M.; KANAFANI, A. (1989): Hubbing and Airline Costs. *Journal of Transportation Engineering*, 115(6), pp. 581-596.

JÄGGI, F. (2000): *Gestaltungsempfehlungen für Hub-and-Spoke-Netzwerke im europäischen Luftverkehr – Ein ressourcenbasierter Ansatz*. Bamberg: Difo-Druck.

JANIĆ, M. (1997): Liberalisation of European Aviation: Analysis and Modelling of the Airline Behaviour. *Journal of Air Transport Management*, 3(4), pp. 167-180.

KAHN, A.E. (1988): Surprises of Airline Deregulation. *American Economic Review*, 78(2), pp. 316-322.

KRÄMER, A.; LUHM, H.-J. (2002): Peak-Pricing oder Yield-Management? Zur Anwendbarkeit eines Erlösmanagementsystems im Fernverkehr der DB AG. *Internationales Verkehrswesen*, 54(1), pp. 19-23.

LAWTON, T.C. (2002): *Cleared for Take-Off. Structure and Strategy in the Low Fare Airline Business*. Aldershot, Burlington: Ashgate.

LENKE, H. (1994): Die europäischen Eisenbahnen – Substitution oder Nischenverkehr? *Schriftenreihe der Deutschen Verkehrswissenschaftlichen Gesellschaft e.V. – DVWG –, series B, no.171: Luftverkehr im Wandel – Chancen und Risiken der Zukunft*, Bergisch Gladbach, pp. 75-84.

LEVINE, M.E. (1987): Airline Competition in Deregulated Markets. Theory, Firm Strategy, and Public Policy. *Yale Journal on Regulation*, 4(2), pp. 393-494.

MCSHAN, S.; WINDLE, R. (1989): The Implications of Hub-and-Spoke Routing for Airline Costs and Competitiveness. *The Logistics and Transportation Review*, 25(3), pp. 209-230.

MEFFERT, H.; PERREY, J.; SCHNEIDER, H. (2000): Grundlagen marktorientierter Unternehmensführung im Verkehrsdienstleistungsbereich. In: Meffert, H. (Ed.): *Verkehrsdienstleistungsmarketing. Marktorientierte Unternehmensführung bei der Deutschen Bahn AG*, Wiesbaden: Gabler, pp. 1-55.

MORRISH, S.C.; HAMILTON, R.T. (2002): Airline Alliances – Who Benefits? *Journal of Air Transport Management*, 8(6), pp. 401-407.

NERO, G. (1999): A Note on the Competitive Advantage of Large Hub-and-Spoke Networks. *Transportation Research Part E*, 35(4), pp. 225-239.

OUM, T.H.; PARK, J.-H. (1997): Airline Alliances. Current Status, Policy Issues, and Future Directions. *Journal of Air Transport Management*, 3(3), pp. 133-144.

OUM, T.H.; PARK, J.-H.; ZHANG, A. (2000): *Globalization and Strategic Alliances. The Case of the Airline Industry*. Amsterdam, Lausanne, New York et al.: Pergamon.

POMPL, W. (1998): *Luftverkehr – Eine ökonomische und politische Einführung*, 3rd Ed., Berlin et al.: Springer.

PORTER, M.E. (1980): *Competitive Strategy. Techniques for Analyzing Industries and Competitors*. New York: Free Press.

PORTER, M.E. (1985): *Competitive Advantage. Creating and Sustaining Superior Performance*. New York: Free Press.

SCHMIDT, S. (1993): *Strategische Allianzen im Luftverkehr - Erfolgsorientiertes Management europäischer Flug-Carrier*. Trier: FTM.

SCHNELL, M. (2000): Zur Effektivität möglicher Kooperationsformen im liberalisierten europäischen Luftverkehr – eine empirische Analyse. *Zeitschrift für Verkehrswissenschaft*, 71(3), pp. 242-270.

VIEREGG, M. (1995): *Effizienzsteigerung im Schienenpersonenfernverkehr*. Munich: Akademischer Verlag.

WHITE, L.J. (1979): Economies of Scale and the Question of "Natural Monopoly" in the Airline Industry. *Journal of Air Law and Commerce*, 44, pp. 545-573.

About the Editors

Prof. Dr. Dr. h.c. Werner Delfmann

Prof. Delfmann (born 1949) graduated in mathematics and business administration (Dipl.-Math.) at the University of Munster, Germany. After being senior professor at the universities of Osnabruck and Frankfurt/Main, he became Full Professor of Business Administration and Director of the Department of Business Planning and Logistics at the University of Cologne, Germany in 1988. From 1999 till 2001, he was Dean of the Faculty of Economics, Business Administration and Social Sciences. His main research activities focus on strategic and international management, logistics, e-commerce and information management, controlling and operations research.

Prof. Delfmann has been a visiting professor and invited lecturer at several European universities, e.g. HEC Paris, Stockholm School of Economics and Copenhagen Business School, as well as at universities and business schools overseas like National University of Singapore (NUS), Asian Institute of Management (AIM) Manila. In 1997, he held the ASEAN-EU chair in Management at the AEMC in Brunei Darussalam.

From 1999 till 2003, he was chairman of the Community of European Management Schools (CEMS), the network of 17 leading European management schools and more than 50 leading companies. Furthermore, he has been acting as a lecturer and scientific coordinator at several institutions for executive management education, e.g. at GSBA Zurich, ISA, Paris and USW, Germany as well as in the international Executive MBA programme (GEM).

In 2004 CORVINUS University, Budapest, Hungary conferred a doctor honoris causa in Business Administration on Prof. Delfmann.

Prof. Delfmann is founder and head of national and international working-groups in Strategic Management and Logistics with scientists and senior executives. He is a member of numerous scientific organizations and management associations, e.g. of the research committees of the European Logistics Association (ELA), the German Logistics Association (BVL) and the German Society for Business Administration (SG-DGfB). Furthermore, Prof. Delfmann has close relationships with leading companies in industry and trade by holding mandates as counsellor, consultant and member of the supervising board, as well as in a broad scope of cooperative research projects.

Dept. of Business Policy and Logistics,
University of Cologne, 50923 Cologne, Germany
Tel: +49 221 470 4316,   Fax: +49 221 470 5007
Email: delfmann@wiso.uni-koeln.de; http://www.spl.uni-koeln.de

Prof. Dr. Herbert Baum

Prof. Dr. Herbert Baum has global experience in transport research and consulting for more than 20 years. Following his academic education, he has been chair owner for transport economics at several German universities (Hamburg, Bochum, Essen, and Cologne). Since 1990, he has been Professor of Economics and Director of the Institute of Transport Economics at the University of Cologne.

He has released more than 200 publications in the field of transport economics both in German and English. He is the editor of the German *Zeitschrift für Verkehrswissenschaft (Journal of Transportation Economics)*.

Furthermore, Prof. Baum is a member of the Advisory Board of the German Federal Ministry of Transport, member of the Enquête-Commission of the Parliament of Northrhine-Westfalia, expert for the Conference of European Ministers of Transport, and member of the Airport Commission of the German Federal Ministry of Transport.

In the topic of air transport he carried out several research projects. Focal points, thereby, were quantitative measures for transport and environmental policy. These research works were realized as efficiency analyses, cost-benefit-studies and goal-achievement analyses. His main research areas are prices and taxes for air transport, demand reaction analyses, modal shift effects, potentials for the substitution of air transport by railways, traffic avoiding strategies for air traffic, monetary evaluations of instruments, employment effects of air traffic, effects of night-flight regulations, environmental effects of air transport policy.

Institute and Seminar for Transport Economics,
University of Cologne, 50923 Cologne, Germany
Tel: +49 221 470 2312,   Fax: +49 221 470 5183
Email: h.baum@uni-koeln.de; http://www.ifv-koeln.de

Stefan Auerbach Dipl.-Kaufmann

Mr Auerbach is Managing Director at Lufthansa Consulting and advises clients in aviation and transport industry in service quality improvement, traffic development, business planning, performance management, privatization and process optimization. Mr Auerbach is heading Lufthansa Consulting's airport practice.

He gained extensive experience in Lufthansa Consulting's Management Consulting Department in many assignments for airline and airport clients worldwide.

Mr Auerbach has spoken at several international airport and airline conferences, on the subject of network management within an alliance. He has written several publications on aviation related subjects like logistics, network planning, recovery management and revenue management.

He joined Lufthansa Consulting in 1994. He started his professional career with a professional training at Bensberger Bank in Germany, receiving a certificate with honors from the German Chamber of Commerce. He worked in several positions, ultimately as a team leader for risk analysis in industrial finance.

Mr Auerbach graduated with honors at the University of Cologne, earning a Degree in Business Administration. He specialized in Corporate Planning and Logistics, Management Accounting and Computer Science.

Lufthansa Consulting GmbH
Von-Gablenz-Strasse 2-6, 50679 Cologne, Germany
Tel: +49 221 88 99 634,   Fax: +49 221 88 99 660
Email: stefan.auerbach@lhconsulting.com; http://www.lhconsulting.com

Dr. Sascha Albers

Sascha studied Business Administration with majors in Strategic Management, Logistics, Marketing and Transport Economics at the University of Cologne, Germany, as well as at HEC School of Management in Jouy-en-Josas (Paris). In 1999 he graduated at the University of Cologne earning his degree in Business Administration (Dipl.-Kfm.). In 2000 he additionally graduated as CEMS Master in International Management, a pan-European Masters degree. From 1999 until 2005 he worked as research and teaching assistant at the Department of Business Policy and Logistics of the University of Cologne, receiving his Doctorate (Dr. rer. pol.) in February 2005. During his doctoral studies Sascha spent a semester as a visiting scholar at the Management and Organizations Department of the

Kellogg Graduate School of Management, Northwestern University, USA. He currently (summer 2005) enjoys New Zealand, following an invitation from the Department of Strategic Management and Human Resource Management of the University of Waikato Management School to teach a course on Alliances and Networks.

His research interests are Cooperative Strategy, Aviation and Supply Chain Management.

Dept. of Business Policy and Logistics,
University of Cologne, 50923 Cologne, Germany
Tel: +49 221 470 6193,   Fax: +49 221 470 5007
Email: albers@wiso.uni-koeln.de; http://www.spl.uni-koeln.de

LIST OF AUTHORS

Robert J. Aaronson

Director General of Airports Council International (ACI). Twenty five years of international aviation experience: as CEO of major airports, e.g. The Port Authority of New York & New Jersey; with passenger and cargo airlines as President and CEO of Air Transport Assoc. of America; on the policy side, as top US official in the FAA responsible for nationwide airport standards, safety and development; worldwide airport development and operations in the position of Exec. VP and General Manager of Lockheed Air Terminal (later to become Airport Group International). Own consulting firm Strategies For Airports, Inc. which became part of Lufthansa Consulting. Chairman of Airports Operators Council International (predecessor to ACI), President of the (US) National Association of State Aviation Officials, and President of the Wings Club (NY).

ACI World HQ
P.O. Box 16, 1215 Geneva 15-Airport, Switzerland
Tel: +41 22 717 8585,   Fax: +41 22 717 8888
Email: aci@aci.aero; http://www.aci.aero

Sandro Agosti lic.oec.HSG

Born in 1979, studies in Business Administration/ Economics, majoring in Business Administration, area of specialization: Tourism and Transport Economics. Since 2002 research associate at the Institute for Public Services and Tourism. CEMS Master Program assistant and assistant for different consulting and research projects (within the fields of airline industry, railway industry, tourism management, destination management and personal services management).

Institute for Public Services and Tourism, University of St.Gallen
Dufourstrasse 40a, CH-9000 St.Gallen, Switzerland
Tel: +41 (71) 224 25 25, Fax: +41 (71) 224 25 36
Email: sandro.agosti@unisg.ch; http://www.idt.unisg.ch

Carsten Bermig Dipl.-Volkswirt, Master in Advanced European Studies

Born in 1970, studies in Economics with majors in Competition Policy, Public Finance and Transport Economics in Cologne (D), Bonn (D), Rotterdam (NL) and Budapest (H). Research assistant at the Cologne Center for Public Finance. Studies and teaching assistant for economics at the College of Europe Natolin. Since 2000 official at the Competition Directorate-General (DG COMP) of the European Commission and since 2002 case handler in DG COMP's transport unit.

Research interest: Competition policy, Transport, Public Finance.

European Commission, Competition Directorate-General
1049 Brussels, Belgium
Tel: +32 2 295 69 85,   Fax: +32 2 296 98 12
Email: carsten.bermig@cec.eu.int;
http://europa.eu.int/comm/competition/index_en.html

Prof. Dr. Thomas Bieger

Born in 1961, full professor for Business Administration, specialization in Tourism, dean of the department Business Administration, Secretary General of the International Association of scientific experts in Tourism (AIEST).

Research and teaching priorities: management of personal services, destination and location marketing, tourism development and planning, transport and network economies. Current publications: *Destinationsmanagement* (Oldenburg, Hamburg, 2001), *Dienstleistungsmanagement* (Haupt, Bern, Stuttgart, 2000), *Finanzierung im Tourismus* (together with Beat Bernet, Haupt, Bern, Stuttgart, 1999), *Zukünftige Geschäftsmodelle* (Springer, Hamburg, 2002), *Dienstleistungskompetenz* (together with Christian Belz, Thexis, St.Gallen, 2000), *Air Transport and Tourism* (St.Gallen, 2002).

Institute for Public Services and Tourism, University of St.Gallen
Dufourstrasse 40a, CH-9000 St.Gallen, Switzerland
Tel: +41 (71) 224 25 25,   Fax +41 (71) 224 25 36
Email: thomas.bieger@unisg.ch; http://www.idt.unisg.ch

Sven Budde Dipl.-Kaufmann

Born in 1978, studies in Business Administration with majors in Strategic Management and Logistics, Retailing, Wholesaling and Distribution Management and Economic and Social Psychology in Cologne. Practical experience during his studies at Ford Motor Company and Lufthansa Consulting. Since 2003 personal assistant of the Chief Executing Officer of Deutsche Bahn, DB Regio AG in Frankfurt am Main.

Research interest: Aviation.

DB Personenverkehr GmbH, DB Regio AG
60326 Frankfurt am Main, Germany
Tel: +49 69 265 61820, Fax: +49 69 265 61804
Email: sven.budde@bahn.de; http://www.bahn.de

Renato Chiavi

Born in 1940, diploma in commerce and apprenticeship in forwarding. Joint 1995 Danzas from Panalpina after 38 years where he was COO worldwide operations at the time and had several positions in Italy, Nigeria, Switzerland and the USA. 1996 Executive VP and Member of the Executive Board of Danzas in charge of Business Unit Intercontinental which under his management became the largest Intercontinental Freight Forwarder in the Industry (No.1 AFR, No. 2 OFR). 2001 CEO of the Danzas Group. Since 2003 COO of DHL Danzas Air and Ocean and member of Global Executive Committee of DHL worldwide, a daughter company of Deutsche Post World Net.

DHL Danzas Air & Ocean
Peter Merian Strasse 88, 4002 Basel
Tel: +41 61 274 71 00, Fax: +41 61 274 71 07
Email: renato.chiavi@dhl.com

Hans G. Fakiner Dipl.-Volkswirt

Born in 1945, studies in political economics and business administration in Cologne and Frankfurt/Main. For 12 years, researcher at Battelle Institute e.V., Frankfurt am Main, in the fields of economics, traffic and environment. Since 1987, within Fraport AG (the former Frankfurt Airport AG), manager in various positions in the divisions of market research, marketing planning and aviation strategies. Commissioner for Intermodality since 2000, responsible for (1) the development of strategies (in fulfilling Fraport's business mission) and (2) the coordination of multilateral intermodal projects, i.e. the realization of products and services.

Research interest: Strategic Management, Aviation, Intermodality.

Dept. of Traffic & Terminalmanagement, Fraport AG
60 547 Frankfurt am Main, Germany
Tel: +49 69 6 90 7 11 46, Fax: +49 69 6 90 5 97 02
Email: h.fakiner@fraport.de

Kim Flenskov cand.merc.jur, M.Sc.

Born in 1963 in Copenhagen, studies in Business Administration and Commercial Law with emphasis on business strategy, industrial organization, competition law and trade regulation from the Copenhagen Business School. From 1991-1992 Trainee with the European Commission Directorate-General IV for competition – Transport Unit; from 1992-1995 Head of Section in the Danish Ministry of Transport – Aero-political Affairs; from 1995-1996 Legal Officer, Danish CAA – Safety Regulations Dept; from 1996-1998 Senior Manager Commercial with British Airways franchise carrier SUN-AIR of Scandinavia; since November 1998 employed at Lufthansa Consulting in Cologne, from January 2001 as Managing Director.

Lufthansa Consulting GmbH
Von-Gablenz-Strasse 2-6, 50679 Cologne, Germany
Tel: +49 221 88 99 698, Fax: +49 221 88 99 660
E-Mail: kim.flenskov@lhconsulting.com; http://www.lhconsulting.com

Thomas Fritz Dipl.-Kaufmann

Born in 1975, studies in Business Administration with majors in Strategic Management, Logistics, Finance and Statistics in Cologne (D) and Stockholm (S). Since 2001 associate with McKinsey & Company in Cologne consulting consumer goods and logistics companies with a focus on marketing, operations and organization. Currently doctoral candidate at the University Witten/Herdecke specializing in sports economics.

Research interest: Aviation, Sports Economics, Consumer Goods.

Subbelrather Straße 128, 50823 Cologne, Germany
Tel: +49 221 589 7939, Fax: +49 221 208 7109
Email: thomas.fritz@gmx.de

Michael Garvens

Michael Garvens has headed the company that runs Cologne Bonn Airport since 1 February 2002. Before this, he was managing director of Globe Ground Berlin GmbH, a joint venture established by the Berliner Flughafengesellschaft and Lufthansa. He held several different posts at Lufthansa over a period of ten years. He was, among other things, managing director of Lufthansa Airport and Ground Services in Friedrichshafen and Leipzig. Born in 1958 and having completed his degree in business administration at the Wirtschaftsakademie Hamburg (Hamburg Academy of Business Administration) in 1984, Garvens gained his first experience in the aviation sector as a controller when he joined Hapag Lloyd in Hanover in 1986. From here he moved to Lufthansa in 1990.

Flughafen Köln/Bonn GmbH
Postfach 980120, 51129 Köln, Germany
Tel: +49 (0)2203 40-4051,   Fax: +49 (0)2203 40-5750
Email: michael.garvens@koeln-bonn-airport.de;
http://www.koeln-bonn-airport.de

Björn Götsch Dipl.-Kaufmann, CEMS MIM

Born in 1977, studies in Business Administration with majors in Strategic Management, Logistics, Finance and Transport Economics in Cologne (D), Helsinki (FIN), and Pennstate (USA). Since December 2003 doctoral candidate, research and teaching assistant at the Department of Business Policy and Logistics of the University of Cologne.

Research interest: Aviation, Strategic Alliances.

Dept. of Business Policy and Logistics,
University of Cologne, 50923 Cologne, Germany
Tel: +49 221 470 6663, Fax: +49 221 470 5007
Email: bjoern.goetsch@uni-koeln.de; http://www.spl.uni-koeln.de

Mark Goh PhD, BSc. (Hons), MBA

Colombo Plan Scholar 1980 – 1984, Adjunct Professor at University of South Australia 1997-2000, Visiting Research Fellow UMIST 1998, Commonwealth Fellow to UK 1998, Citibank International Fellow USA 1998, Visiting Professor Chulalongkorn University 1999-2000, Associate Senior Fellow at Institute of Southeast Asian Studies 2000-2003, Regional Director Asia/Middle East at APL

Logistics 2002-2003, Visiting Scholar Guanghua School of Management at Beida 2004. Listed in Who's Who in Asia and the Pacific Nations, and etc.

Research interest: Supply Chain Management, Transport Policy and Management.

Dept. of Decision Sciences, NUS Business School,
National University of Singapore, 117591, Singapore
Tel: +65 68743014,   Fax: +65 67792621
Email: mark_goh@nus.edu.sg.

Dr. Matthias Graumann

Born in 1964, studies in Business Administration with majors in Organization Design, Social Psychology and Insurance in Hanover and Cologne until 1993. From 1994 to September 1997 employed by Koelnische Rückversicherung AG (now named GenRe Cologne) in Cologne (D) as well as Milan (I). In October 1997 change to the Department of General Business Administration and Organization Design at the University of Cologne. Since October 2003 Associate Professor at the University of Cologne. Since July 2005 Ordinary Professor at the Europa Fachhochschule Fresenius (University of Applied Sciences).

Research interest: Organization Design, Risk Management and Corporate Governance

Europa Fachhochschule Fresenius – Hochschule für Wirtschaft und Medien,
Im MediaPark 4c, 50670 Cologne, Germany
Tel: +49 221 973199 88,   Fax: +49 221 973199 20
Email: graumann@fhwm.de

Peter Grönlund

Peter Grönlund was born in 1954 in Copenhagen, Denmark. After receiving training as a Forwarding Agent he joined the SAS in 1976. During his career he has hold several management positions both within as well as outside SAS. In 1993 he was appointed Vice President of Space Control Services/Revenue Management in SAS. In 1996 Peter Grönlund was appointed Vice President and General Manager of SAS Cargo. When SAS Cargo Group A/S was incorporated in 2001 he was appointed President and CEO. Peter Grönlund holds the following position of trust: Business Leaders Association Denmark (VL52), and he is a member of IATA Cargo Committee.

SAS Cargo Group
Postbox 151, DK-2770 Kastrup, Denmark
Tel: +45 3232 3888
http://www.sascargo.com

Caroline Heuermann Dipl.-Kauffrau, CEMS MIM

Born in 1976, studies in Business Administration with majors in Strategic Management, Logistics, Trade and Distribution as well as Transport Economics in Cologne (D) and Milan (I). Since 2001 research and teaching assistant at the Department of Business Policy and Logistics of the University of Cologne. Summer 2005 visiting scholar at Hitotsubashi University, Tokyo.

Research interest: Supply Chain Management, International Logistics Management, Transport Economics and Management.

Dept. of Business Policy and Logistics,
University of Cologne, 50923 Cologne, Germany
Tel: +49 221 470 3983,   Fax: +49 221 470 5007
Email: heuermann@wiso.uni-koeln.de, http://www.spl.uni-koeln.de

Sin-Hoon Hum PhD (Management), B. Commerce, B. Engineering (Hons)

Studied at the University of Newcastle (Australia) and the University of California, Los Angeles. Served as Faculty member at the National University of Singapore (NUS) Business School since 1988. Also previously served as Dean of the NUS Business School.

Research interest: Operations Strategy, Supply Chain Management, and Balanced Scorecard Applications

NUS Business School, The National University of Singapore,
1 Business Link, Singapore 117592, Republic of Singapore
Tel: +61 65 6874 3025, Fax: +61 65 6779 2621
Email: bizhumsh@nus.edu.sg; http://www.bschool.nus.edu

Christian Kaufhold Dipl.-Kaufmann, CEMS MIM

Born in 1976, studies in Business Administration with majors in Strategic Management, Logistics, Organizational Behavior and Design and Transport Economics in Cologne (D) and Paris (F). Joined Henkel KGaA in 2003 in the field of International Supply Chain Management for the Consumer Adhesives Business.

Professional domains: Supply Chain Design, Outsourcing, New Product Introduction, Quality Assurance, Complexity Management.

Henkel KGaA,
Henkelstr. 67, 40191 Düsseldorf, Germany
Tel: +49 211 797 2474,   Fax: +49 211 798 3680
Email: christian.kaufhold@henkel.com; http://www.henkel.com

Dr. Thorsten Klaas

Born in 1967, studies in Business Administration with majors in Strategic Management, Logistics, Operations Management and Economic Geography at the University of Cologne (D). From 1996 to 2002 PhD student, research and teaching assistant at the Department of Business Policy and Logistics of the University of Cologne. Winner of the German Science Award 2003 with doctoral thesis "Logistik-Organisation". From 2002 to 2004 consultant / senior consultant at Droege & Comp. GmbH (Düsseldorf) and Simon, Kucher & Partners (Bonn). Since 2004 freelance consultant, specialization in strategy and logistics.

Research interest: Supply Chain Management, Supply Chain Design, Logistics, Strategic Management, Organization Theory.

Rurstraße 19, 50937 Cologne, Germany
Tel: +49 221 9434683,   Fax: +49 221 470 5007
Email: klaas@logistik-organisation.de; http://www.logistik-organisation.de

Dr. Joachim Klein

Member of the Eurowings Board since 2000 where he is responsible for the sectors Information Technology as well as Ground Services and since April 2001 Managing Director of Eurowings Flug GmbH which became Germanwings in 2002. Born in 1957 in Namborn, in the Saarland region, he studied business management and commercial IT at the Saarland University and obtained his PhD there at the end of 1991. In 1992 he started working for Rhenus AG in the logistics department in Dortmund and three years later became Head of Logistics and part of the management team. At the beginning of 1998, he took over Product Management Logistics Services at Schenker AG in Essen where he was responsible for the global development and implementation of the logistics activities of the company.

Germanwings GmbH
Terminalstrasse 10, 51147 Cologne, Germany
Tel: +49 2203 1027 400, Fax: +49 2203 1027 300
Email: joachim.klein@germannwings.com; www.germanwings.com

Dr. Christoph Klingenberg

Born in 1961, Dr. Klingenberg studied Mathematics and Computer Science in Hamburg, Bonn Princeton, Harvard and Cologne. He joined McKinsey & Comp. in 1990 as a consultant in various industries with a focus on restructuring and strategic repositioning. Since 1996 with Lufthansa AG in different functions like head of corporate strategy, leading a project on operational excellence and ontime performance and EVP infrastructure.

Since April 2003 EVP "Future European Operations" at Lufthansa AG.

Lufthansa AG
FRA JA, Bldg. 356, 60546 Frankfurt, Germany
Tel: +49 69 696 8990
Email: christoph.klingenberg@dlh.de, http://www.lufthansa.com

Benjamin Koch Dipl.-Kaufmann

Born in 1976, studies in Business Administration with majors in Strategic Management, Logistics, Corporate Finance and Public Finance in Cologne (D) and St. Gallen (CH). Studies in Spanish, Italian and French in New York (USA). Since 2000 consultant with Lufthansa Consulting GmbH in Cologne (D), working in the fields of Aviation Finance, Air Transport Market and Traffic Development and Airport Management and Strategy. Since 2003 doctoral candidate at the Department of Business Policy and Logistics of the University of Cologne.

Research interest: Corporate Finance, Aviation, Supply Chain Management.

Lufthansa Consulting GmbH
Von-Gablenz-St. 2 – 6, 50679 Cologne, Germany
Tel: +49 221 88996 17, Fax: +49 221 88996 75
Email: benjamin.koch@lhconsulting.com; http://www.lhconsulting.com

Andreas Kraus Dipl.-Kaufmann., Dipl.-Geograph

Born in 1968, studies in Business Administration and Economic Geography with majors in Marketing, Logistics, Strategic Management, and Transport Geography in Cologne (D). From 1995 to 1996 research and teaching assistant at the Department of Economic Geography of the University of Cologne (D). 1996 to 2000 Manager Aviation Policy of the Federation of German Industries and Secretary General of the German Air Shippers' Council in Berlin (D) and Brussels (B). Since 2000 Director Market and Sales Development of Lufthansa Consulting GmbH in Cologne (D), responsible for Airline & Airport Marketing, Traffic Forecasting and Aviation Market Studies. Doctoral candidate at the Department of Economic Geography of the University of Cologne.

Research interest: Aviation Sales and Distribution Strategy, Aviation Policy, Network Management, Supply Chain Management.

Lufthansa Consulting GmbH
Von-Gablenz-Str. 2 – 6, 50679 Cologne, Germany
Tel: +49 221 88996 838,   Fax: +49 221 88996 74
Email: andreas.kraus@lhconsulting.com; http://www.lhconsulting.com

Marcus Niedermeyer Dipl.-Kaufmann

Born in 1967, studies in Business Administration with majors in Organization Design, Marketing and Economic Geography in Cologne. Since 1996 working for Lufthansa Cargo AG in different positions like assistant to the Chairman of the Executive Board, Project Manager for implementing a new LH Cargo Strategy, Manager Planning & Steering, Project Manager for value based management and Project Manager Redesign and Re-engineering Corporate & Shared Services.

Since July 2003 General Manager for Product Management at Lufthansa Cargo AG.

Lufthansa Cargo AG
FRA F/MP, Gate 25, bldg. 451, 60546 Frankfurt, Germany
Tel: +49 69 696 92265
Email: marcus.niedermeyer@dlh.de; http://www.lufthansa-cargo.com

Dr. Andreas Otto

Dr. Andreas Otto was born in Mettmann, Germany in 1962. After obtaining a degree in business administration at the University of Cologne, he was awarded a doctorate in political sciences in 1994. In the same year he joined Rhenus AG & Co KG in Dortmund, where he later joined the Executive Board with responsibility for Marketing and Sales. The Supervisory Board of Lufthansa Cargo AG has appointed Dr. Otto to the Executive Board with effect from 1 April 2000.

Lufthansa Cargo AG
60546 Frankfurt/Main
http://www.lufthansa-cargo.com

Andrea Pal MSc.

Born in 1961, Andrea Pal studied Energy Engineering at Bucharest Polytechnical Institute in 1985. Mrs Pal has been Senior Vice President of the Department Global Investments & Management of Fraport since 1 February 2001. Prior to this, she was Vice President of Multi Utility Center RWE AG in Essen responsible for the international strategy of the group and Chief Technology Officer of TESSAG, Frankfurt. She was also Head of Lahmeyer Energy Solutions involved in the development of international BOT projects.

Business interest: Project Financing, Change Management and Reengineering Process in Aviation Companies.

Fraport AG Frankfurt Airport Services Worldwide Frankfurt Airport
D-60547 Frankfurt, Germany
Tel: +49 69 690 20130, Fax: +49 69 690 59901
Email: a.pal@fraport.de; http://www.fraport.de

Kathryn Pavlovich PhD (Waikato), BA (Auckland), MMS (Waikato)

Born and lives in New Zealand. Assoc. Prof. at the Waikato Management School in Hamilton, New Zealand. Teaching interests include Strategic Alliances and Networks, Strategic Partnerships and Collaboration, Business Policy and Strategy, and Entrepreneurship. Current research is futuristic: how will the increasing network interdependencies and global connectedness impact on and influence our social systems as we currently know them.

Research interest: Network Structure, Strategic Alliances, Quality of Relationships, Tourism, Small Business, Phenomenology.

Dept. of Strategic Management, University of Waikato Management School
Private Bag 3105, Hamilton, New Zealand
Tel: +64 7 8384837,   Fax: +64 7 8384356
Email: kpav@waikato.ac.nz; http://www.waikato.ac.nz

Kornelia Reifenberg Dipl.-Kauffrau

Born in 1970, studies in Business Administration. She is Director at Simon-Kucher & Partners, where she has been working since 1997. With Simon-Kucher & Partners, Mrs Reifenberg is member of the Competence Center Logistics. Her main foci are on the development of price strategies, innovation management, strategic marketing management and sales controlling in the logistics industry. She has led and carried out numerous projects for leading international logistics carriers, including Deutsche Post World Net, Lufthansa Cargo AG, TNT and DHL Worldwide Express.

Simon-Kucher & Partners, Strategy and Marketing Consultants
Haydnstrasse 36, D-53115 Bonn, Germany
Tel: +49 (228) 9843 315, Fax +4 (228) 9843 320
Email: kreifenberg@simon-kucher.com; http://www.simon-kucher.com

Dr. Jan Remmert

Born in 1969, studies in Computer Science & Economics, Ph.d. in Business Administration. He is Senior Consultant at Simon-Kucher & Partners, after a multiyear freelancing activity at Lufthansa Consulting. His project and research priorities are in the area of pricing for logistics service provider, especially aircargo business. Current publications are *Referenzmodellierung für die Handelslogistik* (Gabler, Wiesbaden, 2001), *Der Prozess der informatorischen Auftragsbearbeitung* (together with P. H. Klee and H. Makowski, in: Controlling von Logistikprozessen, Schäfer Pöschel, Stuttgart 2003), *Supply Chain Management* (with T. Engelsleben, in: Das große Buch der Strategieinstrumente, Campus, Frankfurt a.M. 2002).

Simon-Kucher & Partners, Strategy and Marketing Consultants
Haydnstrasse 36, D-53115 Bonn, Germany
Tel: +49 (228) 9843 315, Fax: +49 (228) 9843 320
Email: jremmert@simon-kucher.com; http://www.simon-kucher.com

Jens Rühle Dipl.-Kaufmann, CEMS MIM

Born in 1978, studies in Business Administration with majors in Strategic Management, Logistics, Controlling and Statistics in Cologne (D), Copenhagen (DK) and Vancouver (CAN). Since 2003 doctoral candidate, research and teaching assistant at the Department of Business Policy and Logistics of the University of Cologne.

Research interest: Planning Processes, Strategies in Dynamic Environments, Railway and Aviation Management.

Dept. of Business Policy and Logistics,
University of Cologne, 50923 Cologne, Germany
Tel: +49 221 470 3897, Fax: +49 221 470 5007
Email: ruehle@wiso.uni-koeln.de; http://www.spl.uni-loeln.de

Robert Skoog Fil. Kand. (B.A.)

Born in 1952, studies with degree in Administration, Logistics and Transport Economics. Additional studies in Law, Business Law and Business English. Management positions in SAS Cargo since mid 1980s, comprizing inter alia General Administration, Marketing & Sales, Product Development, IT, Training, Management Support, Process Management, Distribution & Industry Affairs, Network Management, Alliance Development etc.

SAS Cargo, Alliances & Corporate Support, Dept. STOFQSK
SE-195 87 Stockholm-Arlanda, Sweden
Tel: +46 8 797 1909, Fax: +46 8 797 4480
Email: robert.skoog@sas.se

Swee-Koon Tan MSc.

Studied Accounting and Finance at the London School of Economics and Political Science, and Financial Engineering at the National University of Singapore. Research Associate at the National University of Singapore. Previously, an MBA tutor in Finance at INSEAD, Asian Campus, and Class Teacher at the London School of Economics and Political Science.

Research interests: Capital Structure, Mergers and Acquisitions, Asset Pricing.

PO Box 182, Singapore 915807, Republic of Singapore
Email: sktan@alumni.nus.edu.sg

Wolfgang Weil Dipl.-Kaufmann

Born in 1957, Wolfgang Weil studied Business Administration with majors in Organizational Theory, IT in ecomomics and Capital Market Theory at Frankfurt's Johann-Wolfgang-Goethe-University in 1982. Mr. Weil joined Fraport AG in 1989 as Manager Quality Management and Deputy Head of Operations in Fraport's Aircraft handling department. He was Managing Director System Development and Consulting in the Aviation Ground Services & Logistics department before he worked as an expat for Fraport as Managing Director Frankfurt Airport Asia, with offices in Bangkok, Thailand. Currently he is Project Manager in Fraport's Global Investments & Management department. Mr. Weil represented Fraport in various working groups of such organizations as Airports Council International (ACI) and the International Air Transport Association (IATA).

Business interest: International Aviation Management.

Fraport AG Frankfurt Airport Services Worldwide Frankfurt Airport
D-60547 Frankfurt, Germany
Tel: +49 69 690 78556;   Fax: +49 69 690 59866
Email: w.weil@fraport.de; http://www.fraport.de

INDEX

Note: Page numbers in **bold** refer to primary references to this topic.

ACMI 500
aerodynamic design 475
African Nations 507
Air Andaman 148
Air Asia 351
Air Berlin 41, 58, 60, 120, 122, 129, 130, 167, 187, 203, 202–5, 207–10, 214–15, 363
Air Canada 100, 102, 210, 259, 330
air cargo 452, 456, 451–66, 473–81, 492, 495
  airmail 452, 461, 523
  alliances 477, 479, 480, 485
  bazaar **539**
  belly 474
  bookings 468
  business facts 456
  capacity 453, 454, 461, 462, 470, 480, 481, 508, 513
  competitiveness 454
  consolidation 261
  cooperation 476
  customer value 478
  demand 454–56, 506, 511, 512
  development 501, 504, 505, 506, 507, 508, 509
  express 460, 492
  facilities 578
  forwarders 380, 457, 502, 503, 541
  freighter 455
  global logistics companies 490
  history 490–98, 511
  hub 531
  joint activities 416
  liberalization 508
  management 459, 470
  market 451, 453, 469, 470, 513, 520, 541
  market share 510
  network 523
  pricing 454, 475, 512, 539, 540–41, 541, 542, 544, 548, 555
  products 465
  quality management 503
  rate card 549, 552
  restructuring 474
  role of 462, 510
  security 469
  strategy 490, 511, 512
  supply chain 489, 512, 513
  traffic 535, 536
  transport chain 456, 469, 518
  value chain 457
  variable costs 541
  yield management 551

Air Cargo Tariff 475
Air Dolomiti 59
Air Express International 501
Air France 4, 29, 30, 58, 61, 99, 110, 113, 120, 174, 182, 306, 310, 423, 495
Air France/KLM 30, 424
Air New Zealand 259, 328, 329
Air Paradise 148
Air Polonia 205
air traffic control 179, 364
air traffic rights 579
Air Transport World 313, 314
air transportation networks 519

air transportation services 558

air travel 33

AIRail **433**

AirAsia 102, 148, 149, 150, 151, 155, 156, 157

Airborne Express 503

Airbus 175–76, 200, 353–54, 494, 502, 506, 509, 524
  A300 176, 494, 498, 500, 509, 524, 525
  A310 500, 509
  A319 110, 121
  A319/320 132
  A320 114, 150–51, 154–56, 176
  A330 509
  A380 9, 60, 353, 354, 509, 567, 571

aircraft
  airport criteria 365
  assignment 525, 526, 527
  belly 474, 475, 521
  capacity 72, 77, 182, 200, 363, 528
  costs 232
  development 493
  drop out 181
  innovation 512, 513
  intercontinental 530
  jet 493, 525
  labelling 136
  maintenance 113
  manufacturers 49, 92, 353, 494, 502, 559, 613
  movements 529, 578
  noise 356
  parking position 566
  private jet 106
  productivity 72, 77, 88, 177
  quality 234
  range 525
  seating 174
  selection 103, 524
  size 114, 175, 176, 353, 363, 370, 432, 475, 529, 530, 531, 610
  standardization 127
  technology **9**, 110, 353, 354, 440
  terrorist attacks 349
  turnaround 144, 174, 531, 609
  type 72, 74, 128, 149, 151, 156, 181, 257, 524, 525, 613, 617, 620
  used 200
  utilization 51, 81, 108, 114, 127, 128, 257, 527, 617
  widebody 110, 495

airfreight *see* air cargo

airline
  alliance 306, 478
  belly 459, 461, 470, 509
  brand 111
  charter 41, 54, 55, 57, 58, 122, 130, 202, 363, 518, 524
  choice of 110, 543, 544
  competition 22
  cooperation 257, 325, 327
  costs 73, 543
  dependency 513
  domestic 337
  dominant 79
  express 518, 520, 524, 525, 527, 528, 529, 530, 535, 536, 537
  feeder 53
  flag 42, 97, 99, 102, 103, 121, 307, 312, 326, 331, 362, 364, 365, 393, 458, 475, 567, 608
  full service 112, 147, 158, 159, 160
  global 307
  hub 52, 80, 83, 165, 175, 430, 613
  network 20, 21, 24, 26, 31, 41, 50, 54–56, 58, 59, 61, 65–76, 79–86, 93, 97–109, 122, 126, 129, 137, 182, 200, 204, 205, 211, 520, 523, 612, 613, 619
  profitability 311, 312, 315
  regional 42, 53, 55, 56, 57, 59, 104, 125, 325, 363, 423
  traditional 160, 166, 167, 172, 182, 362, 518
  yield 458

airport
  acquisition 383, 384
  alliance 413, 424
  cargo 368, 498, 502, 537
  challenge 429
  choice of 114, 172, 517, 519, 535, 537
  competition 386, 394, 432, 572
  congestion 21, 29, 160
  cooperation 366, 411, 412, 413, **414**, 415, 416, 417
  costs 346, 348, 351
  direct effects 590
  european 181
  facilities 415, 529

INDEX 657

fees 21, 24, 36, 73, 81, 92, 172, 432, 433, 617
functions 529, 565
hub 160, 351, 352, 410, 418, 421, 430, 431, 439, 441, 529, 530, 617
impact on economic welfare 586
impact on the environment 355
indirect effects 590
induced effects 590
industry 346, 347, 348, 352, 353, 354, 356, 357, 358, 359
industry structure 350, 352
industry technological change 353
industry transformation 345, 347
infrastructure 71, 353, 529
international 387, 565, 570
internationalization 382, 384, 405
management 345, 358, 377, 380, 384, 391, 393, 409, 414, 419, 421
non-aviation sector 366, 372, 373, 374, 424
operator 414
planning 353
primary 121, 130, 131, 134, 138, 159, 172
privatization 2, 5, 397, 411, 420, 422, 424
regional 37, 57, 393
secondary 36, 37, 99, 103, 123, 127, 128, 130, 134, 144, 146, 159, 160, 161, 172, 200, 351, 352, 353
security areas 465
spoke 70, 71, 526, 527, 528, 529, 530, 531, 534, 609
strategy 395, 396, 567, 575, 576, 612
transportation chain **379**

Airport Council International 415

Airport Express 428, 575

Airports 585

Airports Authority of Thailand 417

Airports Council International 415

Alitalia 29, 36, 99, 310, 376, 495

All Nippon Airways 259, 330, 434, 437

alliance
 agreement 209, 210, 211, 214, 309, 318
 cargo **261**, 464, 477, 478, 485

global 42, 50, 327, 333, 337
life cycle 423
low fare 209
management board 259, 265
multi-partner 310
network 209, 238, 306, 614
performance 319
scale and link 327, 339
strategic 121, 244, 255, 256, 265, 306, 325, 411, 423, 478, 482
structure 6
systems 61, 259

American Airlines 99, 100, 180, 181, 210, 309, 310, 329, 332, 333, 338, 434, 436, 491, 542

Amsterdam Airport 358, 364, 365, 386, 415, 428, 430, 434, 435, 438, 443

Ansett 328, 330, 331, 333, 334, 335, 336, 339

Antitrust authority 204

Article 81 22, 26

ASEAN 145

Asia 233, 576

Asiana Airlines 330

Australian Commerce Commission 336

Austrian 52, 53, 59, 61, 205, 259, 330, 389

Austrian Airlines *see* Austrian

B2B 414, 416, 466, 499

B2C 499

BAA 356, 357, 364, 378, 400, 417, 421

baggage
 conveyor system 433, 434, 439, 445
 handling system 71, 81, 439, 565, 566, 568, 569, 571, 574
 security screening 566
 waiting time 566

Bahrain 354, 497

Bangkok International Airport 417

bargaining power 236

behavior 193, 194, 613

benchmark 169, 179, 180, 305, 308, 310, 318, 319, 320, 502

Berlin-Schönefeld 202, 207

bilateral agreements 20, 42, 103, 364, 465, 501, 502

board product 138, 172, 173, 174

board service 114, 126, 171, 173

boarding system 149

Boeing 9, 60, 144, 149, 156, 176, 200, 326, 335, 340, 353, 452, 464, 481, 500, 502, 506, 508, 509, 524
  B707 493, 509
  B727 497, 498, 509
  B737 60, 149, 176
  B747 354, 495, 497, 500, 503, 509
  B757 495, 497, 509
  B787 9
  forecast 560

brand value 155, 157, 466

British Airways 4, 29, 34, 35, 61, 99, 102, 104, 110, 113, 120, 130, 146, 169, 176, 309, 310, 312, 316, 319, 332, 333, 338, 428

Brussels Airport 365, 376

Brussels-Zaventem 532

budget travellers 148

bunching effect 159, 179

Bundeskartellamt 31, 32

business and/or leisure destination 412

Business customers 616

business cycle 61, 99, 108, 311, 490, 494, 498, 501, 505, 506, 511, 513, 562

business lounges 121

business mission 431

business model 24, 27, 41, **42**, 44, 45, 46, 47, 50, 53, 55, 56, 57, 58, 59, 60, 62, 97, 110, 123, 165, 166, 175, 214, 224, 619
  air cargo 459, 462, 470, 518
  airport 359, 420
  alliance 411
  cargo 455
  charter airline 122
  low cost 53, 58, 60, 116, 119, 120, 123, 124, 125, 129, 130, 131, 134, 136, 137, 138, 139, 140, 141, 143, 146, 178
  network 61, 121
  regional airline 59
  Ryanair 365
  Tiger Airways 155
  traditional 58, 99
  ValuAir 150, 151

business partner program 467

business strategy 135, 226, 409, 424, 427, 518, 561

business traveller 127, 147, 149, 152, 157, 159, 362

business-class 104, 172, 174, 328

Buzz 120, 133, 146, 617

C2C 43

cabotage 186

call centers 468, 618

Canada 24, 54, 69, 210, 350, 385, 390, 391, 497, 499

Canada 3000 54

capacity
  allocation 625
  overcapacity 200, 211, 495, 501
  planning 181, 512
  rationalization 25
  utilization 443, 445, 482, 522, 547, 551, 628

capital
  intensity 607
  investment 492, 507, 561, 562

Cargo 2000 469, 503

Cargo Community Systems 468

Cargolux 461, 495, 496, 500, 503

carrier *see* airline

catchment area 26, 149, 178, 201, 365, 368, 369, 371, 376, 411, 429, 430, 431, 432, 437, 438, 441, 481
  enlarging 431

category assignment 551

Cathay Pacific 111, 309, 310, 312, 315, 316, 333, 465, 575

Cebu Pacific 148

central control system 565

Centralized Airport Operation Center 565

Centralized Management Information System 565

Charleroi Airport 36, 352, 365

Chiang Mai 147, 152, 157

China 145, 151, 155, 161, 351, 354, 416, 423, 453, 465, 507, 510, 558, 559, 572, 575, 576, 577, 579

China Eastern 423

China Southern 423

Civil Aeronautics Act 495

Civil Aviation Administration of China 8

class configuration 144

codesharing 25, 234, 237, 258, 259, 262, 309, 310, 328, 331, 332, 333, 334, 338, **410**, 428, 434, 436, 614

Cologne 134, 428, 433, 434, 435, 436, 437, 438, 439, 443, 445, 531

Cologne/Bonn Airport 75, 76, 132, 170, 187, 201, 203, 204, 207, 361–76, 532, 535

competition authorities 22, 26

competitive
  advantage 33, 60, 124, 141, 186, 197, 209, 227, 240, 537, 562, 614, 628
  disadvantage 458, 623
  dynamics 186
  dynamics research 186, 189, 192, 214
  forces 124
  gap 212, 214
  interdependence 189
  position 24, 186, 190, 206, 207, 227, 235, 236, 237, 256, 310, 382, 577, 607, 619
  response 190, 191
  strategies 146, 194, 605, 607, 621, 626

computer reservation system 21, 24, 37, 38, 80, 99, 169, 178, 179, 328, 330, 433, 437, 446, 613, 614, 620, 626, 633

concentration 125, 191, 198, 210, 223, 226, 231, 235, 241, 248, 365, 410, 411, 424, 458

Condor 54

congestion 37, 58, 354, 356, 415, 522, 569

connecting
  flights 149, 167, 230, 234, 259, 607, 618, 624, 631
  times 178, 180, 434, 609

consolidation 20, 21, 41, 42, 105, 116, 182, 185, 186, 211, 212, 214, 215, 232, 234, 248, 261, 325, 327, 328, 337, 339, 375, 397, 398, 406, 424, 461
  phase 213

consolidation hub 458

Continental Airlines 99, 165, 171, **212**, 310

control mechanisms 179

control system 416, 571

Copenhagen 52, 56, 357, 358, 364, 444, 476, 532

corporate
  cultures 265, 269, 482, 485
  cultures of alliance members 269
  growth 44
  network 44, 45, 46, 47

credible threats 195, 198, 213

CRM 62

Croatian Airlines 52

Crossair 53, 58

cultural differences 387, 485

CUSS 442, 443, 446

customer
  loyalty 32, 139, 149, 431, 480, 544, 545, 549, 554
  relationship-management 468
  requirements 44, 483, 518
  segmentation 110, 467, 549
  service 62, 111, 112, 124, 468, 499
  service quality 531
  specific differentiation 552
  value 62, 109, 131, 135, 137, 139, 140, 237, 476, 478, 484

customization 545, 546, 550

Danzas 501

Dassalt Falcon 495

DB 428, 433, 434

dba 129, 130, 167, 170

Debonair 617

Delta 99, 102, 181, 423, 575, 576, 578, 579

de-peaking 88, 612, 624, 632

deregulation 2, 6, 224, 340, 496, 511, 519
  USA 610

deterrence 193, 194, 195, 196, 197, 198, 208, 210, 211, 213, 214, 215, 216

deterrence mechanism 210

Deutsche British Airways 167

Deutsche Flugsicherung 180

Deutsche Post 458, 461, 501, 503, 511

Deutsche Post World Net 458, 461, 503, 511

DHL 375, 461, 474, 481, 496, 497, 501, 503, 511, 519, 520, 525, 528, 531, 532, 533, 534, 535, 536

Diosdado Macapagal International Airport 537

discount 27, 29, 32, 35, 178, 366, 541, 542, 544, 545, 553, 554, 629
  criterion 549, 554
  guidelines 545, 550
  system 546, 547, 549

dis-intermediating 149

distribution 21, 192, 383
  channels 100, 138, 210, 631
  direct or "online" 139
  indirect or "offline" 138
  process 158
  system 145, 160, 631

domestic services 177, 352

door-to-door transportation 457

downstream effects 559

dumping 541

Dusseldorf airport 366

East Midlands 398, 532

Eastern Airlines 491

Easyjet 61, 102, 129, 187, 200, 202, 203, 205, 206, 207, 208, 210, 213, 215, 368, 617, 618

e-business 499, 510

INDEX

EC Treaty 20, 21, 22, 34, 36

e-commerce 139, 414, 416, 468, 469, 565

economies of
   density 31, 60, 77, 223, 230, 232, 236, 241, 607, 615, 622
   density and scale 236
   density and scope 60
   scale 50, 59, 60, 76, 77, 92, 192, 230, 232, 233, 236, 239, 247, 261, 327, 328, 339, 340, 399, 410, 411, 413, 415, 417, 442, 518, 522, 607
   scale, scope and density 50
   scope 77, 78, 79, 85, 86, 230, 607, 615

economy class 113, 172, 173, 174

Edelweiss Air 54

EDI 499

electronic booking channels 468

Emirates 12, 58, 113, 337

Europe 106, 203, 233, 524

European
   air transport market 606
   aviation 19, 20, 21, 23, 24, 37, 120, 185, 186, 187, 188, 199, 200, 201, 203, 205, 206, 211, 215, 362, 364
   Civil Aviation Conference 497
   Commission 19, 22, 23, 24, 25, 26, 27, 28, 29, 30, 31, 33, 34, 35, 36, 37, 38, 168, 536
   Council 24
   Economic Community 491, 494
   Monetary System 494
   Union 20, 21, 22, 23, 24, 34, 37, 98, 99, 132, 168, 354, 428, 443, 445, 446, 499, 501, 508, 519, 522, 527, 535, 606

external coordination 264, 269, 270

External Costs 599

FedEx 461, 474, 481, 495, 496, 497, 500, 501, 503, 519, 520, 531, 537

FedEx Trade Networks Transport & Brokerage 503

feeder

airlines 52, 121, 212, 215
   flights 60, 82, 429, 430, 431, 432, 437, 439
   services 105, 125, 363

Finnair 58, 444

Finnair/SWISS 444

First Cambodia Airlines 148

fleet 175, 455, 458, 462, 524, 615, 630, 633
   aircraft 81, 82, 88, 175, 261, 495, 524, 525, 613, 615, 617
   assignment 51, 99, 524, **525**, 526
   capacities 609
   short-haul 176
   size 108, 214
   structure 92, 607, **613**, 617
   utilization 527

flight
   attendant 182
   connecting 178, 233
   direct 112, 178
   frequencies 206, 257
   information system 440
   operations 121, 232, 257, 523, 524, 535
   related services 559
   schedule 122, 170, 177, 202, 203, 205, 209, 214, 229, 232, 233, 238, 239, 240, 243, 244, 246, 247, 248, 262, 410, 419, 436, 443, 533, 534
   to-flight connection 440

Flying Tiger Line 497

Flying Tigers 475, 491

fortress hub 178, 632

forwarder 416, 461, 464, 466, 469, 470, 480, 485, 492, 493, 502, 503, 541, 542, 543, 544, 546, 547, 549, 551, 553

France 29, 35, 102, 310, 386, 410, 428, 531, 629

Frankfurt 31, 32, 53, 171, 172, 177, 179, 180, 183, 201, 204, 260, 358, 368, 371, 372, 385, 386, 388, 399, 416, 418, 421, 427, 428, 429, 430,

431, 434, 435, 437, 438, 439, 440, 444, 445, 446

Frankfurt Airport 88, 112, 176, 415, 417, 427, 428, **429**, 430, 431, 432, 433, 434, 435, 437, 438, 439, 441, 446, 590

Frankfurt Hahn Airport 201, 365

Frankfurt-Berlin 31, 32

Frankfurt-Berlin/Tegel 31

Fraport 180, 366, 378, 385, 386, 387, 388, 399, 403, 415, 416, 417, 418, 419, 420, 421, 427, 430, 431, 432, 433, 435, 438, 439, 440, 441, 446, 447
strategy 395, 418, 429

frequent-flyer program 19, 27, 31, 32, 33, 34, 37, 68, 72, 81, 121, 149, 151, 236, 237, 239, 258, 413, 452, 467

GATT 453, 499

GCC 497, 501

GDP growth 312, 504, 507, 509

German Air 495

German Airports Association 415, 429

German Rail 427, 428, 431, 433, 434, 435, 436, 438, 440

Germania 31, 133, 205

Germania Express 167, 204

Germanwings 102, 119, 120, 123, 129, 131, 132, 133, 134, 135, 136, 137, 138, 139, 140, 141, 167, 169, 170, 187, 203, 367, 368, 617
corporate culture 135
corporate design 136
corporate identity 135
strategic positioning 133
strategy 133
yield management 137

Germany 20, 35, 102, 119, 120, 132, 169, 170, 173, 174, 182, 187, 202, 204, 208, 352, 361, 363, 365, 368, 369, 371, 376, 385, 386, 389, 410, 418, 428, 429, 430, 432, 435, 445, 523, 531, 532, 536, 629

Global Airport Monitor 570, 578

global logistics 457, 489, 490, 491, 511, 512, 513
companies, air cargo 490

global network 233, 259, 463, 470

global partnerships 329, 330

global sourcing 43

globalization 21, 120, 224, 239, 382, 470, 489, 490, 499, 506, 508, 511, 513

Go 102, 104, 120, 133, 146, 499, 617

ground handling 36, 49, 181, 232, 257, 261, 364, 380, 381, 386, 395, 403, 416, 432, 468, 469, 503, 559

Gulf Air 58, 110

Haag protocol 475

handling 523, 609

handling capacity 567

handling facilities 439, 479, 481

Hanover 203, 371, 386, 418

Hapag-Lloyd 58, 140, 169, 170, 187, 203, 204, 207, 208, 210, 363, 367

Hapag-Lloyd Express 120, 129, 140, 167, 170, 187, 203, 204, 205, 207, 209, 210, 215, 363, 367, 368, 617

harmonized services 477

Hat Yai 152, 155

Helvetic Airways 48, 60

high-speed trains 428, 430, 435, 437

INDEX 663

Hong Kong 147, 150, 151, 155, 157, 317, 354, 386, 416, 558, 559, 563, 572, 573, 575, 576, 577, 578, 579

Hong Kong International Airport 416, 559, 563, 572, 574, 577

Hong Kong's Chep Lap Kok Airport 151

hub
  air cargo 531
  airport 70, 78, 80, 88, 90, 434, 522, 525, 527, 530, 612
  logistics 576
  operation 77, 522, 534, 561
  secondary 52, 56, 60
  selection 524
  sub 528, 530, 532, 533, 534
  system 429, 441

hub and spoke 108, 110

hubbing 42, 50, 59, 60, 348, 353, 355, 559, 560, 612

IATA 37, 98, 100, 352, 428, 437, 469, 475, 492, 493, 495, 496, 501, 503, 541, 570, 578
  annual conference 159
  forecast 560
  rate framework 502

Iberia 4, 29, 61

ICE 371, 428, 430, 435, 445

incumbents 37, 104, 159, 187, 606, 611, 614, 615, 617, 618, 619, 624, 629

India 102, 110, 393, 498

indirect productivity effects 587

Indonesia 148, 159

Institutional Airport Ownership 358

Institutional Economics Perspective 228

Instrument Landing Systems 575

integrator 54, 457, 461, 462, 470, 474, 481, 497, 498, 517, 518, 519, 520, 523, 524, 527, 528, 529, 531, 535
  network strategies **531**
  networks 519

intensity of competition 145, 165, 191, 192, 193, 196, 201

interconnectedness 325

interface management 255, 256, 258, 260, 261, 263, 264, 265, 266, 267, 268

interline agreements 410

intermodal
  accessibility 529
  competition 615, 621, 623
  integration 444
  product 432, 433
  services 430, 432, 441, 446, 447
  system 576
  transportation 439

intermodality 369, 427, 428, 430, 431, 433, 439, 446

internalization 52

International Air Rail Organisation 444

International Civil Aviation Organization 90, 92, 320, 492, 508, 571

International Federation of Freight Forwarders' Association 493

international routes 173, 235

Intersky 58, 59

intramodal 606, 621, 623
  competition 606, 623

Jakarta 147, 150, 151

JAL Cargo 464

Japan Airlines 261, 310, 315, 329

JFK 417

Johor Baru 149, 156

just-in-time 497, 499, 500

key account organization 467

Khon Kaen 152

KLM 4, 29, 30, 102, 120, 146, 174, 176, 182, 306, 307, 309, 310, 319, 337, 423, 466

Köln/Bonn Airport 599, 600

Korean Airlines Cargo 461

Kuala Lumpur 147, 149, 156, 351, 354, 558, 559, 564, 567, 568

Kuala Lumpur International Airport 351, 559, 563, 564, 568

Kuoni 54

Kuwait 110, 497

La Guardia 417

Lan Chile 334, 338, 339, 465

landing
fees 143, 146, 419, 432, 559
rights 159, 203, 335

landside facilities 414

Langkawi 147, 157

Latin America 384, 392, 507

lead times 356, 476, 478, 524

Leipzig/Halle 203, 536

LH Airport Express 434

liberalization 19, 20, 21, 23, 35, 97, 98, 99, 101, 107, 186, 224, 231, 235, 308, 321, 354, 364, 378, 491, 497, 498, 499, 501, 508, 511, 519, **520**, 523, 613, 614, 622
global 321
process 20

Liège 531, 535, 537

Lion Air 148

load factors 31, 72, 74, 257, 539, 541, 609

location factor 585

Logair 492

London
Gatwick 202, 417, 496
Gatwick Express 444
Heathrow 112, 356, 417, 421, 428, 558
Heathrow Express 435, 444
Luton 187, 202
Stansted 53, 112, 202, 203, 352, 366, 368, 372, 417

long-distance train 428, 432, 434, 435, 439, 441, 443, 446, 606

long-haul services 30, 112, 116, 523, 537

LOT 259, 330

low cost airline 11, 20, 32, 42, 48, 57, 65, 72, 75, 76, 86, 100, 101, 102, 103, 107, 109, 116, 123, 119–41, 143–62, 167, 170, 175, 178, 182, 186, 187, 199, 200, 203, 204, 205, 208, 209, 211, 212, 214, 224, 329, 350, 351, 352, 353, 362, 363, 364, 365, 367, 369, 372, 386, 399, 418, 419, 567, 607, 613, 614, 631
business model 41, 53, 55, 58, 59, 60, 61, 97
challenge **115**
network 56
state aid 36
yield management 144

low cost airport 361, 368

loyalty
aspect 554
program 24, 32, 48, 50, 141, 234, 616, 620, 629, 633

Lufthansa 4, 31–32, 32, 33, 56, 58, 60, 61, 102, 106, 113–14, 120, 129, 165, 167, 169–77, 173, 179, 180, 182, 204, 211, 212, 258, 259, 309, 310, 330, 332, 363, 364, 397, 427, 428, 430, 433, 434, 435, 436, 438, 439,

441, 443, 444, 454, 464, 465, 477, 493–95
Airport Express 434

Lufthansa Cargo 261, 451, 461, 464, 465, 466, 467, 468, 469, 470, 477

maintenance 44, 69, 77, 82, 92, 110, 113, 149, 155, 161, 181, 231, 232, 257, 411, 423, 500, 528, 559, 569, 609, 611, 618
    aircraft 71, 571
    costs 613
    services 559

Malaysia 156, 159, 351, 559, 564, 567, 568, 569
    Airport 564

Malaysian Airlines 148

Malév 4

Malpensa Express 444

market
    barriers to entry 191
    barriers to entry and exit 191
    concentration 191, 196, 198, 201, 210
    coverage 206, 209, 210, 248, 320, 520
    dependence 189
    development 208, 455, 519, 554
    entrant 537
    entrants 101, 124, 172, 606, 611, 621
    entries 84, 192, 206, 208, 212, 213, 420, 617, 622, 629
    niche 53, 55, 125, 132, 140, 161, 200, 212, 215, 362, 419
    position 97, 237, 327, 479, 520, 621
    power 27, 28, 29, 30, 79, 80, 101, 615, 622, 626
    share 20, 28, 29, 33, 52, 54, 55, 120, 146, 166, 169, 170, 197, 198, 199, 236, 306, 312, 335, 366, 501, 510, 544, 623, 629
    spot 543, 547, 552

market segments/airline competition 121–23

market-based view 46, 223, 224, 226, 227, 235, 237, 238, 239, 235–39, 241, 247, 248

marketing
    activities 186, 445, 479
    mix 133, 136, 137
    strategy 32, 136, 340, 467

maximum take-off weight 9

mediation process 593

memorandum of understanding 438, 442

Memphis 537

merger
    cross-border 410
    regulation 22

mergers and acquisitions 223, 224, 225, 235, 236, 237, 238, 245, 246, 248

modal split 431, 437

mode of transport 363

modular construction 613, 625

monopoly 110, 310, 605, 606, 619, 625

multimarket
    competition 187, 192, 193, 195, 196, 198, 199, 201, 210, 212, 213, 216
    contact 187, 192, 193, 194, 195, 196, 198, 199, 208, 209, 213, 215
    degree of contact 210

multi-modal transport 577

multimodality 431

mutual forbearance 193, 194, 196, 197, 198, 199, 208, 210, 212, 213, 214, 215

National Skyway Freight Corporation 491

network
    businesses 44
    configuration 224, 532
    connectivity 612
    economy 43, 46
    effects 29, 43, 44, 52, 53, 55
    management 42, 48, 49, 50, 51, 52, 56, 59, 62, 233, 365

666 INDEX

operation **525**
player 606, 608, 627, 632
relationship 328
services 621, 622
strategies integrator **531**

New Global Cargo 261

New Institutional Economics 228

Newark 417, 444

NIKI 203, 205, 209, 210, 214

no-frills
airline **165**, 166, 167, 168, 169, 170, 171, 172, 173, 175, 182, 369, 512

NokAir 148, 152, 153, 154, 156, 157

North American Free Trade Agreement 499, 508

Northwest **212**, 307, 309, 310, 577

Norwegian Competition Authority 34

oil embargo 494, 495

oligopolistic collusion 191

Oman 497

one-stop shopping 477, 484, 489, 490, 500, 503, 510, 513

one-stop-service 531

one-time customer 553

oneworld 121, 307, 309, 310, 319, 320, 321, 333, 334, 338, 423

open skies 20, 24, 42, 159, 327, 329, 331, 333, 334, 354, 364, 455, 492, 572

opportunistic behavior 228, 229, 241, 244, 245, 246, 247

organizational slack 189

organizational structure 196, 198, 210, 263, 264, 390, 402, 403

outsourcing 49, 498, 499, 518

over-booking 234

Pan American World Airways 493

Pantares 386, 415, 416

Paris Charles-de-Gaulle 435, 437, 441, 531

parking
aircraft position 354
bays 566
facilities 414, 432
revenues 431
services 559

passenger
business 452, 456, 475, 476, 477, 523
capacity 493, 571
charges 432
handling capacity 577
traffic 34, 138, 454, 477, 491, 497, 576, 577
travel habits 348

personal trust 485

Phitsanulok 152

Phuket 147, 152, 155, 157

planning
crew 99, 181, 609
fleet 609
rotation 51, 181
uncertainty 628

point-to-point 70, 177
network 160
relations 122
routes 53, 127
services 128, 353, 607
traffic 113, 352, 353, 362, 432

Polar Air Cargo 503

postal services 461, 519, 520

premium segments 631

premium services 459

price 31, 33, 99, 172, 398
authority 543
bundling 548
competition 109, 127, 200, 611, 617
conscious 172

Index

deterioration 555
differentiated seat categories 121
differentiation 544
earning ratios 60
elasticity 127
level 32, 129, 230, 542, 543, 546, 548, 552, 553, 554
lists 541, 546, 547, 548, 551
low price 58
minimum 542, 554
optimal 542, 543
policy 365
service option 171
setting 30, 475, 543, 554, 555, 628
setting process 543
structure 137, 545, 546, 547, 554
system 555
variations 186, 547
war 105, 159, 161, 212, 337, 540, 541, 632
worthiness 172

pricing 30, 98, 628
air cargo 540, 544
guideline 543
individual 555
instruments 541
marginal 484
operating and marketing strategy 155
predatory 30
scheme 551
standardized 542, 546
strategy 32, 552
structure 137, 158

PrivatAir 60

private-public-partnership 160

privatization 224, 239, 358, 378, 382, 384, 386, 388, 391, 394, 395, 396, 397, 398, 400, 411, 420, 421, 424, 498
airport 412, 420

procurement 232, 309, 365, 416
joint 257

productivity 23, 24, 114, 167, 174, 175, 181, 358, 497, 500, 506, 510, 512, 513, 568
aircraft 177
infrastructure 179, 180

profitability 36, 61, 104, 120, 125, 131, 133, 226, 231, 305, 306, 307, 308, 311, 312, 313, 314, 315, 318, 319, 320, 321, 388, 390, 416, 432, 459, 482, 528, 539, 544, 616
index 305, 311, 313, 314, 315, 320
of Airline Operations 527

punctuality 112, 130, 135, 140, 157, 177, 202

Qantas 145, 159, 309, 310, 329–34, 335–39

Qatar 497

Qualiflyer Group 49, 52

quality 403
customer service 531
management 503
seat 113
standards 469

Radio Frequency Identification 443, 510

Rail Air Intermodality Facilitation Forum 428, 443, 446
Commission 443, 446

rail link 371

rail operator 442

rate agreement 543

rate card system 548, 549, 551

Reagan 496

real estate development 414, 419

regulation barriers 615

regulatory framework 24, 98, 414, 513, 518

relationship management 328

resource-based view 46, 197, 223, 224, 226, 227, 241, 239–41, 247, 248

response lag 190, 191

response timing 190

retail
  management 414
  outlets 153, 416, 443

retaliation 194, 195, 196, 197, 207, 210, 213

revenue management 48, 51, 155, 539, 543

Rhein-Main Verkehrsbund 440

risk diversification 237

route network 32, 42, 51, 87, 91, 126, 128, 138, 150, 152, 154, 167, 202, 223, 229, 233, 234, 236, 239, 241, 246, 247, 259, 374, 495, 529, 605, 607, 608, 610, 615, 617, 619, 622, 627, 629, 632
  design 607
  planning 51

Ryanair 36, 61, 101, 102, 113, 115, 120, 123, 129, 130, 133, 140, 145, 146, 149, 154, 155, 166, 168, 172, 187, 200, 201, 202, 206, 307, 312, 365, 399, 617

Saarbruecken 418

Sabena 362, 376

safety 62, 135, 138, 140, 148, 153, 155, 157, 415, 454, 455, 481, 493, 508, 564, 618

SARS 5, 100, 349, 350, 367, 375, 502, 569, 570, 576

SAS 33, 34, 52, 102, 332, 485
  SAS Cargo 464, 476

SAS/Lufthansa 33

Saudi Arabia 497

SBB 444

Scandinavia 34

Scandinavian Airlines 259, 261, 309, 330

scheduling 37, 51, 161, 224, 499, 526, 577, 617

Schiphol Group 415, 416

Seaboard World Airlines 491

seamless travel 111, 258, 330, 431, 435

seat inventory control 51

seating 103, 111, 113, 127, 128, 131, 140, 144, 151, 154, 167, 173, 174, 177, 182
  comfort 174

security 110, 111, 127, 138, 149, 173, 348, 366, 386, 395, 402, 404, 416, 445, 446, 455, 465, 466, 468, 469, 470, 480, 481, 502, 508, 510, 512, 569, 625
  standards 127, 445, 446, 455

self-coordination 264, 265, 266, 269, 270

September 11th 5, 57, 60, 65, 66, 307, 347, 367, 375, 421, 422, 445, 455, 468, 502, 520, 576

service 482
  fee 131, 182
  in-flight 452
  organizations 484
  packages 468
  quality 79, 129, 130, 137, 569

service system 519

Shigeo Shingo 497

Show Cause Order 495

Signage systems 416

Silkair 151

Singapore 56, 145, 149, 150, 151, 155, 156, 157, 159, 330, 332, 336, 352, 354, 501, 558, 563, 566, 580
  airline industry 560
  airport 157

INDEX

Singapore Airlines 9, 145, 147, 148, 149, 151, 154, 156, 159, 160, 259, 261, 310, 333, 335, 336, 423

Singapore Airlines Cargo 464, 466

Singapore's Changi Airport Terminal 1 151

single-carrier connection 234

single-hub structure 523

Sky Team 53, 423

slot 30, 37, 161, 229
    capacity 431, 438
    issue 629
    shortages 29, 31
    substitution effect 432

Soekarno-Hatta Jakarta International Airport Terminal 2 151

Southwest
    strategy 123

Southwest Airlines 101, 102, 108, 111, 123, 129, 130, 162, 166, 307, 352, 617

Spanair 112, 259, 310, 330

specificity 45, 132, 228, 242, 243, 245, 246

spot rates 541, 542

stagflation 496

Standard Agency Agreement 492

standardization 62, 424, 446, 492, 518, 544, 545, 546, 550
    aircraft 127

Star Alliance 52, 56, 121, 209, 211, 258–**60**, 261, 263, 265, 268, 307, 309–10, 319–21, 330, 334, 337, 410, 423, 429, 430–31, 435, 438

Star Alliance Service GmbH 260, 265

state monopolies 159

strategic

challenges 481, 606
decision areas 607
effectiveness 557, 558, 559, 560, 561, 563, 580
goals 171, 431
manufacturing 561
moves 186, 190
partnerships 325, 627
planning unit 43, 45
Planning Units 46
positioning 119, 123, 130, 131, 371
progress 563, 580
resource 50, 239
scenarios 327
scope 2
similarity 197, 214
success factor 50
success factors 42, 50, 53, 54, 62
trends 1
triangle framework 125

strategy
    cost leader 115, 124, 607, 612, 617, 618, 629, 630
    cost leadership 109, 119, 123–28, 131, 140, 226, 463, 607, 617, 629–32
    differentiation 97, 102, 107–14, 119, 123, 124, 125, 131, 146, 161, 171, 172–75, 226, 230, 236, 237, 463, 464, 466, 469, 555, **606**, 607, 619–22
    focus 119, 124, 125
    global logistics company 489
    intermodal 429
    theory 46

structure
    coordination 52
    ownership 52, 385

Subic Bay 537

supply chain 124, 158, 414, 457, 489, 490, 500, 503, 510, 512, 513, 577
    management 457, 510, 577

Suvarnabhumi International Airport 559, 563, 570

Swearingen Metro 496

Swiss 41, 58, 59, 61, 211, 376, 393, 444, 501

Swissair 52, 58, 310, 362, 364, 376, 393, 423

switching costs 32, 33, 616

synergies 52, 225, 233, 243, 264, 366, 399, 412, 458, 463, 479, 615, 626

tacit collusion 193, 212

TACT rules 541

Taiichi Ohno 497

Taipei 147, 331

Taiwan 145, 155

TAMS 564, 565

TAP Portugal 434, 437

target market 136, 157, 383

tariff structure
 air cargo 555

Tasman Empire Airways Ltd 329, 331

terminal
 capacity 90
 developments 356
 flow patterns 348
 mega 566, 567
 no-frills 567
 satellite 566
 Y-shaped 574

Terror prevention 468

Terrorism 347, 502

Texan regional carrier 166

TGV 435, 444

Thai 152, 153, 498, 571

Thai Airways 153, 156, 160, 259, 309, 310, 315, 319

Thai Airways International 152, 159, 330

Thai Aviation Industries Company 571

Thailand 148, 152, 156, 159, 160, 559, 570, 571, 572

Thailand Airport 417

Thalys Air 444

Third party logistics providers 457

Thomas Cook 54, 58

ticket 31, 33, 120, 121, 128, 131, 137, 147, 151, 153, 157, 158, 169, 171, 230, 330, 444, 476, 613, 616, 618
 commission 147, 157
 flexibility 173
 flexible 171, 173, 175
 price 33, 120, 121, 128, 131, 137, 169, 170, 171, 172, 174

ticketless reservation system 123

Tiger Airways 145, 148, 154–56, 158

time-definite services 466

TNT 461, 474, 481, 519, 520, 524, 531, 532, 535

tour operator 41, 54, 55, 57, 59, 122, 130

tourism 45, 46, 159, 162, 203, 326, 331, 333, 337, 340, 359, 365, 558, 564, 571, 576, 579
 companies 362
 hub 572

Tower Group 503

Toyota Production System 497

tracking 82, 439, 478

tracking and tracing 439

Tradewinds 496

traffic
 continental 170, 176
 densities 610
 network 105, 478
 waves 177, 179

train
 high-speed 371, 437, 441, 446
 station 368, 369, 371, 431–37, 439, 440, 442–45, 629
 tickets 628

training 136, 144, 176, 181, 234, 389, 480, 577, 618
  facilities 181

transaction costs 51, 52, 100, 228, 229, 240, 242, 245, 246, 247, 248

transatlantic flights 475

transit handling 478

transportation chain 518

transportation processes 490

travel agent 19, 25, 32, 34, 35, 37, 122, 130, 138, 145, 158, 169, 182, 559, 614, 620
  commission 35, 34–35, 158, 182

travel time 26, 110, 112, 115, 178, 179, 180, 430, 433, 436, 443, 618, 623

TRIPS 453

trust 29, 31, 37, 108, 229, 240, 265, 325, 339, 372, 483

TUI 120, 203, 205, 363

turnaround
  capacity 609
  time 53, 127, 144, 148, 153

TWA 491

Tyrolean 52, 53, 59

U.S. Airline Deregulation Act 410, 495

Udon Thani 152

UK 410

ULD 495

unions 102, 108, 114, 182, 336, 481, 619
  influence 632

unique selling-proposition 470

United Airlines 100, 102, 211, 258, 309, 310, 314, 334, 445, 491

United Airways 259

United Arab Emirates 497

UPS 375, 461, 474, 495–96, 497, 501, 503, 519, 520, 524, 531, 535, 537

US 24, 79, 90, 98, 102, 104, 106, 111, 112, 120, 132, 143, 145, 147, 159, 166, 170, 178, 186, 197, 212, 231, 235, 312, 329, 331, 334, 352, 354, 358, 375, 444, 452, 455, 491, 492, 493, 495, 496, 497, 498, 499, 501, 502, 503, 506, 509, 510, 511, 523, 546, 608, 617

US Airways 69, 99, 100, 111, 259, 330

ValuAir 145, 148, 150–52, 156–58

value adding services 518

value chain 54, 55, 91, 93, 100, 226, 232, 257, 327, 380, 381, 399
  cargo 457
  primary activities 607
  supporting activities 608

value-added activities 576

VARIG 259, 330

VAT 446

Virgin 34, 101, 145, 335, 337, 352
  Blue 330
  Express 617

Virgin Atlantic 113

Vitoria 532

Volareweb.com 205

volumes
  aircraft 475
  airport 441
  baggage 439
  business 402
  cargo 479–82, 505, 520, 521, 529, 543, 549, 564, 577, 578
  export 504
  financial 215
  forwarding 493
  passenger 80, 85, 363, 430
  production 77
  purchase 91, 257, 479

railway 633
traffic 79, 232, 349, 386, 419, 430, 437, 522, 532
transport 518

VRIO analysis 239

warehousing facilities 578

Warsaw convention, 475

well-balanced market situation 211

willingness to pay 84, 131, 135, 138, 141, 542, 546, 621

Wings 307, 309–10, 318–21, 423

win-win situation 576

WOW 260–63, 263, 265, 268, 464, 465, 485

WTO 453, 499, 501, 504, 507, 579

yield 44, 78, 80, 81, 83, 104, 458, 544, 606, 614
decline 168, 212, 362, 363, 364, 454, 458
management 51, 83, 174, 230, 231, 234, 239, 240, 547, 551, 555, 616, 620, 628, 633

Zurich Airport 365, 376